Green Industrial Restructuring

Springer

Berlin
Heidelberg
New York
Barcelona
Hong Kong
London
Milan
Paris
Singapore
Tokyo

Manfred Binder • Martin Jänicke •
Ulrich Petschow (Eds.)

Green Industrial Restructuring

International Case Studies and Theoretical Interpretations

With 96 Figures and 40 Tables

Springer

EDITORS:

Manfred Binder
Forschungsstelle für Umweltpolitik
Ihnestraße 22
14195 Berlin

Professor Dr. Martin Jänicke
Forschungsstelle für Umweltpolitik
Ihnestraße 22
14195 Berlin

Ulrich Petschow
Institut für Ökologische Wirtschaftsforschung GmbH
Potsdamerstr. 105
10785 Berlin

ISBN 3-540-67467-5 Springer-Verlag Berlin Heidelberg New York

Library of Congress Cataloging-in-Publication Data
Green industrial restructuring: international case studies and theoretical interpretations; with 40 tables / Manfred Binder; Martin Jänicke; Ulrich Petschow.- Berlin; Heidelberg; New York; Barcelona; Hong Kong; London; Milan; Paris; Singapore; Tokyo: Springer 2001
ISBN 3-540-67467-5

Springer-Verlag Berlin Heidelberg New York
a member of BertelsmannSpringer Science+Business Media GmbH
http://www.springer.de

Printed in Germany

Cover Design: Erich Kirchner, Heidelberg
Typesetting: Camera-ready by the editors

SPIN: 10660878 30/3130xz – 5 4 3 2 1 0 – Printed on acid free paper

Preface

This book presents the major results of two research projects carried out in the years 1995–2000. Their common subject was the decline of dirty industries in industrial countries, the driving forces of this decline, and its social management. Such a decline is one of the major challenges to any strategy of sustainable development, since reductions in energy and materials consumption almost inevitably lead to intersectoral structural changes at the cost of industries which provide materials or energy. Unfortunately, the economic, social and political costs of such changes are rarely considered in the environmental debate, let alone properly addressed in environmental policy proposals.

In the first project, financed by the German Science Foundation, the Environmental Policy Research Unit at the Free University Berlin (FFU) developed the analytical framework and the research design for case studies on declining dirty industries and analysed decreases in German steel, aluminium and chlorine production.

The second project, financed by the European Commission and conceived jointly by the FFU and the Institute for Ecological Economy Research Berlin (IÖW), sought to combine this approach with new theoretical perspectives which have gained prominence in economics in recent decades – namely evolutionary economics and theories focusing on institutions, networks and firms. In the project, working within a common research framework, researchers in Austria, Denmark, France, Germany and the Netherlands studied cases of (at least temporarily) declining dirty industries. We invited proponents of the aforementioned theories to discuss these case studies at two workshops. The first of these, in Berlin in October 1997, focused on different views of structural change, while the second, in Vienna in October 1998, focused on policy questions. With this design we hoped to set in motion an itinerate process in which empirical case studies and theoretical approaches would increasingly influence each other. As always, this interdisciplinary work proved difficult. The outcome, however, shows that such an approach is fruitful for the advance of research, as well as for the development of green policies which seek to overcome environmental problems through structural change. The major contributions to the research projects – case studies as well as theoretical considerations – went through a process of presentation and interdisciplinary discussion at the workshops and are now presented in this book.

The editors wish to thank the European Commission and the German Science Foundation for their support and all participants for their stimulating, profound and, more often than not, underpaid input. We are also grateful to Sylvia Theodos for letting us forget the administrative side of research management. The texts were translated into English with the help of Alex Sawyer and Michael Dills. The editors nevertheless take responsibility for any remaining errors or inadequacies. Our special thanks go to Klaus Jacob and Jan Nill, who helped co-ordinate the projects and edit this book and whose ideas shaped this research to a great extent.

MANFRED BINDER
MARTIN JÄNICKE
ULRICH PETSCHOW

Contents

B. The Case Studies

Contributors

Samir Allal
C3ED, 47 Boulevard Vauban, 78047 Guyancourt Cedex, France

Natacha Amorsi
C3ED, 47 Boulevard Vauban, 78047 Guyancourt Cedex, France

Manfred Binder
Environmental Policy Research Unit (FFU), Free University of Berlin, Ihnestr. 22, 14195 Berlin, Germany

Sylvie Faucheux
C3ED, 47 Boulevard Vauban, 78047 Guyancourt Cedex, France

Julia Haake
C3ED, 47 Boulevard Vauban, 78047 Guyancourt Cedex, France

Walter Hüttler
IFF, Schottenfeldgasse 29, 1070 Vienna, Austria

Klaus Jacob
Environmental Policy Research Unit (FFU), Free University of Berlin, Ihnestr. 22, 14195 Berlin, Germany

Matthias Jahncke
Institute for Ecological Economy Research, Potsdamer Str. 105, 10785 Berlin, Germany

Martin Jänicke
Environmental Policy Research Unit (FFU), Free University of Berlin, Ihnestr. 22, 14195 Berlin, Germany

Frank Jöst
Ruprecht-Karls University Heidelberg, Grabengasse 14, 69117 Heidelberg, Germany

René Kemp
Maastricht University, MERIT, P.O. Box 616, 6200 MD Maastricht, The Netherlands

Patrick Kenis
Free University Amsterdam, De Boelelaan 1081c, 1081 HV Amsterdam, The Netherlands

Børge Klemmensen
Department of Environment, Technology and Social Studies, 4000 Roskilde, Denmark

Ans Kolk
University of Amsterdam, Roetersstraat 11, 1018 WB Amsterdam, The Netherlands

Franz Lehner
Institute of Work and Technology, Munscheidstraße 14, 45886 Gelsenkirchen, Germany

Jan Nill
Institute for Ecological Economy Research, Potsdamer Str. 105, 10785 Berlin, Germany

Harald Payer
CULINAR, Mayerhofgasse 4/11, 1040 Vienna, Austria

Ulrich Petschow
Institute for Ecological Economy Research, Potsdamer Str. 105, 10785 Berlin, Germany

Annette Piening
Environmental Policy Research Unit (FFU), Free University of Berlin, Ihnestr. 22, 14195 Berlin, Germany

Winfried Ruigrok
University St. Gallen, FIM, Bodanstr. 6, 9000 St. Gallen, Switzerland

Simone Schucht
Ecole des Mines de Paris, Cerna, 60 boulevard Saint-Michel, 75272 Paris cedex 06, France

Jan van der Straaten
Tilburg University, P.O. Box 90153, 5000 LE Tilburg, The Netherlands

Stefan Zundel
Fachhochschule Lausitz, Großenhainer Str. 57, 01968 Senftenberg, Germany

Introduction

Manfred Binder[1], Martin Jänicke[1], Ulrich Petschow[2], Klaus Jacob[1], Jan Nill[2]
[1]Free University of Berlin, FFU, Ihnestr. 22, 14195 Berlin, Germany
[2]Institute for Ecological Economy Research, Potsdamer Str. 105, 10785 Berlin, Germany

1 The Challenge of Green Industrial Restructuring

Structural change, the rise and decline of industries, is a vital characteristic of economic development and growth. New "sunrise" industries emerge while older "sunset" industries lose their former importance. From the point of view of environmental policy and sustainable development, structural change has become an important issue: "green industrial restructuring" at the expense of "dirty industries", with their high consumption of energy and raw materials, is now seen to be necessary, but is likely to cause social and economic problems and could also be blocked by vested interests. It is therefore crucial to analyse the forces behind environmentally friendly industrial change and of the possibilities of a respective industrial policy.

There are four possible ways in which the environmental burden from production can be reduced:

1. add-on measures: reducing damage to the environment by adding technologies, generally "end-of-pipe" equipment, rather than upgrading the existing plant
2. integrated measures: changing or substituting harmful processes without changing the output
3. structural change: a change in the composition of the output without affecting its value in terms of gross domestic product ("green industrial restructuring")
4. limiting growth: reduced GDP or at least its growth

Until now, environmental benefits from industry have come mainly from *end-of-pipe measures*. These measures are attractive to private enterprise and environmental policy-makers, because they do not entail changing existing technologies, organisational structures or the division of labour. But they are of no help in solving many of the most urgent environmental problems (greenhouse effect, waste arisings, etc.), insufficient to alleviate many others (e.g. noise), and in some cases they merely shift pollutants to other media or create new problems (e.g. the creation of highly toxic filter dust) or are economically or environmentally inefficient. The introduction of *clean technologies*, the second strategy listed above, has therefore become increasingly significant. The most promising integrated technologies bring savings in energy and materials, reducing demand for the supplier sectors, causing intersectoral structural change at their expense.

Nonetheless, this indirect form of *green industrial restructuring* might not prove sufficient from an environmental point of view. In addition, sectoral crises in the energy and materials supplier sectors are forcing businesses to seek new markets for their products – sometimes with success. If the crises take them by surprise, sunk costs in existing factories and machinery might make it sensible for them to start selling at below production costs. Since these basic sectors are themselves resource and pollution intensive, contributing a great deal not only to overall materials and energy consumption but also to emissions, solid waste production, transport volumes etc., reducing the activities of these "dirty industries" would benefit the environment in any case. For this reason, green industrial restructuring must be placed on the political agenda. It should be noted that a decline in basic industries in the industrialised world will not in itself guarantee sustainability, if these industries are simply relocated to less developed countries. Although research in this respect has not in fact been able to identify this tendency in many countries, and certainly not in industrialised countries as a whole, this potential should not be ignored.

Converting natural resources into products, solid waste and emissions across the globe must be considered a critical issue for the long-term health of the environment. Postulates that a *"dematerialization"* could take place, at least in highly developed countries, have become an important component of environmental policy. The focus of this research project has not been on dematerialization in general, but on the specific consequences of dematerialization for certain sectors of the economy and therefore on sectoral change there. As mentioned above, such a development could bring severe *social and economic consequences* in its wake and might be blocked by the potential loser branches. Dirty industries often play a rather important role as employers and political pressure groups, at least regionally. This is a problem not just for green industrial restructuring but for structural change as a whole. From the point of view of environmental policy, the problem has been highlighted in the debate on CO_2 taxes.

The issue of green industrial restructuring is intimately connected with *"autonomous" structural change*, its determining factors and the opportunities for influencing them. And a relative – in some cases absolute – decline in production by some environmentally intensive industries has indeed been taking place in industrialised countries since the 1970s. This has to some extent been due to economic factors such as energy price rises and market saturation.

But some important factors are not directly related to market forces – industrial restructuring is frequently influenced by *industrial policy*, which is in turn influenced by the aforementioned vested interests. The resulting policy is often defensive and structurally conservative (as for coal and steel, for example). Since the 1970s, however, environmental and energy supply problems have also led to government intervention at the expense of dirty industries. One example is the "anything but oil" strategy which – in conjunction with the explosion in oil prices – has led to a decline in mineral oil production in several OECD countries. In Japan, primary aluminium production, among other things, has even been curbed by law

(1978). Generally, however, several "dirty" industries still benefit a great deal more from the current industrial policies of the European Union and its member states than sunrise industries do, especially with regard to subsidies and state protectionism. Finally, environmental policy could itself be a determinant of industrial restructuring, and move towards a "green" industrial policy.

Due to the serious economic and social consequences, an inefficient industrial policy may even put obstacles in the way of innovation or restructuring which favours new industries. Thus, before there is any structural change, "green" or other, these obstacles may have to be eliminated first. And if "automatic" green industrial restructuring is insufficient, policy may need to challenge the might of economically successful industries.

The key question is therefore: what conditions bring about a more or less painless decline of dirty industries on the one hand, and support for the development of cleaner "sunrise" industries on the other? And what part, if any, could green industrial policy play in this restructuring process?

Unfortunately, there is a serious *lack of research* in this area. Economists have not been especially interested in the non-economic aspects of structural change, nor in most aspects of sectoral decline in general. There are very few detailed theories of structural change which can also account for decline. Until recently, they have been limited to a rather general, top-down analysis, with scant reference to the evolution of technologies, companies and institutions across sectors. Nevertheless, the crucial importance of these factors has been suggested, and is now increasingly being acknowledged.

The same is true for the political aspects. The traditional industrial policy debate focuses on the question of whether and how a shift from sunset to sunrise industries should be supported, in order to promote economic development and the competitiveness of national economies. This neglects the older basic industries: if some industries are growing or are expected to grow, others are shrinking or must shrink, at least in relative terms. Recent analytical approaches, however, are shedding a somewhat different light on the political implications.

Environmental economists have usually ignored the role of economic structural change in environmental policy, focusing instead on environmental modernisation and economic instruments, leaving out the structural implications. We do not therefore know much about the management of industrial restructuring from an environmental perspective, and very little indeed from the perspective of potential loser branches.

The empirical basis for analysing environmental aspects of industrial change is somewhat thin on the ground, too. In fact, the study of global environmental change in terms of industrial change still lacks a comprehensive research strategy. We do not even know exactly what is happening in many areas. Many people believe for example that "dirty industries" are being relocated to the developing countries, something flatly contradicted by the existing – albeit few – empirical studies.

Another popular view is the hypothesis of an end to the "era of materials", which suggests saturation in the demand for most basic materials in highly developed countries. If this were the case, it would give reason to hope for an environmental "Kuznets curve", i.e. a trend where, once a certain threshold value is passed, further economic growth no longer aggravates the state of the environment but tends rather to improve it. However, the available statistics are far from sufficient to support such an optimistic view.

2 A Brief Outline

Given this unsatisfactory state of affairs, we have developed a broad approach to the question of green industrial restructuring, by tackling empirical as well as theoretical aspects and by approaching the political aspects based on a broad analysis of structural change. On the theoretical side, we have evaluated and drawn upon different approaches, exploiting their insights for the analysis of industrial restructuring and its implications for policy, while on the empirical side we have conducted case studies. We focused on certain basic industries in Europe where at least a temporary decline in physical and economic production has been observed since the beginning of the 1970s. The results of these studies were then interpreted by experts, using approaches which explicitly consider the (usually neglected) roles of technology and economic actors, in these cases the affected companies and their networks. These interpretations are divided into some which focus on restructuring, whilst others highlight the political dimension.

We therefore not only stress the importance of the environmental dimension in current scientific and political debates on structural change and industrial policy, but also explore some new and widely neglected aspects of the discussion.

The report starts with an introduction to the case studies by *Binder*. These case studies focus on products of strategic importance to industries with above-average environmental problems, e.g. through high intensities of resource consumption, emissions, solid waste, transport, or through the risk of accidents. In the cases chosen, the output of these products declined in the country studied, at least temporarily. Since the decline in production took place against a backdrop of overall economic growth, this suggests green industrial restructuring as defined above (a change in the composition of national output to the detriment of dirty industries).

The cases were analysed by researchers in the corresponding countries:

- chlorine in Germany by *Jacob*
- fertilisers in Austria by *Payer*
- coal and steel in Germany and Luxembourg by *Binder and Schucht*
- aluminium in Germany by *Schucht*
- cement in Austria by *Hüttler*
- cement in Denmark by *Klemmensen*
- petroleum products in France by *Allal, Amorsi, Faucheux and Haake*
- petroleum products in the Netherlands by *Van der Straaten*

After the termination of the project, *Annette Piening* added a case study about nuclear energy in Germany, which is not included in the summary and could not be discussed by the other contributors to this book.

The full case studies form the final part of the report. Research started with a common questionnaire on the main features of the respective markets, the patterns of structural change, the forces driving the developments, the political and decision-making processes, and the possibilities for environmental intervention. Statistics are provided of international developments and profiles of change affecting materials or sectors.

The engines of decline in the use of specific materials, and consequently in the corresponding industries themselves, are many and varied. No general laws governing these changes can be discerned at present. The forces behind reduced material consumption, such as increased process efficiency, market saturation and substitution, have been important for green industrial restructuring; in our case studies this is especially true for cement and steel. In some cases, energy policy (the mineral oil industry) and environmental policy intervention (fertilisers in Austria, chlorine in Germany) have had some effect, although its role was typically a minor one. Other driving forces, such as loss of competitiveness or relocation to other industrialised countries were sometimes active, but can only explain changes in terms of production itself, whereas change in the actual demand for materials is crucial from the point of view of the global environment.

Jänicke assesses the plausibility of an end to the "era of materials", that is a decline in the basic industries of developed countries. A number of long-term developments may be contributing to this alleged trend, such as:

- the shift towards a "post-industrial", service-based society
- the rise of "post-materialism" which, among other things, may favour consumer behaviour which is more oriented around "non-material" criteria
- the increasing material productivity of manufacturing industry, due to input-saving modernisation and the shift towards more flexible and specialised forms of production
- the lower material intensity of basic industries brought about by improvements in product quality, e.g. the strength of steel

Nonetheless, a general dematerialization hypothesis can explain current trends in developed countries only to a limited extent: market forces, as well as political factors, are often working against this decline. Influences such as the rise in the costs of oil and other raw materials have had only temporary significance. The intensity of materials use in terms of GDP has generally fallen in advanced OECD countries such as the USA, Japan, Germany, the Netherlands and Austria. But the rate of the fall in specific materials consumption has not generally exceeded growth in GDP. An absolute fall in the total material input to production processes has not occurred anywhere. The curve described by total material input in highly developed countries is not a broad inverse U curve, but more often a sort of an N curve. Consumption of some of the materials covered in this volume, as well as the total materials requirement of countries like the USA or Japan, exhibit this kind of up-down-up trend. The decline started with the oil crisis, but materials use increased again in the mid-1980s as economic growth picked up and the fall in materials intensity slowed down or halted altogether.

Following on from this, the present report assesses some economic theories of industrial restructuring. The contribution by *Zundel* offers an overview of traditional economic theories of structural change and their contribution to explaining green industrial restructuring. In order to do so, the complex relationship between economic and environmental structural change mentioned above is first unravelled and a number of caveats are made. Traditional economic theories of structural change, inasmuch as they exist at all, generally consider relative change between industries. To oversimplify for a moment, if these theories are applied to absolute declines, phenomena which they were never conceived to deal with, they give cause to hope for some kind of autonomous green industrial restructuring, predicting either a shift towards manufacturing and services or a relocation of old basic industries to developing countries. But their ability to account for the facts is very limited, mainly because they often operate at too high a level and completely ignore "political" forces, actors, and institutions.

There are some recent theoretical developments which are more open to these factors, and which could be helpful in analysing structural change and its political implications. These approaches are in some respects rooted in conventional economic theory, but their scope extends well beyond, borrowing from related disciplines, especially political science and business approaches. These theoretical developments and their possible applications to industrial restructuring are presented and evaluated in the contribution by *Nill and Petschow*, which covers evolutionary economics, neo-institutional economics, network theory and company-based approaches. At the same time, it places the interpretations of the case studies in a broader theoretical context.

Given its emphasis on change, evolutionary economics seems an appropriate approach to industrial restructuring, in particular when seen as a discontinuous process. Structural regularity and structural change are considered specific to countries or technologies. The application of this perspective to industrial restructuring seems promising.

From a neo-institutional perspective, institutions restrict the behaviour of agents, but are themselves influenced by actors and organisations. Public choice theory highlights the fact that collective action, e.g. by powerful interest groups, reduces flexibility and slows down industrial restructuring. A more dynamic view, along the lines of North, emphasises the interaction between technical progress and the institutional environment.

Network theories point out that networks and similar forms of coordination are very important to the course of economic and political processes. These include both innovation and diffusion networks, as well as networks which resist change, and they all determine to some extent the course of industrial restructuring.

The behaviour of firms can be viewed as the micro level of the observable phenomenon of industrial restructuring. The strategies and rationales of companies and their determinants are important, because they provide a sizeable part of the impetus towards structural change. In this interpretation, the major role of multinational enterprises must be emphasised.

Some generalisations are possible, particularly because these approaches can complement one another, for example the evolutionary and institutional or company and network perspectives. In these terms, industrial restructuring can be understood as the interaction between factor markets, technologies and institutions, and it should be stressed that "history matters", and will therefore have a significant influence on future trajectories.

Typically, the resulting structural change will be unpredictable. However, further research is both necessary and desirable. The potential inherent in many recent approaches to analysing industrial restructuring and industrial decline or stagnation has yet to be fully exploited. However, the above approaches do provide a means including such questions in an analysis. Three of these approaches – evolutionary, network and company-focused analysis – are demonstrated practically in three interpretations of the cases studies, by Jöst, Kenis and Ruigrok respectively.

An assessment of the environmental aspects of these approaches is not as straightforward as it might seem. They do not point to a kind of automatic decline, stressing instead restrictions but also opportunities for environmentally beneficial structural change. Hence, any permanent industrial decline must be organised, and at least some basic industries do or will need a *green* industrial policy. The need for environmental policy to act in this way is emphasised by Lehner.

In analysing the processes and actors involved in structural change, and in partially integrating the "political" determinants, some of the new approaches already go some way towards building a bridge between analysis and its political implications, especially in terms of the scope of industrial policy (see Nill and Petschow). At any rate, they acknowledge the need to include the institutional setting in any analysis, and to identify any specific characteristics, actor constellations and the regimes which are at work. Green industrial policy needs to take into account a multitude of factors and restrictions. From an analytical point of view, it should start from the dynamic interaction between markets, actors, networks and technologies, and therefore the political implications are further explored. Subsequently, Kolk and Kemp use a company-based and an evolutionary technology approach to the political aspects of the case studies.

Jöst presents a rather special evolutionary framework, integrating the theory of joint production and neo-Austrian capital theory. Industrial restructuring is brought about by innovation. The description of the development of economies using concepts from evolutionary economics shows that some elements of structural change are not predictable without analysis of past developments. He identifies two essential factors determining structural change in the case studies: on the one hand, scarcity of resources and environmental problems, to which companies and societies responded with a change in production techniques and environmental regulation, and on the other hand, the two oil crises at the beginning of the 1970s and 1980s, which provided a strong incentive for changing technologies in order to reduce energy consumption. However, as his theoretical analysis shows, structural change does not automatically, i.e. for purely economic reasons, lead to a reduction in environmental pollution.

Kenis takes a sectoral network perspective on industrial restructuring. The existence of sectoral networks, of which he identifies several examples, affects the course of restructuring, as is found in the case studies. In the absence of sectoral networks, structural change is leading to faster and more radical decline in employment levels and production capacity in the aluminium industry, for example. The presence of sectoral networks has a far more ambivalent effect. The outcome is contingent on the type of structural change produced by the existing sectoral network. Structural change with sectoral networks can lead to outcomes which better safeguard production capacities and thus also often employment, because sectoral networks are often effective devices for internalising or externalising contingencies or uncertainties. This phenomenon of atypical decline due to the presence of sectoral networks can be observed in the cases of the French oil refineries and German coal-mining. Thirdly, sectoral networks need not necessarily concentrate on conserving existing production facilities and employment levels, but may develop collective strategies which help facilitate a minimisation of the economic and social consequences of structural change, especially if the sector is relatively "open".

Ruigrok combines a company-based approach with network theory. In his analysis of international industrial restructuring, he points to the importance of "industrial complexes", which develop around large core enterprises and imply different concepts of control. They constitute a kind of hierarchical network. The strategies of these complexes are shown to be decisive for industrial restructuring. On the one hand, industrial complexes may act comprehensively across sectors while, on the other hand, competing complexes may exist within a single sector. Different intersectoral strategies are a possible, albeit not inevitable, result of the activities of industrial complexes. The driving forces behind industrial restructuring do not therefore necessarily emerge from a single sectoral level.

None of these approaches assume any longer that autonomous market forces alone can explain or predict structural change. They clarify some of the conditions which any environmental strategy or "green" industrial policy must take into consideration: it must address not only the technical and organisational situation but also contradictory, to some extent endogenous, political intervention. The possibilities and limitations of such a green industrial policy are further explored in the subsequent articles.

Kolk focuses on the role of multinational enterprises (MNEs) in environmentally beneficial structural change and their interaction with governments. Core companies may have a heavy impact on regulatory authorities and other companies. Kolk takes a strategic environmental management approach, differentiating between outside-in, inside-in and inside-out perspectives. She highlights the importance of MNEs in most of the cases, noting that they choose different environmental strategies. Referring extensively to the cases of petroleum and cement, the often restrictive impact of MNEs on environmental policies is demonstrated. Nevertheless, different regulatory traditions also partly explain the diversity of the cases. Given the importance of core businesses, the explicit promotion of strategic environmental management within these firms is seen as a way to further green industrial restructuring.

Kemp explores the challenges of eco-restructuring from the political side, using insights from technology studies and evolutionary system approaches. He states that little experience has been gained from policies aiming at "creative destruction" or system renewal. Starting from the production chain, Kemp shows that the window of opportunity for making integrated process change is limited. This is due not only to the influence of economic, policy and innovation networks, but also to the technological trajectories which make radical change difficult. Thus it is not surprising that the case studies show few examples of major product and process change, and it is hard for policy to make industry adopt environmental measures that entail financial loss. Kemp proposes and explores strategic niche management as a probe-and-learn method for exploring the potential for environmentally friendly innovation and managing the introduction process by providing a mixture out of initial protection and selective market pressure. Government acts mainly as a facilitator of new approaches.

The starting point of the contribution by *Lehner* is the current lack of competitiveness of the European economy. He points out that the problems of unemployment and weak growth rates have their main causes in a late structural shift towards knowledge intensive economies. He states that delayed structural change is also a major reason for current environmental problems and that a radical environmental policy could be a key force for innovation and structural change towards a knowledge-based and service-oriented economy. According to his analysis, the case studies show that there are rather large similarities within Europe, in that environmental policy has had no effect on structural change. It is a type of policy which is integrated into the dominant industrial structures and fits in with incremental innovation and change, the dominant pattern in most European industry. As a result, he concludes that there is a need for a more radical environmental policy, as an innovation policy to overcome existing structures and structural deficits in the European economy.

3 Conclusion: Towards a Green Industrial Policy?

The interpretations offer some advice on strategies for coping with the difficulties a green industrial policy will encounter. For a long time, mainstream economics did not consider the influence of government action on the structure of the economy. "Outsiders", however, have always pointed to the possible benefits to be gained from political action. Recent developments in traditional economic theory are now also suggesting that interventionist industrial policy could in fact promote economic development. While the classic debate on the justification for industrial policy remains, a *green* industrial policy brings a new dimension into the discussion. It does not primarily aim at growth and (short-term) competitiveness, but does integrate the negative "externalities" of such an orientation. With respect to "picking winners" – the specific support of new industries and technologies –, this implies a need to screen these industries environmentally. Green arguments not only point to further disadvantages from strategies to conserve heavily polluting industries such as coal, steel etc.: a green industrial policy should aim at a decline of (parts of) particularly dirty industries ("picking losers").

Such an approach is quite new and research is only at an early stage. The potential difficulties arising from restructuring are often neglected in traditional policy concepts. Political management of structural change is a complex task for regulators. Green industrial policy must consider a range of different influences and restrictions, such as the sectoral or regional conditions and the interaction between actors, technologies and institutions. Screening the cases under this aspect points up some important aspects of green industrial policy.

No common pattern of political influence caused structural change in our case studies. Only in two cases did environmental policy operate partly as a green industrial policy. The example of chlorine in Germany shows that agenda setting by environmental policy actors can contribute to industrial restructuring in the long term. The demonstrability of the environmental consequences of some chlorine uses and a broad discussion of possible regulation encouraged the development of technical alternatives. Doubts about a key process and the threat of regulation led to a restructuring of the sector and a fall in chlorine production at the end of the 1980s, before regulations actually came into effect.

The example also points to limitations: a more far-reaching green industrial policy, as lobbied for by environmental groups, has been successfully blocked by a close political network of the industrial association, the union and the industry ministry. Only an internal restructuring of the industry, with limited effects on employment and the region was feasible.

Environmental policy played a part in the decline of Austrian fertiliser production, too, but not a decisive one. Between 1986 and 1994 a fertiliser tax, and later a limit on the use of fertiliser in agriculture, contributed to falling demand. The industry, which is essentially shared between two oligopolies, could partially compensate for declining demand with increased exports. At the end of the 1980s, the profitability of reinvestments by one of the two firms was only assured by a state subsidy from environmental funds. Environmental pressure groups, the ministry of the environment and some economists were loath to pay this subsidy, but the company and the ministry of industry won through, and investment was subsidised, even though the effects on employment were minor.

Other cases also demonstrate the limitations of far-reaching environmental policies. In the Danish cement industry, the Danish-English monopolist used extensive lobbying and the threat of relocation to obtain an exemption from the new CO_2/energy tax in 1993. Incidentally, a similar power of veto for MNEs was also evident in the Dutch case, where environmental modernisation was blocked by a single company, in spite of an existing proven technological alternative (the modernisation took place some years later, for economic reasons).

As long as strong economic actors and networks impose limits on politically induced change, the dynamics for overcoming them are important. Here, market liberalisation policies have sometimes succeeded in destabilising existing network structures and thereby functioned partly as green industrial policies.

Closed networks and core companies proved important in many of the cases. However, in some cases, European integration has contributed to the destabilisation of tight networks, as take-overs by multinationals have fragmented political networks. Sometimes technological innovation has played a part by destroying the homogeneity of interests. Nevertheless, reducing capacity is anything but easy, especially if there are strong regional employment interests. In the case of coal and steel, the affected sector and its regional political networks were able to mobilise considerable resources to slow down the rate of change. Difficulties also increase when green industrial policy has to act against market forces.

Nor is there a common pattern to the management of the social and economic impact of structural change in the different cases. It ranges from hardly any public policy at all (e.g. the case of German aluminium production) to the multitude of actions taken in response to the decline in coal production.

The most important conditions for successful change seem to be:

- the end of an investment cycle in the affected sector
- satisfactory technological alternatives
- an innovator in the branch to promote the diffusion of alternative technologies
- clear environmental targets
- a high degree of political integration between different government actors
- availability of funding for compensation to lessen the social and regional disruption caused by change

Very often, political actors fail to formulate strategies for a green industrial policy. Some strategic restrictions and possible measures to overcome the deficits are shown in Table 1.

This list of restrictions and possible remedies does not claim to be complete. Most of the proposals aim to enhance the ability to set strategic targets, rather than their regulatory implementation. It should not be forgotten that a green industrial policy must be developed and implemented with a long-term perspective. Huge investment costs, regional impact and a large number of dependent jobs are the main problems faced by an ambitious environmental policy.

Table 1. Obstacles to green industrial policy and their remedies

obstacles	remedies
lack of agreement between various political actors on the need for structural change; lack of acceptance for structural change	• research into the environmental risks from a sector • developing ambitious policy targets for a sector's environmental performance • developing and promoting policy networks, including relevant actors and consuming industries, to support environmental targets • improving interdepartmental policy co-operation, e.g. integrated policy planning
lack of alternative technologies for a sector or for its specific requirements	• stimulating R&D activity through subsidies • stimulating competition to innovate by setting policy targets • promoting R&D networks involving the consuming sectors • strategic niche management (see Kemp)
lack of acceptance for technological alternatives	• finding innovators in the consuming sectors to promote diffusion
lack of information about the sector	• setting up information systems on sectors which are potential targets of an ecological industrial policy • identifying best practice internationally • tracking investment activities • identifying strategically important technologies
expected social disruption (unemployment, regional problems, etc.)	• compensatory measures (training, social plans, promoting disadvantaged regions, etc.)
strong opposition from affected industries and its allies	• ex ante estimation of the bargaining power of these industries • identifying potential innovators in a sector and in the consuming industries • promoting acceptance for change outside the affected branch • compensatory policies

Dirty Industries in Decline: An Introduction to the Case Studies

Manfred Binder
Free University of Berlin, FFU, Ihnestr. 22, 14195 Berlin, Germany

1 The Significance of Dirty Industries

Our empirical research on structural change focuses on industrial sectors which have a particularly heavy impact on the environment. The most important industries are

- extraction of natural resources
- transformation of primary energy such as crude oil or coal into final fuels such as petroleum products or electricity
- production of metals (steel and aluminium), based primarily on processing metal ores
- production of non-metallic mineral products, such as cement, glass or ceramics;
- the chemicals industry
- production of wood pulp and paper products

These industries are the main polluters among all producing sectors. This is not only due to their economic scale but also to their high intensities of pollution, i.e. pollution caused per unit of added value. This means, that a change in the composition of the economy at the expense of these industries would be beneficial for the environment. Because many environmental problems are external costs, the share of these industries in total gross domestic product might be too large, and industrial policies to correct for this kind of market failure justified.

There is no simple way of assessing the environmental impact of an industrial sector, and no single indicator can encompass the many dimensions of pollution and consumption of natural resources. For an easy, readily available approximation, we have used energy consumption as a "quick and dirty" indicator for the importance of these industries.

The significance to energy consumption of the energy sectors themselves – i.e. the extraction of fuels, the refinement of petroleum and the production and distribution of electricity – need not be shown in detail. Fig. 1 therefore looks only at the non-energy sectors from the list above, which are all manufacturing subsectors: non-metallic mineral products, chemicals, paper production and basic metals industries (divided into ferrous and non-ferrous metals in IEA data). The lower half of the figure shows the contributions of these "dirty" industries and of the manufacturing sector as a whole to the gross domestic product (GDP) of the 24 OECD countries in 1990 (OECD–24). The upper half of the figure shows the

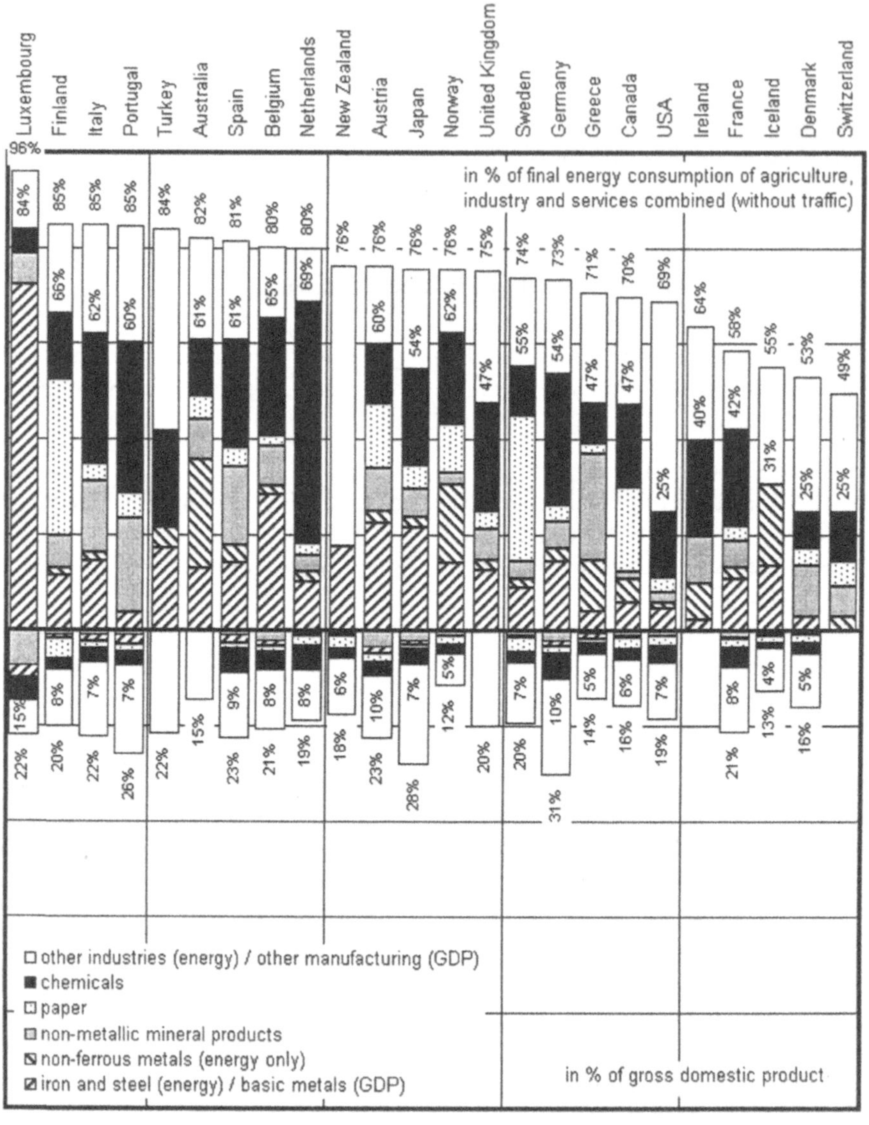

Source: IEA Energy Balances 1997, OECD: National Accounts II 1997
German GDP data refer to Western Germany only. All GDP shares are based on local currencies. No percentages with insufficient data. Final energy consumption does not include energy producing industries, but GDP does. Industrial final energy consumption includes manufacturing (without petroleum refineries), non-energetic mining and construction. GDP data do not distinguish ferrous and non-ferrous metal production.

Fig. 1. Proportions of GDP and total final consumption for dirty industries and the manufacturing sector in OECD–24 countries in 1990

shares of the same industries in final energy consumption used for producing GDP, i.e. used in the agricultural, industrial and service sector. Not included in these calculations is the energy used for traffic purposes, because IEA statistics do not distinguish commercial and private traffic.

These sectors used 51% of total final energy in these countries in 1990. For most of the countries, no more than 6–8% of GDP comes from "dirty" sectors, but about half of final energy consumption in GDP producing sectors is used here. The predominance of these industries in energy consumption by the manufacturing sector overall is also striking: in most of the countries, they consume two or three times as much energy as the other manufacturing sectors combined, but yield only half or a third as much economic value.

It should be noted, however, that the situation in the United States, which alone account for 43% of the OECD–24's final energy consumption, 31% of its GDP and 27% of its population, is less extreme: according to the International Energy Agency, "dirty" manufacturers are responsible for only 25% of the USA's final energy consumption in GDP producing sectors.

The data used here should be taken with a large grain of salt. For several countries there are no data available for value added by economic sector. For many other countries, the time-series are rather short, indicating that they are less than reliable. Furthermore, price structures differ between countries, which alone can often explain the sectors' different GDP shares when compared internationally. Moreover, the IEA's energy balances are far from perfect: for instance, the minor importance of dirty industries in New Zealand's final energy consumption is due mainly to a complete lack of data on energy consumption for non-metallic mineral products and non-ferrous metal production, even though New Zealand actually specialises in highly electricity-intensive primary aluminium production.

Due to a lack of data, it is not possible to produce a comparable graphic for the 1960s or 1970s. But it can be safely said that the predominance of the industries selected was far more pronounced then than it is today, after two decades of declining materials intensity (see the chapter by Martin Jänicke in this book and later this chapter) and environmental modernisation, in particular energy savings, in these sectors (Jänicke et al. 1993, 1997).

2 The Significance of Single Products for Dirty Industries

In contrast to monetary data, long time-series of physical data are quite readily available for most of the important products of dirty industries. Moreover, they indicate the environmental impact of these industries more properly than production value or value added.

Again, energy consumption serves as a proxy for environmental impacts. The following scatter plots show the importance of differences in per capita production of major "dirty" outputs for energy consumption by dirty industries in the OECD–

24 countries. In addition, linear regression results are also shown, weighted according to the populations of each country, in order to prevent small countries from having too much influence.

Figure 2 shows the very strong correlation between crude steel production and energy consumption by the iron and steel industries in the OECD–24 countries in 1990. Iceland, Norway and New Zealand are the only countries where high sectoral energy consumption cannot be explained by the amount of crude steel produced. But the energy data for Iceland, at any rate, are highly dubious: Iceland produces neither crude nor finished steel, but apparently ranks third in per capita energy consumption for iron and steel making in the OECD. For all other countries, sectoral energy use is fairly proportional to steel production, with almost 400 g oil equivalent (OE) required per kg crude steel. In a weighted linear regression using population for the countries in the sample, the differences in crude steel production explain 91% (= adjusted R^2) of the variance in sectoral energy use. This does not mean that energy intensities do not make a difference: energy intensities in steel production have decreased tremendously in recent decades. But the data show that energy intensities do not differ that much between countries in a given year.

Other dirty industries are not as easily described by a single production indicator as the steel industries are by crude steel. The most important non-ferrous metal is aluminium, and specialists in primary aluminium production – Iceland, Norway, Australia and Canada – also use far more energy in non-ferrous metals production

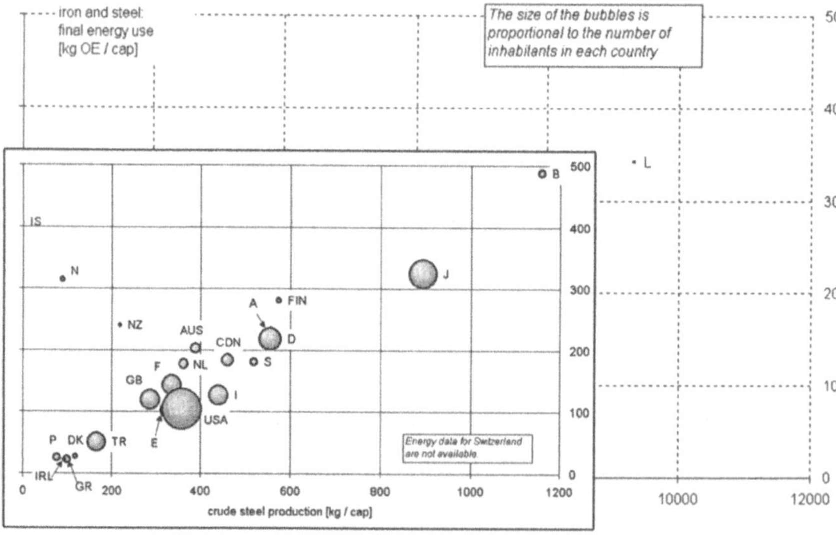

Sources: ECE, UN, IEA, US

Fig. 2. Crude steel production and energy consumption for iron and steel production in the OECD–24 in 1990

than other countries (Fig. 3). But the differences in sectoral energy consumption between these other countries cannot be explained by differences in the scale of primary aluminium production (Fig. 4). The regression statistics corroborate this result: differences in production levels explain 69% of variance in sectoral energy use. But if Iceland, Norway, Australia and Canada are removed from the sample, only 4% explained variance remains.

Specialists in producing paper and paperboard – Finland, Sweden, Canada, Norway and Austria – show extremely high energy consumption figures as well (Fig. 5). Nevertheless, their energy intensities – approximately measured as sectoral energy consumption per unit of paper production – differ widely, from 360 g OE/kg paper production in Austria to 890 g OE/kg in Canada. Countries with less paper production per inhabitant usually have lower and more similar energy intensities, mostly under 300 g OE/kg paper (Fig. 6). 76% of variance in energy use is explained by differences in production levels. If Finland, Canada, Sweden and Norway are excluded, 77% of variance in the remaining sample is still explained, but the slope of the regression line is no longer as steep (210 g OE/kg paper compared to 580 g OE/kg paper with all countries).

Outliers are less important with respect to non-metallic mineral products (Fig. 7). Although this sector produces a wide array of products including glass, ceramics and building materials, 56% of the variance in sectoral energy use per capita can be explained by differences in cement production per capita alone.

Chemicals are the most heterogeneous of the sectors analysed, and no single production indicator is sufficient to model its energy consumption. Moreover, the data available for chlorine production are not comprehensive, and maybe not so reliable either: data are missing for some small producers – Ireland (annual capacity 6 kt), Austria (60 kt) and Switzerland (104 kt) – but also for the Netherlands, whose capacity amounts to 647 kt, more than 40 kg/cap (Eurochlor 1999). Nevertheless, weighted linear regression produces a remarkably high R^2 of 44% with a slope of 3.5 kg OE/kg chlorine (Fig. 8), although this result is not robust: excluding Norway, the USA and Japan, R^2 rises to 95% and the slope to 15.0 kg OE/kg chlorine. Of course, neither 15 nor 3.5 kg OE, but rather 0.13 kg OE, are needed to produce 1 kg of chlorine (Haas 1990, p. 47). Chlorine production could serve as an indicator for the size and the stage of development of a country's chemicals industry (Henseling 1992, p. 249), and it might indicate energy consumption inasmuch as it correlates with a lot of other energy intensive processes. But this correlation may not last, e.g. due to the development of chlorine substitutes.

All in all, the significance of single products for the selected industries is quite high, and it makes sense to focus on them in the analysis of dirty industries. But one should also include other materials as well, namely non-ferrous metals other than aluminium, chemicals other than chlorine and the core energy sectors (mining, electricity, petroleum).

Astonishingly, the data suggest that, in many cases, international differences in the scale and structure of production are more important in determining energy consumption *in a given year* than differences in energy intensities. This does not

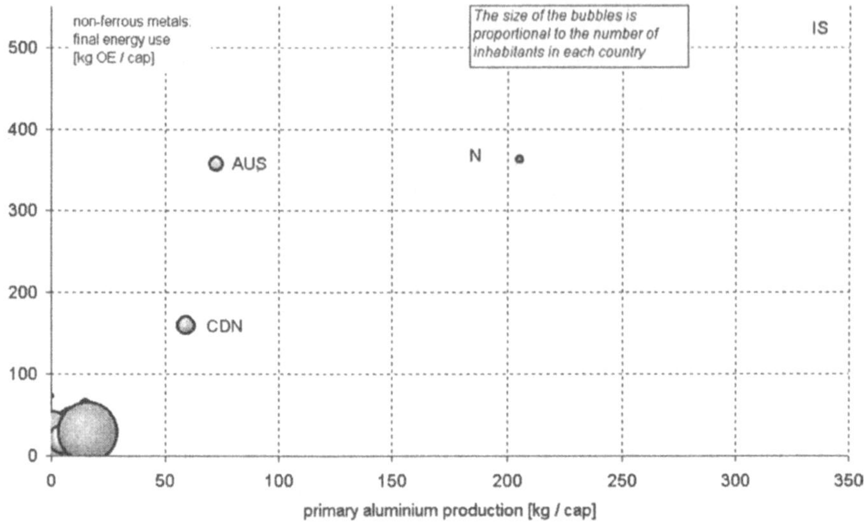

Sources: Metallgesellschaft, IEA, US

Fig. 3. Primary aluminium production and energy consumption for non-ferrous metals production in the OECD–24 in 1990: specialists Iceland, Norway, Australia and Canada

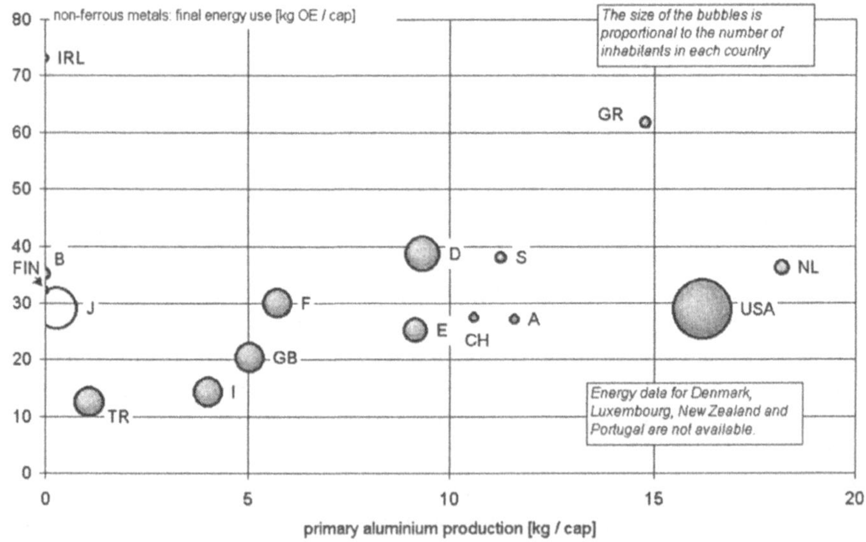

Sources: Metallgesellschaft, IEA, US

Fig. 4. Primary aluminium production and energy consumption for non-ferrous metals production in the OECD–24 in 1990 without Iceland, Norway, Australia and Canada

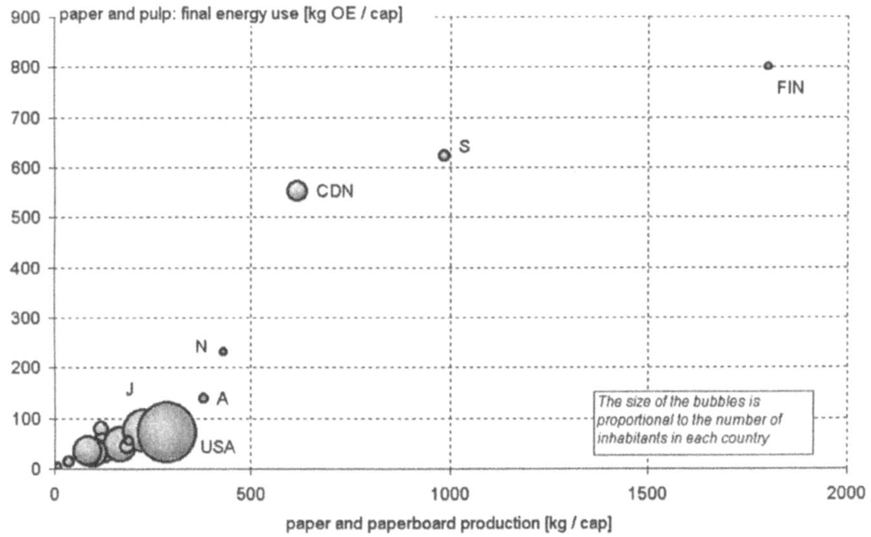

Quellen: FAO, IEA, US

Fig. 5. Production of paper and paperboard and energy consumption by the paper and pulp industries in the OECD–24 in 1990: specialists Finland, Sweden, Canada, Norway and Austria

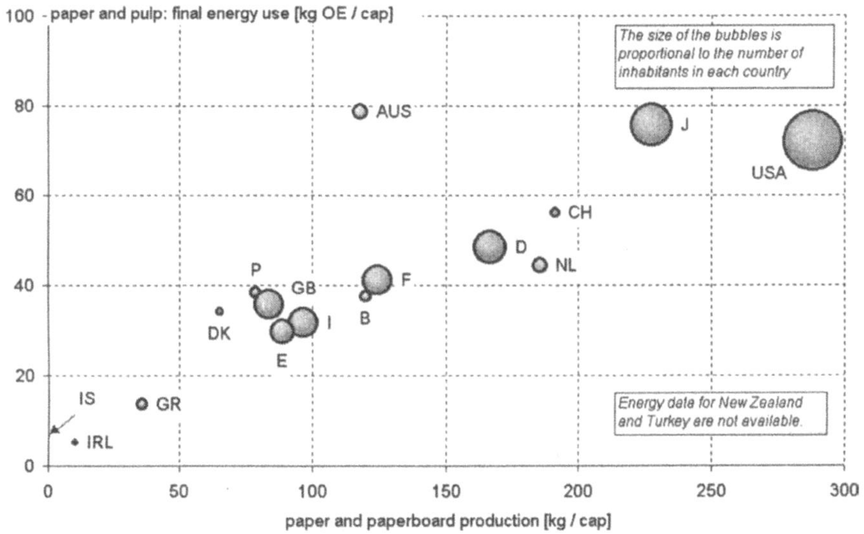

Quellen: FAO, IEA, US

Fig. 6. Production of paper and paperboard and energy consumption by the paper and pulp industries in the OECD–24 in 1990 without Finland, Sweden, Canada, Norway and Austria

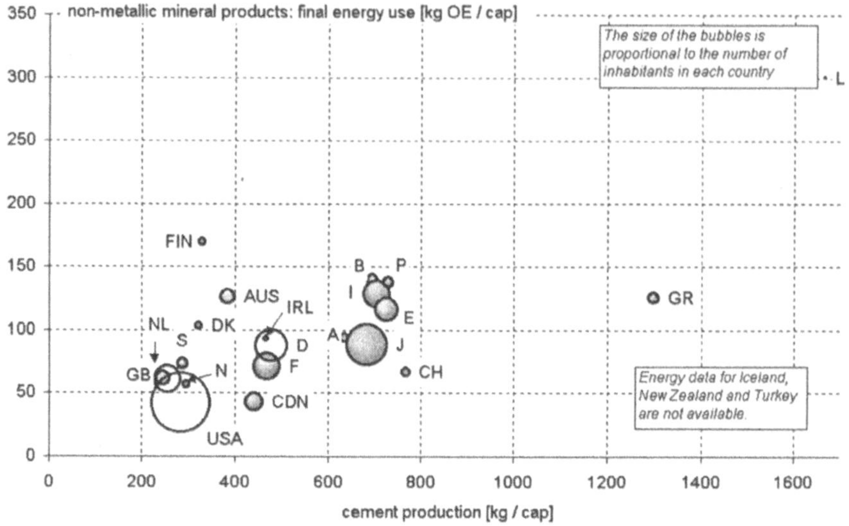

Sources: UN, IEA, US

Fig. 7. Production of cement and energy consumption for non-metallic minerals production in the OECD–24 in 1990

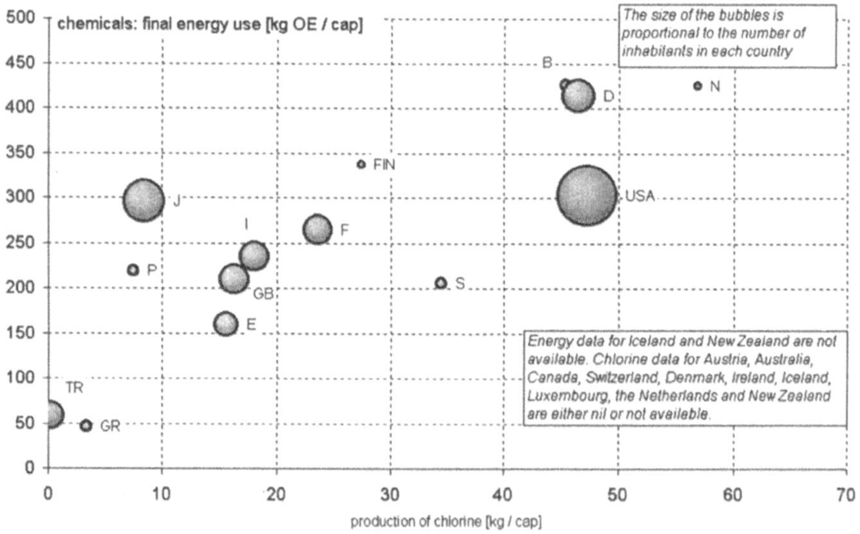

Sources: UN, IEA, US

Fig. 8. Production of chlorine and energy consumption for chemical industries in the OECD–24 in 1990

mean that structural *change* has been more important than technological change affecting energy intensities. On the contrary, almost all statistical studies on the subject have shown that the impact of intersectoral change on energy consumption has been limited compared to intrasectoral, i.e. mostly technological change (see e.g. Binder 1993, Greening et al. 1998, Sun 1998). However, this is due not to the supposed insignificance of economic structures but to their rigidity: structure matters, but it does not change very quickly.

3 The Cases

In this book, most of the dirty industries are analysed as national case studies. In order to throw some light on the conditions and problems associated with green industrial restructuring, we selected the cases according to the following criteria:

- The material output of these industries has declined at least temporarily in the country in question (environment).
- Such a decline in material output usually implies a decline in an industry's share of gross domestic product which, as a rule, causes economic and social problems. Thus, the decline of these industries must also entail negative economic and social consequences due to restructuring (economy).
- We tried to select cases where government policy or the behaviour of the general public has contributed to the emergence and/or management of structural crises (policy).

We therefore selected the following case studies: industrial production of chlorine (in Germany) and fertilisers (in Austria), coal mining (in Germany), production of iron and steel (in Germany and Luxembourg) and by far the most important nonferrous metal, aluminium (in Germany), cement production (in Denmark and Austria) and petroleum products (in France and in the Netherlands). The most important dirty industries ignored in this collection are electricity and paper production. Neither have experienced substantial declines in production in any industrialised country, making them unsuitable for the purposes of this book.

The cases were analysed by researchers in the corresponding countries on the basis of a common questionnaire, which provided

- *basic information*, such as the number and size of companies, the degree of concentration, competition or co-operation between producers, the physical output and value added by the industry, the number of employees, its importance to regional and national economies, its relationship to supplier and consumer industries, initial state regulations, technological maturity and the key environmental impact
- a *description of the structural change*, including the duration and extent of the decline in material output (in comparison with other countries), the resulting economic and social problems, and the environmental benefits of the decline, taking into account the environmental impacts of dirty substitutes and imports

- *driving forces* of the development, supply side reasons (such as innovation, factor price changes, changes in management strategy, regulation or foreign competition) as well as demand side reasons (such as market saturation, substitution by other materials, decline of consumer industries)
- *political information*, including a description of the actor network(s), their respective interests, opinions, strategies, actions and interaction
- *possibilities for intervention* for the benefit of the environment

Before trying to bring out tentative results and lessons for policy makers, we first present and compare the individual cases briefly, using a common scheme:

- For each case a scatter plot shows the development of each industry in each of the OECD–24 countries. Per capita production of the material in question is plotted against GDP per capita for each country for 1960–97 (all available years). In order to distinguish the developments in different countries, we have added lines between the average values in 1970–75 and 1990–95 for each economy. Also shown is a cubic regression[1] computed from these data, as well as the coefficient of determination R^2 as a standard measure for the goodness of fit. This is the simplest – and most common – way to test for a relationship between wealth and environmental indicators in research on the environmental Kuznets curve (see the contribution of *Jänicke* in this book; Ekins 1997): a graph with such a regression can be shaped like an inverted U or an N (among others). Nevertheless, there are many more possible forms of Kuznets curves, or its plausible alternatives, than can be detected with such a simple approach (Binder 1999). And another important shortfall of such a form is that it can never model a development which leads to a steady state in the respective market, with eventually constant production over time: whatever the data, the curve has to go either up or down at the end, due to the equation it is built from.
- The development of production and consumption of the material at issue is shown for each country and for the OECD–24 for the years 1960–97. Thus we can assess whether the developments in each case – i.e. the timing as well as the scale of the decline in production – was a typical one.

3.1 German Chlorine Production

Chlorine production is a key sector in the chemicals industry and a central issue in the environmental debate on chemicals, owing to its human and environmentally highly toxic final products, and its risky and highly energy intensive production, among other things. Countries differ enormously in their levels of chlorine production (Fig. 9). Germany's per capita production ranks amongst the highest, with an absolute production second only to the United States. While it is definitely a

[1] Cubic regression tries to estimate the per capita production data according to the formula: $P_{ij} = a_0 + a_1 Y_{ij} + a_2 Y_{ij}^2 + a_3 Y_{ij}^3$ with P_{ij} = estimated per capita production and Y_{ij} = per capita GDP, both in country i in the year j. If R^2 were 1, all points on the scatter plot would be situated exactly along the regression graph. If it were 0, no regression line could fit the data.

sector for rich countries (adjusted R^2 = 41% with a decreasingly rising slope), chlorine production seems to have passed its zenith: in several important countries – notably Sweden, Finland, France, and Great Britain – less chlorine is produced today than in the early 1970s. Elsewhere, growth has simply slowed since then, and only in recent years could an absolute decrease be noticed.

Germany is pretty typical for the development in the OECD–24 as a whole (Fig. 10[2]). Chlorine production peaked in 1986, at a level surpassed only slightly in 1988 – a time of stagnation – followed by a fall of a quarter between 1989 and 1992, mainly due to less production of PVC, CFCs and chlorinated solvents. While the decline in the production and use of PVC proved temporary, the process of chlorolysis, which produced the solvents TRI and PER as well as TETRA, a pre-product of CFCs, was virtually abandoned. This case is therefore atypical within this book, since the – albeit short-lived – decline in production was caused primarily by environmental issues, namely the ozone layer and public health. Unlike most of the materials discussed in the other case studies, many products of the chlorine industry are toxic, carcinogenic (or at least suspected of being so).

Many environmentalists are therefore opposed to any kind of chlorine industry, but as yet without much success. A powerful coalition between chemicals companies and their industrial associations, the union of chemical workers and the ministry of economic affairs has clearly dominated decision making on chlorine issues, while the environmental and health authorities have usually been divided and indecisive. Nonetheless, they have put chlorine on the agenda by commissioning a host of research studies, thus channelling resources and attention to this product group. Several times, this has resulted in the development of chlorine free processes or substitutes, which has begun to weaken the coalition in this respect.

There were no major social problems arising from sluggish growth and falling production, since the process is not labour intensive and no major financial losses have been entailed.

3.2 Austrian Fertiliser Production

Fertiliser consumption in industrialised countries is not expected to rise further, given the stagnation in the agricultural sector and the quantities of agri-chemicals already used. Furthermore, restrictions on the use of fertilisers have increasingly become an explicit goal of environmental policy. But again, the differences in scope and speed between the highly industrialised countries are tremendous (Fig. 11). They are so large that there is no reasonable fit for the regression line, whose decrease in production above a suggested turning point of about 20 000 US-$ GDP per capita is mostly due to the – continuous – low per capita production of high-income Luxembourg, Switzerland and Japan. Austria's low per capita pro-

2 No consumption data are available for chlorine. But domestic consumption and production do not differ greatly, especially in Germany, where about 92% is processed further on the site where it is produced (see *Jacob* in this book).

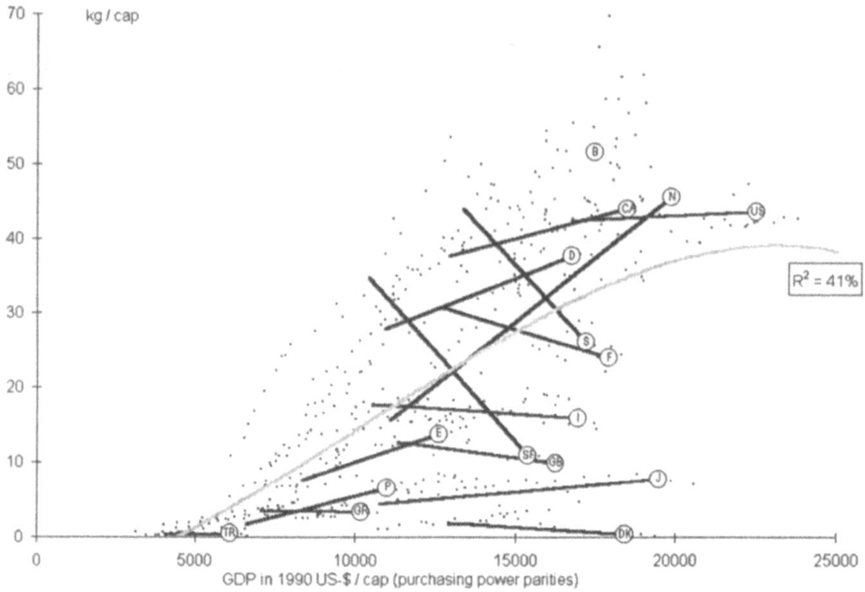

Sources: UN, IEA

Fig. 9. Chlorine production and GDP per capita in OECD–24 countries 1960–96 and national developments 1970/75–1990/95

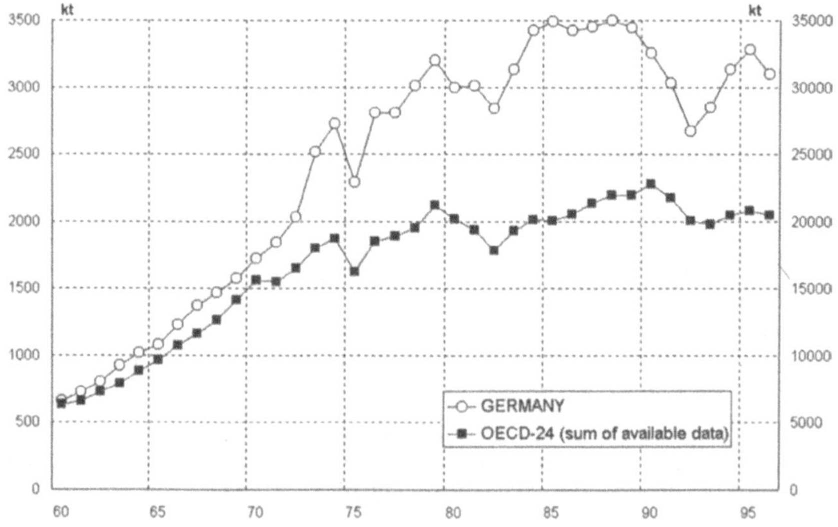

Source: UN

Fig. 10. Chlorine production 1960–96 in Germany and the OECD–24

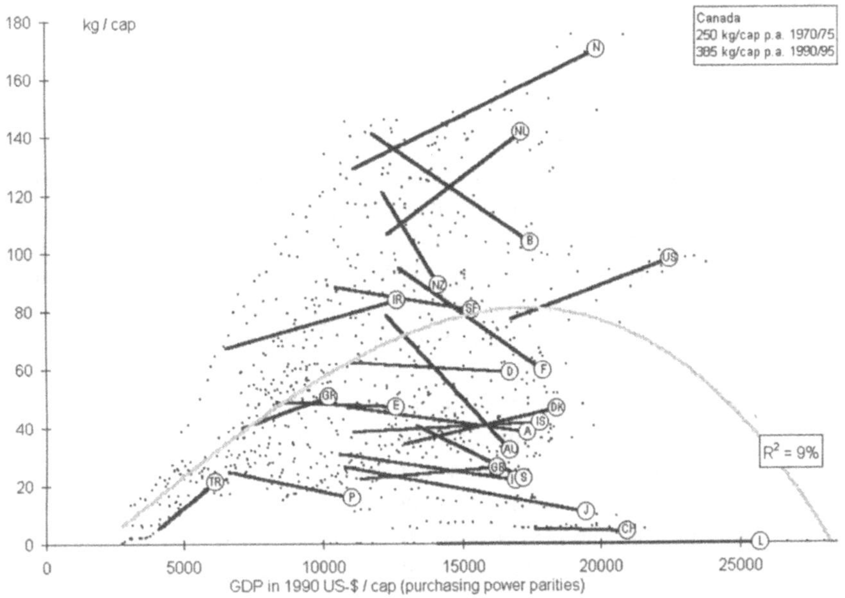

Sources: FAO, IEA

Fig. 11. Fertiliser production and GDP per capita in OECD countries 1960–96 and national developments 1970/75–1990/95

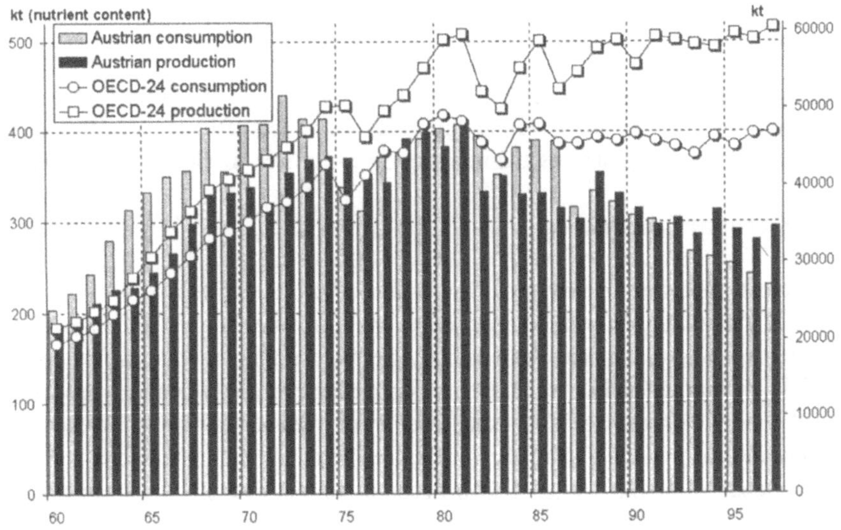

Source: FAO

Fig. 12. Fertiliser production and consumption 1960–96 in Austria and the OECD–24

duction has fallen but not extraordinarily so: half of the sample has more or less the same or lower per capita production figures, and the decrease in France, Belgium, Australia, Sweden and Japan has been much more pronounced than the one in Austria.

In the OECD–24, the high growth era, which even survived the first oil crisis and the subsequent recession for some years, was followed by remarkable stability after 1980, with no sign of an absolute decrease (Fig. 12). The OECD–24 is still a net exporter of fertilisers, but the European Union has become a major net importer. Not so Austria: until the early 1970s, Austrian farmers were consuming more fertiliser than Austrian industry produced, but exports and imports have subsequently more or less balanced each other out[3]. Recently, constant production and falling domestic use have led to net exports. Austrian fertiliser use has decreased by a third since the mid–1970s, and production has also fallen by some 20% since the early 1980s. Austrian producers sell more fertiliser to foreign than to domestic customers, and Austrian farmers also prefer to use imported products. The market has become increasingly competitive due to deregulation (EU, WTO) and the opening up of the former socialist countries. Thus, price competition is seen as the most important determinant of structural change, although there are only three domestic producers, with Magrolinz as undisputed core actor and one dominant wholesaler (RWA). The restructuring did not cause major problems, because the industry is small, with only some 500 employees.

The reasons for declining consumption were mainly

- over-production and corresponding agricultural set-aside policies
- increased efficiency in fertiliser use, to save costs and decrease environmental stress
- the rise of organic farming

While the impact of public policies on over-capacity in agriculture has been ambiguous at best, they have clearly supported increased efficiency by funding research, imposing regulations and introducing a fertiliser tax (levied 1986–94).

3.3 German Coal Mining

The most important mining sector in industrialised countries is still coal mining. Coal is a relatively harmful energy source, both in its extraction and its use. Substituted by oil, gas and nuclear energy, the once dominant coal is now present mainly in market niches, such as iron ore reduction and electricity production. Furthermore, for geological reasons, coal from overseas is now far cheaper than domestic coal. By the end of the 1950s, freight rates had decreased to a level at which coal from the United States became competitive on European markets.

[3] The Austrian statistics used in the case study in this book measure fertiliser by weight, while the FAO statistics used here measure net nutrient content. Thus, the domestic statistics show gradually rising net exports from 1973.

Other export countries, like Australia or South Africa, have followed suit and increased their mining capacity enormously. Thus, European coal production has long been in decline and some countries have shut down all their pits. Differences in per capita GDP do not correlate with differences in mining activity (adjusted R^2 = 2%; Fig. 13).

There is not much of a difference between the coal production and consumption figures for the OECD–24 (Fig. 14). After a period of stagnation, when coal was substituted by oil in almost all applications, coal use recovered in the aftermath of the first oil crisis, stagnating again in the 1990s. But Germany has faced a continuous decline since the late 1950s, changing from a small net exporter to a net importer. Although not at all competitive, West German coal mining has not been abandoned, because of opposition from organised suppliers' interests and the severe social problems this would entail. Instead, subsidies and strict import restrictions have helped the coal industry to stay in business. Although totally dependent on public support, the coal industry has remained private, but it has merged into a single company, with the former owners and major users as shareholders, originally mainly the steel producers, and today the electricity generators. With the decline of the industry, its importance and its power have declined as well, and by the early 21st century, German coal mining could eventually become almost negligible – after half a century of subsidised restructuring.

3.4 Iron and Steel production

Coal mining and steel production are closely related, since coke and iron ore are the two major inputs for pig-iron production in blast-furnaces. The "natural" production sites for steel producers were therefore in the vicinity of coal and ore deposits. Because of the aforementioned fall in freight rates, European coal and iron ore both lost out to competition from resources from North and South America as well as Australia (among others) and as a consequence the traditional steel sites became less attractive than sites at the coast. Additionally, major innovations in post-war decades made newly built steel plants far more cost effective than existing ones and induced a tremendous capacity build-up. Given the decade-long life of capital intensive iron and steel producing equipment, this provoked a crisis, further aggravated by an unexpected slackening in demand from the 1970s on, and the astonishing success of several newly industrialised countries in steel import substitution. Only a few countries increased steel production during this period of crisis (poorer countries, Italy, Finland and the smaller producers Switzerland, Denmark and Norway). The others reluctantly reduced capacity and production volume more or less in unison (Fig. 15). Germany is still one of the largest steel producers, in absolute terms and per capita. At the start of the steel crisis, Luxembourg had a per capita steel production ten times as high as that of second-placed Belgium, and still has by far the highest rate, even after halving production. Again, differences in income hardly correlate to variance in crude steel production, and the regression line makes no sense.

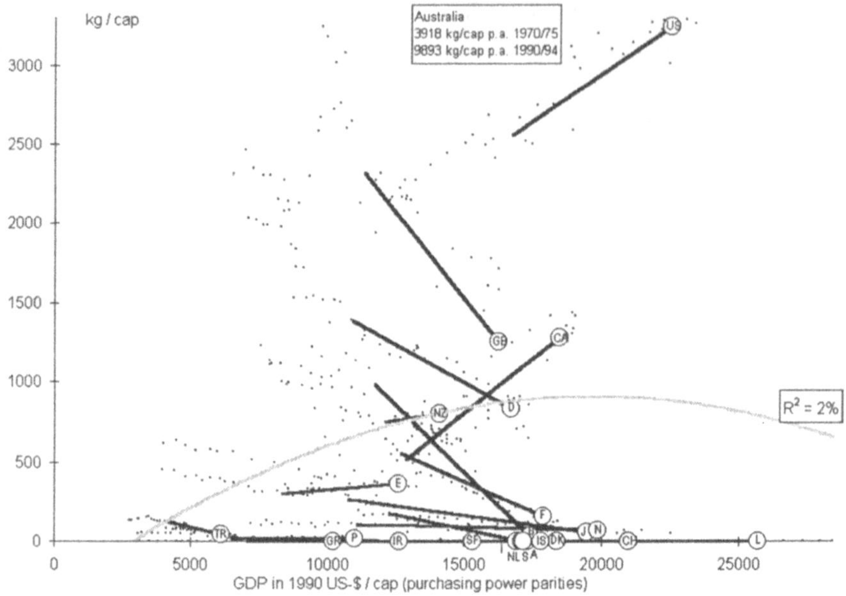

Source: IEA

Fig. 13. Coal mining and GDP per capita in OECD–24 countries 1960–94 and national developments 1970/75–1990/94

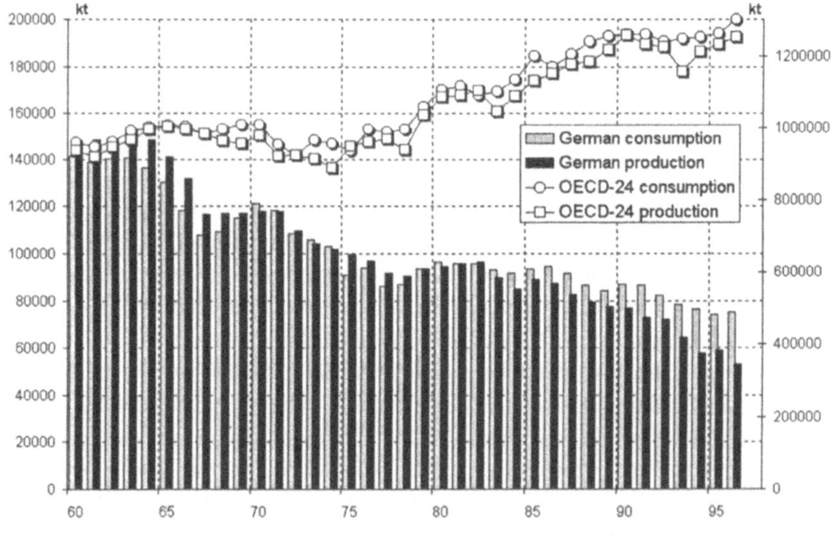

Sources: IEA

Fig. 14. Coal mining and consumption 1960–94 in Germany and the OECD–24

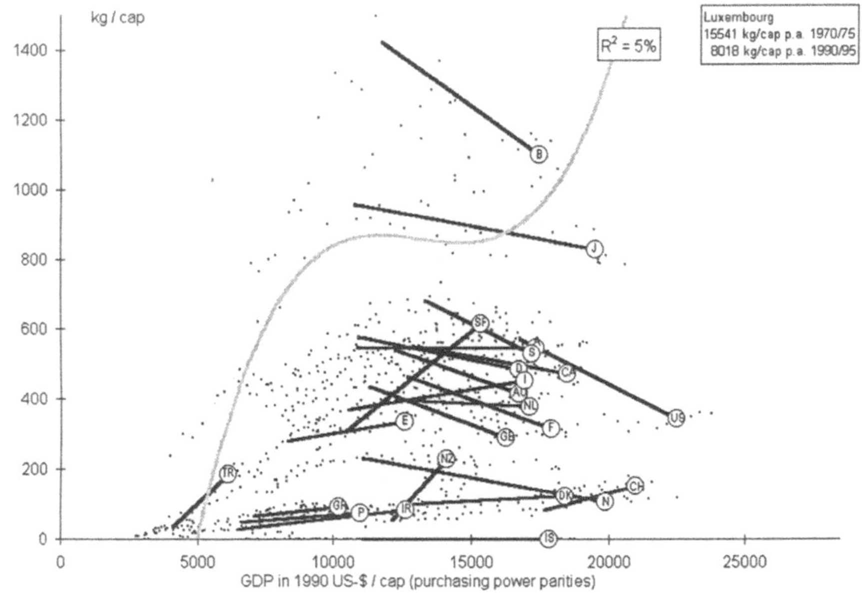

Sources: ECE, UNCTAD, IEA

Fig. 15. Crude steel production and GDP per capita in OECD countries 1960–96 and national developments 1970/75–1990/95

Germany shows a typical development for the OECD–24 countries (Fig. 16), with the exception of slower growth rates in the 1960s, due only to the meteoric rise of the Japanese steel industry from nowhere to the top of world markets. Most traditional German steel producers were lucky in that the huge coal deposits where they were located were also close to the river Rhine, along which – after more than a century of rapid industrial development and the corresponding build-up of infrastructure – resources from overseas could be transported almost as cheaply as in maritime waters. Although the switch to resources from abroad is no major problem, they have actually used hardly any coal from abroad, in order to support the local coal industry. But prices were set according to what they would have paid for imports, the difference being paid by government. Besides, production facilities were more efficient than those of most of their competitors at the beginning of the crisis, giving them a good chance to survive in a free market. However, producers in the Saarland on the French border and many smaller producers were not as lucky; and with high subsidies in other countries and a highly regulated European market, German businesses had to face European Community bargaining resulting in a more coordinated and equal reduction of capacity across countries than a – politically illusory – free market solution would have caused.

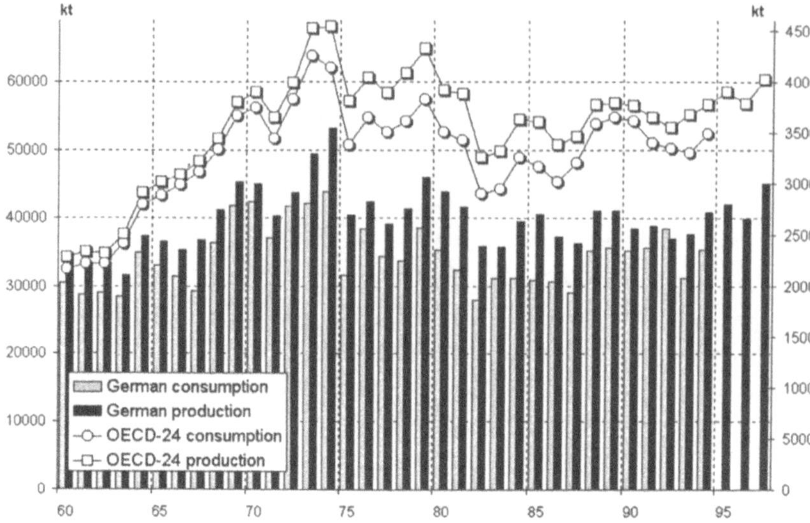

Source: ECE, UNCTAD

Fig. 16. Crude steel production and consumption 1960–96 in Germany and the OECD–24

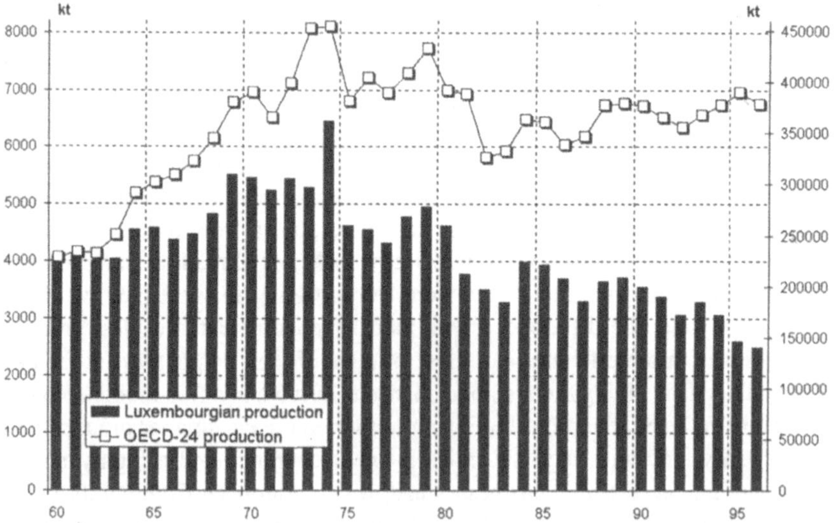

Source: ECE, UNCTAD

Fig. 17. Crude steel production 1960–96 in Luxembourg and the OECD–24

The study of Luxembourg's steel industry illustrates the far more dramatic results of the steel crisis in what was an almost mono-sectoral industrial area (Fig. 17[4]), and which suffered from all the disadvantages a traditional production site could have: local but suddenly (or at least unexpectedly) uneconomic resources, no access to cheap waterways, and fairly old production facilities to boot. Luxembourg has nevertheless managed to become one of the most prosperous industrialised countries in recent decades. Slowly – but far faster than most other producers – Luxembourg's steel industry has completely abandoned the early stages of production – ore mining and pig-iron production – and now specialises in steel recycling. Of course, this could not prevent a major decline in employment by the steel industry, but several social policies, new manufacturing industries and, most of all, the rise of financial enterprises and other service sectors like media and the international bureaucracy, have more than made up for the loss. Much of this success may be due to the special situation of such a very small country with traditionally (and naturally) close links to neighbouring countries and international institutions. The most important financial sector takes advantage of very favourable legal regulations to attract international money. Besides this, a large share of employees are not Luxembourg citizens, making it far easier to switch from one industrial sector to another, i.e. from one group of commuting workers to another. Nevertheless, the economic and social success of the transition remains admirable.

3.5 German Aluminium Production

Production of aluminium, especially *primary* aluminium, is more concentrated than any other material discussed here. This is typical for non-ferrous metals, since production is mostly located where the rare natural resources are concentrated. GDP does not correlate to the level of production (adjusted R^2 is only 7%) and therefore the corresponding regression line is useless. Non-ferrous metal ores usually contain only little of the metal, making it rather uneconomic to transport a mostly useless raw material over large distances. For aluminium production, it is not so much the ore (bauxite) which is decisive, but the availability of cheap electricity. Thus, countries with power generation based on hydroelectricity (Iceland, Norway, New Zealand and Canada) or other unusually cheap sources (coal in Australia) have extraordinarily high per capita production data. Once, nuclear energy seemed a promising candidate for providing cheap electricity. Of course, the rise in energy prices in the wake of the two oil crises made differences in electricity prices even more important. Germany's per capita production is below average even ignoring the aforementioned specialists (Fig. 18).

With such differences between countries, only summary statistics can help to identify the trend (Fig. 19): contrary to the recently stable development in steel in industrial countries, the use of primary aluminium is still rising, but its former

[4] Almost all steel produced in Luxembourg is bound for export. Because of the close relationship to Belgium, crude steel consumption data refer only to both countries combined.

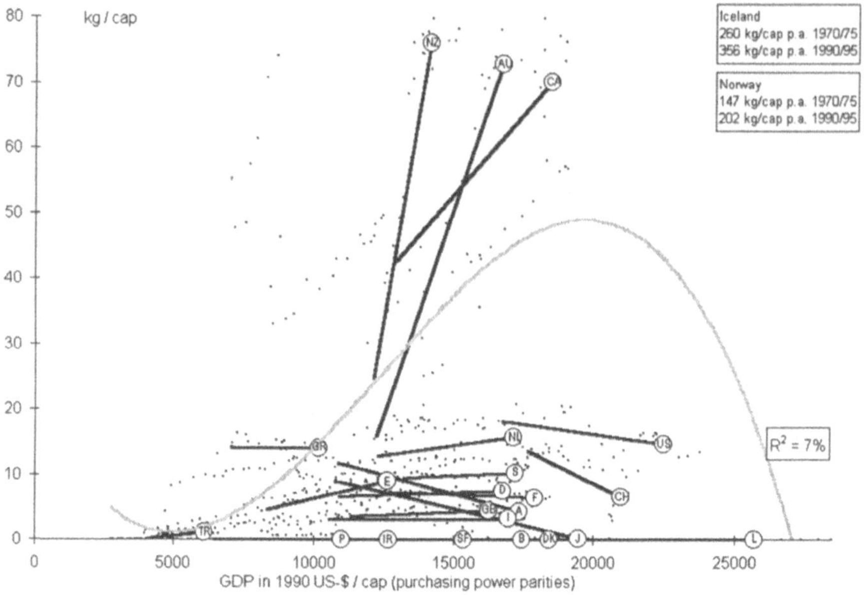

Sources: Metallgesellschaft, IEA

Fig. 18. Primary aluminium production and GDP per capita in OECD–24 countries 1960–96 and national developments 1970/75–1990/95

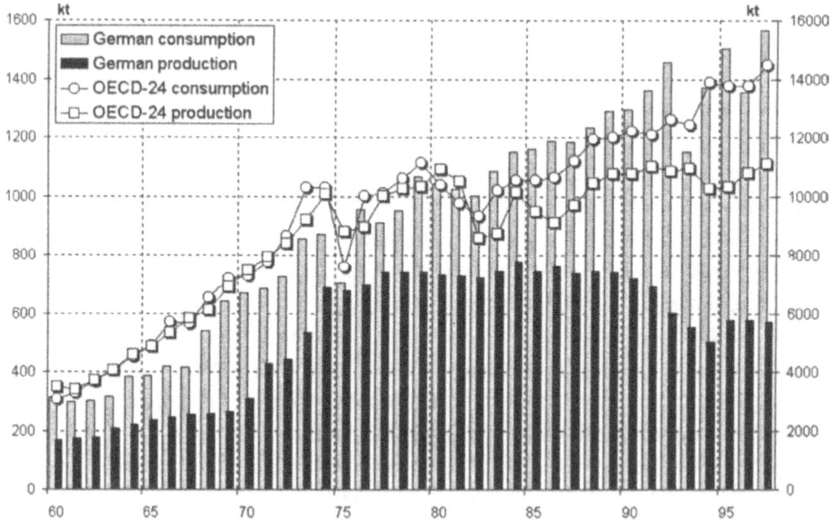

Source: Metallgesellschaft

Fig. 19. Primary aluminium production and consumption 1960–96 in Germany and the OECD–24

tremendous growth rate of almost 10% annually in 1960–73 was interrupted by a decade of crises and stagnation. Since the mid–1980s, aluminium consumption in the Western industrialised countries has recovered, but production has remained fairly constant, the difference being made up by imports from third countries.

Germany has been a net importer of aluminium during the entire period in question, but its import dependency has increased in recent years: aluminium production has fallen by more than 20% since the end of the 1980s, due to the decommissioning of four aluminium works. At the same time, domestic demand has risen by 20%. Mainly because of high electricity costs and strong competition from the Commonwealth of Independent States (CIS), the future, if any, of the German aluminium industry remains doubtful. This decrease has not been accompanied by any social problems or loss of capital worth mentioning, nor has public policy been involved. Although the possibly disastrous consequences for the aluminium industry of even slight increases in electricity prices have always been used as an argument against, for example, decommissioning nuclear power plants, the current decline of the industry – which is set to accelerate – has not received any public attention.

3.6 Austrian and Danish Cement Production

Cement producers are struggling in the face of falling demand due to reduced building activity. With the exception of the increasingly export oriented Luxembourg and the poorest countries – Turkey, Greece and Portugal – only Japan, Italy, Belgium and Spain have reported – sluggish – growth in production. All other countries show the same downward trend, albeit at different levels. The cubic regression leads to a modest R^2 of 24% and a regression line with a sharp increase for low, a slight decrease for medium and an – absurdly – dramatic increase for high GDP/cap (Fig. 20).

For the OECD–24 countries together, there has been only a temporary decline in production in the early 1980s followed by a recovery. Nevertheless, cement production is rising far more slowly than GDP, and may be entering long-term stagnation (Fig. 21). Typically for most countries, Austrian cement consumption and production were more or less equal until the early 1990s. Both fell by about 30% between 1972 and the mid–1980s. Afterwards, consumption rose substantially, but fell again after 1994. In recent years, imports from Eastern Europe have increased and forced domestic production to new lows. The massive job losses were addressed by social compensation plans. The close network between administration, private management, trade unions, researchers and the interested general public provide innovative management of the social as well as the environmental problems of this industry.

Denmark has experiencd the same 50% fall in cement consumption as all Scandinavian countries since the early 1970s (Fig. 22). At that time, Scandinavian per capita cement production was around the average for industrialised countries; now, only New Zealand, Great Britain and the Netherlands have lower per capita

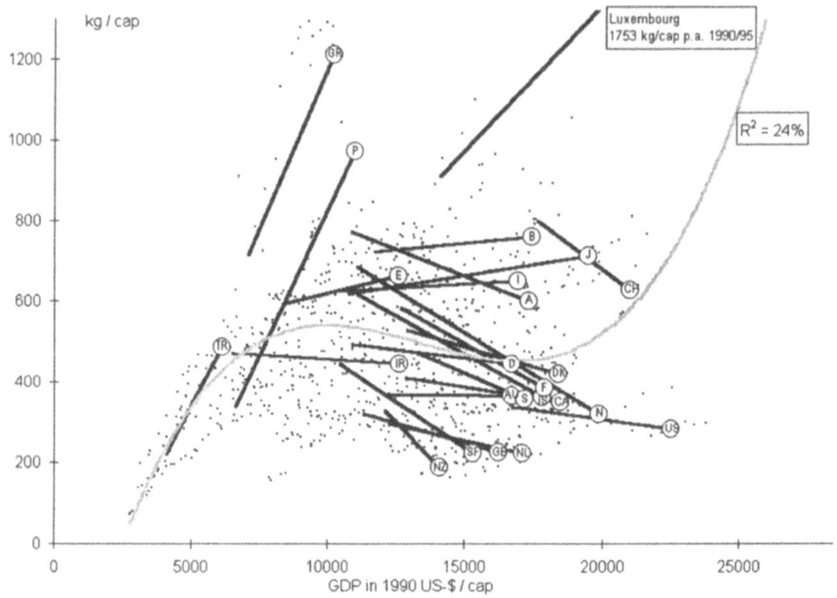

Sources: UN, IEA

Fig. 20. Cement production and GDP per capita in OECD–24 countries 1960–96 and national developments 1970/75–1990/95

production. The decrease in cement consumption was mainly a consequence of the slowdown in construction activity, especially for residential housing, and the 20% reduction of the amount of cement per unit of concrete. Nevertheless, the Danish cement industry – consisting of a single company, Ålborg Portland – has recovered, by expanding exports in the 1990s, and is now producing at almost as high a level as before the crisis. This was made possible by modernising production and focussing on an export market niche, white cement, a third of the world market for which is produced in Denmark. Modernisation cut the number of jobs by two thirds from 2 200 in 1978 to 750 in 1997, but the social impact was limited and could be further alleviated by regional policy. In recent years, energy policy, espe cially the introduction of a CO_2 tax, has had great impact on the industry, and caused Ålborg to lobby furiously, gaining special regulations and tax exemptions for energy intensive industries.

3.7 French and Dutch Petroleum Production

As a result of the two oil crises, energy conservation became an issue and petroleum products were substituted with other energy sources (mostly nuclear energy and natural gas). While crude oil extraction in Europe was negligible before the exploitation of North Sea oil, there have always been refineries located in the

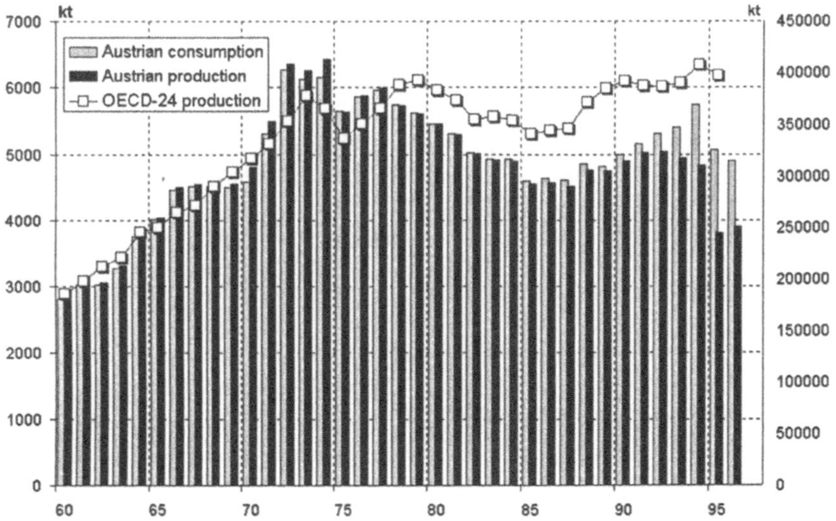

Source: UN, ÖSTAT, VÖZ

Fig. 21. Cement production in Austria and the OECD–24 and consumption in Austria 1960–96

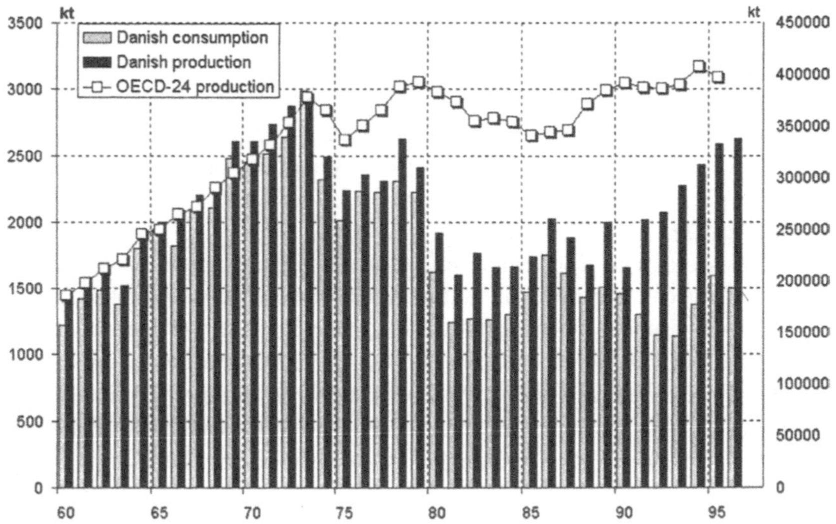

Source: UN, ECE

Fig. 22. Cement production in Denmark and the OECD–24 and consumption in Denmark 1960–96

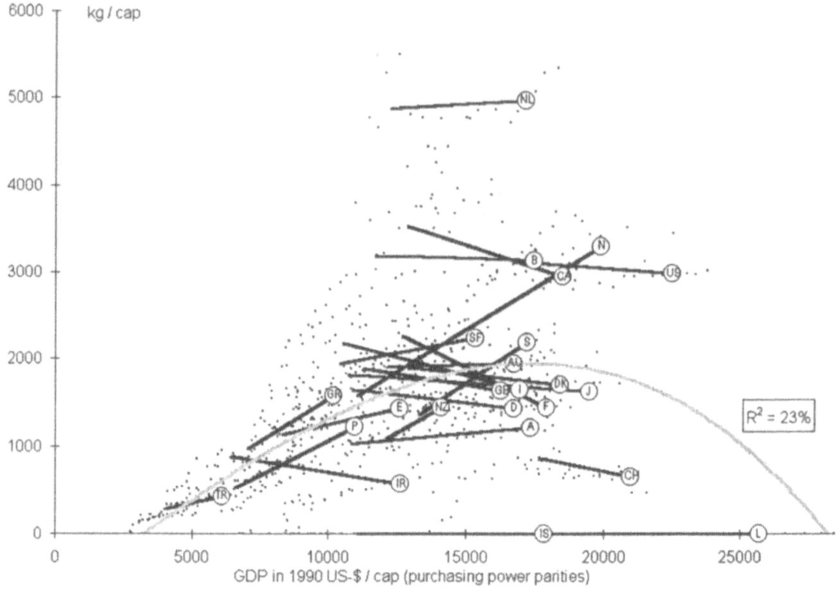

Source: IEA

Fig. 23. Output of petroleum products and GDP per capita in OECD–24 countries 1960–96 and national developments 1970/75–1990/95

industrialised countries. Multinational companies, such as British Petroleum or Royal Dutch Shell, are decisive in extraction, refining and distribution. Although these companies represent some of the most powerful economic organisations in the industrialised world, and are by no means unimportant to employment, the industrialised countries have generally practised a policy of explicit abatement. In a number of developed countries, refinery output has fallen slightly since the early 1970s (Fig. 23).

Nevertheless, in most of the countries, petroleum production and consumption has risen again, as energy prices have fallen since the mid-1980s. In France, however, consumption has not risen as noticeably, although the previous decline was sharper than in most other countries, as was the rise in the 1960s (Fig. 24). Furthermore, petroleum production has been partly substituted by imports from neighbouring and Arab countries. In contrast, export-oriented Dutch petroleum production has increased more than domestic consumption (Fig. 25).

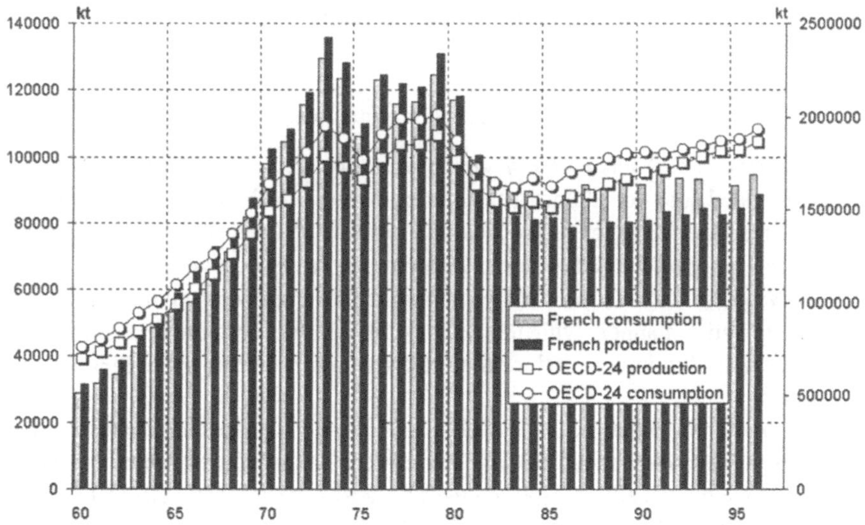

Source: IEA

Fig. 24. Output and consumption of petroleum products in France and the OECD–24 1960–96

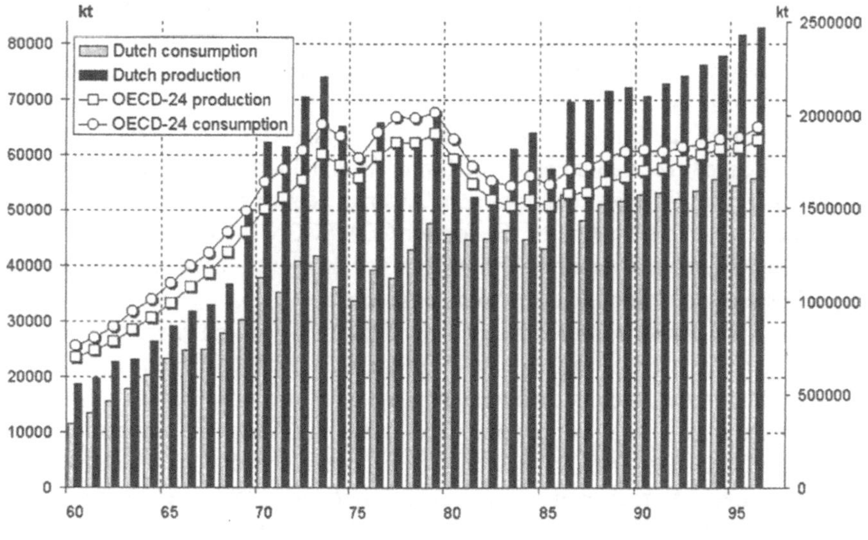

Source: IEA

Fig. 25. Output and consumption of petroleum products in the Netherlands and the OECD–24 1960–96

4 Comparative Results and Conclusions for Green Industrial Policy

The industries selected differ in many important respects:

- *Economic and social significance.* While the German coal and steel industries provided jobs for almost a million employees in the early 1960s, the significance of sectors such as the Austrian fertiliser and cement industries has always been far more limited (600 and 2000 employees respectively).
- *State-business relations.* Major decisions in German coal mining are made in a corporatist manner, with the nation-state as the most important actor, while German aluminium producers are almost totally free to make their own decisions.
- *Extent of the markets.* 92% of German chlorine is processed further on-site and only final products are traded. The cement industry's raw materials and products are not usually transported over distances greater than 100–150 km. On the other hand, the markets for steel and crude oil are essentially global markets, where regional markets have only limited independence.
- *Reasons for structural change.* The driving forces behind changes in the use of specific materials, and consequently the corresponding industries, are quite different. At present, no general laws governing this change are evident.

Nevertheless, the selected industries also have much in common. Usually, most national production is organised within a small number of companies. But foreign trade, mostly with neighbouring countries, restricts the influence of single companies or national authorities, and has become increasingly important in recent years. Deregulation of the European market and the expansion of the European Union have challenged the predominance of former national monopolies, cartels, or oligopolies (e.g. in German coal mining or Austrian cement production). Moreover, with the economic transition of the former COMECON countries (and their successor states), new competitors – as well as new markets – have emerged nearby (which is especially important for both Austrian cases).

Most of the industries were hit especially hard by the two oil crises and the subsequent economic crises, which usually triggered a decline in production and/or consumption in the cases studied here, because all of the industries in question are highly energy intensive. Furthermore, the slump in investment reduced demand for buildings and machinery, resulting in a substantial reduction in demand for basic materials like metals or cement. With lower energy prices and the recovery and stabilisation of economic growth in industrial countries since the mid 1980s, these factors have lost most of their impact on the development of basic industries. In the long run, other driving forces will be more important.

The impact of national GDP per capita on the size of these industries remains dubious at best, as the regression results and summary time-series indicate (see also the contribution of *Jänicke* in this book): for most indicators, the cubic regression could find hardly any relationship to GDP in the samples chosen (fertilisers, coal, steel and aluminium have R^2 below 0.10), for others the calculated regression lines were rather misleading when compared with the actual data in the scatter plot (extremely so with respect to cement production, to a lesser extent also for petroleum products and chlorine).

The technologies for producing and using these materials have been well-known for decades or even centuries. It is therefore highly unlikely that significant new uses and markets will emerge nowadays, and many established markets in industrial countries are more or less saturated. Innovation focuses on increasing the efficiency of production and consumption in established markets. Since materials and energy consumption contribute substantially to total costs, rising economic efficiency is usually accompanied by environmental benefits from declines in material and energy intensities. Thus integrated measures are often win-win solutions, at least if they are timed according to the investment or re-investment schedules of the affected companies. Companies which modernise early often have a competitive advantage over more conservative companies with a more hostile attitude to environmental issues, e.g. Dutch Esso compared to Dutch Shell in the case of petroleum refining in the Netherlands.

Environmental problems have sometimes contributed to a decline in production and consumption. In Austria, a tax on the use of fertiliser prevented a fall in fertiliser prices which might have led to a renewed rise in consumption. In Germany, chemicals businesses, under pressure from the public and facing imminent regulation, were eager to find alternatives to the most problematic uses of chlorine. But in most cases, environmental issues were not a major driving force. Whether recent, more strategic approaches to environmental protection, such as resource management (recycling requirements, life cycle analysis, or eco-auditing), "green" taxation or planning will have an impact on materials use remains to be seen.

This may lead to economic, social and political problems. From the point of view of private companies, they can only be successful if they adapt as fast as possible to the decline of their market. Adaptation might mean

- specialising in niche products (like Danish cement)
- speeding up modernisation or the decommissioning of unprofitable plants (e.g. Dutch Esso vs. Shell)
- using producers' expertise to sell less intensive use of their products (e.g. steel producers selling lighter products)
- diversifying investment

But it is always important to know *whose* problems are to be solved: the investors', the companies', their suppliers' or customers', their workers', the regional or national economy's, etc. For instance, diversification to more profitable sectors might sometimes be a solution for a troubled company or for an old industrial region like Luxembourg, but not necessarily so for skilled workers like well-paid miners or steel workers, if their skills are only of value to the now declining industry.

Problems mostly arise when the resources used by declining sectors cannot easily be reapplied:

- money invested in machinery which suddenly proves to be out-dated or simply superfluous (e.g. over-capacity in steel and petroleum)
- the advantages of traditional locations vanishing in the wake of technical change (e.g. easy access to European coal and iron ore resources becomes irrelevant because of falling freight rates) or institutional change (e.g. increasing competition for Austrian cement and fertiliser producers due to deregulation)
- human resources based on the specific skills, experience and values in the workforce
- institutions, networks, etc. designed or evolved to smooth traditional pathways of development and innovation in times of prosperity, but which now block new ones (e.g. coal and steel)

On the other hand, in those cases where these problems are not severe, the decline in production occurs less spectacularly, sometimes almost without public attention or even without political assistance. This is especially true if only a few jobs are involved and the skills are not restricted to the industry in question (e.g. fertilisers, chlorine, cement), and if the cuts can be managed simply by not replacing decommissioned machinery (e.g. German aluminium and chlorine).

All these problems are examples of "sunk costs" (in a very broad sense, including transaction costs). Sunk costs are a problem if and only if the "sunk" investment proves to be bad business, i.e. if decision makers made wrong decisions. Structural change does not necessarily mean financial loss and unemployment. If the decline in demand (or the rise of new, more profitable competitors) is anticipated correctly, workers may turn to other occupations (including re-training) instead of specialising in a doomed sector, investors may invest only in necessary capital (which then will be profitable even in declining markets) etc.

But especially in sectors in which machinery is used for decades (e.g. electricity, coal and steel, petroleum) it is next to impossible to forecast the market situation for the entire relevant period. In fact, in every single case in this book, neither public nor private actors predicted the falls in production, not even shortly before they occurred. Myopia is in this respect very often less a sin than a fact of life. This has two major implications:

- The losers under these developments are not really to blame for unavoidable losses. It is only fair to share this kind of burden from structural change across society – without, of course, prolonging the necessary economic restructuring and without undermining possible improvements in forecasting.
- Flexibility of economic resources is crucial for restructuring, in order to minimise the losses from wrong decisions: e.g. small scale, factor mobility, diversification of regions and companies, multi-functional skills, no streamlining of expectations, technologies, etc.

There are only few cases where the fall in production was a result of environmental issues or other political goals. In each of these cases, consumption and not the production of materials was criticised or even regulated, and producers were forced to react to the decrease in demand. This is the most likely scenario for all kind of dematerialization or green restructuring. Indeed, it would be difficult to achieve dematerialization by simply reducing production, since in free markets, demand could easily be satisfied by increasing imports, with probably no environmental benefit whatsoever. In several of our cases, a decline in production was accompanied by correspondingly higher imports (e.g. German aluminium and Austrian cement) or lower exports (e.g. steel in Luxembourg) as a consequence of deregulation. Either consumption would have to be directly coerced to a parallel decrease (e.g. by introducing marketable permits for the use or sale of materials) or foreign trade would have to be restricted (e.g. by import quotas).

Restrictions on foreign trade would scarcely be unfamiliar to European industrial policy, but they were usually placed as part of a strategy to protect domestic materials producers (for instance in agriculture, textiles, coal and steel). Political networks which address industries directly are likely to be dominated by these very industries: if a decline in steel *production* is the goal, steel producers will be invited to discuss the issue and outline the problems on reaching this goal. If a decline in steel *consumption* is the goal, the most important industrial actors may be the construction sector, the producers of vehicles and transport equipment etc. who are interested in saving steel. Industrial policy should therefore help to organise "demand" for green restructuring (agenda setting, formulation of targets, informational and market stimuli), leaving the "supply" side to private actors.

But well organised and powerful dirty industries, like energy, metals and chemicals producers, will oppose any industrial policy which is detrimental to their market position, as shown by the politics of global warming and especially the discussion of green taxes (see the French and Danish cases for details). Green industrial policies are more likely to succeed if they concentrate on a few strategic points in material flows, and if they support innovators or outsiders.

References

Binder, M. (1993): Energieverbrauch in der OECD. Berlin: FFU (Forschungsstelle für Umweltpolitik) rep 93–2.

Binder, M. (1999): Wachstum, Strukturwandel und Umweltschutz. Berlin: FFU (Forschungsstelle für Umweltpolitik) rep 99–5.

ECE (Economic Commission for Europe): Annual Bulletin of Housing and Building Statistics for Europe and North America. Genf.

ECE: The Steel Market in ... and Prospects for ... Genf.

Ekins, P. (1997): The Kuznets curve for the environment and economic growth: examining the evidence. in: Environment and Planning, Vol. 29.

Eurochlor (1999): Location of Chlorine Production Plants. in: http://www.eurochlor.com.

FAO (Food and Agriculture Organization): Yearbook Fertilizer. Rom.

FAO (Food and Agriculture Organization): Yearbook Forest Products. Rom.

Greening, L.A., Davis, W.B., Schipper, L. (1998): Decomposition of aggregate carbon intensity for the manufacturing sector: comparison of declining trends from 10 OECD countries for the period 1971–1991. Energy Economics 20, 43–65.

Haas, U. (1990): Standortperspektiven stromintensiver Produktionen in der Bundesrepublik Deutschland. Mögliche Konsequenzen umweltschutzinduzierter Strompreise bei Aluminium und Chlor. Göttingen: Vandenhoeck & Ruprecht.

Henseling, K.O. (1992): Ein Planet wird vergiftet. Der Siegeszug der Chemie: Geschichte einer Fehlentwicklung. Reinbek: Rowohlt.

IEA (International Energy Agency): Energy Statistics and Balances. Paris.

Jänicke, M., Mönch, H., Binder, M. (1993): Ecological Aspects of Structural Change. Intereconomics. 28 (4), 159–169.

Jänicke, M., Binder, M., Mönch, H. (1997): "Dirty Industries": Patterns of Change in Industrial Countries. Environmental and Resource Economics 9 (4), 467–491.

Metallgesellschaft AG: Metallstatistik. Frankfurt/Main.

OECD (Organisation for Economic Co-operation and Development): National Accounts, II. Paris.

Sun, J. W. (1998): Changes in energy consumption and energy intensity: A complete decomposition model. Energy Economics 20, 85–100.

UN (United Nations): Industrial Statistics Yearbook. New York.

UN: Statistical Yearbook. New York.

UNCTAD (United Nations Conference on Trade and Development): Yearbook of International Commodity Statistics. New York.

ÖSTAT (Österreichisches Statistisches Zentralamt): Außenhandelsstatistik. Wien.

ÖSTAT: Industriestatistik 1. und 2. Teil. Wien.

VÖZ (Vereinigung der österreichischen Zementindustrie) (1994): 100 Jahre Vereinigung der österreichischen Zementindustrie. Wien.

VÖZ: Jahresberichte. Wien.

A. Theoretical Approaches

Towards an End to the “Era of Materials”? Discussion of a Hypothesis

Martin Jänicke
Free University of Berlin, FFU, Ihnestr. 22, 14195 Berlin, Germany

1 Introduction

The postulate of reducing the rate at which the world converts material resources into emissions and waste has been part of the environmental discourse from the outset. Interestingly enough, even in the 1970s this prescriptive and somewhat pessimistic formula for preventing a global environmental crisis had received a more optimistic response: the hypothesis that advanced industrialised countries are progressing “beyond the era of materials”.

Both of these “schools” have supported their arguments with long-term statistical trends, and both have great importance for the debate on global industrial change and environmental restructuring (ecorestructuring). A general trend towards “dematerialization” would be accompanied by innovation and the rise of high-tech, information and service sectors, but it also would create losers in the mining and basic industry sectors, both of which not only play a strategic role in materials use but also display a high environmental intensity in their production processes. This change is essentially what ecorestructuring means.

In the following text I will attempt to show that the optimistic vision of an end to the era of materials remains an important one, even if it has thus far failed to be borne out by the facts. At present, there are more “N curves” than inverted “U curves” to be observed in per capita use of materials (shadowing the development of GDP). So where has the optimist’s hypothesis gone wrong? And how can the environmental postulate of dematerialization (Herman et al. 1989; Schmidt-Bleek 1994) be realised in spite of this? Perhaps the hypothesis failed to take the powerful structural rigidity of the older industrial countries sufficiently into account. If this is the case, a “green” industrial policy would gain even greater importance than before. The key task is to transform the de facto industrial policy in favour of basic industries into a strategy which makes structural change possible and acceptable.

2 The Development of the Hypothesis

The hypothesis of a general decline in materials consumption by advanced economies was explicitly proposed by Larson, Ross and Williams in the mid-1980s (Larson et al. 1986). It was based on an analysis of long-term (1890–1983) trends in US consumption of cement, steel, ammonia, paper, chlorine, aluminium, and ethylene. It had parallels with a Swedish study on the long-term development of the steel and cement sectors (Johansson et al. 1983). Trends in Germany, France and the UK also reflected this picture in the period between 1950 and 1983. The studies were repeated in the famous study "Energy for a Sustainable World" (Goldemberg et al. 1988). This too came to the conclusion that "a broad-based trend away from basic materials use in economy" is taking place, not only with respect to steel or cement, but even in "modern" materials such as aluminium, chlorine, or ethylene (Fig. 1).

The inverted U curve for intensity of materials use had been described before by Malenbaum (1978, p. 18). It shows materials intensities of use increasing parallel to a country's industrial development up to a certain point, after which it falls. This curve was subsequently also dubbed the "environmental Kuznets curve" – a usage which is now also applied to other environmentally significant indicators of long-term economic development (Grossman and Krueger 1994; WWF 1996).

In 1986, independently of the study by Larson et al., the Free University of Berlin conducted a similar study of steel, cement, primary energy, and the weight of

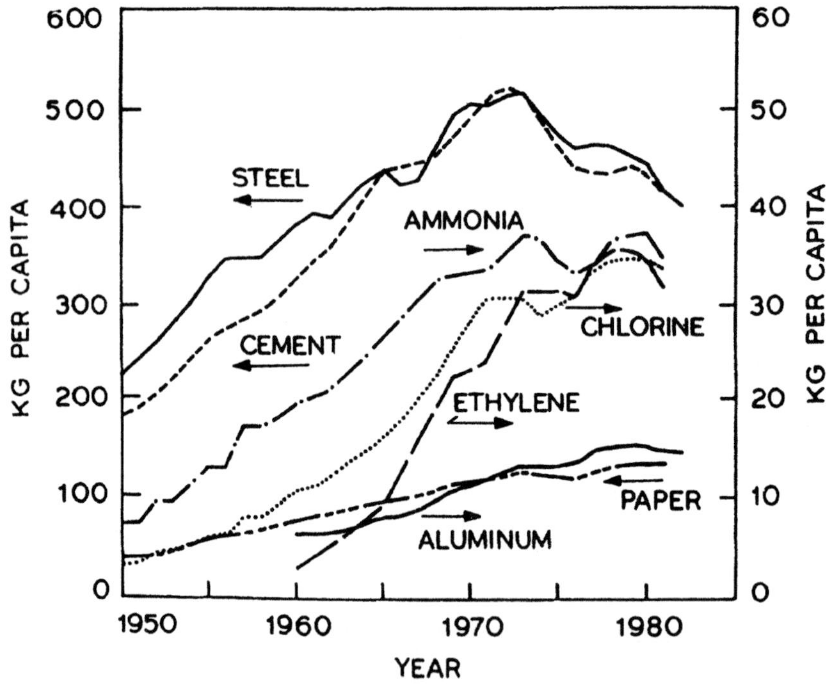

The data are for aggregated apparent consumption in Germany, France, and the U.K., averaged over running 3–year periods and plotted against the mid–year. The ammonia curve is for an aggregate of data for Germany and France only. Source: Goldemberg et al. 1988, p. 98.

Fig. 1. Per capita use of selected basic materials in Germany, France and the UK

transported goods in 32 industrialised countries (Jänicke et al. 1987; Jänicke and Mönch 1988). The study was influenced inter alia by general hypotheses about the long-term cycle of steel production (Prognos 1983). Comparing less developed industrialised countries with advanced OECD countries, we discovered that these factors were also described by inverse U curves (Fig. 2).

Theories of an at least relative decline in the importance of materials to highly developed countries date back to the 1960s and early 1970s.

Daniel Bell's theory of post-industrial society is worth mentioning in this context (Bell 1967, 1973; Kahn and Wiener 1967; Clark 1940). Although only partly concerned with materials and energy consumption, it describes tendencies towards a relative decrease in industrial production. In some senses it could be understood as the theory of a long-term tendency to at least partially substitute physical factors with services, knowledge and of a continuous increase in efficiency (including recycling). The theory consequently contains some grounds for environmental optimism (Bell 1973).

Incidentally, a similar theoretical debate took place early on in Eastern Europe. In the well-known Richta Report (1968), Czech reformers – citing Daniel Bell and other Western authors – described "a new ('postindustrial') type of growth..., based not on the growing quantity of the means of production and the labour force, but on the changing quality and the degree of using new production forces".

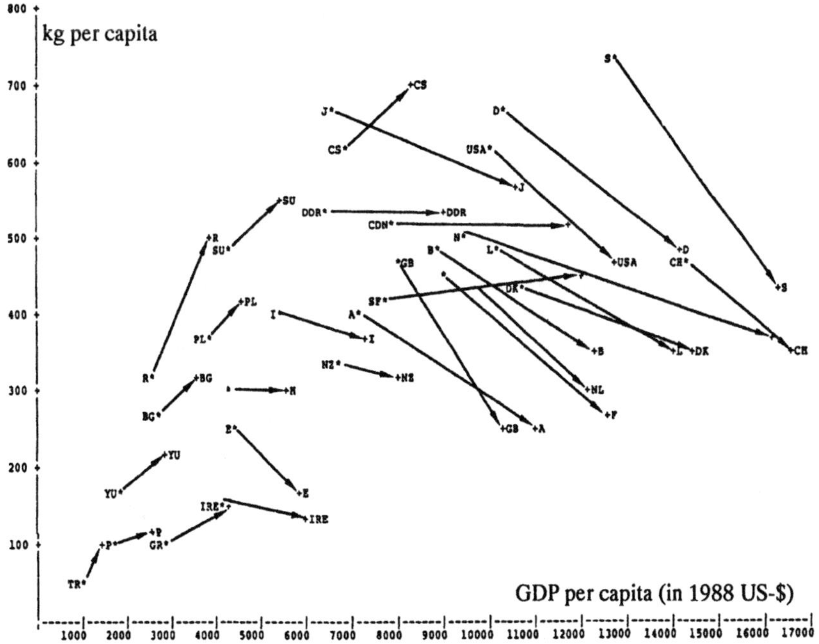

Source: Jänicke and Mönch 1988, p. 397

Fig. 2. Crude steel consumption in industrial countries 1970–85

Compared with "extensive growth" during the era of industrialisation, they emphasised the advent of an "intensive economic growth" based on high R&D input and combined with a growing service sector (Richta et al. 1968). Emphasis on innovation in this direction was well justified, in the light of extreme inefficiency of materials use, especially in the USSR – albeit without much success. The environmental impact of this technological revolution was not mentioned in the Richta Report, but its conclusions were clearly moving in the direction of "qualitative growth", which was subsequently to play an important role in the environmental debate.

Other, more value-based theories concerned with the intensity of materials use could be found in the USA, Japan, and Eastern Europe. Most of their criticism was on environmental grounds, arguing against existing trends in industrial development and seeing nothing in the way of an automatic improvement.

In the USA, authors like Boulding, Ayres and Kneese had been postulating the need for closed materials cycles since the late 1960s – stressing the importance of the "fundamental law of the conservation of mass". In 1974, a comprehensive empirical analysis of material flows relative to population and GDP in the US was published (Kneese et al. 1974; Fischer-Kowalski 1997). The global models of Forrester (1971) and Meadows (1972) demonstrated the potentially disastrous consequences – not least for the environment – if the era of materials and industrial growth were to persist in the long-term.

Japan was the first country to develop a more technocratic economic doctrine based on these new ideas. As early as in 1971, the MITI published its vision of a "knowledge-intensive industrial structure": branches with high energy and material intensities would be substituted with knowledge-intensive forms of production (Foljanty-Jost 1995, p. 159). In 1978, Japan was also the first – and only – country to introduce a law providing precise targets for reduction in material- or energy-intensive branches such as fertilisers (-45%), primary aluminium (-32%), shipbuilding (-35%), or steel (-14%) (Weidner 1996, p. 485).

Even in Russia, economists in the early 1970s linked the intensification of materials use to the natural environment (Fedorenko and Gofman 1973). The first complete audit of national material flows in the Soviet Union was conducted in 1974 – in parallel with Ayres and Kneese in the US – by Gofman, Lemeschew and Reimers, demonstrating (at least for the experts) the extreme wastefulness of materials use at all levels of production (Streibel 1990). Members of the East German Academy of Science stated in May 1973 that "the greatest proportion of environmental pollution is a direct consequence of an underdeveloped materials economy", and therefore that the goal of a "closed material cycle" should be set (Maier and Roos 1974, pp. 32, 35).

Later on, in the debate on economic reform in Eastern Europe, the intensification of production – i.e. higher energy and materials productivity – was regarded as almost a substitute for environmental protection. The low modernisation capacity of the Eastern communist systems, however, did not permit taking the necessary strategic steps.

3 The Hypothesis is Highly Plausible

Thus the hypothesis of an end to the era of materials was developed in two steps. Firstly, Larson et al. and Goldemberg et al. demonstrated empirically the inverted U curve for materials use intensity relative to growth in GDP. On this basis, the conclusion was then drawn that this development would also lead to an absolute fall in materials use. This hypothesis was based on arguments and observations which – at least at the time – appeared more or less plausible.

It was not the hypothesis on intensity of materials use as such, but rather the second part of the argument which has proved problematic today, and which will be discussed in the following.

Larson et al. and Goldemberg et al. mention the following driving forces as causing at least a "saturation in the use of basic materials" (Goldemberg et al. 1988, p. 98):

- substitution of traditional materials
- greater materials efficiency
- greater efficiency of industrial processes
- "more services and less material-intensive products": saturation of markets for products with high material intensity and the development of new markets for products and services with lower material intensity (Larson et al. 1986)

Indeed, there are many plausible theoretical reasons for an inverted U curve in the intensity of materials use to accompany the economic development of a country. To summarise the most important of them:

- the increasing importance of the service and information sectors and an associated decrease in the economic relevance of industry, especially mining and the basic industries
- the rise of post-materialism, or at any rate shifts in consumer preferences towards non-materialistic, "green" products
- the falling demand for basic materials within processing industries, brought about by increased efficiency and "intra-sectoral ecological modernisation" (Jänicke et al. 1993)
- the decrease in the materials intensity of the basic industries themselves
- improvements in the quality of materials (e.g. steel), reducing their relative mass
- political regulation raising the costs of environmentally intensive materials, or directly influencing demand or the efficiency of materials use

The trend towards lower intensity of materials use was enhanced by more temporary factors, such as the oil price rises in 1973 and 1979 and the subsequent fall in industrial growth (see below). Unfortunately, this constellation of driving forces was short-lived. Thus it might be useful to look at the statistical trends in materials use, including more recent developments. What do they tell us?

4 Disappointing Statistics

Larson et al. and Goldemberg et al. did indeed demonstrate a slight reduction in the use of selected materials between 1973 and 1985 (Goldemberg et al. 1988, p. 98). In our comparative study, mentioned above, we found similar trends for some materials in 1986. But we were loath to generalise upon such limited results (Jänicke et al. 1987): "An empirical argument against such theories could be that structural change in the 1970s was mainly due to the increase of prices of raw materials (not only energy prices). As the prices of raw materials have tended to decline since the beginning of the 1980s, the process of structural change could be expected to come to an end. Moreover, the appearance of new 'hard' raw materials could have a similar effect." In a later, more comprehensive study, we found statistical confirmation of this cautious position (Jänicke et al. 1993). At present the statistical trends can be summarised as follows:

1. The hypothesis of a long-term reduction in materials use intensity by highly developed countries remains statistically plausible. A general de-coupling of GDP and the use of certain materials has been observed since 1973. Fig. 3 demonstrates that this is the case for crude steel, cement, primary aluminium and fertilisers in OECD countries, although not for paper/paperboard. Furthermore, in five countries – USA, Japan, Germany, the Netherlands, and Austria – total materials consumption is similarly de-coupled from GDP (Adriaanse et al. 1997; Hüttler et al. 1997). Figure 4 shows this trend for USA and Japan. In addition, the OECD talks of a general "declining primary resource intensity of OECD output" between 1985 and 1995 (OECD 1997, p. 34).
2. *However, no general reduction of materials use can be found in the OECD countries.* This is not only true for most of the materials discussed here. In general, material flows are even growing in absolute terms, at least in the five countries mentioned above (Adriaanse et al. 1997; Berkhout 1997). Total materials use in Japan and the USA had noticeably stabilised between 1975 and the mid–1980s. Thereafter, however, they have experienced a renewed rise (Fig. 4).
3. *Instead of inverted U, or environmental Kuznets curves, we often found an up-down-up trend, which might be termed an "environmental N curve"* (Jänicke 1979, p. 111). There is no stable or general trend in materials use. In the mid–1980s – after the oil crises – a rising tendency in OECD countries could be observed, even for cement, although not for steel or fertiliser. The increased materials use in the USA and Japan has already been mentioned. Even in the case studies of this present book, some of the "declining basic industries" covered have experienced a revival since about 1985.
4. *Up to now, there is also no significant observable trend towards delocation of basic industries into developing countries,* in the sense of production being substituted with material imports (Jänicke et al. 1997). The OECD countries are still seen as the main net exporters of e.g. iron and steel. The share of primary

products in total exports by the "Big Five" developing countries fell by one half, to 29% in the period 1985–1995 (OECD 1997, pp. 34–35).

5. In general, *the engines of demand for materials "are deep-seated and persistent"* (Berkhout 1997, p. 288).
6. *Most of the material is consumed and emitted rapidly;* the amount of recycled materials is of a rather small order of magnitude, usually ranging from 2–5% – but 9% in Japan (Berkhout 1997, p. 291; Foljanty-Jost 1995, p. 124). Even the most highly developed industrial countries are therefore a long way from achieving "sustainable development" in terms of a closed materials cycle.
7. There is, however, an open question as to the effects of recent policy developments. It may be assumed with some plausibility that a more rapid increase in the efficiency of materials use will be brought about by recent waste management and recycling strategies, eco-auditing etc. The debate on factor 10 (Schmidt-Bleek 1994) or factor 4 (von Weizsäcker et al. 1995) may have at least some additional effects. Countries like the Netherlands, Germany (Adriaanse et al. 1997), or Sweden (where factor 10 was proposed by the government "Eco-Cycle Commission" in 1997) may be effectively influenced by the new policy paradigm.

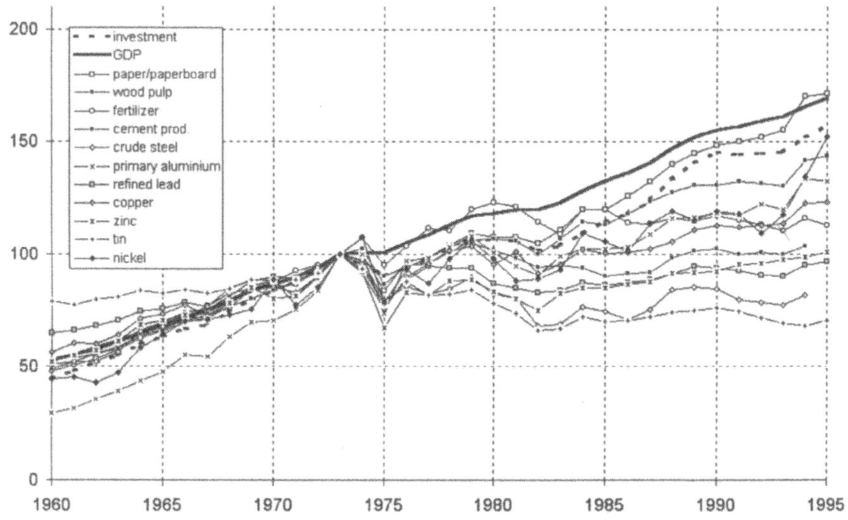

Sources: OECD, IEA, FAO, Metallgesellschaft UN, ECE/UN; figure by Manfred Binder.

Fig. 3. OECD materials use 1960–95 (1973=100)

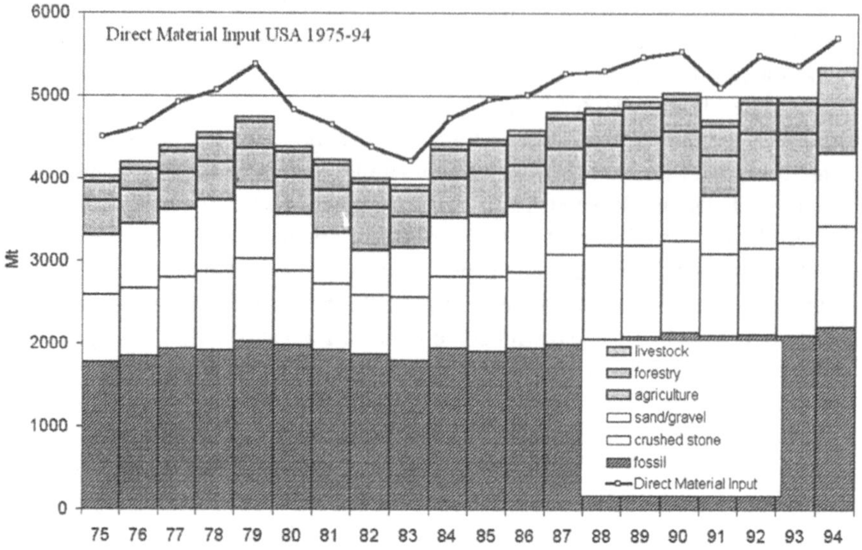

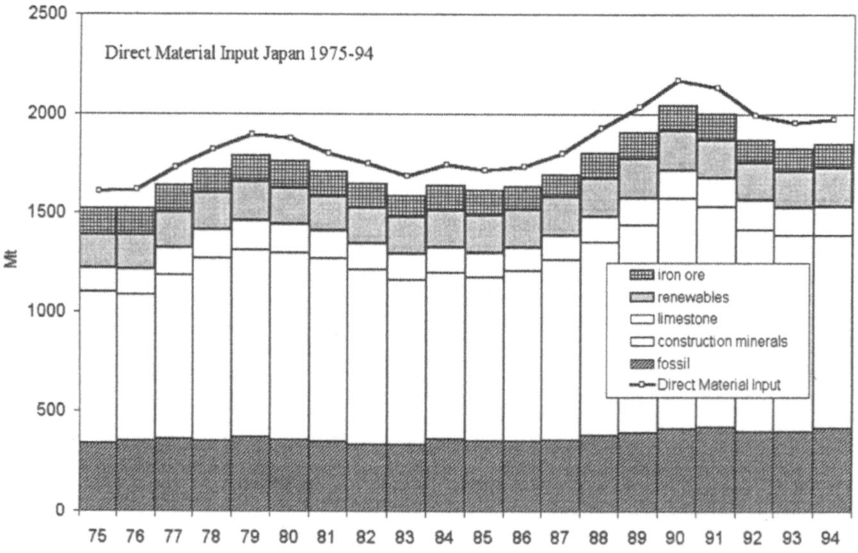

Source: Adriaanse et al. 1997. The DMI was recalculated by Manfred Binder

Fig. 4. Direct material input, USA and Japan 1975–94

5 Inverted U curves vs. environmental N curves

In the following section I discuss the weaknesses of the "end of the era of materials" hypothesis. Firstly, there is the issue of which questions a theory or hypothesis of the decline of materials use in highly developed countries should answer. At least four groups of driving forces should be analysed:

1. long-term technological and socio-economic factors which cause a reduction in materials use intensity
2. exceptional economic factors which cause a reduced demand for materials
3. normal competition within branches for new markets
4. political factors which promote or obstruct structural change

The hypothesis of an end to the "era of materials" is based solely on analysis of the first group: increased process and material efficiency, substitution of materials, shifts towards information and services in consumer preferences, the knowledge and service intensity of production itself, etc. All of these are facts, except that the new knowledge-based service economy has generated new forms of materials use and has been accompanied by a rapid "industrialisation" of private households (Gershuny and Miles 1983).

The second group refers to the rapid rise in oil (and energy) prices post–1973, but also to the subsequent slowing down of growth and investment (lower demand for machinery, buildings and the corresponding materials), which led to overcapacity, especially in the basic industries. This trend reversed itself in the mid–1980s, which at least partly explains the aforementioned N curves.

The third group of driving forces may provide an additional plausible explanation for the N curves. It would be unrealistic to assume that a sector will react passively to a fall in demand, caused in this case by modernisation in purchasing branches and other factors associated with the aforementioned first group. It can be safely assumed that it will fight to overcome the crisis, find new solutions and conquer new markets. Why should it fail?

The fourth group comprises political factors: government or regional political actors where the basic industry sector is concentrated play an important role as to whether or not structural change is accepted. This factor cannot be ignored. Political actors may help the branch to survive and find new markets. Possible options include direct public-sector demand, public investment (in the cases of cement or steel) or subsidies. On the other hand, policy can make the decline of a basic industry more acceptable through subsidies for alternative investment, retraining, social measures etc. Policy has sometimes even been able to force the phasing out of an environmentally harmful product.

The environmental N curve is commonly found in environmental research (Jänicke 1979, 1985). Whenever the causal factor of a problem increases and the counter measures address only the symptoms, there is no permanent solution to the problem. The same holds for a sector's environmental problems: during growth they exhibit a tendency to recur in time, unless additional measures are taken (which may result overall in a long-term stabilisation of problem indicators). This is a crucial point in environmentally sustainable development.

End-of-pipe measures in a growing industry (and also the transport sector) have typically been associated with environmental N curves – or with additional measures which stabilised the de-coupling of growth in output and pollution. The OECD has recently predicted that environmental indicators connected with motor traffic will follow an N curve (OECD 1997a, p. 104).

Within fast-growing industries, even ecological modernisation may be followed by a renewed upturn of environmental indicators. We have shown this for the case of Japan, where impressive industrial modernisation between 1973 and 1985 was in the end balanced out by rapid industrial growth (Jänicke et al. 1993). Even this extremely advanced country has achieved only a weak level of ecorestructuring in basic industries.

There are therefore often N curves where the hypothesis of an end to the "era of materials" would have suggested Kuznets curves. As regards materials use, however, such N curves do not have the same argumentative force as those associated with end-of-pipe treatment or economic growth. They result from the last three groups of driving forces listed above, which means that they are open to very different influences.

One point is nonetheless particularly important in the context of environmental policy. There is a contradiction between ecological modernisation and the response of basic industries to being affected by this as suppliers. Efficient use of materials, energy savings or phasing out the use of a toxic substance creates loser branches, who may react by competing for new markets, as is the rule in market economies. Electricity-efficiency measures, for example, are often responded to by the power industry (and its allies) through strategies to find new markets and, quite possibly, new ways to waste energy (such as the stand-by mode for electrical appliances). Action taken against the use of chlorine has been successfully counteracted by the target industry's efforts to increase consumption of PVC, etc.

More generally speaking, if structural change is not accepted (or effectively enforced), and if the "struggle for revival" remains a driving factor, the environmental improvements from modernisation within branches will be limited, if not balanced out by the supplying industry's search for new markets. An environmental N curve may then be the result (many of them typically tending towards stagnation).

6 The Importance of Political Factors

If this is the case, the stable forces pushing towards an end to the "era of materials" may be seen as *necessary* conditions for the corresponding ecorestructuring. But the *sufficient* condition will be a strategy which makes structural change acceptable to the loser branches (especially mining and basic industries). This provides a special argument in favour of "green" industrial policy, as opposed to a policy of structural conservatism.

The political factor – positive or negative – in structural change which, together with more temporary influences, has been ignored by those hypothesising the end of the "era of materials", may therefore prove crucial. The modern state currently exhibits a strong tendency to increase demand for materials through its policy on sectors such as construction, transport or agriculture, or through its general industrial policy. Only recently has a new strategy of policy integration emerged, and tried to influence those "other" policy areas as well. This has sometimes gone hand in hand with a target group policy, using the innovatory potential in industrial branches to achieve more environmentally sustainable development. As mentioned above, policy measures such as life-cycle analysis, material flow analysis, eco-auditing, a waste policy based more on prevention, or a shift towards environmental taxation may also influence materials use in the long-term. But it is too early to measure their effects.

It may even be too early to formulate a detailed strategy for material flow management and ecorestructuring on the basis of dematerialization. The knowledge base is still insufficient, and the goals of such a green industrial policy still require concrete design and development. For example, we must differentiate between the quantities of a sub-group of materials and its environmental importance. A distinction between *cleaning* and *reducing* materials flows would be a first step. As a second step, it would be important to formulate different strategies for different groups of materials: metals could (and should) be recycled as far as possible, but fossil fuels, which cannot be dealt with in this way, appear to be candidates for radically reduced consumption. The role of construction materials – the other major group of consumed materials – may be better discussed in association with sustainable land use, because even perfect recycling would solve only a small proportion of the environmental problems caused by the spread of built-up areas. The group of biotic materials on the one hand, and salts on the other hand, again imply very different forms of resource management and structural policy.

Many questions, therefore, still need to be answered. The question remains open as to whether the hypothesis of an absolute decrease in materials use by highly developed countries could – in the long run – be borne out in reality. From the point of view of sustainable development, we know only that there is an urgent need to move in this direction. It seems that the realisation of the empirical vision of an end to the era of materials depends upon the successful implementation of the prescriptive goal of dematerialization. Malenbaum, Larson, or Goldemberg have shown empirically that there are powerful driving forces which can be used for a corresponding strategy for the sustainable use of materials.

References

Adriaanse, A. et al. (1997): Resource Flows: The Material Basis of Industrial Economies. Washington, Wuppertal, The Hague, Tsukuba.

Bell, D. (1967): The Reforming of General Education. New York.

Bell, D. (1973): The Coming of the Post-Industrial Society. New York.

Berkhout, F. (1997): Policy Implications of Substance Flow Analysis. In: Bringezu, S. et al. (ed.): National and Regional Material Flow Accounting: From Paradigm to Practice of Sustainability, Proceedings of the ConAccount Workshop 21-23 January, 1997, Leiden (Wuppertal Special 4).

Clark, C. (1940): Conditions of Economic Progress. London.

Fedorenko, N. and Gofman, K. (1973): Rationelle Gestaltung der Umwelt als Problem der rationellen Planung und Leitung. Sowjetwissenschaft. Gesellschaftswissenschaftliche Beiträge, No. 5, 1973.

Fischer-Kowalski, M. (1997): Society's Metabolism. On Childhood and Adolescence of a Rising Conceptual Star. In: Redclift, M. and Woodgate, A. (eds): International Handbook of Environmental Sociology. Cheltenham: Edgar Elgar.

Foljanty-Jost, G. (1995): Ökonomie und Ökologie in Japan. Politik zwischen Wachstum und Umweltschutz. Opladen: Leske & Budrich.

Forrester, J.W. (1971): World Dynamics. Cambridge, Mass.: Wright–Allen.

Gershuny, J. and Miles, I. (1983): The New Service Economy. London

Goldemberg, J., Johannsson, T.B., Reddy, A.K.N., Williams, R.H. (1988): Energy for a Sustainable World. New Delhi et al: Wiley Eastern.

Grossman, G. and Krueger, A. (1994): Economic Growth and the Environment. Working Paper No 4634 of the National Bureau of Economic Research, Cambridge MA.

Herman, R. et al. (1989): Dematerialisation. In: Ausubel, J. and Sladowich, H.E. (eds.): Technology and Environment. Washington D.C.

Hüttler, W., Payer, H., Schandl, H. (1997): Gibt es eine Entkopplung von Wirtschaftswachstum und Ressourcenverbrauch? In: Fischer-Kowalski, M. et al. (eds.): Gesellschaftlicher Stoffwechsel und Kolonisierung von Natur. Amsterdam: Verlag Fakultas.

Jänicke, M. (1979): Wie das Industriesystem von seinen Mißständen profitiert. Opladen: Westdeutscher Verlag.

Jänicke, M. (1985): Preventive Environmental Policy as Ecological Modernisation and Structural Policy. Wissenschaftszentrum Berlin, IIUG dp 85–2.

Jänicke, M. and Mönch, H. (1988): Ökologischer und wirtschaftlicher Wandel im Industrieländervergleich. In: Schmidt, M.G. (ed.): Staatstätigkeit. International und historisch vergleichende Analysen. Sonderheft 19 der PVS, Opladen: Westdeutscher Verlag.

Jänicke, M., Mönch, H., Ranneberg, T., Simonis, U.E. (1987): Improving Environmental Quality through Structural Change. A Survey of Thirty-one Countries, Wissenschaftszentrum Berlin, IIUG Discussion paper 87–1 (German edition 1986: IIUG Discussion paper 86–1).

Jänicke, M., Mönch, H., Binder, M. (1993): Ecological Aspects of Structural Change. Intereconomics 28 (4), 159–169.

Jänicke, M., Binder, M., Mönch, H. (1997): "Dirty Industries": Patterns of Change in Industrial Countries. Environmental and Resource Economics 9, 467–491.

Johansson, T.B., Steen, P., Bogren, E., Fredricksson, R. (1983): Sweden Beyond Oil – The Efficient Use of Energy. Science 219, 355–361.

Kahn, H., Wiener, A.J. (1967): The year 2000: A framework for speculation on the next 33 years. New York.

Kneese, A., Ayres, R.U., D'Arge, R.C. (1974): Economics and the Environment: A Materials Balance Approach. In: Wolozin, H. (ed.): The Economics of Pollution. Morristown: General Learning Press, pp. 22–56.

Larson, E.D., Ross, M.H., Williams, R.H. (1986): Beyond the Era of Materials. Scientific American, 254 (6), 24–31.

Malenbaum, W. (1978): World Demand for Raw Materials in 1985 and 2000. New York: McGraw-Hill.

Maier, H. and Roos, H. (1974): Die Mensch-Umwelt-Beziehung als politökonomisches Problem. In: Reproduktion der natürlichen Umweltbedingungen, Berlin: Akademie-Verlag.

Meadows, D., Meadows, D., Zahn, E., Milling, P. (1972): The Limits to Growth, New York: Universe Books.

OECD (1997): The World in 2020. Towards a New Global Age. Paris.

OECD (1997a): Sustainable Development. OECD Policy Approaches for the 21st Century. Paris.

Prognos AG (1983): Euro-Report 84. Vol. A, Basel.

Richta, R. et al. (1968): Zivilisation am Scheideweg (Richta-Report). Prag.

Schmidt-Bleek, F. (1994): Wieviel Umwelt braucht der Mensch? Das Maß für ökologisches Wirtschaften. Berlin, Basel, Boston: Birkhäuser.

Streibel, G. (1990): Reproduktion und Nutzung der natürlichen Umwelt. Forschungsstelle für Umweltpolitik, FFU rep 90–13.

Weidner, H. (1996): Basiselemente einer erfolgreichen Umweltpolitik. Eine Analyse und Evaluation der Instrumente japanischer Umweltpolitik. Berlin: Edition Sigma.

von Weizsäcker, E.U., Lovins, A.B., Lovins, L.H. (1995): Faktor vier. Doppelter Wohlstand – halbierter Naturverbrauch. München: Knaur.

WWF 1996: Dangerous Curves: Does the Environment Improve with Economic Growth? Gland, Switzerland.

Structural Change in Basic Industries and the Environment – Theoretical Foundations and Classic Economic Theories

Stefan Zundel
Fachhochschule Lausitz, Großenhainer Str. 57, 01968 Senftenberg, Germany

1 Introduction

Modern industries, if we think of computer manufacture, software development or telecommunications, appear to be clean industries. Traditional industries, on the contrary, appear to have harmful effects on the environment, if we take as examples the iron and steel industry, coal mining or chlorine industry. This gives birth to the idea that economic structural change in the sense of the growth of modern industries and decline of traditional, especially basic, industries will have a positive effect on environmental quality. If there is a double dividend of modernising the economy, then it is only a small step to the next idea of a political alliance between the social forces interested in a modernised economy and those interested in environmentally sound economic development. But is there really a natural partnership between active industrial policy and environmental policy?

If the reality were as simple as that described above, it would need no further examination. Unfortunately, economic development is not a priori environmentally sound development. There are a lot of empirical findings which show that modern industries create new environmental problems. And even if the thesis of an end to the era of materials (cf. Larson et al. 1986) were correct, there is empirical evidence that a fall in the output of an old basic industry in a developed country is not always good for the environment. For example, the total effect on the environment of structural change can be worsened by the rise of new branches with new environmental problems. Since the relationship between economic and environmental development is far more complicated, this offers an interesting opportunity for further investigating the environmental effects of economic structural change. This is the starting point of my article.

As structural change is understood in a number of ways, we should start our discussion by introducing three different definitions of structural change: economic structural change, change in industrial metabolism and change in the environmental performance of an economy. In the first part of our article we use these concepts of structural change to shed some light on the relationship between economic structural change and environmental structural change from a systematic point of view. This part concludes by sketching out a scheme of intermediary factors. Each of them represents a hurdle or a hindrance on the path from economic structural change to environmental structural change, so we need to keep them in mind if we are to draw conclusions from an economic structural change on the basis of our empirical findings.

We can use the term "economic structural change" as a simple description, meaning that the share of an industry of the GNP has changed. However, from a theoretical point of view, not every change of this kind is structural. Some is caused by the driving forces of economic structural change, some by accidental circumstances. If we are going to distinguish between driving forces attached to economic structural change and driving forces attached to particular developments, we need a theory of structural change. Although there are a lot of theories *concerning* economic structural change – evolutionary, institutionalist, neoclassical, theories of globalisation and so on – the discussion about theories *of* economic structural change is almost completely lacking in the recent literature, leaving aside the German "Strukturberichterstattung" (structural reporting). In the second part of our article, therefore, we will look at some older theories of economic structural change and sketch out the driving forces behind structural change from the perspective of these theories (especially Hoffmann 1931; Chenery 1960; Chenery and Taylor 1968; Fourastié 1966; Rostow 1971; Schumpeter 1975).

The article concludes with an explanatory guideline of environmental structural change looking at the driving forces behind economic structural change. It can be used as a tool for the empirical analysis of economic development in basic industries and its environmental implications.

2 Economic and Environmental Change – a Complicated Relationship

There are certainly many possible ways to describe economic activities and their relationship to the environment. We believe that the three concepts mentioned above – economic structural change, change in industrial metabolism, change in an economy's environmental performance – represent the most common types of description and are useful for this discussion. At first glance, we can see that economic structural change influences the industrial metabolism and that the industrial metabolism influences the environmental performance of an economy. This is presumably the basic idea behind the "Larson hypothesis" (cf. Larson et al. 1986). Examining this thesis more closely raises an interesting question as to whether this is indeed the case and, if not, which significant intermediary factors are present. In order to answer this question, we need to clarify our concepts of structural change.

Structural change in an economic sense describes simply how an industry grows relative to growth in gross national product (GNP). The relative economic weight of an industry is measured by its net product in relation to GNP. In this context we must be aware that the question of what an industry is is not a trivial one. In principle, we can distinguish between functional or institutional definitions of an industry. Functional definitions are based on the main products or services of an industry. For example, the steel industry is defined simply by its production of steel. In fact, in almost all EU countries, the firms which belong to the steel industry produce a great many products which have no connection with steel production. Furthermore, they incorporate services which can be also found in any other firm outside the steel industry, for example accounting. Both aspects are institutional, because here the ownership is the significant discriminating factor. If a service is provided by a firm which belongs to the steel industry, this service increases the net product of the steel industry. If the same service is provided by a firm which belongs to coal mining, this service increases the net product of coal mining.

As it is almost impossible to give a clear-cut functional definition of an industry, all definitions used in statistics are based on a functional definition mixed with conventional elements. It is obvious that our empirical findings on economic structural change depend to some extent on our definition of that industry. For example, the chemicals industry in Germany is growing mainly because of a growth in services and not because of a rise in physical output. In general, if we define industries in an appropriate way, we can almost always find economic structural change.

There is another problem with economic structural change. If the level of aggregation is very low, a great deal of structural change can be found and vice versa (if the level is high, structural change will be less or will have simply disappeared). The reason is trivial. With a high level of aggregation, the structural changes in different industries neutralise one another. In general, if we aggregate the industries of companies to sectors appropriately, we can almost always find any mixture of intersectoral and intrasectoral change we wish.

Industrial metabolism can be defined as "...the whole integrated collection of physical processes that convert raw materials and energy, plus labour, into finished products and wastes ..." (cf. Ayres 1994) We should emphasise two points only. It is very difficult to quote accurate figures which represent industrial metabolism, because existing national accounting systems collect figures differently. Even if the size of the industrial metabolism is large, there is a significant difference between physical inputs and outputs on the one hand and the environmental effects which affect the quality of human life on the other. Owing to the fact that the status of environmental reserves and receptors are important, too, we need a further concept: the *environmental performance of a national economy*.

At first sight, the environmental performance of an economy can be defined by the status of resources and receptors and their feedback on the quality of human life. In this sense we can speak of environmental aspects as we speak of other objective standards of living, such as the number of doctors or schools relative to the size of the population. For example, we can measure the level of soil desertification or reserves of drinking-water. But almost every change in industrial metabolism will have two consequences: some environmental aspects of human life will improve, some will worsen. This is obvious, but the crucial question which follows on from this asks: under which conditions has our environmental balance sheet been improved? In other words, under which conditions do we have *envi-*

ronmental structural change, if we define this phenomenon as a positive net effect on the environment by economic structural change? In answering this question we need to integrate values into our definition of environmental performance. As nature evaluates its creatures only by their ability to survive, one has to insert human values. The question as to which human values these should be belongs to a discussion we cannot go into here.

Even if we think of all these problems as solved or soluble, there remain a few important intermediary factors between economical structural change and environmental structural change, shown in Fig. 1.

Under what circumstances will economic structural change entail environmental structural change? The intermediary factors listed above can be seen as reasons for economic structural change without environmental structural change. To be precise:

- *Changed environmental problems.* It is possible for the decline of a basic industry to be combined with the rise of modern industries which generate new environmental problems. For example, the substitution of iron or wood by synthetic materials is not always beneficial to the environment.
- *Technological change.* In many cases, changed production technologies (without a change in the physical design of the products) is associated with positive

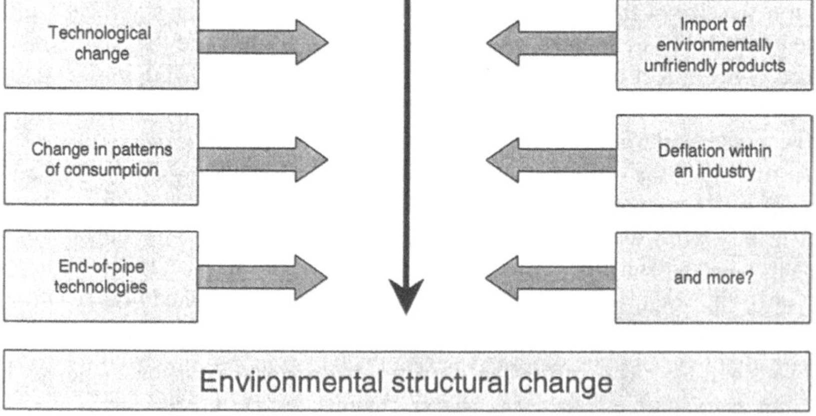

Fig. 1. Relationship between economical structural change and environmental structural change

effects on the environment, but this is not necessarily so. For example, production technologies in the agricultural sector have changed considerably throughout the industrialised world over the past fifty years and, at the same time, this sector's share of GNP has been falling steadily. The environmental problems caused by agriculture have also risen dramatically, if we think of nitrification of the soil and ground water, desertification of the soil or the shortage of water resources caused by modern farming technologies.

- *Change in the proportion of services within an industry.* The widespread practice of outsourcing services manifests itself – all things being equal – in economic structural change. In the statistics we then see the industry shrinking, but this is not associated with real change in the industrial metabolism.
- *Import of environmentally unfriendly products.* In principle, it is possible for the decline of an industry to be combined with an increase in imports of this industry's products. This case can be empirically eliminated by examining statistics on consumption. Far more complicated is a case involving imports of secondary and final products from a basic industry. For example, if there is a reduction in national steel production and consumption, but steel is nevertheless imported in cars bought from abroad.
- *Price cuts in an industry* will cause economic structural change if the fall in prices is not mirrored across all industries. It is obvious that this entails no change in industrial metabolism. In general, where there is a difference in pricing between industries, there will be economic structural change without a change in the physical framework of the economy, unless these differences are eliminated by absolute deflation in the net product of every single industry.

To summarise the first section in the form of a warning: we have to be very careful when linking economic structural change to environmental structural change. This is certainly not impossible, but it requires a few – and maybe sometimes deceptive – simplifying assumptions.

3 The Normal Pattern of Economic Structural Change

Three empirical findings present themselves in this area: an increase in per capita income over time, a relative decrease in the net output of basic industries over time and a decline in the environmental damage caused by industry relative to growth in its net output. The increase in income indicates economic development. The decrease in net output indicates economic structural change and the relative decline in environmental damage indicates environmental structural change. These findings constitute the empirical ground on which we stand with the hypothesis that economic structural change is a driving factor of environmental structural change.

For the sake of argument, we will assume that there is a significant connection between the economic decline of basic industries in developed countries, expressed in production value or net product, and the relative decline in environmental damage. We can then ask whether theories of economic structural change can explain environmental structural change. In using the term "theory" we have to consider an important implication: there must be regularity in structural change. This is not necessarily so. Three possible cases can be made out:

- A country's economic development is independent of that in any other country and follows no general pattern.
- A country's economic development is similar to that of other countries at a particular time.
- Each country's economic development follows a certain pattern.

It is obvious that general theories are not necessary if the economic development of each country is independent and follows no general pattern. In other words, if every case is a particular one, we do not need a theory and we cannot generalise about any single development. This would have major implications: if it were so, we could not use one case to make generalisations about others and we would therefore be unable to create general political rules for designing industrial policy.

Although there has up to now been an interminable debate about the appropriate pattern (for an overview cf. Meißner/Fassing 1989, pp. 104–126), there is a great deal of empirical evidence in the economic literature supporting the third possibility. In principle, economic development follows a so-called "normal pattern", which is shown in Fig. 2.

The idea of a "normal" pattern is based on a distinction between three types of industries, called simply early, middle and late industries (cf. Chenery 1960): Early industries are normally defined by two features. They meet basic needs and use simple technologies. On the other hand, late industries are the opposite. They meet advanced needs and use high technology. Middle industries meet ordinary needs and use middle level technologies. These distinctions are somewhat artifi-

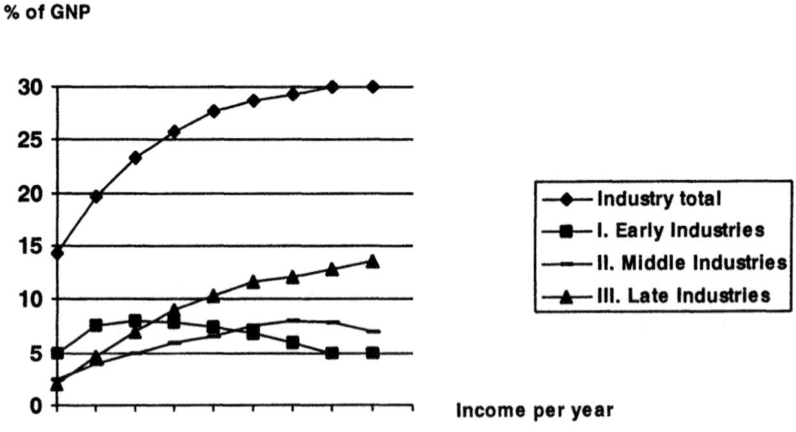

Fig. 2. The "normal" pattern

cial, but they fit our purposes. It is obvious that they reflect two theoretical ideas on economic change: that it is influenced by technological change and by changing needs.

As one can see from the idealised graph above, these three types of industries follow a particular development pattern over time. Early industries peak at a low level of per capita income. Middle industries peak at a medium level of income. Late industries grow steadily over time. The driving factors behind these trends, as normally found in older literature, are shown in Fig. 3.

The main force behind economic structural change is technological progress. It enables increases in productivity and entails higher incomes. On the other hand, it also generates new products, sometimes even new industries. An increase in income is normally accompanied by a change in consumer preferences and a subsequent change in demand. Increased productivity and the creation of new products cause a change in supply. Changing supply and demand conditions produce economic structural change.

The modern German system of structural reporting, "Strukturberichterstattung", paints a far more complicated picture of these driving forces (Fig. 4). In principle, this overview shows the same forces as above, but in a far more subtle and differentiated form. An important exception should be mentioned: subsidies and regulations. In general, industrial policy is itself a driving force of structural change. It

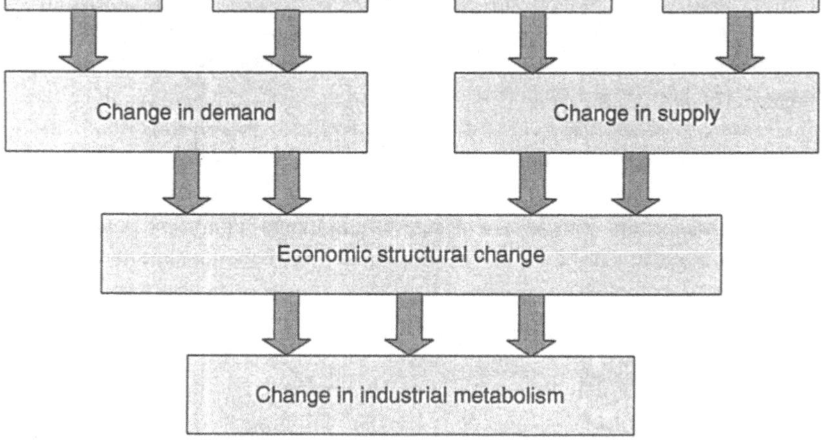

Fig. 3. Driving forces behind economic structural change

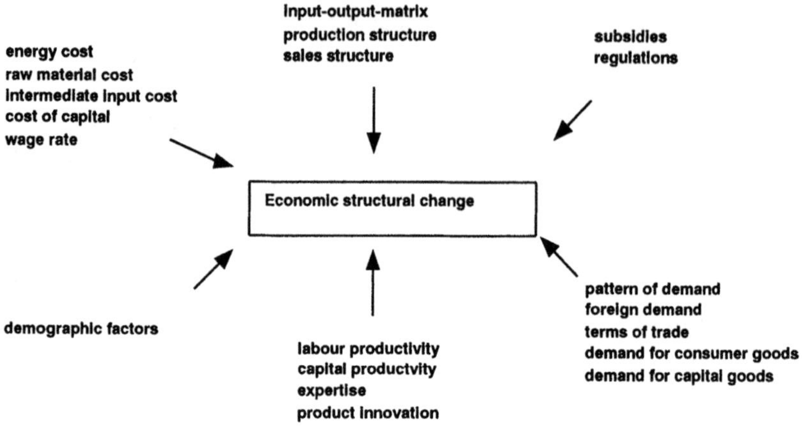

Fig. 4. Driving forces of economic structural change in the light of the "Strukturberichterstattung"

can shift economic development in one direction or another and can boost it or slow it down. Traditional economic theory is weak in this respect, because it assumes that policy normally reacts to event.

Even from a traditional point of view, not every country need necessarily follow the normal pattern. There are well accepted reasons for particular developments:

- If a country is rich in natural resources there can be development oriented around raw materials. That means that the primary sector of the economy is dominant for longer than it would be under the normal pattern.
- If the national markets are large, for example in the United States, national peculiarities in its economic development can persist for longer.
- If there is a great deal of international trade, the proportion of export-oriented industries is relatively high. This is one of the reasons for the underdevelopment of the service sector in Germany.
- If the infrastructure and human capital are well-developed, industries based on such inputs are also well-developed.
- National investment conditions are important for the rate of economic development and thereby for the rate of structural change. Persistent political turbulance can in particular be a great hindrance to economic development.

4 Different Approaches to Explaining Economic Structural Change

With respect to Fig. 2 and Fig. 3, we can make out a number of approaches to explaining economic structural change.

The first can be traced back to Hoffmann (cf. Hoffmann 1931) and has been popularised by Chenery and Taylor (see above). From their point of view, the main driving forces behind structural change are elasticity of demand, which expresses different consumer preferences for different types of goods – for example essential or luxury goods, nondurable and durable consumer goods – and technological progress, which produces a continuous change in the relative prices of labour and capital, entailing a substitution of labour by capital.

A similar approach, but at a far higher level of aggregation, is followed by Fisher, Clark and Fourastié (cf. Fisher 1939; Clark 1957; Fourastié 1966). It is well known as the "three sector hypothesis". As the name suggests, the protagonists of this approach distinguish between only three sectors of the economy. The primary sector contains branches such as agriculture and mining. Branches such as mechanical engineering, chemistry or electronics belong to the secondary sector. Branches which deliver services are in the tertiary sector, for example insurance or banking. The hypothesis claims that if income rises over time, the primary sector will decline over time, the secondary sector will first grow and then decline, the tertiary sector will grow. Like Hoffmann, Chenery and Taylor, Fisher, Clark and Fourastié also assume that elasticity of demand and technological progress are the main driving forces behind economic structural change. Consequently, the rise of the tertiary sector is explained by a combination of high elasticity of demand and low technological progress. On the other side, the rise and fall of the secondary sector is explained by good technological progress and changes in the elasticity of demand over time. The hypothesis has been confirmed by the past histories of industrialised countries.

The picture for the future is considerably more complicated, and for two main reasons. Firstly, it is unclear whether the assumption of low technological progress in the service sector can hold, with the widespread implementation of information technology in this sector. Secondly, as mentioned above, many services are closely linked with industrial activities and classification of these services in statistics is often only a question of ownership.

We can explain the fate of basic industries by this approach if we suppose that basic industries belong to the primary or secondary sectors. We must also decide the level of development which a country has achieved. But it is not so easy where basic industries are placed within the secondary sector, because we cannot then say a priori that there will be no unexpected changes in the rate of technological progress and in the elasticity of demand. All explanations which use these kinds of arguments are weak in this sense, because they extrapolate existing trends – a given rate of technological progress, a given elasticity of demand – into the future of individual industries. Inasmuch as we have no better explanation, this kind of analysis is nevertheless useful. And if we suppose furthermore that an economic decline of basic industries will be beneficial to the environment, we have here an argument for the existence of autonomous environmental structural change in developed countries.

Both approaches provide a necessary but insufficient raison d'etre for industrial policy. On the basis of a normal pattern of development we can compare the actual development of a country with its ideal development in relation to income levels. If there is a significant difference, there may be an opportunity for stimulating growth by supporting underdeveloped branches or sectors. For example, it has been assumed that the service sector in Germany is underdeveloped relative to that in similarly highly developed industrialised nations. We therefore have a rationale for active industrial policy but it is insufficient, because it is unclear whether government is capable of conducting industrial policy effectively and, with respect to neoliberal beliefs, appropriately.

The third popular approach is Vernon's life cycle hypothesis (cf. Vernon 1966). Figure 5 shows the typical life cycle of a product. Vernon assumes that the typical phases of the life cycle are specifically linked to allocation between developed and developing countries. In the first phase we see labour-intensive research and development activities in industrialised countries. In the second phase, standardised mass production is typical, and can be reallocated to less developed countries to take advantage of cheap labour. In the third phase, production is switched again to industrialised countries, because it is now capital-intensive automatic production. This approach emphasises technical progress as a driving force behind economic structural change.

This hypothesis has not been as broadly confirmed as the three sector hypothesis of Fourastié et al. There are branches which follow this pattern, but we also have data on branches which do not. This is therefore probably no general rule and every branch or basic industry must be analysed individually. Especially in basic industries, we can often observe varied developments, simply because most of the interesting branches produce not a single product, for example steel, but a range of products, for example different kinds of steel: sheet steel for cars or special steels for construction and so on, and each of these products has its own life cycle.

The idea of a product life cycle can be used to justify industrial policy as well. From this point of view, each branch has its ideal location, dependent on its technological development. If there is significant deviation from the ideal allocation of

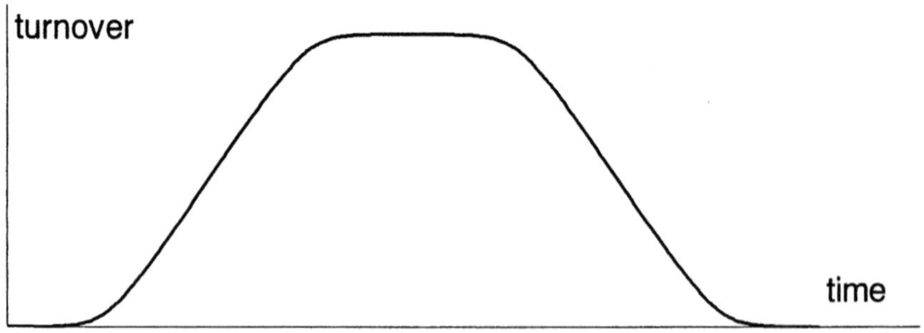

Fig. 5. Life cycle of a product

industries, the mobility and flexibility of production factors are probably restricted. And it can therefore be a task for governments to facilitate change, especially social change.

Last but not least, mention should be made of approaches in the tradition of Schumpeter (for example Hauff and Scharpf 1975). The fundamental idea is that new basic innovations can create new key branches. The life cycle of these basic innovations constitutes an industrial era. These eras are often identified with Kondratieff cycles, which last about 50–60 years. Owing to the fact that there is a first mover advantage for firms or branches which develop new products, innovations stimulate growth. Here also, the main driving force behind economic structural change is technological progress.

The main political message of this approach can be formulated as a simple rule: increase the rate of economic structural change. The instruments which can be used for this purpose depend partly on attitudes towards political intervention in the market. From an ordoliberal or neoclassical point of view the best method for speeding change is the reduction of subsidies for old branches and deregulation of new ones. But subsidies for innovators and for research and development in new products are also possible instruments.

5 Conclusions

Figure 6 represents an amalgamation of Fig. 1 and Fig. 3. It shows how we can explain environmental structural change by using the arguments which explain economic structural change. What can we learn about the fate of "dirty industries" from this particular economic perspective?

1. The first lesson, mentioned above, is simple but crucial: be careful! The relationship between economic structural change and environmental structural change is not a simple one. When researching the development of basic industries we must carefully investigate intermediary factors – developments which fit the general scheme are often only accidental.
2. There are certainly a great many advanced theoretical studies of structural change which use better theoretical frameworks than the one described above. Nevertheless, the main burden in explanation of structural change rests on traditional arguments, as we can see from the following example:

> It is a long observed fact that steel intensity (it can be seen as an indicator for the size of industrial metabolism caused by the steel industry, the author) – the comparative rate of growth of steel consumption and of GNP – changes according to the stage reached in a country's development. When the economy "takes off" steel consumption advances faster than GNP. Once the industrial equipment and the infrastructure have been installed, steel consumption slows down to the same rate of growth as GNP. Finally, once the market for industrial is saturated the rate of steel consumption growth falls behind that of GNP, which is increasingly diverted to sophisticated industries and services which are less intensive users of steel. (Keeling 1992, p. 154)

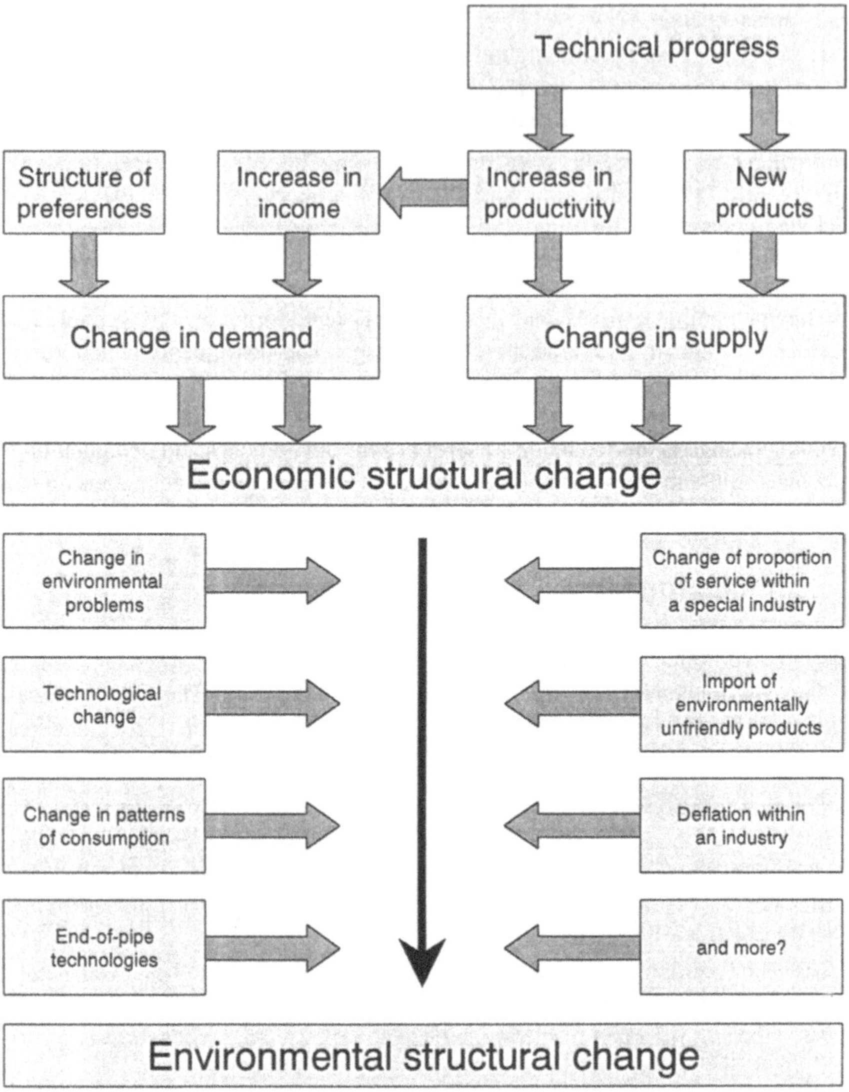

Fig. 6. Explaining environmental structural change with the driving forces of economic structural change

This is a typical example of the explanation of a specific aspect of industrial metabolism using the driving forces behind economic structural change. In this case, steel can be seen as a typical middle industry as defined above. The driving forces are provided by the saturation of important steel markets. In all case studies in this volume, it is likely that factors taken from the general framework developed above are used as an explanation for the relative decline of a basic industry.

3. But the crucial question from an environmental point of view is whether steel consumption in fact slows down in absolute terms. If replacement demand stabilises steel consumption at a high level, resources can nevertheless become exhausted, receptors overburdened, because the industrial metabolism has now simply become too large. In this case, economic structural change does not in fact imply environmental structural change
4. Last but not least, we cannot be sure a priori that basic industries will not experience a revival in the future if we classify them as early or middle industries. Certain indicators show that some basic industries in developed countries do recouple with the growth rate. Figure 7 shows this problem.
   The future of a basic industry must therefore be investigated if we are to be sure that economic structural change creates benefits for the environment in the long run. This point is very important because, from a theoretical point of view, we can not be sure if economic modernisation really yields a double dividend.
5. Traditional approaches to explaining economic structural change are not very closely linked to theories of government regulation. There is therefore no clear-cut position as to whether government interventions are useful and, if so, which

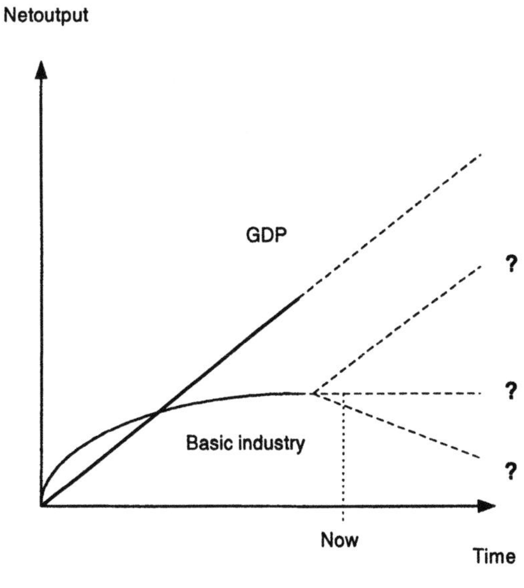

Fig. 7. Linking up with the growth rate again ?

form of government intervention is required. Nevertheless, theories of economic structural change represent a great temptation for political actors, even those who generally believe that political intervention in the market is not very useful. Consequently, we can observe throughout all industrial countries similar political attempts to subsidise research and development in new branches which are believed to be technologically strategic. Unfortunately, there is little difference between these national policies, and it therefore begins to appear that these kinds of policy create an international zero game.

6. Traditional approaches completely ignore the importance of existing government intervention for the progress of basic industries. From a traditional point of view, economic development is autonomous and policy reacts to this development. In fact, the fate of basic industries depends on government subsidies. Thus the crucial question is not whether there should be political intervention – in many cases there is political intervention like it or no. The crucial question is which kind of political intervention we prefer.

References

Ayres, R.U (1994): Industrial Metabolism: Theory and policy. In: Ayres, R.U. and Simonis, U.E (eds.): Industrial Metabolism. United Nations University Press. Tokyo/New York/Paris. p. 3.

Chenery, H.B. (1960): Patterns of Industrial Growth. The American Economic Review 50, 624–654.

Chenery, H.B. and Taylor, L (1968): Development Patterns: Among Countries and over Time. The Review of Economics and Statistics L (4), 391–416.

Clark, C. (1957): The Conditions of Economic Progress. 3rd edn (1st edn. 1940). London.

Fisher, A.G.B. (1939): Production, Primary, Secondary and Tertiary. Economic Record 15, 24–36.

Fourastié, J. (1966): Le progrès technique et les activités économiques. Revue économique 1, 115–126.

Hoffmann, W.G. (1931): Stadien und Typen der Industrialisierung. Jena.

Hauff, V. and Scharpf, F.W. (1975): Modernisierung der Volkswirtschaft. Frankfurt a.M., Köln.

Keeling, B. (1992): Structural Change in the World Steel Industry: A North-South Perspective. In: van Liemt, G. (ed.): Industry on the Move: Causes and Consequences of international Relocation in the manufacturing industry. Geneva, International Labour Office.

Larson, E., Ross, M.H., Williams, R.H. (1986): Beyond the Eve of Materials. Scientific American. 254 (6), 24-31.

Meißner, W. and Fassing, W. (1989): Wirtschaftsstruktur und Strukturpolitik. München.

Schumpter, J.A. (1975): Kapitalismus, Sozialismus und Demokratie. 4th edn. München.

Rostow, W.W. (1971): Politics and the Stages of Growth. Cambridge.

Vernon, R. (1966): International Investment and International Trade in the Product Cycle. Quarterly Journal of Economics 80, 190–207.

New Theoretical Perspectives on Industrial Restructuring and their Implications for (Green) Industrial Policy

Jan Nill and Ulrich Petschow (with Matthias Jahncke)
Institute for Ecological Economy Research, Potsdamer Str. 105, 10785 Berlin, Germany

1 Introduction

Experience has shown that changes in the industrial structure of an economy are a prerequisite for, as well as a side-effect of, economic development and growth. Notwithstanding, this mesoeconomic level of the economic process has rarely been addressed by the economics literature, which usually focuses on either micro- or macroeconomics, whereas an analysis of industrial restructuring must start with industries or appropriately defined sectors of an economy. The increasing incidence of sectoral crises afflicting entire industries in industrialised countries, and the course these crises took, have indicated determinants of industrial restructuring which cannot be explained, or at least not sufficiently, within the framework of traditional economic theories. It has been observed that, in certain sectors of the economy, the expected exits from and entries into the market have not taken place, that there has been massive and repeated political intervention, that the lifespans of production technologies and product cycles have shortened and that economic structural change has generally been discontinuous.

Given these new features of industrial restructuring, the evolution of firms, technologies and institutions in different industrial sectors is increasingly becoming a focus of interest, and precisely these factors have been taken up by more recent theoretical approaches. Evolutionary economics highlights technological progress as an engine of economic change, with an essential influence not only on markets, but also on the industrial structure of the economy. With the emergence of new institutional economics, a return to the analysis of the institutional setting which determines the field of possibilities for market processes can be observed. However, other varieties of the numerous institutional approaches are also gaining importance, for example public choice and network analysis. Finally, the part played by companies is also increasingly moving into the field of vision.

All these approaches are held to be of considerable importance in explaining the process of industrial restructuring, not least because their analysis emphasises the role of the actors in this process. This is also important with regard to the political implications. Industrial policy aims to influence industrial restructuring, but its implementation is faced with a number of difficulties, so that there is a rather large gap between theoretical optima and the empirical evidence which calls for explanations and strategies to overcome this gap. This holds also for policies which attempt to foster green industrial restructuring. These approaches may be able to shed new light on these problems.

However, there is still a considerable gap between their potential explanatory power and their practical application to industrial restructuring. This observation is the starting point of this chapter. It is our aim to offer an exploratory survey of the theoretical developments outlined above and to examine their contribution to the explanation of industrial restructuring and their implications for policy. At the same time, we attempt to place in a broader theoretical context the contributions by Frank Jöst, René Kemp, Patrick Kenis, Ans Kolk and Winfried Ruigrok, which test the productivity of some of these approaches in interpreting the case-studies.

We focus on industrial restructuring in quite a narrow sense: the evolution of the output of entire industries relative to other industries and the overall economy. Of course the definition of an industry varies between differing theoretical approaches. In this book, industries are equated with sectors such as steel, aluminium, chlorine etc. The phenomenon analysed is also often referred to as structural change. However, the emphasis is on the development of single industries. We do not give detailed consideration to restructuring within these industries nor to restructuring on a macro level (e.g. the slow-down of growth or the debate about postfordism). Finally, we limit our scope to industrialised countries.

In Sect. 2, we begin with a brief recapitulation of the deficiencies of classical approaches to industrial restructuring. Four important theoretical approaches will then be introduced: evolutionary economics (Sect. 3), new institutional economics (Sect. 4), network analysis (Sect. 5) and company-based approaches (Sect. 6). In Sect. 7 we summarise and demonstrate the relationship between industrial restructuring and industrial policy. The chapter closes with an analysis of the political implications of the different theoretical approaches, with a particular focus on "green" industrial policy aimed at green industrial restructuring (Sect. 8).

2 The Prevailing Economic Perspectives on Industrial Restructuring

Theories which consider or try to explain processes of industrial restructuring have been somewhat at the margins of economic research for some time. Processes of this kind have not been viewed as an economic problem. This is understandable if one bears in mind the neoclassical perspective on industrial restructuring. Conventional theories of economic structural change, on the other hand, only deal with this phenomena at a high level of aggregation. Moreover, their analysis is mostly cut off from political processes.

2.1 Industrial Restructuring and Neoclassical General Equilibrium

The perspective of the dominant neoclassical paradigm in economic theory does not allow for the existence of long-term economic problems on a mesoeconomic level. The operation of (competitive) markets and the price mechanism ensure a balance between supply and demand. Consequently, the results of the market are generally efficient, and thus the concrete industrial structure resulting from this process is economically of lesser importance. According to the basic model of perfect competition, the influence of companies as central economic agents on economic development, whether through market power or participation in (policy) networks, is usually a factor to be ignored. However, these results are usually obtained at a static level. Factors such as preferences, income, technology, and those institutions relevant to changes in economic structure are viewed as exogenous and thus treated as constant parameters. They can therefore be analysed only comparatively and staticly, but changes in these factors themselves cannot be analysed (Biervert and Held 1993, p. 8). However, as price flexibility ensures that the market adapts to changes in these parameters, the restructuring process itself is of minor interest. From a dynamic perspective, only the conditions for balanced growth, as analysed by traditional neoclassical growth theory, are deemed interesting. The steady-state framework assumes that there is no structural change and implies that demand grows proportionally (Pasinetti 1981, p. 69). Technological progress appears as a residual factor[1].

2.2 New Growth Theory and Industrial Restructuring

In the models of new growth theory, technical progress is not only an (exogenous) function of time but also an endogenous result of activities by economic actors. At the same time, (positive) externalities are incorporated into the theory. A theoretical understanding of the mechanism of knowledge production as a conscious activity of individuals with constant marginal returns and the possibility of achieving monopoly rents constitute the most important extensions, in comparison with the traditional approach. This underpinning of the growth process at micro-level, however, does not lead to a disaggregation of sectors. Thus different national and regional growth rates can be explained, but they are usually not yet linked to the phenomenon of industrial restructuring. Such a link seems possible only if forms of positive externalities (e.g. technological spillovers) are incorporated in strategic industries with increasing economies of scale[2].

2.3 Structural Change in Pasinetti's Classical Growth Model

Pasinetti (1981) develops a dynamic multisectoral model of growth along both classical and keynesian lines. In apparent contrast to neoclassics the model represents pure production, which is to a great extent independent of and neutral with regard to the institutional organisation of society. Structural change, i.e. structural

---

1 For a comparative analysis of neoclassical and evolutionary approaches see *Jöst* in this book.

2 A model which can be interpreted in this way has been developed by Grossman and Helpman (1991).

and sectoral dynamics, is modelled explicitly as a necessary attendant circumstance of the growth process. The driving force is not the relative scarcity of resources, but the human process of learning, which is expressed in terms of technological progress and changing consumer preferences (Pasinetti 1981, pp. 23–27). "Declining Industries" are explained as a natural consequence of the discrepancy between increased productivity, stimulated by technological progress, and change in demand. They are the necessary corollary of rising industries.

However, above and beyond the aforementioned points, this theory offers little which can be used in empirical study. Firstly, no prediction is made of the direction taken by structural change, except for strong emphasis on the importance of rising incomes for disproportionally rising demand. Secondly, Pasinetti defines sectors as "vertically integrated" only in respect of final products. He does not consider, at least not in his basic model, intermediate goods, and sectors are therefore not to be equated with "industries" (Pasinetti 1981, pp. 2–3, 68 and 110). Thus the theory is difficult to apply empirically because it agrees only partially with the conventional statistical definition of a "sector". Hence, especially for basic industries, successful application of this theory in practice seems unlikely.

2.4 Traditional Economic Theories of Structural Change

Most of these theories, which are presented in greater detail by *Zundel* in this book, deal with the phenomenon of structural change at a more general level than that of individual industries. Here, they do identify important potential determinants of industrial restructuring, namely changes in demand, technological progress and changes in the international division of labour. However, they are only able to deal with individual industries and specific changes to a limited extent. Structural change is usually understood as a continuous process, and explained with only a few strictly economic factors. Institutional factors, as well as possible political determinants, are left out of the analysis completely. Thus, although these theoretical approaches have a significant potential to explain, "the latter cannot be considered sufficient to explain the actual development of industrial restructuring" (Hartwig 1990, p. 2, our translation).

3 Evolutionary Economics

Evolutionary economics is a fairly young branch of economic theory, although its origins go back to Schumpeter at the turn of the century. Theories of economic evolution deal with the gradual change and development in an economy in terms of the intensity, structure and quality of economic processes over time (Feldmann 1993, p. 54). The causes, development and consequences of endogenous, innovation-led change are particular points of interest, as is the motivation of the relevant economic agents. The concept of economic evolution, the central elements of which are the development of innovations (mutations), their survival (selection), and the interaction between them, is usually developed in analogy to the biological evolutionary theory of Charles Darwin [3].

However, above and beyond this more or less pronounced paradigmatic reference to biology as opposed to mechanics, there is no unified theoretical framework. It consists rather of the works of individual economists, such as Schumpeter (e.g. 1912) or Kenneth Boulding (e.g. 1981), and a common range of topics. It is therefore somewhat difficult to separate it from institutional approaches (Seifert and Priddat 1995, p. 20). Its main starting-points are the influence of innovation and technology on change in aggregate economic output, reference to darwinian ideas in analysing the behaviour of companies and industries in innovation-related competition, the evolutionary perspective on market mechanisms and the path dependence of historical economic developments (Witt 1995, pp. 154–155).

In general, evolutionary economics takes an actor-based – microeconomic – approach in discovering the characteristics of a system. Given the constitutive character of uncertainty, it understands rationality as "bounded rationality". The behaviour of the heterogeneous actors springs from the interaction between ignorance, uncertainty and learning, which from an evolutionary perspective includes innovation (Witt 1995, pp. 164–165). The latter is stimulated by, for example, a general dissatisfaction with the present situation among actors or failure to reach the desired degree of satisfaction. The principle of profit maximisation is replaced, due to the presence of uncertainty, by the concept of "satisficing" (e.g. Nelson and Winter 1982).

The dynamics of the system result from interaction between heterogeneous actors. In modelling micro-macro-interaction, concepts are often imported from the natural sciences, e.g. synergetics or theories of self-organisation. Innovation is understood as a search process characterised by discontinuity and uncertainty. The market not only serves as decentralised processor of information, but also, along the lines of Friedrich von Hayek, as a selective institution which generates and evaluates novelty (Witt 1995, p. 167). One constitutive element of evolutionary theory is therefore that it explains innovation ex-post. As a corollary, as soon as the latter invokes changes in the genotype, it becomes methodologically impossible to develop deterministic prognoses about the results of the system's future evolution.

3.1 A Basic Model by Nelson and Winter

One important contribution is the model developed by Nelson and Winter (1982). It focuses on the behaviour of firms and industries in innovation-related competition, i.e. intra-industrial restructuring. Based on Schumpeter, economic development is sketched out as a dynamic, unbalanced process driven by the successful market introduction of innovations. In analogy to biological evolution, a distinction is drawn between two different levels of change:

[3] The basics of the theory are presented in more detail by Dosi and Nelson (1994).

- changes in the genotype (repeated sequences of actions within a firm – routines – and the total stock of existing technological knowledge)
- changes in the phenotype (central economic parameters such as quantity, quality, price, number of technical procedures used)

The prototype change is the change in phenotype in response to changes in the genotype, i.e. when research results are developed to marketability. But in economic evolution there is also feedback between the two types: frequently the mere application of a certain production technique expands the base of technical knowledge (learning by doing, learning by using) (Hartwig 1990, p. 31). Thus the separation between the two levels is not as clear-cut as in biological evolutionary theory.

There are two basic forms in which novelty appears: the rearrangement of existing elements or the emergence of completely new elements (mutation). Both processes lead to a broadening of technological knowledge and hence to a change in the genotype. Unlike in the biological world, where mutation and recombination are purely random, every developed economic system comprises a multitude of private and public efforts which are deployed in the conscious development of new products and processes. Thus evolutionary economics entails an actor-oriented perspective.

Nevertheless, not every new invention is implemented in the economy, because economic selection criteria operate and determine the direction of technical evolution. Internal selection works through choice between different production techniques, external selection through market competition. Thus the model is well-suited for analysis of the diffusion of innovations, i.e. the feedback from genotypic change to phenotypic factors. However, within the framework of this model, little can be said about the characteristics of the innovation itself or the creation of new alternatives. Given the assumptions of satisficing and bounded rationality, the only systematic force seems to be competitive selection at this level also, i.e. competition between routines (Witt 1995, p. 163).

3.2 Technological Change and Evolutionary System Dynamics

Technological change is another focus of evolutionary economics. Technologies can be seen as typical cases of evolutionary development (Dosi and Nelson 1994, p. 160). Technological knowledge is the basis of innovative activity. Firms are the main carriers of technologies. At the same time, technologies are closely associated with their organisational application. In order to highlight these facts, Dosi (1982) has introduced the notion of the "technological paradigm", i.e. a common idea which directs the action taken. In combination with corresponding hypotheses on learning and the diffusion of information, this produces a framework for the analysis of regularity (the "technological trajectories"). Such regularity is frequent at the level of single industries and it shares features with the well-known life-cycle model of industries. At the beginning, there are many firms and technologies, but in the course of evolution a technological paradigm emerges and a stable industrial structure of larger firms develops. However, this does not hold for all sectors. Moreover, the question of the influence of new technologies on industrial structure becomes important. "Competence destroying technical advance" can lead to discontinuous processes (Dosi and Nelson 1994, pp. 164–166).

Path dependence, and often also dynamic increasing returns, play an important role in almost every evolutionary model, particularly for technological trajectories. The recent discussions of this topic are thereby easy to integrate in an evolutionary perspective. Nevertheless, the central contributions have frequently used a variant of the neoclassical framework for analysis. In the models developed by Arthur (1989), the introduction of the law of increasing returns with respect to adoption and the acceptance of stochastic elements (small "historical" events as selection mechanism) are sufficient to produce path dependence. The main result is the occurrence of technological "lock-in": One technology can advance on competitors without being necessarily better or more efficient. Increasing returns on adoption can be caused by technology-specific learning, which is also important in evolutionary theories (so-called cumulative technologies) or by network externalities. The latter form the basis of the paradigmatic case of path dependence analysed by David (1985), the success of the QWERTY keyboard.

Irreversible, self-reinforcing changes to the system during the diffusion process, a multitude of possible equilibria and the dependence of results on environmental conditions (the central argument of selection) fit well with the evolutionary approach. However, deducing the phenomenon of persistent technological lock-in is a theoretical challenge, as e.g. the model developed by Silverberg, Dosi and Orsenigo shows. Here, assuming the adopters are sufficiently heterogeneous, it is highly likely that a better new technology will succeed after successful learning processes (Dosi and Nelson 1994, p. 167). A solution is proposed by Witt (1997). He models both lock-in as well as the conditions for overcoming it through technological change in an evolutionary framework. The main driving force is the effect of frequency dependency, i.e. the importance of a critical mass of actors or innovators for the respective result. In his model, it is crucial that a critical mass of diffusion agents organise themselves in order to ensure the breakthrough of the better technology and to overcome negative network externalities (Witt 1997, pp. 768–769).

Hence, the system's evolution is not only characterised by the influence of chance but also dependent on decisions taken in the past (path dependence), and so evolutionary economics is naturally able to explain macro-level regularity. Technical change is identified as a motor of growth and development processes. Different technological regimes, i.e. complexes of firms, research and education institutions and regulatory structures lead to differing growth rates (Dosi and Nelson 1994, p. 167)[4]. Up to now, such technological regimes are often related to different national innovation systems (e.g. Nelson 1993, Edquist 1997).

[4] For a more detailed presentation cf. Cimoli and Dosi (1995).

3.3 Industrial Restructuring in Evolutionary Economics

Given its emphasis on change, evolutionary economics seems an appropriate approach to analysing industrial restructuring, particularly discontinuous processes. Structural regularity and structural change are seen as specific to countries or technologies. The application of this perspective to industrial restructuring seems perfectly possible. Nevertheless, this potential has been little exploited up to now, especially with respect to the relationship between path dependency and industrial restructuring, although certain real phenomena, such as reactions to the oil crisis, should be relatively easy to explain[5].

One exception is the analysis of Faber, Manstetten and Proops (1996). They develop a specific evolutionary framework by integrating the theory of joint production and neo-austrian capital theory. This approach is then applied to the analysis of industrial restructuring (Faber et al. 1996). In his contribution, *Frank Jöst* presents this approach in greater detail and applies it to the case studies.

A different approach to explaining industrial restructuring is taken by Berg (1995), with the integration of additional institutional factors (see Sect. 4.3), which appears logical in the light of the importance of the selection environment. Alchian (1950) had already suggested extending the methodological concept of individualism with a description of the selection environment in order to gain useful insights into the evolution of an economic system. Institutions may constitute an important factor here. Moreover, the technological lock-in focused on above and cumulative effects are precisely the factors which lead to interaction with institutions. Taking this into account, the scope of the argument for economic selection in a strict sense of the word is still open to debate. The integration of political and social forces may then be necessary for a long-term analysis as well (Nelson 1994, pp. 51–52).

4 New Institutional Economics

Institutional economics is characterised by somewhat diverse approaches. Traditional institutional economics is based primarily on socio-economic factors. In contrast to technological progress, institutions constitute rather a retardant factor within societies (Reuter 1996). New institutional economics is based on neoclassical analytical methods for identifying the rationale of economic institutions.

4.1 Foundations and Approaches of New Institutional Economics

The approaches of new institutional economics aim to fill the vacuum of neoclassical theory without renouncing its theoretical basis. This common basis is the

---

[5] One example could be the interaction of price changes and energy intensity in time. Even neoclassical authors like Klodt (1994) state important deficiencies of neoclassical explanations in this domain.

model of rational agents maximising individual benefit within given restrictions (maximisation of benefit, bounded rationality, opportunistic behaviour, methodological individualism) (Furubotn and Richter 1991, p. 4). Thus the dynamics of the emergence and alteration of institutions, as well as the impact of institutions on the behaviour of economic actors are traced back to the cost-benefit calculations of (more or less) rationally acting individuals. Both aspects, institutional change and the impact of institutions as restrictions on behaviour, are closely related in reality. Berg (1995, p. 121, our translation) speaks of a "circular causal relationship between regular behaviour on the one side and the regulatory framework on the other side".

Institutional economics therefore looks for explanations of the choice of organisational form and deals with the emergence, development and decline of economic institutions. Institutional economics is in this respect strongly influenced by evolutionary economics (Hartwig 1990, p. 8). Important approaches in new institutional economics are agency theory, the theory of property rights and transaction cost economics. The latter is of particular interest for analysing industrial restructuring.

One theoretical branch which is sometimes thought of as belonging to new institutional economics in a wider sense is public choice. This theory applies neoclassical methods such as utility maximisation to political processes. One stream within this broad collection of models deals with the logic of collective action and deduces institutions as a result of this logic. The founding father of this reasoning is Mancur Olson (1965). He explains the influence of interest groups on decision processes by their (differing) capacity to organise their special interests. Quite small groups can thereby exert considerable influence and enforce a result which is far removed from the optimal result for a society.

4.2 Transaction Cost Economics

Transaction costs play a major role within new institutional economics. In contrast to orthodox economic models, in which the exchange of goods is considered in an ideal form, without the input of resources, these transactions are in reality associated with costs. These costs arise above all because of imperfect information, in the initiation and conclusion of contracts and in the costs of monitoring compliance (Weise et al. 1993). New institutional economics focuses mainly on the comparison of existing institutional arrangements. In contrast to traditional welfare economics it is not based on the assumption of zero transaction costs in a perfect, universalised ("nirvana") situation (Hartwig 1990, p. 8). An important characteristic of transaction cost economics is that the theoretical basis has some dialogue with neighbouring disciplines. Its concepts can therefore be applied in sociology, economics, political and administrative studies (Schneider and Kenis 1996, pp. 14–15).

The transaction costs approach pioneered by Oliver Williamson (1975, 1985) is oriented towards analysing these institutional arrangements and their effects of reducing transaction costs. To put it in simplified terms, there is the institutional supply side (authorities, ministries and so on), and the demand side (firms, associations). Research aims to explain the development of forms of organisation and agreement within a given institutional framework, constant technological knowledge and stable orders of preferences. Its main focus is consideration of efficiency at a static and comparative static level. It should be pointed out that the institutional development path is mainly oriented around potential for reducing transactions costs within given institutional constraints (Williamson strategy), and therefore the long-term evolution of institutions is neglected.

Douglas North (1990) introduces dynamics into the transaction costs approach. The focus of his research is the interaction between technological progress and institutional development and especially the possibility of exerting strategic influence on the institutional framework. In contrast to other institutional economists, such as Williamson, who are interested in comparing existing institutions, North focuses primarily on attempts by firms to change the constraints of the institutional framework (North Strategy). North identifies two main influential factors: changes in relative prices, influenced mainly by technological progress, and changes in the structure of preferences, influenced mainly by society's values or "ideology" (Berg 1995, pp. 134–135).

Furthermore, North starts out by assuming that economic growth is only possible through increasing specialisation or the increasing division of labour. This specialisation increases transactions costs, which therefore have to be more than offset by gains through specialisation. Just as innovative technologies can reduce productions costs through learning processes, institutional innovation is able to reduce transaction costs. Institutional change can be explained in an extended analogy to the production process (Berg 1995, pp. 137–138).

4.3 Institutions as Determinants of Industrial Restructuring

According to the argument developed by public choice theory, collective action can lead to political influence impacting on industrial restructuring and growth. Using a neoclassical efficiency framework, Olson (1982) links the rise and decline of nations to the power of organised special interests, which he explains to some extent endogenously as a result of the corresponding economic situation. Authors at the Kiel Institute for World Economy apply a similar approach to the process of industrial restructuring (Klodt 1988). The emergence of institutional change, e.g. powerful interest groups, is assumed to reduce flexibility and thus to lead to insufficiently rapid industrial restructuring. A reactive policy of subsidies is identified as a particular brake to industrial restructuring. The policy itself is partly explained as the result of rent-seeking by firms (and not necessarily as collective action) (e.g. Lenway et al. 1996). But even if the emergence of political factors is partly endogenous, this has no influence on the notion of (economic) structural change; the latter is still regarded as an independent process; institutions occur as an exogenously determined disruptive factor in industrial restructuring.

Institutions restrict the behaviour of agents but, from a dynamic perspective, institutional constraints are not exogenous, but influenced by actors and organisations. In particular there is feedback between technical progress and the institutional environment, e.g. the incentive structure of the institutional structure. Referring to North, Christoph Berg therefore tries to explain industrial restructuring by linking evolutionary and institutional economic considerations. Berg (1995, pp. 169–184) thus develops a scheme of phases in sectoral development. The basis of this scheme is the evolution of a particular technological system consisting of four phases: development, expansion, maturity, and stagnation. The development phase is influenced decisively by the institutional framework, which provides the necessary incentive mechanisms and financial support, as well as by procedures in science which direct the realisation of genotypic potential (science and technology push).

In the growth phase, there is interaction between technology push and demand pull. At this early stage the institutional structure, and especially the state, can act as diffusion agent. However, the organised interests of established industries can also constitute a constraint. At the same time, promoted by increasing returns on adoption, specific technological trajectories evolve via positive network externalities, pronounced learning effects and adapting expectations (Berg 1995, p. 176).

Consolidation of the development path is reached in the maturity phase, which includes slackening of technical progress, as well as stagnating or diminishing returns on adoption. Thus firms in mature industries have no interest in investing further in technologies with diminishing returns, but would rather invest in gaining the appropriate influence on industrial policy measures instead. The reason for this is that the structure of interests on the supply side is homogeneous in this phase and transaction costs can therefore be reduced significantly.

In the stagnation phase, the markets are becoming increasingly more concentrated and restriction of competition becomes more probable and attractive because of increasing elasticity of supply and decreasing elasticity of demand. The restrictive character of the institutional framework can be reduced by means of successful collective action (and, in extreme cases, the formation of cartels) and can become a strategic variable, although this is limited by institutional competition on the part of other industries. Technological progress becomes more and more expensive and co-operation in innovation sometimes becomes necessary. Technical homogeneity increases and market shares show a tendency to be determined by the political process. There is also a conservative tendency with respect to structures. Two points are important for the end of the stagnation phase (be it a decline or a recovery) and especially the speed with which it comes: successful institutional competition and in particular the adaptive behaviour of the industry as a whole with respect to the possibilities resulting from other technological economic paradigms (such as substitution or diversification).

5 Network Analysis and Industrial Restructuring

For a long time, the theoretical debate has been dominated by the opposition between markets and organisations (or hierarchies and state) as forms of social organisation. Within the past few years, networks have gained importance as an intermediary "mode of governance". The underlying concept of networks is increasingly used by new institutional economics as well as in other social sciences.

5.1 The Economic View of Networks

New institutional economics offers some explanation for the emergence and modus operandi of networks and similar structures. According to Wolff and Neuberger (1995), there are two, mutually compatible, lines of argument: one follows the lines of transaction cost economics, the other follows a perspective based on the theory of incomplete contracts. However, both patterns of explanation are associated with the theory of property rights. The following presents an overview of the transaction costs approach.

Socio-economic activity can be related to three forms of coordination: market-oriented, co-operative and hierarchical. According to the respective characteristics of the transactions, the minimum transactions costs must be identified and, in the light of this, the most appropriate form of institution must be selected. Under increasing division of labour and thereby the increasing specificity and strategic importance of single transactions, the costs of market transactions are increasing. This provides an incentive for internal hierarchical coordination of production.

These forms of organisation can create a stable, long-term framework. Once this framework is established, it makes possible production at reduced transaction costs with respect to protection against, for example, opportunistic behaviour.

Networks and similar structures thus spread through the economy (and of course through the polity), becoming even more important in the light of steadily improving information and communication technologies (Wolff and Neuberger 1995, pp. 74–75).

Even here, the broad spectrum of conceivable network structures, from market-based to hierarchical coordination (co-operative, operative or strategic networks) already becomes apparent. This also results in close parallels to (some) company-related approaches (cf. Sect. 6).

5.2 Network Approaches in Political Science

In political science, the main focus of network analysis in policy analysis is to demonstrate the limits to centralised regulation of societies and the importance of systems of negotiation (König 1995; Scharpf 1991). The importance of non-state actors has been shown empirically. Network analysis in political science speaks against a clear separation of state and society (Mayntz 1993; for a survey see

Kenis and Schneider 1991). Furthermore, the "state" is no longer seen as a centre of social control.[6] Policy networks are mostly understood as informal, but fairly stable forms of interaction between actors with different, but mutually dependent interests (Kenis 1991, p. 299). Problems of action are treated decentrally and not on a hierarchical level.

A policy network is characterised by (relative) stability. The actions of the network members are based on relative autonomy. This does not, however, exclude an asymmetrical balance of power between the members (Héritier 1993, pp. 432–433).

5.3 Networks as Determinants of Industrial Restructuring

All the above approaches assume that network-like forms of coordination are very important to the course of economic and political processes. They include both innovation and diffusion networks and networks of resistance to change. The following section presents two approaches which may be appropriate for analysing industrial restructuring.

Gernot Grabher (1993) points out the importance of regional networks for industrial restructuring, emphasising socio-economic factors as explanatory variables. He concentrates mainly on cases involving industrial districts in which economic sectors are strongly tied to regions. His approach contrasts with the aforementioned approach by Berg (1995), who places the emphasis on technological factors. Explanations for the rise of particular regions usually emphasise factors such as economies of scale, R&D and population density on the one hand, and on the other hand, factors often neglected in other approaches, such as the "industrial atmosphere" (Marshall), interfirm linkages and strong support by regional institutions. However, factors which are supportive during emergence can, in the course of development, prove to be no longer appropriate and can become a serious impediment to innovation. If this is the case, the consequence of these socio-economic conditions can be a lock-in. Grabher (1993, pp. 255–256) calls this a "rigid specialisation trap".

The relationship between industrial restructuring and those forms of networks which are characterised by tight interfirm linkages and close relationships between leading industries and politics is thus ambivalent. To demonstrate this with an example, Grabher analyses the Ruhr Area, which was (and still is) dominated by the steel and coal industry. He points out the fairly negative influence of the existing sectoral structure and interlinkages on the development of new industries. One main reason for this is the close connections and interdependencies among firms in the core and supplier industries, which can be recognised as an advantage at the outset but which can constitute a problem for a region's long-term adaptability. Investment in stability of company networks considerably reduces the costs

[6] This is for example rather important for the German discussion because of the point of view of the ordoliberals in German economics.

of market transactions, but at the same time it creates serious weakness in "boundary-spanning functions" (Grabher 1993, p. 260). These functions are important for perception of the economic surroundings and the availability of external information. Any advantage from these networks is strongly linked to the economic development of the core sectors. Disadvantages appear when demand for these products falls.

This pattern is reinforced by close, consensual and co-operative, relationships between the core industries, trades unions and polity. At first, these relationships create and secure mutual advantages, such as monopoly power, secure wages, social security and an electoral base. These advantages can become disadvantages when the development of the core industry has reached its peak. They can prevent prompt restructuring not least because the firms within the network have adapted perfectly to the specific economic conditions, at the same time reducing their ability to re-adapt to further change. This is another indicator of the ambivalence of networks.

In contrast to Grabher, Patrick Kenis orients his concept of networks more around the mode of governance. He therefore contrasts (private) policy networks with market control and industrial policy control. Policy networks are understood as horizontal systems of coordination, and constitute fairly stable groups of organisation. These represent collective interests through alliances and coalitions. The relative stability of these groups is ensured by shared standards, mutual trust, shared knowledge and by group socialisation. Opportunistic behaviour is thereby discouraged (Kenis 1991, p. 299). In his contribution in this book, *Kenis* analyses the influence of sectoral networks on industrial restructuring in more detail.

6 Firms as Actors in Industrial Restructuring

The behaviour of firms can be understood as the micro level of the observable phenomena of industrial restructuring. The role of firms is emphasised not only by evolutionary economics but also in some other approaches. In contrast to most neo-classical approaches, they take firms into account as "political" actors as well. The strategies and rationales of action by firms and their determinants are important because they provide a sizeable share of the impulse for industrial restructuring. Giovanni Dosi argues that the history of economy and technology, and thereby the process of industrial restructuring, can be interpreted as the history of firms. In this interpretation, the role of multinational enterprises must be emphasised as one of the major actors in the industrial restructuring of various sectors. In the following, we present several theoretical approaches which focus on firms as actors in sectoral industrial restructuring, focusing particularly on shrinking sectors. However, there is another important question to be answered: against the background of increasing global mobility of capital and diversified multinational enterprises the question arises as to whether industrial restructuring should be analysed on a sectoral level at all, in the light of the increasing difficulty of separating sectors and the emergence of economies of scope.

6.1 Industrial Restructuring as a Mainly Exogenous Phenomenon

The usual management literature does not consider the process of industrial decline as a central topic (Porter 1980, p. 254). The shrinking of sectors is usually considered as an exogenous constraint on the development of the firm. Appropriate adaptation strategies are then analysed within this framework.

These characteristics of the management-oriented approach are to be found in Porter's analysis of competitive strategies in declining industries (Porter 1980; Harrigan and Porter 1983). The starting point of Porter's analysis is the absolute decline of an industry (in unit sales) over a sustained period. Against this background, he develops possible end-game strategies. Porter (1980, p. 267) identifies four different strategic options: leadership, niche, harvest and quick divestment. The course of the end game is mainly determined by the level of exit costs, therefore the focus is on long-term specific activa, the level of fixed exit costs, as well as on strategic and information barriers (Porter 1980, pp. 259–265). The following theoretical debate focuses on questions about the appropriate strategy and the question of whether a combination of strategies must necessarily fail[7]. Furthermore, the issue has been analysed in terms of the question of whether large or small firms shrink or leave the market first. Game-theoretic analysis points to the first option (Ghemawat and Nalebuff 1990). However, empirical findings either contradict this result or are ambiguous (Liebermann 1990).

According to the management-oriented approach, the firm's strategy has an important influence on the internal structure of industries and their long-term viability (e.g. Porter 1985, pp. 22–23, 296–297). In an open economy, this alone may influence the course of industrial restructuring in individual national market segments[8]. In a later publication, Porter also points out that firms have some influence on the size of industries and hence on sectoral industrial restructuring. The point of departure is the economics of substitution, i.e. strategic decisions over the substitution of products and technologies (Porter 1985, pp. 273–314).

6.2 Firms as Designers of Industrial Restructuring

However, the fundamental view that external framework conditions are the most important factor – the resource-based view of the firm (Baden-Fuller and Stopford 1994, p. 43) – is called into question by the management literature itself. The groundwork was laid by an empirical research by Rumelt (1991), based on data from the 1970s. Rumelt (1991, p. 168) contradicted earlier studies and showed that the situation of an industry ("stable industry effect") accounts for only about

[7] For a survey see e.g. Parker and Helms (1992).

[8] One of the few contributions highlighting this point is Daems (1990).

8% of the success of firms and that 46% of this success can be put down to "stable business unit effects". Baden-Fuller and Stopford (1994, p. 23) in particular support the thesis that "the firm matters more than the industry" and that there is therefore no life cycle of businesses but only of products. Management strategy is now the decisive variable. This is the case for the success of the individual enterprise but can become decisive for the success of the whole sector, too. With an adequate management strategy, an innovative firm can influence the development of a whole industry. The British steel industry is, according to Baden-Fuller and Stopford (1994, p. 31), an example. In this case, even an end to falling demand is attributed explicitly to the behaviour of one actor.

6.3 Linking of Firm and Network Perspectives

The management literature focuses mainly on the perspective of single firms. Another theoretical point of view emphasises the interaction of firms with one another and with political actors. In this case, the focus is similar to that of network analysis. Michael Best, for example, points out that in the age of mass production, multinational enterprises are seeking stability. Thus these firms tend towards interfirm organisation and to work towards regulation of their sector. Best (1990, p. 74) distinguishes between three approaches: "dominant firm regulation", "collective inter-firm action" (formally as cartels or informally) and "government regulation". These forms of regulation have different outcomes with respect to the long-term dynamics of sectors and the success of industrial restructuring. Best argues that "cartels of rationalisation" have positive effects, whereas spontaneous market coordination and price cartels are important causes of decline (Best 1990, pp. 76, 104).

Winfried Ruigrok and Rob van Tulder (1995) attempt a synthesis of the perspectives of firms and networks. They emphasise that today, firms act internationally. In their analysis of international industrial restructuring, Ruigrok and van Tulder distinguish between two levels:

- the level of the "industrial system", applying to an entire country
- the level of "industrial complexes", which develop around large core enterprises and imply different concepts of control; they constitute a kind of hierarchical network

The rationale of the second level is held to be decisive for industrial restructuring. On the one hand, industrial complexes can act comprehensively across sectors while, on the other hand, competing complexes may exist within a single sector. Different intersectoral rationales are a possible, but not inevitable, result of the activities of industrial complexes.

The driving forces of industrial restructuring are thus not necessarily bound to a sectoral level (Ruigrok and van Tulder 1995, p. 272). In his contribution to this book, *Winfried Ruigrok* presents this perspective in greater detail.

7 From Industrial Restructuring to Industrial Policy

7.1 New Approaches to Industrial Restructuring – a Summary

The approaches presented above can certainly be described as diverse, but they nevertheless share a common basis. This consists mainly of a focus on actors and institutions, in contrast to the approaches presented in Sect. 2 which analyse economic development where the main influences are held to be exogenous. From this perspective the new approaches permit deeper and more detailed insights into industrial restructuring, bringing out a variety of specific influential factors. In this context, special mention should be made of the emphasis on discontinuous developments.

Some of the approaches presented above do not exclude one other and can sometimes complement each other instead. This can be seen in the analyses by Berg and Grabher, for example. Against this background, industrial restructuring can be understood as the interaction of factor markets, technology and institutions, whereby it should be stressed that "history matters", and will therefore have a significant influence on future trajectories.

The four general approaches presented above constitute important contributions to the analysis of industrial restructuring, as can be seen from the work of Faber et al., Berg, Grabher, Kenis, Ruigrok and van Tulder, for example. They can be seen as appropriate basis for further analysis of industrial restructuring. Three of them are presented in further detail and applied to the case-studies in this book (see the contributions by *Jöst*, *Kenis* and *Ruigrok*). It is likely that combinations of these approaches will further improve our understanding of these processes.

Nevertheless, it must be stressed that the theoretical potential of many of the recent approaches to analysis of industrial restructuring has not yet been fully exploited. This is particularly the case with the question of the decline and stagnation of industries. However, the approaches do provide for the inclusion of these questions in an analysis. And such questions have not only been neglected by these approaches. They represent a general failing in economic theory and discussion of industrial policy.

There is a research gap between the development of economic and related theories and their application to real world politics. This also holds for many of the newer approaches, and there is much room for corresponding research in the future. Nevertheless, by focusing on the interaction between actors, markets, technologies and institutions, they offer interesting insights into the possibilities and limitations of industrial policy.

7.2 Industrial Policy and Industrial Restructuring

There is no uniform concept of industrial policy in the literature. Depending on the theoretical background, the definitions and justifications for industrial policy vary considerably; for a survey see e.g. Audretsch (1996). Following the Anglo-Saxon line in the debate, Klodt proposes defining industrial policy as "specific political influence on the sectoral production structure in the economy" (Klodt 1992, p. 1592, our translation). In particular, industrial policy encompasses sectoral economic policy, R&D and technology policy, as well as trade policy. Thus industrial policy (IP) can explicitly encourage industrial restructuring ("active" IP), or merely act in response to structural changes ("passive" IP) (Nicolaides 1993, p. 4). While reference to sectors is generally accepted, there is a lively debate as to whether a general or cross-sectoral policy aimed at growth and increasing competitiveness can also be called industrial policy.

The relationship between industrial restructuring and political regulation can be understood, albeit rather simplistically, by two extreme points of view:

On the one side, politics can be seen as an external actor capable of influencing industrial restructuring. Under this assumption, the question to be answered is why and with what goal there should be political intervention in industrial restructuring. The answer depends on the evaluation of endogenous economic processes and therefore, in essence, on the identification of market failure (e.g. Gual 1995).

On the other side, one can ask first to what extent the process under analysis is already influenced by politics, or whether the internal conditions of industrial restructuring offer any opportunity for political intervention. Industrial restructuring should then be understood as a process of interaction between economy and polity or between markets, technologies and institutional structure. Only in this context is the question of evaluating political intervention raised.

However, traditional discussion of industrial policy is dominated by the former approach and focuses largely on justifying political intervention. While the issue of the justification for a sectoral industrial policy is controversial (e.g. Nicolaides 1993), industrial policy measures are a fact in most industrial countries. There are usually two types of action:

- support of specific new industries and technologies in order to promote industrial restructuring ("picking winners")
- measures to conserve industries or structural adjustment in industries in order to slow down or to cushion structural change

Approaches focusing on the interaction between markets, actors, technologies and institutional structure, often neglected in the political debate, may help to shed new light on this situation.

8 Implications for (Green) Industrial Policy

8.1 Green Industrial Policy and Traditional Industrial Policy

As shown in previous chapters, green industrial restructuring will not be brought about by the market alone, and a policy is needed to it. Such a policy must cope with new challenges because it must set out to have a selective impact on different industries. Thus, in accordance with the definition of industrial policy given above, we term it a green industrial policy, a term first introduced by Jänicke et al. (1995).

In certain respects, green industrial policy (GIP) refers to classical industrial policy debates, but also highlights new dimensions. Justifications for GIP differ: it does not primarily aim for growth and (short-term) competitiveness, but also integrates the negative "externalities" of such an orientation. In the first type of industrial policy, this implies the need to screen support for new industries in terms of their environmental impact. The second type of industrial policy is applied mostly to heavily polluting industries such as coal, steel etc. Environmental arguments not only point to the further disadvantages of conservation strategies, GIP explicitly aims for a decline in particularly dirty industries ("picking losers"). Depending on the delimitation of industries, these may be whole sectors or only parts of them.

This second type of GIP in particular is fraught with (often neglected) difficulties. GIP creates losers, because existing production structures, technologies and investment are put in question. The losers, however, will not usually accept such political intervention, but will rather try to influence them. Thus, GIP is faced with similar problems to those of classical industrial policy of the second type. The latter is also distinguished by a frequent discrepancy between theoretical optima and political practice. Due to the sectoral conditions and actor configurations, the possibilities for intervention appear limited. It is therefore one challenge for (not only) green industrial policy analysis to integrate the implementation conditions from the outset, in order to clarify the possibilities and restrictions. Some contributions by the new approaches to industrial restructuring presented here may be helpful in this respect.

8.2 Lessons from New Approaches to Industrial Restructuring

One common feature in several of the new approaches is a growing theoretical awareness that the focus on politics as an external actor is questionable. At least the need to include the institutional setting in the analysis is recognised.

A frequent characteristic of evolutionary approaches (Sect. 3) is the implicit view that innovation and support for technological progress may have a positive outcome, with destruction being a creative factor. Political evaluation of an actual process of industrial restructuring remains problematic, however, as a variety of attractors can be imagined without being able to rank one above the other (Biervert and Held 1993, p. 11). With regard to factors such as path dependencies and network externalities the (frequently implicit) identification of economic selection by the market with economic efficiency becomes questionable from a theoretical point of view (Nelson 1994, p. 52). Technological "lock-in", i.e. the stabilisation of an inferior trajectory, may hinder a change of trajectory, as implied by GIP, for example. Such change is difficult and can only succeed under certain conditions (windows of opportunity).

Some evolutionary approaches therefore propose *strategic niche management* as a political strategy for encouraging radical change (e.g. Kemp et al.1998). Strategic niche management is a probe-and-learn-strategy to explore the viability of potential technological alternatives, with a mixture of protection and selection pressure. It consists of five steps, namely choosing a technology, selecting and setting up an experiment, scaling the experiment up through policy, and breaking down protection. The institutional component of the strategy is taken into account from the outset. *René Kemp's* contribution therefore takes an evolutionary approach to the political challenges of green industrial restructuring, in its interpretation of the case-studies.

In the political aspects of his evolutionary and institutional theory of industrial restructuring (Sect. 4), Berg comes to similar conclusions. He distinguishes between a number of phases. With respect to the development phase, Berg (1995, pp. 165–168) takes the view that supportive political intervention in the sense of strategic dynamic regulation is desirable during this phase, but his analysis nevertheless points up its limitations, given the dominance of established industries. For the stagnation phase, Berg makes reference to public choice theory and explains political intervention as the inevitable consequence of strategic influence on the institutional framework by mature sectors, this being more attractive than further investment in technological development. This points clearly to the restrictions on a green industrial policy.

A common feature of all network approaches (Sect. 5) is that industrial policy will have specific results in all sectors characterised by network structures. Political evaluation, however, differs widely between the various approaches. From the point of view of public choice theory, there is a general suspicion that networks serve mainly to assist the pursuit of group interests. The goal is not efficient allocation of resources but rather a distribution which is beneficial to the network members. Anti-trust policy is therefore necessary to combat these influences within an industry, whereas industrial policy intervention should be refrained from, in order to prevent the development of such policy networks.

Other approaches in the corporatist tradition come to the conclusion that those sectors which are shaped by networks and corporatist industrial relations are more successful in coping with industrial restructuring, at least if these processes are the result of international influence. The task of state actors is therefore to support this through a kind of "meta-governance" (e.g. Traxler and Unger 1990, pp. 217–218). This is in line with recent contributions, which emphasise the advantages of network governance compared with top-down intervention or pure market forces (e.g. Kenis 1991; Meyer-Stamer 1995) State industrial policy then has a more indirect task of influencing private policy networks.

Networks are particularly important if the actor configuration is marked by long term stability. This is a typical feature of mature industries. For GIP, the different political aspects mentioned above seem significant, and excessive optimism must be qualified and balanced with realism. It is crucial to differentiate between open and closed networks (e.g. Messner 1995, Jänicke et al. 1999). Open networks facilitate the inclusion of important environmental interests and potential innovation drivers, whereas closed networks may degenerate into simple resistance to change. Destabilising and restructuring closed networks seems important for the success of GIP. However, the structure and importance of policy networks also depends on policy style and the configuration of the underlying economic networks. Grabher (1993) argues similarly. His perspective starts from a close interrelationship between industrial restructuring and politics, caused by endogenous factors. For him it is not the existence of networks which is decisive – from a political point of view – but their exact shape. One implied political recommendation is to support slack networks, as they are important in producing the redundancy which is necessary for adaptability. Grabher points out that against the background of insecure returns on investments, there is an economic temptation to avoid redundancies. The main task of industrial policy therefore should be to prevent just this "streamlining" (Grabher 1993, p. 272).

Limits on the political regulation of industrial restructuring become obvious when the importance of firms as decisive actors is taken into consideration. This is pointed out especially by evolutionary economics and the approach of Ruigrok and van Tulder, focusing on cases when the economic networks have a more vertical structure and are dominated by a core firm or "industrial complex" (Sect. 6). Often mature industries are dominated by one or two major players. Innovative strategies by single companies create opportunities for change and put competitors under pressure. At the same time, the scope of the strategic options available to multinational corporations (i.e. the exit option) limits the possibilities for political intervention. The Interpretation by *Ans Kolk* therefore focuses on the role of multinational enterprise in environmentally beneficial structural change and their interaction with governments.

Limited as their contributions may be, the new approaches point to the fact that a green industrial policy is by no means a simple matter, and offer some advice on strategies for coping with the difficulties. Some aspects of this are explored in the aforementioned contributions from Kemp and Kolk. While the line of reasoning in the approaches presented appears very promising, further theoretical investigation and clarification will certainly be required.

References

Alchian, A. (1950): Uncertainty, evolution, and economic theory. Journal of Political Economy 58, 211–221.

Arthur, W. B. (1989): Competing technologies, increasing returns, and lock-in by historical events. The Economic Journal 99, 116–131.

Audretsch, D. (1996): Industrial Policy. wzb-discussion papers FS IV 96–9, Berlin.

Baden-Fuller, C.W.F. and Stopford, J.M. (1994): Rejuvenating the Mature Business. Boston.

Berg, C. (1995): Technologischer Fortschritt und ökonomische Regulierung. Frankfurt/Main.

Best, M.H. (1990): The New Competition. Institutions of Industrial Restructuring. Cambridge.

Biervert, B. and Held, M. (1993): Das Evolutorische in der Ökonomik: Neuerungen – Normen – Institutionen. Eine Einführung. In: Biervert, B. and Held, M. (eds.): Das Evolutorische in der Ökonomik: Neuerungen – Normen – Institutionen. Frankfurt/Main, pp. 7–22.

Boulding, K. (1981): Evolutionary Economics. Beverly Hills.

Cimoli, M. and Dosi, G. (1995): Technological paradigms, patterns of learning and development: an introductory roadmap. Journal of Evolutionary Economics 5, 243–258.

Daems, H. (1990): Industry and Country Exit: Reflections on International Differences in Exit Behaviour. In: C.W.F. Baden-Fuller (ed.): Managing Excess Capacity. Oxford, pp. 59–74.

David, P.A. (1985): Clio and the Economics of QWERTY. American Economic Review, Papers and Proceedings 75, 332–337.

Dosi, G. (1982): Technological Paradigms and Technological Trajectories: A Suggested Interpretation of the Determinants and Directions of Technical Change. Research Policy 11, 147–162.

Dosi, G. and Nelson, R.R. (1994): An introduction to evolutionary theories in economics. Journal of Evolutionary Economics 4, 153–172.

Edquist, C. (ed.) (1997): Systems of Innovation. Technologies, Institutions and Organizations. London and New York.

Faber M., Jöst, F., Manstetten, R., Müller-Fürstenberger, G., Proops, J.L.R. (1996): Linking ecology and economy. Joint production in the chemical industry. In: Faber, M. et al. 1996a, pp. 263–278.

Faber M., Manstetten, R., Proops, J.L.R. (eds.) (1996a): Ecological Economics. Concepts and Methods. Edward Elgar, Cheltenham.

Feldmann, H. (1993): Theorien des Strukturwandels: Welche Orientierungshilfen bieten sie für die Strukturpolitik? Ifo-Studien 39/1993, 40–64.

Furubotn, E. and Richter, R. (1991): The New Institutional Economics: An assessment. In: Furubotn, E. and Richter, R. (eds.): The New Institutional Economics. Tübingen, pp. 1–32.

Ghemawat, P. and Nalebuff, B. (1990): The devolution of declining industries. The Quarterly Journal of Economics, February, 167–186.

Grabher, G. (1993): The weakness of strong ties. The lock-in of regional development in the Ruhr area. In: Grabher, G. (ed.): The embedded firm. London, pp. 255–277.

Grossman, G.M. and Helpman, E. (1991): Innovation and Growth in the Global Economy. Cambridge, MA.

Gual, Jordi (1995): The three policies: an economic analysis. In: Buigues, P., Jacquemin, A., Sapir, A. (eds.): European Policies on Competition, Trade and Industry. Conflicts and Complementarities. Aldershot/ Brookfield, pp. 3–48.

Harrigan, K.R. and Porter, M.E. (1983): End-game strategies for declining industries. Harvard Business Review, July-August, 111–130.

Hartwig, H. (1990): Wirtschaftlicher Strukturwandel und staatliche Strukturhilfen, Frankfurt/Main.

Héritier, A. (1993): Policy-Analyse. Elemente der Kritik und Perspektiven der Neuorientierung. In: Héritier, A. (ed.): Policy Analyse. Kritik und Neuorientierung, PVS Special Issue 24/1993. Opladen.

Jänicke, M., Binder, M., Mönch, H. (1995): Green Industrial Policy and the Future of "Dirty Industries". FFU-Report 95–1, Berlin.

Jänicke, M., Kunig, P., Stitzel, M. (1999): Umweltpolitik. Politik, Recht und Management des Umweltschutzes in Staat und Unternehmen. Bonn.

Kemp, R., Schot. J., Hoogma, R. (1998): Regime shifts to Sustainability through Processes of Niche Formation. Technology Analysis and Strategic Management 10 (2), 175–195.

Kenis, P. and Schneider, V. (1991): Policy Networks and Policy Analysis: Scrutinizing a New Analytical Toolbox. In: Marin, B. and Mayntz, R. (eds.): Policy-Networks: Empirical Evidence and Theoretical Considerations. Frankfurt/Main, pp. 25–59.

Kenis, P. (1991): The Preconditions for Policy Networks. Some Findings from a Three-Country Study on Industrial Restructuring. In: Marin, B. and Mayntz, R. (eds.): Policy-Networks: Empirical Evidence and Theoretical Considerations. Frankfurt/Main, pp. 297–330.

Klodt, H. (1988): Industrial Policy and Repressed Structural Change in West Germany. Kiel Working Paper No. 322.

Klodt, H. (1992): Industriepolitik. In: Gabler-Wirtschafts-Lexikon, 13th edn., Wiesbaden, pp. 1592–1599.

Klodt, H. (1994): Energy Consumption and Structural Adjustment: The Case of West Germany. Structural Change and Economic Dynamics 5, 177–185.

König, T. (1995): Die Koordination in Politischen Verhandlungssystemen. Kölner Zeitschrift für Soziologie und Sozialpsychologie 47, 268–294.

Lenway, S., Morck, R., Yeung, B. (1996): Rent seeking, protectionism and innovation in the American steel industry. The Economic Journal 106, 410–421.

Lieberman, M.B. (1990): Exit from declining industries: "shakeout" or "stakeout"? RAND Journal of Economics 21, 538–554.

Mayntz, R. (1993): Policy-Netzwerke und die Logik von Verhandlungssystemen. In: Héritier, A. (ed.): Policy-Analyse. Kritik und Neuorientierung. Politische Vierteljahresschrift, Special Issue 24/1993. Opladen.

Messner, D. (1995): Die Netzwerkgesellschaft. Köln.

Meyer-Stamer, J. (1995): Industrial Policy in the European Union: Old Dilemmas and New Options, Series Eurokolleg, Friedrich-Ebert-Stiftung, Bonn.

Nelson, R.R. (ed.) (1993): National Innovation Systems. Boston.

Nelson, R.R. (1994): The Co-evolution of Technology, Industrial Structure, and Supporting Institutions. Industrial and corporate change 3, 47–63.

Nelson, R.R. and Winter, S. (1982): An Evolutionary Theory of Economic Change. Cambridge.

Nicolaides, P. (1993): Industrial Policy: The Problem of Reconciling Definitions, Intentions and Effects. In: Nicolaides, P. (ed.): Industrial Policy in the European Community: A Necessary Response to Economic Integration? Dordrecht, pp.1–19.
North, D.C. (1990): Institutions, Institutional Change and Economic Performance. Cambridge.
Olson, M. (1965): The Logic of Collective Action. Public Goods and the Theory of Groups. Cambridge MA.
Olson, M. (1982): The Rise and Decline of Nations. New Haven.
Parker, B. and Helms, M.M. (1992): Generic Strategies and Firm Performance in a Declining Industry. Management International Review 32 (1), 23–39.
Pasinetti, L. (1981): Structural Change and Economic Growth: A Theoretical Essay on the Dynamics of the Wealth of Nations. Cambridge.
Porter, M.E. (1980): Competitive Strategy. New York.
Porter, M.E. (1985): Competitive Advantage. Creating and Sustaining Superior Performance. New York.
Seifert, E.K. and Priddat, B.P. (eds.) (1995): Neuorientierungen in der ökonomischen Theorie. Marburg
Reuter, N. (1996): Zur Bedeutung von Institutionen für die ökonomische Theorie. In: Priddat, B. P and Wegner, G. (eds.): Zwischen Evolution und Institution – neue Ansätze in der ökonomischen Theorie. Marburg, pp. 113–144.
Ruigrok, W. and van Tulder, R. (1995): The Logic of International Restructuring. London.
Rumelt, R.P. (1991): How much does industry matter? Strategic Management Journal 12, 167–185.
Scharpf, F.W. (1991): Handlungsfähigkeit des Staates am Ende des 20. Jahrhunderts. Politische Vierteljahresschrift 32, 621–634.
Schneider, V. and Kenis, P. (1996): Verteilte Kontrolle: Institutionelle Steuerung in modernen Gesellschaften. In: Kenis, P. and Schneider, V. (eds.): Organisation und Netzwerk. Institutionelle Steuerung in Wirtschaft und Politik. Frankfurt/Main, pp. 9–43.
Schumpeter, J.A. (1912): Theorie der wirtschaftlichen Entwicklung. Leipzig.
Traxler, F. and Unger, B. (1990): Institutionelle Erfolgsbedingungen wirtschaftlichen Strukturwandels. Wirtschaft und Gesellschaft 16 (2), 189–223.
Weise, P. et al. (1993): Neue Mikroökonomik. Heidelberg.
Williamson, O.E. (1975): Markets and Hierarchies. New York.
Williamson, O.E. (1985): The Economic Institutions of Capitalism. New York.
Witt, U. (1995): Evolutorische Ökonomik – Umrisse eines neuen Forschungsprogramms, in: Seifert, E.K. and Priddat, B.P. (1995).
Witt, U. (1997): “Lock-in” vs. “critical masses” – industrial change under network externalities. International Journal of Industrial Organization 14, 753–773.
Wolff, B. and Neuburger. R. (1995): Zur theoretischen Begründung von Netzwerken aus der Sicht der Neuen Institutionenökonomie. In: Jansen, D. and Schubert, K. .(eds.): Netzwerke und Politikproduktion: Konzepte, Methoden, Perspektiven. Marburg.

An Evolutionary Perspective on Structural Change and the Role of Technology

Frank Jöst[1]
Ruprecht-Karls University Heidelberg, Grabengasse 14, 69117 Heidelberg, Germany

1 Introduction

Environmental problems and scarcity of resources have reached a hitherto unknown scale and extent. This is evident if one considers the problem of climate change or the destruction of the ozone layer through CFCs, for example. These problems are caused mainly by the patterns of production and consumption developed in industrial countries over the last 200 years. Nowadays, many fear that global environmental problems are being considerably intensified by a world-wide industrialisation following in the footsteps of Europe and North America, as well as by continuing economic growth in the industrialised world. However, whether such a prediction comes to pass will depend on many factors. It is quite conceivable that economic development and a reduction in raw materials consumption and environmental pollution can be achieved simultaneously, if the heavily polluting industries and the resource-intensive sectors lose importance in the course of economic evolution, and/or if a change in production techniques leads to a reduction in the raw materials used and the associated pollutants[2].

The case studies presented in this book are an attempt to discuss, on the basis of selected basic industries in various European countries, whether such a change in the structure of the economy will improve environmental quality. In addition, these case studies should allow one to identify the determinants of past structural change and to describe its impact on the use of the environment as a supplier of resources and a disposal site for waste. With regard to the consequences for environmental pollution, the case studies are in no way unambiguous. In addition, it seems that only general statements can be formulated on the determining factors of structural change. At first sight, structural change appears to be the result of complex interactions between technical, economic, institutional and ecological factors. The goal of the following analysis is to develop a conceptual framework which allows a more accurate description of the interaction between the factors

[1] I am grateful to Stefan Baumgärtner, Malte Faber, Klaus Jacob, Jan Nill and Ulrich Petschow for their helpful comments.

[2] This is discussed in economic literature as the so-called "environmental Kutznets Curve". Several authors have shown in empirical studies that, for some pollutants, the relationship between per capita income and environmental quality corresponds to an inverted U-curve. Empirical evidence is given by Grossman and Krueger 1993, 1995; Seldon and Song 1994; Shafik 1994 and Stern et al. 1996.

which are important for change in the production structure and consumption patterns of an economy. This should help to answer the following questions:

1. To what extent is structural change caused by scarcity of raw materials and environmental deterioration?
2. Is structural change a process which automatically leads to an improvement in environmental quality or must this be enforced by environmental policy measures?
3. What should be taken into account, if new environmental legislation is necessary for environmentally friendly structural change?

The paper is organised as follows. Sect. 2 contains some general remarks on structural change and economic theory. The problems of neo-classical theory in explaining structural change are briefly discussed. A framework which allows a more appropriate analysis of the dynamics of modern economies is developed in Sect. 3. While 3.1 introduces concepts from evolutionary economics, 3.2 develops a more detailed framework for analysis of technological change. In Sect. 4, this framework is used to describe and explain the main characteristics of the development of two branches of the chemicals industry, namely the chlorine (4.1) and the sulphuric acid industries (4.2). Sect. 5 discusses to what extent the results concerning the nature of structural change in the chemicals industry apply also to the other case studies. Sect. 6 summarises the discussion and looks at the role of environmental policy in changing the structure of an economy.

2 Structural Change and Economic Theory

Industrialised economies are characterised by continuing change in the production techniques used. This change of technologies is one important factor determining the evolution of an economy and has far-reaching consequences for the economic agents living in these economies. Structural change is necessarily accompanied by friction and transitional costs. Growing and contracting sectors cause, for instance, an unavoidable shift in the employment structure. Such a labour-reallocation process brings adaptation costs in the form of training expenses or costs in the form of unemployment. Furthermore, a change in the importance of a sector or in the production techniques used leads to a change in the environment's role as raw materials supplier and pollution sink. It is therefore important for society that economists analyse the conditions which can influence structural change. However, within the dominant economic paradigm, neo-classical theory, it is rather difficult to analyse economic evolution.

The core of neo-classical economic theory is the analysis of the problem of allocating scarce means to achieve given ends. In formal models, conditions are studied which guarantee that the decisions of independently acting individuals lead to a market equilibrium. Furthermore, neo-classical economics deals with the preconditions which must be met so that a market equilibrium is as good a state as possible for society, which means that the market equilibrium is pareto-efficient (cf. for this Malinvaud 1985, Chap. 2–5). A comparative static analysis then examines in which way the equilibrium resulting from the economic process may change if exogenous variables, such as preferences, production techniques or the availability of resources vary. Comparing two equilibria does not consider the processes of adaptation from the old equilibrium to the new one and does not normally investigate the circumstances which will lead to a change in the aforementioned exogenous variables.

A similar procedure is used in the analysis of intertemporal allocation problems. In this context, one usually considers "steady-state" solutions with intertemporal optimisation models to describe the evolution of an economy over time. A "steady state" is a situation in which all the model's variables increase at constant rates (Barro and Sala-I-Martin 1995, p. 19). Such a state represents a long-term equilibrium, which the economy may achieve in response to all the exogenous factors acting on the economy (Azariadis 1993, pp. 11–12). However, in a situation where all variables are changing at constant rates, the structure of the economy is also constant. In addition, even if non-steady state solutions are considered, important aspects of structural change are neglected. Romer (1994, p. 5) has pointed out that conventional neo-classical economists have difficulties in analysing, for instance, how policy affects invention and innovation when it is assumed that new goods can be introduced. From Romer's point of view, new goods include not only tangible products but also the development of new technologies. From such a perspective, neo-classical models are inappropriate for describing and analysing the development of real economic systems. Novelty arises, which leads to a gradual change in what are normally considered to be exogenous variables (cf. Ayres 1991, p. 256; Bollmann 1990, pp. 40–41; Nelson 1987, pp. 14–15, Dosi 1986).

An alternative concept for analysis of the intertemporal transition in an economy has been developed in evolutionary economics (Nelson and Winter 1982). However, in contrast to neo-classical economics, there is at present no unified framework for evolutionary economics. Evolutionary economics is interested above all in the phenomenon of economic change and its regularities when novelty arises. Accordingly, it is assumed that the decisions of economic agents are not based on an assumption of optimisation by exogenous constraints. Rather, it is assumed that patterns of behaviour developed in the past – to deal with the problem of imperfect knowledge – form the basis of future economic decisions and that exogenous constraints such as preferences and technologies are changing over time (Dosi and Nelson 1994, pp. 157–159; Witt 1994, p. 503; Nelson 1987, p. 1). Such a framework allows for a better description of structural change in an economy. The essential elements of an evolutionary theory, its basic concepts and its application to economics are outlined in the following section.

3 The Evolutionary Perspective on Structural Change

There are various approaches to analysing the development of an economy from an evolutionary perspective, as Nill and Petschow have pointed out in their contribution. In interpreting the results of the case studies I shall follow an approach developed by Faber et al. (1996a). The main characteristics of this framework are presented in Sects. 3.1 and 3.2 and may be summarised as follows:

- The authors develop a conceptual framework which allows the unified description of evolutionary processes in different sciences, namely physics, biology and economics. This offers insights into the nature of change over time, in particular with respect to its predictability and hence the possibility of influencing such processes.
- The theory of joint production (Faber et al. 1996b) and the neo-austrian capital theory (Faber 1986) are used to analyse comprehensively the interdependencies between invention of and innovation in new technologies (Faber and Proops 1991) and the use of the environment for raw materials and waste disposal.

3.1 Evolutionary Concepts in Economics

Faber et al. (1996a, p. 137) conceptualise evolution as "...the process of the changing of something over time". This statement is not strictly speaking a definition, but can be operationalised using concepts of biological evolutionary theory. Biology distinguishes between the genotype and the phenotype of an organism or biological system. The genotype represents the potentiality of an organism or biological system given by its genetic structure. The phenotype is the concrete appearance of an organism or system, determined by its potentiality and its environment.

Faber et al. (1996a, p. 154) use these concepts to describe economies in terms of their potentiality – their genotypes – and their realisation – their phenotypes. The economic genotype is described by the following elements: (i) preference rankings by economic agents, including their ethical and religious norms; (ii) production technology and (iii) legal, economic and social institutions.

The economic phenotype contains the following components: (i) the production techniques actually employed, including the types and quantities of capital goods used, (ii) the types and quantities of capital and consumption goods produced, (iii) the prices of all goods produced, (iv) the distribution of consumption, income and wealth among economic agents, and (v) the market structure.

The terms genotype and phenotype make possible a distinction between two categories of economic modification: (i) modifications which are due to changes in the potentiality of an economy and (ii) modifications of the economy which occur with given technology, preferences and social institutions. The former is called "genotypic", the latter "phenotypic" change. Within this framework, structural change can be genotypic as well as phenotypic. For instance, the reallocation of industries which was explained by the theory of comparative advantages is phenotypic change. In this case, there is neither a change in the set of available production processes and the tastes of consumers nor a change in the institutional setting of the economies. The invention of a new technology, which is an example of genotypic change, might lead to an increase in the importance of one sector in the economy and correspondingly to a decrease in that of another.

The distinction between genotypic and phenotypic change helps us to understand the problems which occur in explaining structural change and in trying to control economic development through economic or environmental policy. Analysis of biological systems has shown that the outcome of modifications of an organism's genotype are unpredictable while that of phenotype modifications can be predicted in principle. Faber et al. (1996a, p. 148) mention three reasons why genotypic change in biological systems is unpredictable:

- factors which cause a mutation can not be specified ex ante
- the influence of a mutation on the genetic material cannot be foreseen
- the effects of a genotype modification on the phenotype are not predictable

Similar difficulties arise when we try to predict the development of the economic genotype. It is not possible to identify factors leading to the invention of a yet unknown production process, and it is hardly possible to predict the way an invention will modify the economic phenotype. Consequently, it is very difficult to predict the future development of an economy where such development is influenced mainly by genotypic change.

Faber et al. (1992, pp. 86–88) explain the underlying epistemological reasons for these difficulties in an extensive analysis of the problems of ignorance, novelty and evolution. They distinguish between processes whose description is dependent on time and processes whose description is not. The latter can be explained by analogy to time-independent laws such as the Law of Gravitation in physics or the Law of Supply and Demand in economics. Because of their time-independent structure, these processes are predictable when the underlying regularities are known. But if processes are intrinsically time-dependent, i.e. if they involve novelty, they are unpredictable. This structure is important for the process of mutation in biology. Similarly, the invention of a new production process and its total effect on the economy are not predictable. Faber et al. (1992, p. 88) mention that "...all the consequences of the invention of the steam engine were unpredictable, at least up to the time it was invented."

These considerations on the limits of our knowledge and understanding are of great importance if we want to predict and control the invention of new technology and the effects of fundamental change on the institutions or preferences in our society. This does not mean that it is impossible to examine genotypic modifications to an economy. If production techniques, institutions or the preferences of a society change fundamentally over time, we are able ex post to identify factors, which favoured the invention of new technologies or led to a change of institutions and attitudes in society. And it can be shown afterwards which modifications of the phenotypes occurred due to a modified genotype. This insight is important for the present investigation of structural change in basic materials industries. We have to take into account that essential aspects of structural change can only be investigated by an analysis of the historical development of the sectors. Two examples will be presented in Sect. 4.

3.2 A Framework for the Analysis of Technical Change

The economic history of the last 200 years, particularly in Europe, is characterised above all by alterations in production techniques and changes in the raw materials inserted and the corresponding environmental pollution. The case studies presented in this book demonstrate that this also applies today. Hence, if one wants to understand the long-term development of an economy it is important to look at the interplay between natural boundary conditions and technology.

A framework which permits descriptions of the long-term relationship between natural boundaries and technological change is developed in Faber et al. (1996a, 1998). The basic ideas, developed in several studies, are the concept of joint production and neo-austrian capital theory.

Faber et al. (1998) show that from a thermodynamic point of view, all production is joint production. This means that the manufacture of one product with a combination of inputs is necessarily accompanied by the production of more than one output. From an ecological point of view, solid, liquid or gaseous waste always arises as the joint product of manufactured goods. The phenomenon of joint production can be studied in an ideal form through the development of the chemicals industry, because chemical processes, by their nature, give rise to by-products. Joint production has therefore essentially determined the development of the chemicals industry in the past, and is of central importance for the chemicals industry's ability to adapt to a change in economic and environmental conditions in the future. However, if we define pollution as unwanted joint products of manufactured goods, the following structure is of general importance. When joint products arise, the following three adaptation strategies can be made out:

1. The by-products are released into the environment as waste, waste water, or air pollution. If the capacity of the environment is restricted, there may be bottlenecks, and the enterprises can react with alternatives 2 & 3.
2. It is possible to implement end-of-pipe technologies, such as waste water purification plants or exhaust gas filters, in order to prevent emission of the unwanted by-products, or to reduce emissions by changing production processes.
3. The by-products can be traded or used as inputs to other production processes, processed into marketable products, or transformed into substances which can be disposed of without problems.

If one considers the history of the chemicals industry, its development has a characteristic structure. By-products were very often first disposed of into the environment (1), until environmental problems led to the sue of options 2 & 3. These changes normally presuppose the invention of a new process and its implementation, as well as expansion (2) and restructuring (3) of existing capital stock. The history of the chemicals industry shows that it was often environmental policy which initiated such changes by means of new regulations. Therefore environmental legislation, which was itself a response to environmental problems, has been an important force for structural change. This conclusion can be backed up by looking into the histories of the chlorine alkali and sulphuric acid industries. But before we go into the details of the chemical industry, it is useful to examine the interaction between technology and the environment more precisely. In order to do this, we need to introduce the following concepts from neo-austrian capital theory[3].

- The *technology* of an economy comprises all the techniques (production processes) available to the economy. However, all techniques are not necessarily used.
- *Innovation* is the introduction of a technique which has been available but which has not up to now been implemented.
- *Invention* is the addition of a new production process to the technology.

These concepts help to describe the long-term relationship between invention, innovation and the natural world (These interdependencies are illustrated in Fig. 1; cf. Faber et al. 1996a, p. 158). The starting point is a state of the economy in terms of its phenotype and its genotype. The economy uses nature as a supplier of raw materials and as recipient of pollutants in order to produce goods. Two interdependencies can be distinguished:

1. The continuous extraction of raw materials leads, among other things, to a rise in price of resources and thereby to adaptation with respect to the quantities of the resource used. This is, in our terminology, a phenotypic change in the economy. Such a modification is simultaneously an incentive to develop a new production process which requires less raw materials or uses different resources. If

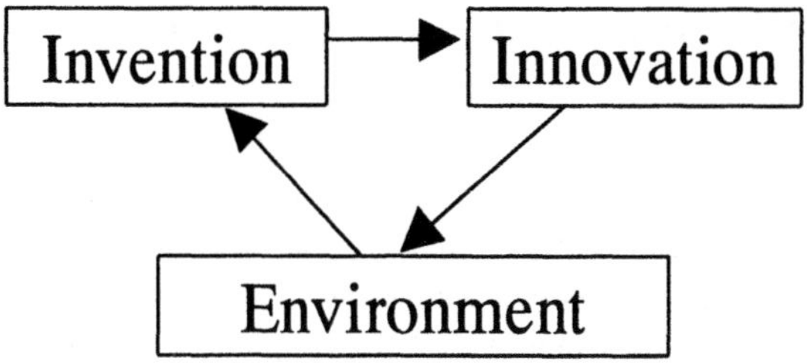

Fig. 1. Long-term interaction between the economy and the environment.

[3] For an introduction into neo-austrian capital theory cf. Faber 1979, 1986; Stephan 1995.

such a process is invented, the genotype of the economy changes and a correspondingly new phenotype emerges.

2. If we produce goods, we have, besides the desired output, a joint product of unwanted emissions. These emissions very often diminish the quality of ecological systems. If society is no longer willing to accept the associated damage, new laws are implemented to protect polluted areas. The implementation of such new institutions is a genotypic change in the economy and it is to be expected that changes in legal institutions will also cause a modification of the phenotype. Either the quantity of goods produced will decline or end-of-pipe technologies are implemented to reduce emissions, or known production techniques are introduced in practice (phenotypic change).

Change in the genotype and phenotype of an economy might be an incentive to invent and implement new production techniques. The initial genotypic change can therefore induce further genotypic and phenotypic modifications. This is one reason why it is often extremely difficult to distinguish between genotypic and phenotypic change in economies. In addition, it is also difficult to distinguish precisely between genotypic and phenotypic change, since the changes in the genotype of an economy, unlike in those of biological systems, can occur very fast. However, the interdependencies between the genotype and the phenotype of an economy can be described if one considers the changes in an economy over a long period of time.

The interdependencies outlined above have played an important role in past economic development. Faber et al. (1996a, p. 163) argue that the industrialisation of Europe in the 18th and 19th centuries can be considered ex post as a substitution of the scarce raw material charcoal by the relatively abundant fossil fuel coal. Charcoal was the essential energy input in the smelting of iron. The rising price of this input was a strong incentive to finding an alternative technique for producing pig-iron. With the invention of a process which transformed coal into coke, it was possible to produce pig-iron considerably more cheaply. This was a precondition for the universal use of iron in the process of industrialisation.

4 Structural Change in the Chemicals Industry: The Evolution of the Chlorine and Sulphuric Acid Industries

The development of the chemicals industry, particularly the chlorine and sulphuric acid industries, is an almost perfect example for using the framework developed above to describe the interdependency between invention, innovation and the use of the environment as a supplier of raw materials and a recipient of pollutants. The most important aspects of structural change, especially in chlorine chemistry over the last decades, are described in the case study by *Jacob* in this book. We shall integrate his results into the conclusions of our historically oriented study on the problem of joint production and the environment (cf. Faber et al. 1995, 1996c).

4.1 Chlorine Alkali Industry

The history of the chlorine alkali industry is a good example of how the necessary treatment of unwanted by-products has influenced the development of an important industrial sector of the economy. At the end of the 18th century, Nikolaus Claus Leblanc invented a process which produced soda by splitting common salt (sodium chloride). As an unwanted output of this production process, the environmentally harmful by-product hydrogen chloride arises. This was at first emitted into the air and caused damage to agricultural land. In response to this, the British parliament passed the 1864 Chlorine Alkali Act. As this new law required, hydrogen chloride was subsequently dissolved in water and discharged into the rivers. This led to tremendous corrosion problems for flood gates and ships. With the 1874 amendment to the Chlorine Alkali Act, this type of disposal was also forbidden, but the chemicals industry was then able to introduce a recently-developed process by which chlorine could be extracted from hydrogen chloride. From this, a chlorine chemistry has evolved into one of the most important branches within the chemicals industry today. Approximately 60% of the revenue of the chemicals industry depends on chlorine as a process chemical or as part of the end product (cf. Buttgereit 1994 and *Jacob* in this book).

However, the production of chlorine, the use of chlorine as a process chemical as well as the production and the use of chlorinated final products are all associated with many environmental problems. The solution of the initial environmental problem, through the emergence of the chlorine industry since the second half of the last century, has therefore led to new problems. As Jacob's case study shows, the chemicals industry has been able to react to these problems. For instance, chlorine paraffin, used as a lubricant in metalworking, was removed from the production palette by the Hoechst AG (Europa-Chemie 1992). Furthermore, integrated environmental protection measures have been added to meet significantly reduced chlorine requirements from the chemicals industry (Faber et al. 1995). New closed loop processes, in which chemicals such as chlorine are used but do not enter the final product, have been developed. In addition, new production facilities are increasingly being introduced, allowing production of the same output with less inputs, and therefore with less waste. Many substitutes for chlorinated final products are now available. Polypropylene, for example, is a substitute for polyvinyl chloride (PVC) in many applications.

4.2 Sulphuric Acid Industry

Sulphuric acid is quantitatively the most important product in the chemicals industry (Müller 1994)[4]. The environmental problems caused by sulphur compounds such as sulphates are mainly the result of the use of sulphuric acid, which was

[4] The economic and environmental problems of the sulphuric acid industry are analysed in detail by Faber et al. 1996c.

therefore often the subject of environmental regulation. The history of the sulphuric acid industry also offers an example of the importance for structural change in the chemicals industry of the interdependence between invention, innovation and the natural world (for a description of the early history of the sulphuric acid industry cf. Droesser 1908; Meyer 1997).

The development of sulphuric acid production occurred in four stages: In the first stage (1774–1838), the British textiles sector required large quantities of sulphuric acid. This acid was necessary for washing raw materials, and it was important for the production of soda in the commonly used Leblanc process. The production of sulphuric acid was based on sulphur imported from Sicily. This was first oxidised, in so-called lead chamber plants, to sulphur dioxide and then to sulphur anhydrite, which was then transformed into sulphuric acid. However, the sulphur combustion in the lead chamber plants also led to joint products, and a considerable quantity of sulphur dioxide was lost through technical inefficiency and caused environmental damage. Thus there was a direct link between the sulphur dioxide emitted, environmental pollution and inefficient production of sulphuric acid.

At the end of the first stage, the king of Sicily had a monopoly in sulphur exports. Because of the sharp increase in the price of sulphur, there was a strong incentive to look for a substitute raw material. This search marks the beginning of the second stage, which lasted from 1838–1900. Within this period, ores containing sulphur were discovered as raw material in the production of sulphuric acid. The metals industry used these ores as raw material, which caused sulphur dioxide emissions during the roasting process. The environmental problems caused by sulphur dioxide from the extraction and roasting of metal ores was a continuous incentive to reduce the unwanted emissions and motivated the processing of sulphur dioxide to sulphuric acid. Due to the utilisation of an unwanted by-product from metals processing by the chemicals industry, both sectors gained advantages: the metals industry had a possibility for disposing of its sulphur dioxide and the chemicals industry had a cheap substitute for the expensive sulphur from Sicily.

In the second half of the second stage, the demand conditions for sulphuric acid changed. Increasing production of fertilisers and coal tar dyes led to increasing demand, in particular for highly concentrated sulphuric acid. The high price of this acid was an incentive for the development of the Winkler contact process (Müller 1994). This technology was improved during the following years, as was the production of chamber acid (the sulphuric acid produced in the lead chambers) after contact printing also became profitable.

The third stage is characterised by the replacement of the lead chamber process with the single contact process. It lasted from 1900–1963. The existing lead chamber plants remained in partial use until they reached the end of their working lives – the average lifespan of an intensively used lead chamber plant was only eight or ten years. Since there was no need to change the existing inventory, this steady technological substitution had no direct capital losses.

The evolution of vanadium contacts as catalysts was the prerequisite for the beginning of the fourth stage in 1963, when Bayer AG implemented the double-contact process, which permits more efficient use of sulphuric acid. At the same time this resulted in a further decrease in sulphur dioxide emissions. The double-contact process is currently the best available technology for producing sulphuric acid and leads to negligible emissions. Since sulphur dioxide is required as a pre-product, the process is even suitable as a use for this environmentally harmful substance. However, the sulphur dioxide must be present in relatively high concentrations. Nevertheless, considerable environmental problems arise from the use and modes of use of sulphuric acid.

4.3 Conclusions

The history of the chlorine alkali industry and the current changes in the use of chlorine in the chemicals industry as well as the history of sulphur chemistry show that evolution in this important economic sector is to a large extent instigated by resource and environmental problems. The rises in resource prices, induced by increasing scarcity, were a strong incentive to invent and implement new production processes. The deterioration of environmental quality due to unwanted by-products was also often the reason for society's introduction of environmental regulations. In addition, we can learn from the history of the chlorine alkali and sulphuric acid industries that a change in production techniques requires the construction and implementation of new capital goods. However, long periods of time were necessary to realise the new environmental friendly solutions. It was by no means certain that solutions could also be found at the time when the problems first arose.

5 Resource Scarcity and Environmental Problems as a Cause of Structural Change in Basic Industries

The two examples, the chlorine and sulphur industries, show that environmental problems and the scarcity of raw materials have in the past been a stimulus to the invention and implementation of new production techniques. In this way, the response to changed environmental conditions contributed considerably to structural change in the chemicals industry. Joint production is a permanent incentive to develop and introduce new production processes, for two reasons: on the one hand, the utilisation of by-products permits an increase in the efficiency of a production process and contributes to an improved economic position in the sector. On the other hand, unwanted by-products often result in damage, and if they are no longer tolerated by society, change becomes necessary.

The history of the chemicals industry shows that modifications in its natural boundary conditions were a stimulus to develop and introduce new techniques. In the language of evolutionary economics, the deterioration of the environment arising from the production of chemical products has led to a change in the genotype and phenotype of the economy. However, it should be pointed out that at the moment when raw material and environmental problems arose, it was by no means certain that solutions could be found. Furthermore, the evolution of the chemicals industry also demonstrates that the availability of new production techniques does not automatically lead to their application. Two other factors are decisive:

1. New capital goods must be financed, produced and installed.
2. There is the problem of what should happen to old capital stock, which cannot be used for new processes.

Examples of these problems can be found in the chlorine-free production process for propylene oxide (cf. *Jacob* in this book). The expansion of propylene oxide production is a consequence of the introduction of a new process in the production of ethylene oxide. One factor in the rapid introduction of the new production process was the possibility of using old plant with only minor changes for propylene oxide production. In the meantime, chlorine-free propylene oxide production processes have become available, but the existing production facilities represent a hurdle for the introduction of the new techniques. Because of such circumstances, discontinuities may arise in adapting to a change in the natural and social boundary conditions of the economic process.

In the other case studies, similarly strong interdependency between scarcity of resources, environmental problems and changes in production technology cannot be found. Nevertheless, an analysis of structural change in the other sectors of basic industry shows that new environmental regulations have had a considerable influence on their evolution in the last decades. Environmental policy often constitutes an incentive for the industry to implement end-of-pipe environmental protection measures. However, we could also observe that enterprises have altered their production process in order to avoid the unwanted by-products (Faber et al. 1996b, p. 271). The introduction of new techniques was a prerequisite for the substitution of more expensive raw materials, as the examples of the steel industry in Luxembourg and the Austrian cement industry show. In the former case, more expensive iron ores were replaced by scrap (*Binder and Schucht* in this book). In the latter example, fossil fuels became more expensive with the oil crisis and were replaced with waste from other sectors (*Hüttler* in this book). In the steel and aluminium industry in Germany, environmental legislation was, among other things, an incentive to change production processes. In the latter example, the changes in the steel industry improved the efficiency of the production processes. It was possible to produce a higher output with the same amount of inputs. This, in turn, has led to a reduction of production residues and, at the same time, to a decrease in specific energy consumption by the steel industry (*Binder and Schucht* in this book). A similar development can be observed in the aluminium industry. In this sector an increased quantity of secondary raw materials is used in order to reduce the energy input to the production process (*Schucht* in this book).

A prerequisite for technological change in the examples considered was that alternative production processes were known and available. It also appears that we may be able to explain ex-post the circumstances which led to the invention and introduction of a new technique. However, it is not possible to predict how structural change will develop in the future, in the sense of predicting the invention of new techniques. The description of structural change in the categories of genotypic and phenotypic change shows the difficulties inherent in such a forecast.

6 Concluding Remarks: Structural Change, Environmental Protection and Policy

The case studies demonstrate that the structure of the economies considered has changed dramatically in the last decades. The change in share of Gross Domestic Product and in employment is largely determined by a change in the production techniques within the sectors or by a change in the production techniques in those sectors which use the outputs of basic industries as input. Nevertheless, on the basis of the available case studies alone, one could not conclude that the importance of basic industries has diminished. What general conclusions can be found, and which answers can be formulated to the following three questions posed in the introduction of this paper?

1. To what extent is structural change caused by scarcity of raw materials and environmental deterioration?
2. Is structural change a process which automatically leads to an improvement in environmental quality or must this be enforced by environmental policy measures?
3. What should be taken into account, if new environmental legislation is necessary for environmentally friendly structural change?

The examples examined above show that structural change is a process which takes place over long periods of time. The analysis of the case studies yields two essential factors determining structural change in the corresponding sectors of the basic materials industry:

1. scarcity of resources and environmental problems to which firms and societies reacted with a change in production techniques and environmental laws
2. the two oil crises at the beginning of the1970s and 1980s were a strong incentive for a change of technologies in order to reduce energy consumption

An essential indicator describing the change in an economy's production structure over time is the substitution of production processes. This brought about by the invention and introduction of new techniques. The case studies have also shown that a change of production technique often leads to a decrease in environmental pollution. However, it cannot be deduced from this that structural change leads automatically, i.e. only via considerations of profitability, to a reduction in environmental pollution. So far, the source of structural change is change in production processes (Faber et al. 1996b, p. 271): The implementation of new technologies can:

1. increase technological efficiency (produce the same quantity of marketable goods from fewer inputs)
2. decrease environmental damage
3. increase efficiency and decrease environmental damage at the same time

The case studies concerning structural change in basic industries show that a lot of technological change is of type 3, i.e. it increases efficiency and decreases environmental damage. In this situation, it is possible to increase profitability and improve environmental quality simultaneously. One might therefore expect that considerations of profitability and the resulting change in technology automatically reduce environmental deterioration.

However, we have already pointed out during the analysis of the chemicals industry that the superiority of a new technique is not sufficient for the introduction of a new production process. Faber and Proops (1991) have shown in a theoretical study that technological superiority is only a necessary condition for technological change. Whether a superior technique is introduced depends, among other things, on the time horizon of the decision-makers. Faber et al. (1996b, p. 276) report that in the chemicals industry, investments in new plant are expected to pay for themselves within four to six years. With such short time horizons, the technological superiority of new techniques over the old one must be very great. From this it follows that many production processes which reduce environmental pollution and simultaneously increase profitability are not implemented. For this reason, efficiency considerations alone do not lead to structural change which reduces environmental pollution. However, if environmental policy measures make existing procedures more expensive, then this is an incentive to introduce environmentally friendly production techniques which, in turn, also allows for an increase in an enterprise's efficiency in the long run (Faber et al. 1995). This is confirmed in the case study by *Van der Straaten* (in this book) of the oil industry in the Netherlands. Without appropriate environmental policy measures we can not expect structural change automatically to improve environmental quality.

Finally, it is important to consider that many environmental problems are global and the damage irreversible. Therefore, we can not wait until unacceptable environmental pollution or a serious scarcity of resources arise and be recognised by the public – as was the case in the historical examples in the chemicals industry, which finally led to a change in production structure or patterns of consumption. However, even if environmental policy is necessary for structural change which reduces environmental pollution, there is no guarantee of success for such regulations.

This description of economic development in terms of evolutionary economics has shown that a part of structural change is not predictable. We should take this into account when implementing new environmental policy measures. For this reason, it is necessary for environmental policy measures to be formulated flexibly, so that adaptation in the light of new findings is possible. Finally, environmental policy requires long-term goals and the corresponding regulations should be announced early and introduced gradually. Such a strategy will reduce the friction which could occur with changes in the existing production structure. The experience of two oil crises at the beginning of the 1970s and 1980s has shown that unexpected economic change can have profoundly negative results for the economy and society.

References

Ayres, R. (1991): Evolutionary economics and environmental imperatives. Structural Change and Economic Dynamics 2 (2), 255–273.

Azariadis, C. (1993): Intertemporal Macroeconomics. Blackwell, Oxford, UK.

Barro, R.J., Sala-I-Martin, X. (1995): Economic Growth. McGraw-Hill, New York.

Bollmann, P. (1990): Technischer Fortschritt und wirtschaftlicher Wandel. Physica-Verlag, Heidelberg.

Buttgereit, R. (1994): Die Chlorchemie auf dem Prüfstand – gibt es Alternativen? Spektrum der Wissenschaft, August, 108–113.

Dosi, G. (1986): Technical Change and Industrial Transformation. 2nd edn. Macmillan, London.

Dosi G., Nelson, R. (1994): An introduction to evolutionary theories in economics. Journal of Evolutionary Economics 4 (3), 153–172.

Drösser, E. (1908): Die technische Entwicklung der Schwefelsäurefabrikation. Verlag Dr. W. Klinkhardt, Leipzig.

Europa-Chemie (1992): Internationaler Informationsdienst für Chemie, Umwelt und Wirtschaft. Verlagsgruppe Handelsblatt, Düsseldorf, Heft 29, 2.

Faber, M. (1979): Introduction into Neo-Austrian Capital Theory. Springer Verlag, Heidelberg.

Faber, M. (ed.) (1986): Studies in Austrian Capital Theory, Investment and Time. Springer Verlag, Heidelberg.

Faber, M. and Proops, J.L.R. (1991): The innovation of techniques and the time horizon: a neo-austrian approach. Structural Change and Economic Dynamics 2 (1), 143–158.

Faber, M., Manstetten, R., Proops, J.L.R. (1992): Toward an open future: ignorance, novelty and evolution. In: Costanza, R., Norton, B., Haskell, B. (eds): Ecosystem Health: New Goals for Environmental Management. Island Press, Washington, pp. 72–96.

Faber, M., Jöst, F., Müller-Fürstenberger, G. (1995): Umweltschutz und Effizienz in der chemischen Industrie – Eine empirische Untersuchung mit Fallstudien. Zeitschrift für Angewandte Umweltforschung 8 (2), 168–179.

Faber, M., Manstetten, R., Proops, J.L.R. (1996a): Evolution in biology, physics and economics: a conceptual analysis. In: Faber, M., Manstetten,R. , Proops, J.L.R. (eds.): Ecological Economics: Concepts and Methods, Edward Elgar, Cheltenham, pp. 136–167.

Faber, M., Manstetten, R., Proops, J.L.R., Jöst, F., Müller-Fürstenberger, G. (1996b): Linking ecology and economy: joint production in the chemical industry. In: Faber, M., Manstetten, R., Proops, J.L.R. (eds.): Ecological Economics: Concepts and Methods, Edward Elgar, Cheltenham, pp. 136–167.

Faber, M., Jöst, F., Manstetten, R., Müller-Fürstenberger, G. (1996c): Kuppelproduktion und Umweltpolitik: Eine Fallstudie zur Chlorchemie und zur Schwefelsäureindustrie. Journal für praktische Chemie, 338, 497–505.

Faber, M., Proops, J.L.R., Baumgärtner, S. (1998): All production is joint production – a thermodynamic analysis. In: Faucheux, S., Gowdy, J., Nicolai, I. (eds): Sustainability and Firms. Technological Change and the Changing Regulatory Environment, Edward Elgar, Cheltenham.

Grossman, G. and Krueger, A. (1993): Environmental impacts of a North American free trade agreement. In: Garber, P. (ed): The U.S. Mexica Free Trade Agreement. MIT Press, Cambridge, pp. 165–177.

Grossman, G. and Krueger, A. (1995): Economic growth and the environment. Quarterly Journal of Economics 110, 353–377.

Malinvaud, E. (1985): Lectures on Microeconomic Theory. Elsevier Science Publishers B.V., Amsterdam.

Meyer, B. (1997): Sulphur, Energy and Environment. Elsevier Scientific Publishing Company, Amsterdam, Oxford, New York.

Müller, H. (1994): Sulfuric Acid and Sulfur Trioxide. Ullmann's Encyclopedia of Industrial Chemistry, VCH, Weinheim. Vol. 25, pp. 635–703.

Nelson, R. (1987): Understanding Technical Change as an Evolutionary Process. Lectures in Economics, Theory, Institutions, Policy. Vol. 8, North-Holland, Amsterdam.

Nelson, R. and Winter, S. (1982): An evolutionary theory of economic change. Harvard University Press, Cambridge.

Romer, P. (1994): New goods, old theory, and the welfare costs of trade restrictions. Journal of Development Economics 43, 5–38.

Selden, T.M. and Song, D. (1994): Environmental quality and development: is there a Kutznets curve for air pollution emissions? Journal of Environmental Economics and Management 27: 147–162.

Shafik, N. (1994): Economic development and environmental quality: an econometric analysis. Oxford Economic Papers, 46, 757–773.

Stephan, G. (1995): Introduction into Capital Theory – A Neo-Austrian Perspective. Springer Verlag, Heidelberg.

Stern, D.I., Common, M.S., Barbier, E.B. (1996): Economic growth and environmental degradation: the environmental Kutznets curve and sustainable development. World Development, 24 (7), 1151–1160.

Witt, U. (1994): Wirtschaft und Evolution. WiSt. 10, 503–512.

Sectoral Network Structures as Determinants of Structural Change

Patrick Kenis
Free University Amsterdam, De Boelelaan 1081c, 1081 HV Amsterdam, The Netherlands

1 Introduction

Sectoral networks can theoretically influence environmental protection in industry in two different ways. Firstly, sectoral networks can influence the direction and pace of structural change in industries (A_1 in Fig. 1 and can consequently have an indirect effect on the environmental stress caused by these industries (A_2 in Fig. 1, i.e. inter-sectoral change). Secondly, sectoral networks can also have a direct effect on environmental stress through their capacity to develop collective strategies to cope with the environmental stress produced in these sectors (B in Fig. 1, i.e. intra-sectoral modernisation). Although the latter is a fascinating topic for analysis in itself, this paper will mainly concentrate on the former (and on A_1 in Fig. 1 in particular). As far as structural change is concerned, it is clear that other factors (such as business strategies, local factors, legal measures, etc.) also play an important role. However, for conceptual reasons, this paper will demonstrate and concentrate on the importance of sectoral networks for structural change.

On the basis of the empirical evidence presented in the case studies of environmentally intensive European basic industries, this chapter will examine whether and, if so, how conclusions can be drawn as to whether sectoral networks provide an explanation for the process of restructuring and decline observed in the industries studied. The following starts by sketching out what is generally understood by "sectoral networks" (Sect. 2). In Sect. 3, the case studies in this volume will be presented in the light of the information they contain on the presence of sectoral networks and the types of structural change they have undergone. In the subsequent section, some speculation will be formulated on the possible relationships between types of networks and the processes of structural change observed (Sect. 4). The paper closes with some general conclusions and suggestions for a follow-up to the project which is more oriented around actors and policy (Sect. 5).

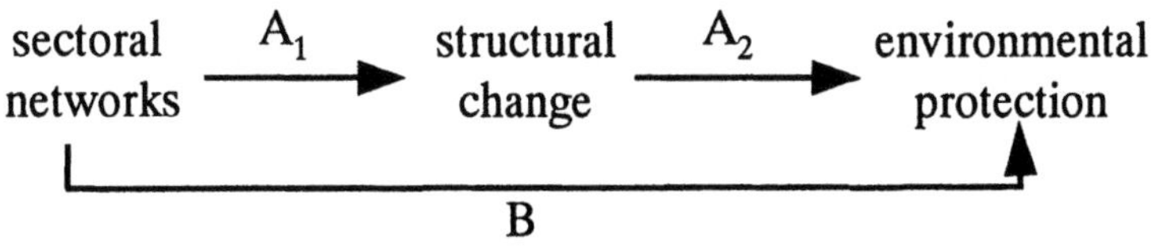

Fig. 1. Possible influences of sectoral networks on environmental protection

2 What are Sectoral Networks?

An extensive body of literature on sectoral networks has appeared, which cannot be reviewed in detail here (Kenis and Schneider 1991, 1996; Alter and Hage 1993; Messner 1995). Generally speaking, three main research themes within the literature can be made out: The first is extensive research which concentrates on the identification, definition, structural and functional *description* of sectoral networks. Secondly, some research attempts to identify and specify the *causes* of their existence and of differences in their structure and functioning. Thirdly, there is research which concentrates on the *consequences* of sectoral networks.

Given the title of the present chapter (sectoral network structures as determinants of structural change), it is the second and third research areas which are particularly relevant here. In the following, the sectoral network is treated as the *independent variable* while structural change is treated as the *dependent variable*. What follows therefore explains in somewhat more detail what sectoral networks are and how they are commonly described in the literature. On the basis of this description, the next section specifies a number of dimensions which are useful for evaluating the degree and type of sectoral networks as identified in the case studies (the first research area). Based on this information it will then be possible to analyse the consequences of the presence and type of sectoral networks on structural change (the third research area). The second research area (the causes of sectoral networks) will only be touched upon in the conclusion.

Structural change can be dealt with by implementing various types of governance structures. In addition to the traditional *market/state* dichotomy, *networks* have been identified as another distinct, significant, and frequently implemented governance mechanism in industrialised countries (see Kenis and Schneider 1996, Hollingsworth et al. 1994, Hollingsworth and Lindberg 1985, Campbell et al. 1991). In the case of structural change[1], governance through sectoral networks means that the parties negotiate among themselves the need for, or distribution of, mutual burdens and benefits from industrial restructuring. Sectoral networks constitute observable and relatively stable groups of organisations formed in alliances or coalitions to promote the collective interests of some or all industrial actors by discussing common interests and by ranking their collective priorities.

Industrial governance by networks differs from state-led industrial governance in the sense that the principal mode of co-ordination is not based on command or direction but rather on negotiation. It also differs from market-led industrial restructuring in that each firm does not merely pursue its own business rationale by concentrating exclusively on "internal restructuring" (see Ruigrok and van Tulder 1995). In contrast, networks are horizontal systems of co-ordination between firms, public administrations, and associations.

---

[1] Sectoral networks are understood here as those affecting structural change and industrial restructuring. It should be noted, however, that a substantial portion of the sectoral network literature deals with other sectors than the industrial, e.g. the health sector, the welfare sector, etc.

They tend to be more stable than market governance, even though this stability is not imposed from above. Instead, shared norms, attitudes of trust, considerable knowledge about and respect for one another stabilise the relationships between the actors. Normative mechanisms, negotiations, and socialisation within the group co-ordinate relationships and discourage opportunism over relatively long periods. Here is a governance mode in which medium- and long-term policies can develop to cope with the structural problems of a specific industry.

Take, for example, the problem of over-capacity in an industry which produces basic goods and is exposed to all the pressures of the global market. Such industries are characterised by low technological complexity and mass production. Large economies of scale and significant fixed costs also tend to be common characteristics. In these industries, full capacity utilisation is the prime condition for competitiveness. Consequently, when firms are confronted with high fluctuations in demand they are almost immediately confronted with high risks. Much is at stake: many workers stand to lose their jobs and investors stand to bear huge losses. During a period of recession, market-led industrial policies in such sectors therefore lead to the closing down of plants – eventually even to the complete disappearance of an industry in individual states.

In order to prevent or overcome the negative consequences of market-led industrial restructuring (such as massive unemployment and large-scale capital depletion), a directly state-led industrial policy is often the preferred option. In most cases, such a state policy amounts to subsidising industries in order to save jobs. However, such actions often increase the risk of impairing the long-term structural adaptability of that particular industry at the same time.

Alternatively, in order to compensate for the shortcomings and self-destructive elements of the market – and, at the same time, pre-empt state intervention – firms within an industry can organise themselves and join together to cope with an impending crisis. In such a situation, they agree upon mutual rules of conduct, linking their own individual autonomy to their medium- and long-term collective interests. Networks put actors into a situation where they are not forced to react so rapidly, more flexible, and less constrained by disruptive developments. This offers a governance mode through which they can better adapt to changing circumstances. Industrial restructuring through networks is based on intensive integration and co-operation between the parties concerned. The actors explicitly negotiate and define rules of conduct, even though they do not necessarily codify them in contracts or other formal agreements. These arrangements make possible the design and implementation of medium- or long-term policies for greater competitiveness and facilitate collective decisions by the firms involved in high-risk investment projects.

In addition to the *functional* properties described above, sectoral networks can also be described in structural terms (Atkinson and Coleman 1989, van Waarden 1992, Messner 1995, Alter and Hage 1993, Kenis and Schneider 1991). In general, they include "a few or not too many" (Marin and Mayntz 1991: 17) *actors* from different backgrounds (private, public, non-profit); they are characterised by horizontal, informal and decentralised *relationships*; they can be distinguished by *rules of conduct, actor strategies* as well as *power structures* between the actors.

3 A "Sectoral Network" Evaluation of Ecologically Intensive European Basic Industries

The case studies in this volume will now be evaluated as to the evidence they contain regarding sectoral networks as a driving force in structural change. As a first step, a table has been drawn up, summarising information on the properties of sectoral networks and on structural change in the ten different sectors. A second stage speculates on possible relationships between sectoral networks and structural change (Sect. 4). The types of *structural change* summarised in Table 1 are primarily those concerning the type and degree of decline observed in the different industries. These types of structural change seem to be the most significant in the context of this volume, as they also directly influence the degree of environmental stress produced by these industries.

As far as summarising the properties of *sectoral networks* is concerned, it should be stated clearly at the outset that it is not possible to provide a comprehensive or exhaustive picture of the various sectoral networks. The main reason is that the data-collection instrument on which the different case studies were based was not explicitly directed at assembling the sectoral network data referred to in Sect. 2. Consequently, evidence on sectoral networks must rather be implicitly deduced and may on occasion be incomplete. In order not to run up against the "law of the instrument"[2], the case studies will be evaluated flexibly, using somewhat general categories. Attention will therefore be paid to some basic and minimal components which every sectoral network comprises: the number and types of actors involved, the extent and types of relationships between these actors and the degree and type of collective strategies produced through these relationships.

Before turning to some speculation about possible relations between certain types of networks and the processes of structural change observed, some general remarks on two points: firstly, the degree and type of structural change observed across the different sectors and, secondly, the importance and significance of sectoral networks across these sectors.

As far as structural change is concerned, it is clear that a fall in production and employment has been observed in all sectors over the last two decades. However, there are some important differences regarding the absolute pace and extent of the decline. What is even more important for the present chapter, however, is that there are significant differences between the rates of decline in different sectors. In some cases, the decline fits well into the general picture of structural change

[2] "Give a small boy a hammer, and he will find that everything he encounters needs pounding. It comes as no particular surprise to discover that a scientist formulates problems in a way in which requires for their solution just those techniques in which he himself is especially skilled" (Kaplan 1964, p. 28).

across Europe (e.g. aluminium in Germany and fertilisers in Austria). This is an indicator that market forces and *internal restructuring* were the principal governance modes at play. In other cases the decline is atypical (e.g. coal in Germany). It is especially these atypical structural changes which point to the fact that there is something else at work. The important question now is whether what is at work here is the sectoral network and if so, how it produces these atypical situations. Before turning to an analysis of this relationship in the next part, the first question is where evidence on the existence of sectoral networks is to be found.

The best way to identify sectoral networks is by first concentrating on the evidence regarding "sector integration" (see Table 1). In cases where specific relationships (such as information exchange, coordination, co-operation, etc) exist between a differentiated set of actors, occasionally leading to collective sectoral strategies, there is strong evidence for the presence and functioning of sectoral networks as a governance structure. Based on these criteria, it can be assumed that sectoral networks exist in the chlorine and cement industries and in the iron and steel sector in Germany, as well as in the oil refineries sector in France.

4 The Relationship Between Sectoral Networks and Structural Change

In what follows, a number of observations are made as to whether and how sectoral networks have been influencing structural change in the cases studied. These observations are of a somewhat general nature and should be treated with some caution, for two reasons. In the first place, as mentioned before, the data available on structural change in the case studies are limited. Secondly, even if such data were available, a thorough analysis would still be problematic, since the study of the consequences of sectoral networks is still a very underdeveloped area in the literature. As a result, few analytical tools or concepts are available to study this relationship. Nonetheless, a number of clear indications can be found.

Firstly, in the absence of sectoral networks, structural change leads to a faster and more radical decline in the number of jobs and in production capacity. This observation is based on the fact that such sectors are governed primarily by the mechanism of the market, which will push towards internal restructuring.

Secondly, structural change in the context of sectoral networks can lead to outcomes which better safeguard production capacity and thereby often employment as well. This can be deduced from the fact that sectoral networks, if they develop, will aim primarily at protecting the industry from certain unexpected events. Sectoral networks are often effective devices for *internalising* or *externalising* uncertainty (see Kenis 1992). This phenomenon, an atypical decline due to the presence of sectoral networks can be observed in the case of the French Oil refineries. Here, co-operation in cutting surplus capacity developed on the sectoral level. Or, in the case of German coal mining, sectoral networks developed which led to an atypical (for Europe) slowing of production.

Table 1. A "sectoral network" evaluation of ecologically intensive European basic industries

sector study	sector differentiation		sector integration		type and degree of structural change
	number of firms	*other actors involved*	*types of relationships*	*collective strategies*	
chlorine Germany		products licensed by the state; extensive state regulation	members of one association; consumer relations	division of labour for final products	drop in production at the beginning of the 1980s and 1990s; minimal drop in employment; production of CFCs and chlorinated solvents halted
fertilisers Austria	very high concentration (3 main firms)	fertiliser tax introduced in 1986 and abolished after Austria's entry in the EU; regulations on fertiliser use			between 1974 and 1995 a 35% fall in production and a general decline in fertiliser use (less agricultural land, increased organic farming, etc.)
aluminium Germany	5 primary and 11 secondary aluminium companies	environmental regulations; no state protectionism			slight decline in production; the number of primary aluminium companies has halved, the number of sec. aluminium companies has fallen slightly since the 1970s
cement Denmark	1 company (since 1980)	zoning regulation, health protection and recent environmental regulations; Northern Jütland County Council (environmental authority)			same level of production in 1996 as in 1960, with falls in 1981 and 1990
cement Austria	9 companies: 3 of them make up 2/3 of the turnover; multinationals have recently entered the market	government import regulations; construction sector; Eastern European cement producers; government environmental regulations (since 1993)	negotiations between government, construction and cement industries and between Austrian and Eastern European producers on export self-restriction	an Austrian cement cartel until recently; cartel was transformed into *Strukturkrisenkartell*	39% decline in production between 1974 and 1996, and a fall in the number of companies and production sites; increased import pressure since the 1990s; increased competition from multinationals; decline in employment (54% since 1974); decline in cement prices

coal mining Germany	3 enterprises in 1997	government import and other regulations; state subsidy of production; regulated by the European Union; strong trades unions; importance of regional actors	close links between companies and regional politics; decisions mostly compromises between all actors	sustained coal production despite its not having been competitive since the 1950s	drastic decline in production since the mid-1950s (by 2/3); drastic decline in employment since the mid-1950s (by 85%); decline in number of enterprises (from 32 in the 1960s to 3 in 1997)
steel Germany	22 enterprises in 1996	state subsidy; political regulation only since the beginning of the steel crisis; environmental regulation	co-operations dating back to the 19th century (e.g. horizontal and vertical integration, specialisation); mergers	co-operation on better utilisation of capacity (since the early 1960s)	decline in production between the mid-1970s and early 1980s; stagnant since then; decline in employment since the 1960s (by more than 70%); decline in the number of enterprises; over-capacity; diversification towards high-quality and high-grade steel
steel Luxembourg	1 company		(not relevant since only 1 firm involved)	(not relevant since only 1 firm involved)	mining abandoned (since 1981); pig-iron: from 25 production sites in 1970 to none in 1997; crude steel: halved, finished steel: slight drop (since 1973); drastic decline in employment since 1974 (by 79%)
oil refineries Netherlands	8 enterprises	a large number of actors have been involved in the formulation of the NEPP (National Environmental Policy Plan)	many actors are traditionally involved in public decisions	climate of tolerance and consultation	some decline in the 1970s, significant decline in the early 1980s; recovery and increased production since 1983
oil refineries France	13 refineries (six of which control the largest share of the market)	government price control until 1985; a manifestly state-directed energy policy; associations (U.F.I.P., Europia, E and P); the EU Council; *L'Institut Français du Pétrole*; French Ministry of Industrial Relations; *Direction des hydrocarbures*; Ministry of Environment; ADAME	agreements on co-operation between companies (e.g. on distribution activities, capacity reduction); negotiations increasingly taking place	a "petrol charta" was in use until 1992; joint distribution activities; joint negotiations and action on reducing over-capacity	employment has halved since 1975; production capacity halved between 1976 and 1989; it has stayed at the same level since 1989; 37 refineries have been closed since 1977

Thirdly, sectoral networks do not necessary have to concentrate on conserving existing production facilities and workforces, but may develop collective strategies which help to facilitate processes to minimise the economic and social impact of structural change. For example, sectoral networks could facilitate the restructuring of the sector in such a way that existing resources (production facilities and personnel) are invested and redirected towards other productive areas.

Finally, there is some evidence that the influence sectoral networks have on structural change depends to some extent on how far the sector is "closed" or "open". Sectoral networks in "closed" sectors are generally thought to be counterproductive for an industry's successful long-term structural adaptability, whereas sectoral networks in "open" sectors facilitate this adaptability (see Traxler and Unger 1990). The reasons are that, in closed markets, institutional sectoral networks will tend to reduce competition between firms and may develop into "quasi-cartels". In open sectors, on the other hand, pressures emanating from the global market play an important role. In such situations, sectoral networks will facilitate modernisation and innovation within their sectors.

These various findings as to the effects of sectoral networks on structural change, although they represent only a few examples, indicate the complexity of the relationship. While the absence of sectoral networks generally seems to lead to a faster decline in production capacity, the effects of the presence of sectoral networks are much more ambivalent. Sectoral networks can stabilise or slow down a decline in production, can lead to a decrease of, or even a shift in, production and can affect long-term structural adaptability positively or negatively, depending on the characteristics of the sector. As a result, on the basis of these findings, it is impossible to give a general answer to the question of whether – in those cases where sectoral networks exist – they lead to a better or worse situation from an environmental perspective. The answer to this question is very much contingent on the type of structural change produced by the sectoral network in place. The fact that no straightforward correlation seems to exist between the presence of sectoral networks and the direction of structural change might be considered rather disappointing, from both a theoretical as well as a practical point of view. If, however, we examine the logic behind this finding in somewhat more detail, interesting new perspectives emerge, with consequences for further research.

5 Conclusions and Suggestions for Future Research

This chapter set out to analyse the relationship between sectoral networks and the direction and pace of structural change in industrial sectors. The assumption was that the direction and pace of structural change ultimately influences the environmental stress produced by these industries. An evaluation of the sectoral networks as well as of the direction and pace of structural change in the industries studied in this book shows that there is indeed variance between industries in respect of these two dimensions. Some industries are characterised by a high pace of industrial change, others by rather moderate change; in some industries, sectoral networks seem to constitute an important governance structure, in others they are not present at all. At the same time, however, another important finding is that the two variables (i.e. structural change and sectoral networks) hardly correlate with one another. At first sight, this conclusion might appear disappointing. A closer look reveals, however, that there is a certain logic to it, and this logic becomes especially important for the formulation of future research and has implications for policy.

The above findings are logical and consistent if one takes into account an important characteristic of sectoral networks, i.e. that they are non-*specialised* governance structures. This means that they do not concentrate on a specific concept of governance but must be seen rather as structures which combine a variety of such concepts. In contrast, markets or power structures are considerably more specialised, and consequently bind the actors to a more limited spectrum of alternatives (e.g. short-term target maximisation in the case of market competition). It is especially during periods of radical change that specialised governance structures can be seen as somewhat disadvantageous. In contrast, the availability of a greater spectrum of alternatives – as is the case with network and similar structures – increases flexibility for actors and allows for alternative paths of action. It is precisely this property of governance forms, i.e. the degree of specialisation, which explains why consistent outcomes can be expected from cases where the market or the state is the principal mode of governance and why outcomes can vary greatly between each other in cases where sectoral networks are the governance structure. This is mainly the result of the flexible character of sectoral networks as a form of governance.

On the basis of this conclusion, two main directions for further empirical and theoretical research can be mapped out. Firstly, far more detailed research must be conducted on the relationship between sectoral networks and structural change. Since sectoral networks clearly have an influence on structural change, but not one which is straightforward, additional research is necessary. This research should concentrate on the relationship between specific characteristics of sectoral networks and the outcomes these produce. It is obviously the characteristics of given sectoral networks which influence the outcomes they produce. In this research, different sectors in different countries must be analysed in detail in the two main dimensions specified, i.e. differentiation and integration. Only such detailed research can give indications as to which specific types of outcome (e.g. increased, decreased or stagnant production) result from which types of sectoral network.

A second line of research might be to concentrate rather on the direct influence of sectoral networks on the environment alone (see B in Fig. 1). As argued above, given the fact that sectoral networks as governance forms are characterised by the fact that they combine a variety of approaches and interests, it seems obvious that they are the ideal device by which such differing and often contradictory demands as efficiency, employment, environmental protection, etc. can be met. There is already a number of studies which have shown that sectoral networks can indeed play an important role in this respect (cf. e.g. Grundmann 1997, Gray and Hy 1986, Rinquist 1994). The cases covered in this volume could also be very valuable tools for research analysing these types of relationships.

References

Alter, C. and Hage, J. (1993): Organizations Working Together. Newbury Park CA: Sage.

Atkinson, M. and Coleman, W.D. (1989): Strong States and Weak States: Sectoral Policy Networks in Advanced Capitalist Economies. British Journal of Political Science 19, pp. 48–67.

Campbell, J.L., Hollingsworth, J.R., Lindberg, L.N. (eds) (1991): Governance of the American Economy. Cambridge: Cambridge University Press.

Grundmann, R. (1997): Politiknetzwerke und globale ökologische Probleme: Der Fall der Ozonschicht. Politische Vierteljahresschrift 38 (2), pp. 274–292.

Gry, B. and Hy, T.M. (1986): Political Limits to Interorganizational Consensus and Change. The Journal of Applied Behavioral Science 22 (2), pp. 95–112.

Hollingsworth, J.R., Lindberg, L.N. (1985): The governance of the American economy: the role of markets, clans hierarchies and associative behaviour. In Streeck, W. and Schmitter, P.C. (eds): Private Interest Government. Beyond Market and State. London: Sage, pp. 221–254.

Hollingsworth, J.R., Schmitter, P.C., Streeck, W. (eds) (1994): Governing Capitalist Economies: Performance and Control of Economic Sectors. New York: Oxford Univ. Press.

Kaplan, A. (1964): The Logic of Inquiry. San Fransisco: Chandler.

Kenis, P. (1992): The Social Construction of an Industry. Frankfurt: Campus.

Kenis, P. and Schneider, V. (1991): Policy Networks and Policy Analysis: Scrutinizing a New Analytical Toolbox. In: Marin and Mayntz (1991), pp. 25–59.

Kenis, P. and Schneider, V. (1996) (eds): Organisation und Netzwerk. Institutionelle Steuerung in Wirtschaft und Politik. Frankfurt, New York: Campus.

Marin, B. and Mayntz, R. (eds.) 1991: Policy Networks. Empirical Evidence and Theoretical Considerations. Frankfurt: Campus Verlag.

Messner, D. (1995): Die Netzwerkgesellschaft – Wirtschaftliche Entwicklung und internationale Wettbewerbsfähigkeit als Probleme gesellschaftlicher Steuerung. Köln: Weltforum Verlag.

Rinquist, E.J.: Policy Influence and Policy Reponsiveness in State Pollution Control. Policy Studies Journal 22 (1), pp. 25–43.

Ruigrok, W. and van Tulder, R. (1995): The Logic of international restructuring. London: Routledge.

Traxler, F. and Unger, B. (1990): Institutionelle Erfolgsbedingungen wirtschaftlichen Strukturwandels – Zum Verhältnis von Effizienz und Regulierung aus theoretischer und empirischer Sicht. Wirtschaft und Gesellschaft 16 (2), pp. 189–223.

van Waarden, F. (1992): Dimensions and types of policy networks. European Journal of Political Research 21, pp. 29–52.

Industrial Complexes as the Centres of Gravity for Green Industrial Restructuring

Winfried Ruigrok
University St. Gallen, FIM, Bodanstr. 6, 9000 St. Gallen, Switzerland

1 Introduction

Many European observers have long regarded restructuring as a passing phase, during which the industrialised economies have to adjust to slower growth and to changing international conditions, before eventually settling in a new equilibrium. Today it appears not only that global competition has intensified, but that restructuring may be a permanent phenomenon, whether companies be knowledge-intensive, capital-intensive or labour-intensive.

However, the driving forces and the forms of restructuring differ widely between industries. In the late 1990s, restructuring in branches as varied as textiles, consumer electronics and insurance seemed driven primarily by business considerations. In the industries discussed in this book, however, restructuring is often *not just a matter of business efficiency, but also subject to the impact of stakeholders* such as environmental groups, organised labour and national and supranational authorities. Green industrial restructuring is a *combination* of optimising internal structures, processes, and market orientation, *and* of adapting or adjusting to the company's economic, social and political relationships to external stakeholders. The term is applied to restructuring in what have been variously termed "dirty" (Jänicke et al. 1997), "declining" (Harrigan and Porter 1983), or, somewhat more broadly, "mature" industries (Baden-Fuller and Stopford 1992). While these phrases are by no means synonymous, restructuring within the corresponding industries involves the need to manage (and hence analyse) a variety of *economic, social* and *political* relationships *simultaneously*. This chapter aims firstly to present an approach which addresses restructuring both in terms of the business rationale *and* in terms of the companies' respective economic, social and political networks. The second part of this chapter attempts to test the plausibility of this approach by applying it to the various cases analysed in this book.

2 At What Level Should Industrial Restructuring Be Analysed?

Industrial restructuring has long been viewed primarily as the result of fundamental, essentially anonymous, economic forces. The dominant analytical approaches here include neo-Schumpeterian economics (cf. Nelson and Winter 1982;

Rosenberg 1982; Freeman 1982; Perez 1983; Dosi et al. 1990; Lundvall 1992), institutional economics (Hodgson 1993; Groenewegen et al. 1995), transaction cost economics (Williamson 1975, 1979; Coase 1937; Teece 1985, 1988), regional economics (Marshall 1961; Piore and Sable 1984), and industrial economics (or "industrial organisation" in the U.S., e.g. Porter 1980, 1985, 1990). These approaches are discussed in *Petschow and Nill's* contribution to this book.

An interesting debate in the literature on industrial economics has focused on the *level of analysis* when examining industrial restructuring. Industrial economists have traditionally preferred a high level of abstraction, in order to be able to apply their results to as large a group of companies or industries as possible. However, one may seriously question whether "industry", a term often used in this book, is a meaningful concept in the first place. From the perspective of industrial economics, the term "industry" is meaningful if company profit levels within a particular sphere of activity tend to converge over a certain period of time. Conversely, the term may lose much of its meaning if profit levels within any group of companies tend to diverge. Using US data from the period 1974–77 to test convergence of profit levels, Richard Rumelt (1991) made a strong case for the idea that the *company* matters, not the *industry*. Many economists have arrived at a similar conclusion after studying the technological revolutions of the 1980s and 1990s, arguing that the emergence of numerous applications and products which are converging on one another from different industrial sectors have made classification by industry increasingly arbitrary.

A second discussion has focused on the issue of whether the concept of a *strategic group* (Hunt 1972; Caves and Porter 1977) may offer a better instrument for grouping companies together. In 1980, Porter defined a strategic group as "the group of firms in an industry following the same or similar strategy along strategic dimensions" (1980, p. 129). Although this basic definition has since been adapted several times (cf. McGee et al. 1995), its idea is still to identify *clusters of similar companies within a given sphere of activity and market*. Porter (1979) provides three explanations for the formation of strategic groups: investment to overcome mobility barriers, (varying) relationships between business units and their parent companies and developments in the nature of demand, production technologies or the characteristics of the product, for example. Research into strategic groups has concentrated on manufacturing industry, including traditional areas such as paints, brewing, producer goods, petroleum, and airlines (McGee et al. 1995, pp. 260, 265). However, there is "no consistent pattern in strategic group characteristics" and the "descriptions of groups lack clarity" (McGee et al. 1995, p. 262). Moreover, industries characterised by high levels of internationalisation and/or horizontal networking pose insurmountable problems for the concept of strategic groups, and research into the implications for performance (the original aim of the research) has produced disappointing results (McGee et al. 1995, p. 263, cf. Fiegenbaum and Thomas 1995).

The discussion above has major implications for studying restructuring in the traditional manufacturing industries. Firstly, it suggests that *attempts to determine trends at the level of an industry, or even at the level of groups within the industry, are risky if the aim is to discover implications for individual companies*: each firm will respond differently to opportunities or threats. Secondly, *policy-makers facing demands for industrial regulation (or even direct intervention) should recognise that the chances of such industrial policy succeeding are prohibitively small.*

As an unintended consequence, these discussions of industrial economics may well have contributed to the increasing legitimacy of the management sciences. The scope and variety of the dominant approaches in this field are certainly too great for a complete list to be given here (cf. Mintzberg 1990). Authors and researchers from widely differing backgrounds have contributed information and perspectives stressing the importance of a company's resources (cf. Peteraf 1993; Wernerfelt 1995), its organisational structures (cf. Bartlett and Ghoshal 1989; Womack et al. 1990, 1995) and its strategic goals (Ansoff 1991; Mintzberg 1991a, 1991b), to name but a few. While many of these approaches provide managers with a wealth of information and/or useful frameworks for conducting internal and external restructuring, the material generated in each situation cannot usually be applied more broadly and has not generally been used as the basis for formulating industrial policy.

3 The Network Approach to Industrial Restructuring

In response to the dominance of mono-disciplinary economic approaches to restructuring, a number of authors has searched for an alternative, interdisciplinary approach. Disciplines as far afield as economics, geography, sociology and political science have all contributed to the rise of what has been referred to as the *network approach*. As it lacks a common disciplinary basis, the network approach has not tended to crystallise into a set of generally shared methods, tools or concepts, diminishing the rigour of network theory. It is also less characterised by any common research agenda.

There has long been great optimism as to what network analysis can deliver. According to Paul DiMaggio, *"The moment for those who come from a network perspective to be the head priests of organizational analysis is now."* (cited in Nohria 1992, p. 16). Nitin Nohria attributes recent interest in network theory to the emergence of what Michael Best (1990) termed *The New Competition* (*i.e.* the challenge to the traditional hierarchical company by smaller, more flexible firms and networks), recent technological developments (in information and production technologies) and to the "maturing of network analysis as an academic discipline" (1992, p. 4). Although some authors have claimed that recent progress in this field is a sign of the rise of the "network paradigm" (Cooke and Morgan 1991), it is doubtful that network theory has actually attained such as status. As several authors have indicated, the literature on networks is a "terminological jungle in which any newcomer may plant a tree" (Nohria 1992, p. 3; Burt 1982).

Networks are omnipresent – which poses a major research problem. According to Nohria, "All organizations are in important respects social networks and need to be addressed and analyzed as such" (1992, p. 4). Network theory views organisations as consisting of *recurrent* links, sets of nodes or interactions, and studies their nature. Most work has focused on inter-organisational research (e.g. Piore and Sabel 1984; Håkanson 1989; Powell 1990; Gerlach 1992; Grabher 1993; Ruigrok and van Tulder 1995). However, network theory is applicable at any level of analysis – from the relationship between two individuals, through sub- and inter-organisation units, to complex global systems (Nohria 1992, p. 4).

Much network research has concentrated on the positions of actors within networks ("structure") (e.g. Burt 1992; Fennema 1982), trying to infer "conduct" from "structure" – not unlike the attempts of industrial economists. As a result, due to the large number of actors and processes involved, network research has often been more or less static (e.g. Axelsson and Easton 1991; Håkanson 1989). In itself, a network perspective would appear to offer every opportunity for more dynamic, horizontal research. However, it is not clear whether or how the dimensions of process (behaviour) and structure (form) may be successfully integrated into a single network approach.

A lot of effort has been put into identifying the relevant actors and unit of analysis, and into defining and operationalising the links and interactions between these actors and units (the "who" and "how" in Richard Burt 1992, p. 59). While the definition of the arena is clearly contingent on the actual subject, the nature of interaction has been approached in terms of the "availability of information" (Burt 1992), redundancies in structural network positions (Burt 1992) and bargaining (Ruigrok and van Tulder 1995; see below).

In principle, network analysis can be applied to any industry, although its significance is greater when it is applied to industries where the nature or frequency of interactions is important. The range of industries covered by network analysis is very wide, covering textiles, shoes and leather, biotechnology, telecommunications and computers. Network analysis has been particularly useful in understanding (both technological and organisational) intra- and inter-firm innovation and diffusion (Freeman 1991; Midgley et al. 1992; Smith et al. 1991), in comparing business systems (Whitley 1992; Whitley and Kristensen 1996) and in assessing the opportunities for and effects of government intervention. The latter clearly represents a major advantage compared with economic approaches to restructuring: network analysis provides many possibilities for analysing the relationships between business organisations, such as companies, and non-business organisations, such as governments or trade unions.

4 Bargaining in Industrial Complexes

Attempts at producing a synthesis of the economic and network approaches to restructuring have been few (cf. Freeman 1991; but for some interesting attempts see Hall 1986; Zysman 1994). In doing so, one needs to decide whether to use the network approach or economic theory as one's basis. While there are good arguments in favour of both, this chapter sets out from a network perspective. Firstly, network analysis is able to address restructuring above the level of the individual firm, taking into account the roles of and interaction between clusters of organisations and non-economic actors. Secondly, network analysis is more flexible in integrating economic principles than economic theory is in integrating a network perspective, and thirdly, network theory can provide useful conclusions for policy-makers, as it is concerned with identifying the significant actors, interaction between them and the outcome.

The synthesis proposed here has been described in greater detail elsewhere (Ruigrok and van Tulder 1995). The central unit of analysis is taken to be the *industrial complex*.[1] An industrial complex has the following characteristics: (1) it is a relatively stable network, i.e. the actors do not meet accidentally, and the purpose of interaction is (2) the exchange of goods, capital, technology, information, skills and/or personnel, and (3) the assignment of *value* (with respect to how the actors should interact) (cf. Galtung 1964, 1971; Wallensteen, 1973, p. 32). The third element deals with the power structures among the constituent parties.

It is assumed here that all large firms with major activities in production and technology aspire to position themselves at the *core* of supply and distribution networks, as well as of political and financial networks, in order to play a leading role in creating added value and in restructuring.[2] A core firm has direct access to domestic and foreign end markets and/or customers, and owes its relative independence to its financial power and its control over a number of core technologies and other strategic competencies particular to the industry or industrial activity. Most importantly, a core firm has an explicit vision of 1) the organisation and management of the value chain, including the internal labour process; and 2) the role that external actors (such as suppliers, banks and governments) should play to facilitate the creation of added value and the (re)structuring of the network.

An illustration of the significance of core firms can be found in the work of Lundvall who, while using different terminology, has documented how, through the 1980s, large technological leaders were able to impose a global restructuring trajectory upon other firms. These core firms, which Lundvall and his fellow researchers labelled *user-producers*, not only brought forth new products and production technologies, but were also among the major users of these technologies. This endowed them with greater power and a better bargaining position than other, especially smaller firms. In this way, these large companies have worsened the

---

1 In principle, the term "industrial complex" also refers to the "service industries". A core firm will often be a manufacturing company, but may also be an independent service provider, such as ATT, IBM or Microsoft.

2 This does not necessarily mean that the *core firm* does all this itself. A core firm aspires to be the spider in an industrial web. A core firm is the principal actor as well as the director of the play. In certain types of networks, a core firm may have to give up its role as sole director, yet it will always retain the leading role, and, if possible, will try to regain control of the action.

prospects for new entrants and have influenced the development of substitutes (Freeman and Lundvall 1988; Andersen and Lundvall 1988).

An industrial complex can thus be defined as a bargaining configuration organised around a core firm, consisting of (groups of) actors which are directly or indirectly engaged in the production and distribution of a given product. An industrial complex is made up of five bargaining relationships for the core firm (cf. Porter 1985):

1. with its suppliers
2. with its workers
3. with its dealers
4. with its financiers
5. with its government(s)

Each of the relationships between the core firm and a bargaining partner can be understood as a *dependency*, representing a *control problem* for the core firm. The management of each core firm has to decide whether, how and to what extent it is willing and able to control its bargaining partners. Usually, a decision to control one group of bargaining partners has direct implications on other bargaining relations.[3] Elsewhere, five basic dependency patterns have been identified between the five bargaining partners and the core firm itself: independence; independence with influence; interdependence; dependence with influence; dependence without influence. The first and fourth represent the two extremes of Williamson's dichotomy between market and hierarchy, the second and third represent two intermediary positions, whereas the fifth represents an extension of the Williamson framework. Table 1 lists these five basic positions (Ruigrok and van Tulder 1995, pp. 112–113).

Since national economies generally contain more than one industrial complex, it is useful to distinguish between an industrial complex and the *national industrial system* (Lundvall 1992; Nelson 1993; de Banville and Chanaron 1991). For example, the German car system consists of the Volkswagen, Ford, Opel, Mercedes-Benz and BMW complexes. If a national system contains more than one industrial complex, different types of bargaining and different concepts of control may co-exist for a brief period. However, within national industrial systems, bargaining patterns have tended to converge around the bargaining rules set by the most cohesive complex, at the expense of the weaker complex(es). Within the EU, convergence around such a common denominator is much less likely to emerge, due to the sovereignty of individual member states. As a result, fifteen industrial systems and dozens of industrial complexes produce a cacophony of agendas competing for dominance (Ruigrok and van Tulder 1996).

[3] For instance, if companies decide to limit their dependence on external suppliers by producing essential inputs themselves, they may by confronted with greater capital requirements (affecting their bargaining relations with financiers) and with the demands of organised labour. This is indeed what happened during the rise of *Fordism*: one uncertainty was replaced by another.

Table 1. Five bargaining characteristics and five concepts of control

core firm towards... suppliers	distributors	employees	financiers	governments	*which are*
division of labour; flexible supply: high level of substitution between suppliers; local delivery; competitive and co-operative relationships; no clear profit hierarchy	risk of domination by distributors is high; limited influence on ultimate price of product	loose relation to (craft) unions or no unions; nationally or internationally organised unions	efforts at self-financing; family company; loose relations to (a limited number of) banks	arms-length relations; government support of "small is beautiful" policies; relatively strong supranational government	independent
minority shareholding; supplier of strategic components, longer term supply contracts favoured; many single sourcing relations; profit for supplier may exceed that of core firm	distributors can hold shares in core firm; core firms are confronted with many mega-dealers; dealers can earn larger profit margins than core firms	national union confederations; representation at company council; national productivity coalitions	industrial bank with unilateral directorships and considerable shareholding in core firm; institutional investors; commercial bank has financed a large part of company's debt	strong domestic governments, making use of a variety of instruments to curb the influence of individual firms; nationalised companies	independent with influence
division of labour in which (a few) suppliers also have relationships with other core firms; local delivery; no clear pattern of profit distribution; single sourcing relations	converging profit margins; important function for importers; importers "partners" of the core company: low global price variation	(multinational) company council, representing sectoral unions; sectoral productivity coalitions	financial group; relationship with first tier bank issuing bonds; or first tier bank issuing stock as head of underwriter syndicate; coalition with bank	strong regional/state home governments; loose coalitions; active procurement policies; subsidies sustaining diversity in the industrial structure	interdependent
vertical integration; large numbers of competing suppliers; international delivery (global sourcing); antagonistic relations; profits tend to diverge, but not according to a hierarchy; low level of black box engineering	core firm seeks to own distributors and importers; low price at production site, higher prices abroad (due to higher transport costs)	internal or sectoral unions; contrasting interests; country by country productivity coalitions (weak co-ordination)	direct control over own finances; large number of small stockholders; high degree of self-financing; underwriter hierarchies; adversarial relations with banks	ad-hoc coalitions with national / federal governments; adaptive/independent from local environment; active "creation" of local environment in company towns	dependent with influence
vertical disintegration; small number of prime contractors; local delivery; control hierarchy with coalition between core firm and first tier suppliers; profits: convergence only among first tier suppliers; high level of black box engineering	control (i.a. via data networks); single franchise dealers; less importance for importers; higher prices in market of production site, lower prices abroad ("export residual")	plant/company unions; firm loyalty; decentralised productivity coalitions (centrally co-ordinated); rigorous selection	collaborative relation with first tier banks; mutual shareholdings in other firms of the same conglomerate; high degree of self-financing without this being used	local player status preferred; focus primarily on local governments; structurally dominant position; no need to control the region directly	dependent without influence

5 Applying the Concept of the Industrial Complex

The perspectives of the case studies presented in this book encompass industries rather than individual companies or networks, and they examine a particular industry's overall patterns of restructuring and the consequent environmental effects. This chapter argues that, when developing, producing and delivering goods or services to the customer, individual firms within their industrial complexes, rather than anonymous "industries", are producing restructuring patterns and making strategic choices which affect the environment. The structures, processes and participants within one industrial complex vary widely, even between competing industrial complexes in the same branch. This variation may provide useful insights into the chances for raising environmental standards, for example. The aim of this section is therefore to assess whether and how an analysis of the industrial complex could contribute to the issues raised in this book and to identifying ways of promoting higher environmental standards.

Firstly, a number of market conditions, product characteristics, regulatory features and contexts may be identified, encouraging core firms to develop long-term strategic visions of the role to be played by non-core actors. The following conditions tend to contribute to the formation of more tightly controlled industrial complexes:

- Non-market allocation mechanisms dominate the mode of economic organisation, or, in Coasian terms, market transactions are too costly. Under such circumstances other modes of economic organisation may emerge, such as hierarchies (Williamson 1979) or networks (Powell 1990). According to Williamson, firms would base their sourcing decisions for their inputs on three criteria: a high asset specificity (high investment in capital equipment, high sunk costs, high exit barriers), a high frequency of transactions and a high level of uncertainty. In the late 1990s, firms in industries such as steel, chemicals or petroleum refining would tend to exhibit the following characteristics:
  - high asset specificity, as capital investment is usually required, fostering larger and vertically integrated firms;
  - medium to high transaction frequency, favouring firms with medium to high levels of vertical integration;
  - medium levels of uncertainty (except for petroleum refining, where political uncertainty clearly plays an important role), leading to a trend towards medium (high for petroleum refining) levels of vertical integration.
- In general, steel, chemicals and refinery firms tend to strive for medium to high degrees of vertical integration, which in turn implies that they would indeed seek to control strategic elements of their industrial complexes.
- There is imperfect competition, providing core firms with incentives to seek long-term ways of organising their business environment, e.g. by seeking to integrate suppliers and customers into one overriding strategic vision, and thus sustain above-normal profits.

- Firms have realistic prospects of using non-market mechanisms to maximise market share, profits or shareholder value, e.g. by persuading a government to adopt particular industrial standards or to establish trade protection.
- Society is dominated by strong principles of solidarity and wealth sharing. Hofstede (1980) has identified four cultural dimensions in the national consciousness: masculinity vs. femininity, high vs. low aversion to uncertainty, high vs. low power, and individualism vs. collectivism. Each of these dimensions shape the nature of bargaining. Large firms based in national cultures that may be characterised by their femininity (such as Scandinavia and the Netherlands) are likely to face pressure to invest into schemes benefiting e.g. the unemployed or the environment. Societies characterised by a relatively high power distance and strong uncertainty avoidance (such as in the German-speaking world) are likely to show well-structured industrial complexes and social bargaining.
- Progression along the product life cycle, leading to scale advantages, concentration and a shake-out of firms, makes it crucial to secure the long-term survival of a company by developing clear strategies vis-à-vis its bargaining partners.

This list (which is not exhaustive) suggests that in the industries tackled in this book external conditions would encourage leading players to actively organise their industrial complexes. How plausible can it be made that they do, and in what ways?

5.1 Petroleum Refining

There can be little question that, even today, the world oil industry is still dominated by a relatively small number of nations and companies. In 1979 the OPEC countries were producing as much as 66% of the world's petroleum and, at the start of the 1970s, what used to be known as the Seven Sisters (five U.S. petroleum giants plus Royal Dutch Shell and British Petroleum) controlled some 60% of the world's crude oil (Harvey 1996). Things have changed since then, and what are today the Six Sisters control only some 30% of the world's crude oil. Nevertheless, the oil majors are still pursuing an integrated strategy, trying to control their own refining, transporting and marketing. Thus the large oil companies decide on the location, specifications and output of oil refineries in the European Union, just as everywhere else. Each of these core companies develops scenarios on price setting, tolerable emission levels and on whether or not to promote the development of substitute products, for example. Faced with excess capacity, the oil majors have rationalised along various lines, such as striking strategic alliances with their competitors, seeking state support or closing down facilities (see *Allal et al.* in this book). At the same time, the oil companies have often been active in "playing the employment card" with national governments, who have been struggling to maintain employment levels during economic recession (see *Van der Straaten* in this book).

The world's largest oil majors have maintained a high degree of vertical integration (although less so in the case of state-owned oil companies) and are operating an oligopoly. Suppliers often have a fairly weak position with regard to the leading oil majors, since the core firms themselves produce much of the oil that they subsequently refine. The oil industry is relatively labour extensive (although employment considerations do play a role, see below) and labour disputes in the industry are rare. Furthermore, the largest oil companies usually control their distribution structures, and are sufficiently cash-rich to limit their dependence on external financiers.

While each of the bargaining partners identified as members of the industrial complex is represented in the oil complex, by far the most important bargaining dimension is that between the core firm and the government. The sheer size and power of the oil majors make them formidable bargaining partners, even vis-à-vis national governments. Shell's delayed investment in the Hycon refinery is a case in point. Van der Straaten has shown how, after an incorrect assessment of the market, Shell effectively blackmailed the Dutch government into accepting additional emissions – rather than acknowledging that it had only itself to blame and redoubling its efforts to catch up with Exxon.

Conversely, the oil majors have also been the object of government policy. Many European oil companies have been, and in some cases still are, part of state-owned oil complexes. Examples include CFP in France (Total and ELF), ENI in Italy, INI in Spain and Statoil in Norway (see Ruigrok and van Tulder 1995, p. 162). Even independent oil companies, such as Shell and Exxon, have had to bargain heavily with national governments on concessions for exploration. Alternatively, governments have tried to leverage their position by raising environmental standards, gaining employment assurances, etc. It is this bargaining game across each dimension, especially with the corresponding national governments, that deserves further attention in follow-up studies. Such analysis could provide insights into the patterns of governments' concessions to national champions, and into the strategies for higher environmental standards which may be pursued.

5.2 Aluminium, Iron and Steel

Although these industries are discussed separately in this book, the aluminium, iron and steel industries may be taken together here, since the metals industries all produce relatively mature products for core firms in *other* industries, especially in the automobile industry, mechanical engineering and construction sector (see *Binder and Schucht*; *Schucht* in this book). Consequently, most companies in the aluminium, iron and steel sectors will *not* have succeeded in positioning themselves as core firms in an industrial complex, but will rather serve as *lower tier suppliers* to other firms. This implies that aluminium, iron and steel producers have only limited opportunities for determining their strategic future. For instance, an aluminium producer is itself unable to promote the production of lighter vehicles, just as it is unable to decide whether to use more aluminium or different materials if lighter vehicles are being manufactured. Analysing the aluminium, iron and steel industries would therefore start at the core firms of the various industrial complexes in which these industries participate, and examine the core firms' visions both of the materials to be used in the future and of their supplier relations management.

5.3 Coal

The role of coal, alongside iron and steel the engine of European recovery after World War II, has declined dramatically (see *Binder and Schucht* in this book). The environmental impact of burning coal has undoubtedly contributed to this, as well as the availability of alternative energy sources. However, from an industrial perspective, two additional reasons may be suggested. Firstly, labour conditions in coal mines have always been notorious, leading to high levels of unionisation. This is likely to have accelerated the search for alternative sources of energy, and will have made it easier for many governments to close down their coal mines (as shown in the confrontation between Thatcher and Scargill in the U.K.). Secondly, governments were often the main shareholders in coal mines as well as the leading consumers of its output. As the innovation literature suggests, such core user-producers (Lundvall 1988) are exceptionally well-placed to influence the fate of an industry, as indeed seems to have been the case.

5.4 Cement

The case of the cement industry shows how changing market conditions (stagnation in the construction industry, increasing concentration of market players, import substitution) and changing price structures (especially for energy as an input) have led to a drastic reduction of cement production in Austria (see *Hüttler* in this book). As Hüttler indicates, a number of large cement producers such as Holderbank and Lafarge Copée have effectively gained control over large segments of the Austrian cement industry. The removal of the cement cartel and the aforementioned market and price conditions have been major reasons behind this concentration. From the perspective of the industrial complex, the question would be to identify what caused the cartel to disappear, and especially how concentration affected the cement producers' bargaining relations with their major clients in the construction industry.

5.5 Chlorine and Fertilisers

Chlorine and fertiliser production are the Cinderella sectors of the chemicals industry. As *Jacob* (in this book) shows, it is the chemical majors such as BASF,

Bayer, Dow, ICI and Solvay who play a leading role in chlorine production, due to the high capital requirements. *Payer* (in this book) shows that the situation is comparable with respect to fertiliser production, with Austria's largest company OMV and Rohne-Poulenc dominating the market in Austria. Although in many cases chemicals companies serve as suppliers to other companies (within and beyond the chemicals industry), the large chemicals companies may be considered core firms in their own right, since their names, for instance in the pharmaceuticals and paints sectors, are known to many end consumers. Such strong, financially independent (see *Jacob* in this book) core firms will be reluctant to surrender a profitable activity until either government regulations force them to do so, or industrial or private customers reduce their demand. While Jacob suggests that government regulation has played a role in limiting chlorine production, the nature of the impact by government and customers might deserve further attention, as a case study of how bargaining processes within industrial complexes may bring about changes which ultimately benefit the environment.

6 Conclusion

This chapter has introduced the industrial complex and its constituents as an alternative unit of analysis when studying industrial restructuring processes. It has argued that industrial complexes may be seen as the centres of gravity for international restructuring. Industrial complexes incorporate the leading economic, social and political actors explicitly engaged in or facilitating the development, production and delivery of goods and/or services to the respective customers. Industrial complexes are held together by the core firm's vision of how to organise its value system, and on the role the various bargaining partners should play in this process.

This was followed by a discussion of a number of market conditions, product characteristics, regulatory features and contexts, which suggested that the concept of the industrial complex could help in analysing the subjects discussed in this book. A preliminary assessment has been presented as to what such an analysis might look like. The conclusion is that there may indeed be real advantages in analysing the various cases in this book from the perspective of the composition and bargaining dynamics of the industrial complexes in the various sectors and that, moreover, this could bring an increased understanding of the chances of campaigning for or against certain environmentally friendly policies.

References

Ansoff, I. (1991): Critique of Henry Mintzberg's "The Design School: Reconsidering the basic premises of strategic management". Strategic Management Journal, 449–461.

Andersen, E.S. and Lundvall, B.Å. (1988): Small National Systems of Innovation Facing Technological Revolutions: An Analytical Framework. in: Freeman and Lundvall (eds.) (1988), pp. 9–36.

Axelsson, B. and Easton, G. (eds.) (1991): Industrial Networks. London and New York: Routledge.

Baden-Fuller, C. and Stopford, J. (1992): Rejuvenating the mature business. London: Routledge.

Barney (1997): Gaining and Sustaining Competitive Advantage. Reading MA: Addison Wesley.

Bartlett, C. A. and Ghoshal, S. (1989): Managing across borders: the transnational solution. Cambridge MA: Harvard Business School Press.

Best, M. (1990): The New Competition: Institutions of Industrial Restructuring. Cambridge: Polity Press.

Bogner, W., Thomas, H., McGee, J. (1996): A Longitudinal Study of the Competitive Positions and Entry Paths of European Firms in the U.S. Pharmaceutical Market. Strategic Management Journal 17, 85–107.

Burt, R. (1982): Toward a Structural Theory of Action. New York: Academic Press.

Burt, R. (1992): The Social Structure of Competition. In: Nohria and Eccles (1992), pp. 57–91.

Caves, R. and Porter, M. (1977): From Entry Barriers to Mobility Barriers. Quarterly Journal of Economics 91, 241–261.

Coase, R. (1937): The Nature of the Firm. Economica 4, 386–405.

Cooke, P. and Morgan, K. (1991): The Network Paradigm: New Departures in Corporate & Regional Development. Regional Industrial Research Report, no. 8, University of Cardiff.

Dosi, G., Pavitt, K., Soete, L. (1990): The Economics of Technical Change and International Trade. Hemel Hempstaed: Harvester/Wheatsheaf.

de Banville, E., Chanaron, J.-J. (1991): Vers un Système Automobile Européen. Paris: CPE-Economica.

Dosi, G., Freeman, C., Nelson, R., Silverberg, G., Soete, L. (eds.) (1988): Technical Change and Economic Theory. London: Pinter Publishers.

Fennema, M. (1982): International Networks of Banks and Industry. The Hague: Martinus Nijhoff Publishers, Studies in Industrial Organization, vol. 2.

Fiegenbaum, A. and Thomas, H. (1995): Strategic Groups as Reference Groups: Theory, Modelling and Empirical Examination of Industry and Competitive Strategy. Strategic Management Journal 16 (6), 461–476.

Freeman, C. (1982): The Economics of Industrial Innovation. 2nd edn. London: Frances Pinter.

Freeman, C. (1991): Networks of Innovators: A Synthesis of Research Issues. Research Policy 20, 499–514.

Freeman, C. and Lundvall, B.Å. (eds.) (1988): Small Countries Facing the Technological Revolution. London and New York: Pinter Publishers.

Galtung, J. (1964): A Structural Theory of Aggression. Journal of Peace Research 2, 95–119.

Galtung, J. (1971): A Structural Theory of Imperialism. Journal of Peace Research 2, 437–482.

Gerlach, M. (1992): Alliance capitalism: the social organization of Japanese business. Berkeley: University of California Press.

Grabher, G. (1993): The Embedded Firm: The Socio-Economics of Industrial Networks. London and New York: Routledge.

Groenewegen, J., Pitelis, C., Sjöstrand, S.E. (1995): On Economic Institutions: Theory and Applications. Aldershot: Edward Elgar.

Håkanson, Håkan (1989): Corporate Technological Behavior: Co-operation and Networks, London: Routledge.

Hall, P. (1986): Governing the economy: the politics of state intervention in Britain and France. Cambridge: Polity Press.

Harrigan, K. and Porter, M. (1983): End-Game Strategies for Declining Industries. In: Harvard Business Review, July/August, 111–120.

Harvey, Colin (1996): The UK downstream industry: commercial challenge and structural change. Speech by Shell U.K. Ltd managing director to the Institute of Petroleum European Retail Conference, Birmingham – June 4.

Hodgson, G. (1993): The Economics of Institutions. Aldershot: Edward Elgar.

Hofstede, G. (1980): Motivation, leadership and organization: Do American theories apply abroad? Organizational Dynamics, Summer, 42–63.

Hunt, M. (1972): Competition in the Major Home Appliance Industry, 1960–1970. Unpublished doctoral dissertation, Harvard University.

Jänicke, M., Binder, M., Mönch, H. (1997): Dirty industries: Patterns of change in industrial countries. Environmental and Resource Economics 9, 467–491.

Katzenstein, P. (1985): Small States in World Markets: Industrial Policy in Europe. Ithaca/London: Cornell University Press.

Kenis, P. (1991): The precondition for policy networks: Some findings from a three-country study of industrial restructuring. In: Marin and Manytz (1991), pp. 297–330.

Kenney, M. and Florida, R. (1988): Beyond Mass Production: Production and the Labor Process in Japan. Politics and Society 16 (1), 121–158.

Krugman, P. (ed.) (1986): Strategic Trade Policy and the New International Economics. Cambridge: MIT Press.

Lundvall, B.Å. (1988): Innovation as an Interactive Process: from User-producer Interaction to the National System of Innovation. In: Dosi et al. (1988), pp. 349–370.

Lundvall, B.Å. (ed.) (1992): National Systems of Innovation: Toward a Theory of Innovation and Interactive Learning. London: Pinter Publishers.

Marin, B. and Manytz, R. (eds.) (1991). Policy networks: Empirical evidence and theoretical considerations, Frankfurt am Main: Campus; Boulder Colorado: Westview.

Marshall, A. (1961): Principles of Economics. 9th edn.. London: Macmillan.

McGee, J., Thomas, H., Pruett, M. (1995): Strategic Groups and the Analysis of Market Structure and Industry Dynamics. British Journal of Management 6, 257–270.

Midgley, D., Morrison, P., Roberts, J. (1992): The Effect of Network Structure in Industrial Diffusion Processes. Research Policy 21, 533–552.

Mintzberg, H. (1990): Strategy formation: Schools of thought. In: Frederickson, J.W. (ed.): Perspectives on strategic management. New York, Harper and Row.

Mintzberg, H. (1991a): The Design School: Reconsidering the basic premises of strategic management. Strategic Management Journal, March, 171–195.

Mintzberg, H. (1991b): Learning 1, planning 0: Reply to Igor Ansoff. Strategic Management Journal, September, 463–466.

Nelson, R. (1993): National Systems of Innovation. Oxford: Oxford University Press.

Nelson, R. and Winter, S. (1982): An evolutionary theory of economic change. Harvard University Press, Cambridge.

Nohria, N. (1992): Is a Network Perspective a Useful Way of Studying Organizations? In: Nohria and Eccles (1992), pp. 1–22.

Nohria, N. and Eccles, R. (eds.) (1992): Networks and Organizations: Structure, Form and Action. Boston, MA: Harvard Business School Press.

Perez, C. (1983): Structural Change and the Assimilation of New Technologies in the Economic and Social System. Futures, October, 357–375.
Perez, C. (1985): Microelectronics and World Structural Change: New Perspectives for Developing Countries. World Development 13, March, 441–463.
Peteraf, M. (1993): The cornerstones of competitive advantage: A resource-based view. Strategic Management Journal 16, 179–191.
Piore, M. and Sabel, C. (1984): The Second Industrial Divide: Possibilities for Prosperity. New York: Basic Books.
Porter, M. (1979): The Structure within Industries and Company Performance. Review of Economics and Statistics 64, 417–427.
Porter, M. (1980): Competitive Strategy. New York: Free Press.
Porter, M. (1985): Competitive Advantage: Creating and Sustaining Superior Performance. New York: Free Press.
Porter, M. (1990): The Competitive Advantage of Nations. New York: The Free Press.
Porter, M. and van der Linde, C. (1995a): Green and competitive: Ending the stalemate. Harvard Business Review, Sept.–Oct., 120–134.
Porter, M. and Van der Linde, C. (1995b): Toward a new conception of the environment-competitiveness relationship. Journal of Economic Perspectives 9 (4), 97–118.
Powell, W. (1990): Neither Market nor Hierarchy: Network Forms of Organization. Research in Organizational Behavior 12, 295–336.
Rosenberg, N. (1982): Inside the Black Box: Technology and Economics. Cambridge.
Ruigrok, W. and Van Tulder, R. (1993): The Ideology of Interdependence: The Link Between Restructuring, Internationalisation and International Trade. Ph.D. dissertation, University of Amsterdam.
Ruigrok, W. and Van Tulder, R. (1995): The logic of international restructuring. London: Routledge.
Ruigrok, W. and Van Tulder, R. (1996): The price of diversity: Rival concepts of control as a barrier to an EU industrial strategy. In: Devine, P., Katsoulacos Y., Sugden, R. (eds.) (1996): Competitiveness, subsidiarity and objectives: Issues for European industrial strategy Routledge, London, pp. 79–103.
Rumelt, R. (1991): How much does industry matter? Strategic Management Journal 12 (2), 167–185.
Smith, H.L., Dickson, K., Smith, S.L. (1991): "There are Two Sides to Every Story": Innovation and Collaboration within Networks of Large and Small Firms. Research Policy 20, 457–468.
Stegeman, K. (1989): Policy Rivalry among Industrial States: What can we Learn from Models of Strategic Trade Policy? International Organization, 43 (1), 73–100.
Teece, D. (1985): Multinational Enterprise, Internal Governance and Industrial Organization. American Economic Review 75 (2), 233–238.
Teece, D. (1988): Technological Change and the Nature of the Firm. In: Dosi et al. (1988), pp. 256–281.
Wallensteen, P. (1973): Structure and War: On International Relations 1920–1968. Political Sciences Association in Uppsala, no. 64.
Wernerfelt, B. (1995): The resource-based view of the firm. Strategic Management Journal 12, 171–174.
Whitley, R. (1992): Business Systems in East Asia. London and Thousand Oaks: Sage.
Whitley, R. and Kristensen, P.H. (1996): The Changing European Firm: Limits to Convergence. London and New York: Routledge.

Williamson, O. (1975): Markets and Hierarchies: Analysis and Antitrust Implications. New York: Free Press.
Williamson, O. (1979): Transaction cost economics: the governance of contractual relations. Journal of Law and Economics 22, 233–261.
Womack, J.P., Jones, D.T., Roos, D. (1990): The Machine that Changed the World. New York: Rawson Associates.
Womack, J. P. and Jones, D. T. (1995): Lean Thinking: Banish Waste and Create Wealth in your Corporation. New York: Simon and Schuster.
Zysman, J. (1994): How Institutions Create Historically Rooted Trajectories of Growth. Industrial and Corporate Change 3 (1).

Multinational Enterprises and Industrial Restructuring: A Strategic Environmental Management Approach

Ans Kolk
University of Amsterdam, Roetersstraat 11, 1018 WB Amsterdam, The Netherlands

1 Introduction

Different perspectives can be taken, when looking at structural change and environmental protection. This book focuses on selected basic industries with a high environmental impact that have experienced a (temporary) decline in material output. These include chlorine, fertilisers, cement, petroleum refining and heavy industries. The economic and environmental pressures on these industries have brought individual, but also often collective responses from governments and companies, at European and/or international level. Companies also have the option of reacting through sectoral organisations at one or more of these levels, and sectoral restructuring – as well as the possibilities for intervention – therefore suggests a focus on companies and governments as two salient types of actors, as driving forces.

This chapter considers the case studies from the perspective of core companies, multinational enterprises in particular. It focuses on how companies' strategic decisions impact on environmental protection in relation to structural change and how they have interacted with governments in particular. To some extent, multinational enterprises bind together the various sectors and countries examined in this book: their activities are cross-border, usually across more than one sector, and they are accustomed to operating within diverse regulatory systems. The following demonstrates the significance of multinationals, paying attention to their persistence and their location as companies in networks, as well as presenting a framework for analysing strategic environmental management. Subsequently, this analysis is applied to the case studies included in this book. The final section deals with some of the policy issues relating to (green) industrial restructuring.

2 Core Companies and Strategic Environmental Management

Whereas industrial sectors may decline, large companies in particular are able to escape this fate by restructuring their activities in time. This is a continuous and dynamic process, with the changes applying more to the activities and the continuity to the companies themselves. A closer look at the largest industrial companies shows that the overwhelming majority is relatively old. While the average age of industrial companies is approximately 40 years, large multinationals are far more stable. Of the twenty-two chemical companies in the 1995 Fortune global 500, for example, two were founded between 1800 and 1850, nine between 1851 and 1900, eleven between 1901 and 1935, and none of them after 1936 (compiled from Hikino et al. 1998, p. 105). If this is broadened to include the 100 largest industrial companies, a similar pattern emerges: Only four of these companies – Daewoo, Digital Equipment, Petrobras and Sunkyong – originate from after the Second World War, 47 were founded between 1901 and 1950, 36 between 1851 and 1900, and 13 before 1851 (compiled from Ruigrok and van Tulder 1995, 239–268). This persistence underlines the importance of large multinational companies and their strategies when dealing with the driving forces of industrial restructuring.

With regard to structural change and the environment, the production processes of the largest industrial multinationals are frequently highly polluting (endogenous environmental risks, see Rondinelli and Vastag 1996). They are confronted with environmental issues to a greater extent than local companies, especially if they produce well-known brand names and have a high visibility. Public attention is therefore often focused on multinationals, and on their large financial and technical resources. For the companies, the exogenous risks are likely to be high; environmental management thus acquire strategic importance more readily.

In addition, multinationals usually occupy crucial positions within networks; they might be designated "core companies" (Ruigrok and van Tulder 1995; *Ruigrok* in this book) or "flagship companies" (Rugman and D'Cruz 1997). Core companies control (part of) a product chain, and strive to obtain or increase this control. By virtue of this position, they may exert a strong influence on regulatory authorities and other companies. They can powerfully affect environmental management in other companies, upstream or downstream in the chain, or competitors in other chains, and may have the capacity to impose environmental conditions on suppliers.

Companies face strategic dilemmas in managing all these relationships – with other companies (competitors, suppliers, customers), regulatory authorities at various levels, non-governmental organisations (NGOs), consumers and local communities. In all of these fields, the environment has become an issue because of its potential to affect image, efficiency and economic performance. With respect to the network approach to industrial restructuring (Ruigrok and Tulder 1995; *Ruigrok*, this book), the number of bargaining relationships is larger in the case of environmental issues, although the urgency for companies' core business activities is likely to be less. Viewed from such an environmental perspective, NGOs and local communities have an indirect interest in the production and distribution of companies' products.

The different components of strategic environmental management are summarised in Table 1, viewed from three interrelated perspectives: outside-in, inside-in and inside-out. Outside-in focuses on the external influences which are strategic and/or environmental, inside-in on the internal aspects of environmental management, and inside-out on the impact of companies on their networks, considering their position and the environmental problems in the product chain. The model presented here, as well as an overview of the relevant literature, is described in greater detail elsewhere (Kolk 2000).

It should be noted that while research at the level of an industry (or strategic groups or clusters within industries) continues to be problematic with regard to industrial restructuring as such (see *Ruigrok's* chapter), an industry analysis can be useful when it serves to uncover overall environmental impact, vulnerability, risks and opportunities (Rondinelli and Vastag 1996; Steger 1993). Environmental costs and expenditure, for example, are very much sector-specific (Blodgett 1997; Jaffe et al. 1995; OECD 1996), and regulation will be targeted similarly (Kolk 2000). Those sectors with the greatest environmental impact are most active in environmental reporting (Kolk 1999) and have the largest number of sectoral self-regulatory initiatives. Nonetheless, and here the parallel to restructuring re-emerges, sectoral characteristics only provide the background for understanding the implications for individual companies. This underlines once more the importance of analysing core companies, their strategies and industrial networks.

The three perspectives defined in Table 1 are interrelated: in particular, inside-out and outside-in cannot always be separated easily. The case in point is self-regulation, where a company takes measures in response to or in anticipation of

Table 1. Salient aspects of strategic environmental management: three perspectives

perspective	aspects
1. outside-in *external influences*	• structure of the industry (competition, markets, products, variety of technologies used, environmental risks attached to the industry) • key (environmental) regulatory influences (home country, host countries international) • other forms of (environmental) pressure (customers, societal organisations)
2. inside-in *company*	• company's economic profile (profits, market share, market strategy, control structure, degrees of vertical integration, diversification and internationalisation) • organisational structure • company-specific resources and capabilities • environmental impact and risks inherent in the company's activities • performance in environmental management
3. inside-out *network*	• network and dependency relationships between the different companies; assessment of core companies • major environmental problems in product chain(s), the specific companies where these major problems can be found and their location (which countries) • environmental requirements placed on suppliers/contractors

Source: Kolk 2000, chap. 4

outside pressure, to avoid regulation, to profit from or further enhance its specific areas of competence, and/or to increase control over its network or product chains. At the same time, voluntary initiatives may raise expectations, serve as input and provide a model for future legislation (for example, regulations on environmental reporting), or subsequently become a market condition.

More generally, regulation involves not simply a one-way process whereby governments devise rules and companies comply (the traditional "command-and-control" approach). Instead, many companies try to influence the form and contents of regulation before it is adopted, in such a way that it harms their interests least or suits their particular cost structure. In this complex lobbying process, some companies are better placed than others. Especially the larger ones are generally more influential, owing to their impact on employment and economic growth. Moreover, and closely related to this point, their managers often have better access to politicians and know how to exploit their position. Nevertheless, the final outcome of this negotiation and lobbying provides the environmental confines within which companies should stay.

Finally, whether an aspect is inside-in or inside-out may depend on strategic decisions taken by companies, for example whether to outsource activities or not. In spite of these interrelationships, however, the perspective is important in making conceptual distinctions: focusing on strategic (environmental) management within a company (inside-in), market and regulatory pressure on a company (outside-in), and the company's influence on its network and chains (inside-out).

The framework in Table 1 approaches the relationship between environmental and strategic management from an unusual perspective. Although all companies are confronted with environmental issues, this does not mean that they automatically act strategically. In some cases, where environmental management neither offers market opportunities nor affects competitiveness, compliance may prevail, involving primarily operational staff. It should also be noted that strategies, whether environmentally-related or not, are not always deliberately planned; they can also emerge over time (Mintzberg et al. 1998). In addition, deliberate or emergent strategies may not necessarily be realised or effective.

With these caveats, the framework presented here may shed further light on the case studies in this book, as it considers companies as the driving forces of restructuring. If large companies begin to incorporate environmental considerations into their strategies and, more importantly, implement the resulting policies, green structural change is already underway. At the same time, industrial restructuring can also be accompanied by environmental benefits if polluting sectors go into decline. As the sectoral studies in this book show, this may sometimes be only a temporary phenomenon. Governments, however, may want to devise policies both to stimulate the incorporation of environmental issues into companies' strategies and to take advantage of decreasing activity in polluting sectors if this occurs. To obtain more insight into the dynamics of this process, elements of the case studies will be analysed in the next section. In view of the design and scope of this book, this will focus on the outside-in perspective.

3 Analysing the Cases From a Strategic Environmental Management Perspective

Core companies can be identified in all the sector and country studies in this book. The markets are dominated by a limited number of multinational enterprises (and sometimes only one). National monopolies can be found in the Luxembourg steel sector with Arbed, a partly state-owned company, and in the Danish cement industry. In Denmark, Ålborg Portland, owned by a Danish holding and the British multinational Blue Circle, is a spider in the industrial web and the world leader in white cement.

Until recently, German coal mining has been almost the exclusive domain of Ruhrkohle Bergbau, owned by RAG (which ranks as number 282 in the 1997 Fortune global 500; cf. Fortune, 3 August 1998). Only Preussag (number 229) and the publicly-owned Saarbergwerke did not belong to Ruhrkohle Bergbau, but concentration is still increasing. Since 1999, all German coal mining has been concentrated in Deutsche Steinkohlen AG, a subsidiary of RAG. The Austrian cement industry is highly concentrated, with three multinational companies accounting for 80% of production: the French Lafarge, the German Heidelberger and the Swiss Holderbank. Two national companies dominate the Austrian fertiliser sector, one in production (Agrolinz Melamin) and one in wholesale retail (RWA).

The other sectors are characterised by a greater variety of core companies, but a few of them still account for the bulk of production, and a further concentration can be expected. The only exception may be secondary aluminium in Germany, but detailed production figures are not available. Four companies produce primary aluminium: VAW Aluminium with two plants, followed by Hamburger Aluminium-Werk, Aluminium Essen and Hoogovens. In German iron and steel, Thyssen Krupp, Mannesmann, Preussag and Arbed are the major companies.

Finally, in both the French and Dutch oil sectors, a considerable number of refiners can be found, but large multinationals dominate the markets: Royal Dutch/Shell (number 5 in the Fortune 500), Exxon (7), British Petroleum (20), Mobil (26), Elf Aquitaine (44) and Total (74). A similar pattern can be seen in German chlorine production: although there are at least ten active companies, most of production is accounted for by the large multinationals Dow (184) and Bayer (80), followed some way behind by Hoechst (92) and BASF (77).

Examining the different cases, a considerable awareness of the environment becomes evident. In the German iron and steel sector, for example, environmental improvements were achieved when the companies adopted end-of-pipe measures, implemented changes in the production process and started to close material flow cycles (recycling and re-use). Similarly, Ålborg Portland has improved energy efficiency substantially, compensating for the increased consumption resulting from the shift to the more energy-intensive production of white cement, it has installed a heat recovery system which supplies energy to the local municipality, and uses industrial and household waste for energy generation. In the largest cement companies in Austria, incineration of waste materials (scrap tyres, waste oil / solvents and paper sludge) accounts for 20% of the total energy input.

A note of caution should be sounded here, however, because there currently is an increasing number of companies who are going on the offensive, explicitly presenting themselves to the public as environmentally responsible. In the international debate on climate change, for example, various oil and car companies are openly supporting action, in spite of the uncertain future in these areas: British Petroleum, Royal Dutch/Shell, Elf Aquitaine, General Motors and Toyota (Kolk 2000). The Austrian fertiliser producer Agrolinz Melamin also presents itself as an environmental front-runner. Likewise, chemical companies are becoming increasingly active; they usually publish extensive environmental reports and subscribe to voluntary industry guidelines, such as Responsible Care. A conspicuous feature of chlorine production is that the largest producers assert that there are insufficient incentives to use the more energy-efficient membrane technology.

To some extent, greater environmental sensitivity is unsurprising. All the companies in this book operate in polluting sectors, in countries where society pays considerable attention to environmental issues. They are therefore exposed to high endogenous and also exogenous environmental risks, and will have taken at least some measures at the level of communications or internally. At the same time however, less proactive, even defensive approaches can also be noted within the same sectors, sometimes for a limited period of time, and within a specific national context. Regulatory, social and competitive pressure plays a role here.

In *petroleum refining*, for example, there is wide divergence between responses. In the Dutch case, which focuses on acid rain abatement in the 1970s and 1980s, Exxon (operating as Esso in the Netherlands) and Shell took the opposite approach (see *Van der Straaten's* chapter in this book). Shell long resisted emissions reduction, in particular using the argument of competitive *dis*advantage, and threatening to abstain from further investment in the Netherlands if strict standards were imposed. In view of the fact that Germany, where most of Dutch production was exported, had stricter standards, this argument was hardly sustainable. Shell, however, even went as far as to invest in a refinery based on traditional technologies, although Exxon already had a new – and profitable – technology up and running. In spite of considerable public attention and the existence of emissions reduction policies, the government did not use Exxon as an example to undermine Shell's arguments and pressurise it to change position. Eventually, however, Shell was forced to invest in the new technology as well – due to continued demand for light oil fractions.

The Dutch government's relatively cautious approach to domestically based multinationals in particular fits into the Netherlands' consensus-oriented, corporatist model (Van Tulder 1998). This "polder model" can also be observed in environmental policy: policy-making results from a long process of negotiation between a large number of parties, encompassing business (associations), ministries, provinces and semi-autonomous government agencies, and usually also NGOs, sometimes in the form of consultation (Kolk 1996; *Van der Straaten* in this book).

In close co-operation with business, the government adopted many negotiated voluntary agreements in the 1990s. While acceptance of and compliance with such covenants seems high, they also suffer from weaknesses, for example limited cost-efficiency, controllability and effectiveness (see Algemene Rekenkamer 1995; cf. Rennings et al. 1997 for the German situation). In addition, covenants tend to lead to a lower common denominator than that which some companies may have been able to achieve otherwise.

The issue of CO_2 emissions, which is not examined in *Van der Straaten's* chapter, has been the bone of controversy for petroleum refiners – and Dutch industry in general – since the mid-1990s. In fear of strict government standards following the December 1997 Kyoto conference, the industry offered – earlier that year – to participate in creating international benchmarks for energy efficiency. As a *quid pro quo*, the government refrained from imposing specific emissions standards on the industry (Dutch Government 1997). In April 1999, the parties agreed on a covenant whereby Dutch companies commited themselves to the goal of becoming the world's most energy efficient companies as soon as possible. If they require substantial measures to achieve this goal, these should be taken by the year 2012 at the latest. For the purposes of the covenant, "world leaders" are defined as companies with comparable installed processes, as assessed by a consultant.

During the whole process, large domestically based multinationals such as Shell played a large role. Host multinationals were less eager to participate in preparing the covenant on benchmarking energy efficiency, and a similar lack of interest can be expected in its implementation as well. Shell has taken a rather different posture than the one it took on reducing acidifying substances in the 1980s. It reflects changing policies and perceptions within Shell following the Brent Spar and Nigeria controversies. Specific national aspects also play a role: within Royal/Dutch Shell, the Dutch and UK petroleum companies have been more active in adopting the new approach than in the US. Shell Oil, for example, long resisted the Dutch and UK parts' insistence on withdrawal from the Global Climate Coalition (which finally took place in April 1998) (Kolk 2000).

Compared to France, the Dutch petroleum case is notable for the steady increase in production since 1985, for technological changes, and the role of NGOs and government agencies. In France, production declined until the late 1980s and has stabilised since, producing continuous pressure to restructure (see *Allal et al.'s* chapter in this book). Environmentally induced technological change, such as that described for the Netherlands, was not evident in the French case. Furthermore, environmental organisations and the role of public opinion are conspicuous by their absence in this case. Compared to other European countries, environmental pressure from society is far lower in France. Accordingly, the majority of companies, even ones with highly polluting production processes, do not explicitly publish environmental information, for example an environmental report. The percentage of French companies who produce environmental reports is far lower than in other (European) countries. In the energy sector, however, large multinationals such as EDF and Elf do practice environmental reporting.

The French political process has been characterised by the close relationship between regulators and industry (Wallace 1995). This originates from traditionally extensive state ownership and highly centralised decision-making – led by technocratic and administrative elites (Michalet 1997). As a result, the process is more or less closed, offering little opportunity for public participation or influence (Andrews 1994). The chapter on petroleum refining in France reflects these special characteristics, depicting industrial associations, ministries and government agencies or institutes as the central arena for negotiation and decision-making.

Although the position of the oil refiners may have been weak compared to that of the public energy companies, particularly Electricité de France and Gas de France, in view of the government's different emphasis, they still seem to enjoy a relatively close relationship. Relevant environmental regulation aimed at regulating the oil sector consists merely of implementing EU directives in France, in spite of the official position claiming to pay more attention to environmental issues. The fact that refiners continue to point in public at their ongoing difficulties in restructuring and low profit margins or rates of return has certainly played its part. The moderate optimism about future developments on the part of both the Ministry of Industry and the oil companies (see the final section of the chapter on France) is interesting in this respect.

Overall, French companies may not have much to fear, because the government, for example, has been rather slow in considering EU proposals for implementing energy taxes; small changes have taken place only after 1997 (Schlegelmilch 1998). In the debate on the car oil programme, France successfully insisted on looser requirements, taking a position between the Northern and Southern countries of the EU. Finally, with regard to climate change, oil companies' willingness to reduce emissions expeditiously generally seems related in part to reaping "low-hanging fruit" – savings which can relatively easily be achieved by, for example, eliminating gas losses by venting and flaring.

À propos energy taxes, Denmark has been very active in adoption since 1991, and implementation since 1992 (see *Klemmensen's* chapter in this book). This has had serious implications for the *cement industry*, i.e. the monopolist Ålborg Portland, which did everything it could to prevent the adoption of a CO_2 tax. The 1991 legislation only applied to consumers and small energy users, but the issue re-emerged with the 1996 review of the law. In a similar debate in the Netherlands in the early 1990s, major industries successfully vetoed the imposition of an energy tax, which was subsequently introduced for consumers and small users only. The problem of industrial emissions has, in the meantime, been addressed by the Dutch voluntary agreement on benchmarking energy efficiency, as explained above.

In Denmark, however, the proponents of a CO_2 tax had entered the government in 1993 and enjoyed a parliamentary majority. Industry, with Ålborg Portland as an important energy user, used all available means to oppose the combined CO_2/SO_2 tax. This included a threat of jobs losses, closure of plants and relocation to countries with friendlier legislation. Although the "environmental flight" argument in particular was considered overkill, Ålborg Portland's emphasis on its existing high energy efficiency did exert an effect. The law was adopted, but with a provision to channel tax revenues back into industry via subsidies on energy saving measures and other kinds of compensation for prior investment. In return, companies must deliver a report on energy management, a plan for future reductions and an energy check on their complete production processes. In the Ålborg Portland case, this cycle of (adapted) taxation, compensation and improved energy management is expected to produce considerable energy savings.

Extending the comparison to Austria, energy and emissions have been the main issues for the cement industry here as well (see *Hüttler's* chap. in this book). Since the 1970s, the Austrian cement industry has begun partial substitution of fossil fuels with secondary energy sources, particularly scrap tyres, waste oil, paper sludge and plastics waste. Similar incineration of industrial and household waste for energy generation also takes place in Denmark, with the exception of scrap tyres, which has been banned by the government. This has led to the curious situation whereby scrap tyres are exported for use in the cement industry from Denmark to Sweden, where incineration is allowed.

Although the cases of Denmark and Austria show some similarities, the market situation is somewhat different. In Denmark, the monopolist Ålborg Portland is the world market leader in white cement, although grey cement is also exported. In contrast, the three large multinational companies which dominate in Austria produce for the national (or even regional) construction sector, and must respond to the entry of cheaper exports from Eastern European countries onto the market.

With regard to the environmental challenges for the sector, waste incineration has been controversial in Austria (and elsewhere), due to the types of emissions it produces and the different standards applied to municipal incineration plants and to the cement industry. According to the EU directive on incineration, which was implemented in Austrian law in February 1999, these discrepancies will have to be phased out. Other legislation on emissions to air has also been enacted, although only after protracted negotiations that risked ending in deadlock. Compulsory emissions standards and time limits have resulted in – mainly end-of-pipe – investment, which has helped to improve the competitiveness of the Austrian cement industry, according to sector representatives. At the same time, it has also fostered concentration.

Improved competitiveness as a consequence of regulation is also mentioned in the other Austrian case, the *fertiliser industry*. *Payer's* chapter covers a larger part of the production chain than the other studies, since it includes both production (Agrolinz Melamin) and wholesale (RWA). The sectors co-operate well, although their ownership and political affiliations differ: Agrolinz Melamin is part of the Austrian public petrol company OMW, traditionally connected to the social-democratic cluster, whereas RWA is a co-operative associated with the conservative party. Fertiliser production has shown a steady decline, in which the growth of organic farming (currently on 8% of available agricultural land) played a minor role when compared to the implications of the overall rationalisation and restructuring process following Austria's accession to the EU.

With regard to competitiveness, the adoption of the 1989 Clean Air Act, with the strictest emissions standards in Europe, induced Agrolinz to invest in a new plant, supported by considerable government funding. The company has since taken a large number of environmental measures. Due also in part to environmental regulation, Donau Chemie, the other main company on the market, with a one quarter market share, solved its massive waste problems related to the production of phosphoric acid. This innovation gave the company a first-mover advantage, until cost considerations and more restrictive emissions limits halted production in Austria in 1993. Production was subsequently relocated to France, to Rhône Poulenc, which owned Donau Chemie at the time. After a management buy-out, Donau Chemie has become more flexible on its own.

Currently, both fertiliser producers perceive flexibility as the major factor in competitive advantage. Legislation in the 1990s, directed largely at water emissions rather than the atmosphere in the preceding decade, is seen generally as a handicap. Opportunities may also arise in the supply of services related to the agricultural industry. As the production of fertilisers is not a core business of the Austrian companies, shifts can be expected in their activities.

4 Concluding Remarks

The preceding analysis of some of the case studies in this book shows their diversity. However, they have all focused on the relationship between government and industry (the outside-in perspective). A few implications can therefore be sketched out, especially from this perspective, dealing with different regulatory traditions, the role of the EU, competitiveness and the prospects for further greening.

To start with, the importance of *different regulatory traditions* has emerged clearly. Although harmonisation in the EU is proceeding at a considerable pace, concomitant with the adoption of common directives, national peculiarities do play, and will continue to play, a role. This not only includes the specific domestic constituencies, such as the importance of environmental organisations and other NGOs in agenda-setting and decision-making, but also the way in which government and industry have traditionally managed relevant environmental issues. The preceding section has mentioned some interesting differences, especially when comparing the approaches of the Netherlands, France, Denmark and Austria towards energy and/or emissions.

One way to characterise these historical traits is Andrew's (1994) analysis of the flexibility of regulatory enforcement and the openness of the decision-making process. A flexible approach to enforcement means that there is room for negotiation; a rigid style, which emphasises strict compliance with laws and standards, creates the opposite. With regard to the decision-making process, openness refers to the extent to which information is available, and public participation is possible and effective. Although the US (lack of flexibility, open processes) and the UK (flexible but closed) are extremes, clear differences also exist within the EU.

In spite of divergence, most studies underlined the *crucial role of EU regulation* as impetus or pressure for domestic laws. Although fears of a "race to the bottom", towards the lowest common denominator, are frequently expressed, evidence from the US and Europe suggests that this is not necessarily true. The cases in this book also reveal more incentives than disincentives for adopting environmental protection measures on the part of government and business as a result of European integration. Nevertheless, EU directives are compromises and frequently represent watered-down versions of the highest standard available in one of the member states. They help upward harmonisation and stricter enforcement in countries with laxer approaches, but may entail a downward effect for the others if they lack an incentive to over-comply.

A related and repeated issue in government-business relationships has been *competitiveness*. This is the argument used most often by companies and business associations to oppose environmental measures, almost regardless of their strictness and national or EU-wide application. Because the negative implications are cited so frequently and prominently, the positive impact of regulation on competitiveness tends to be overlooked. As pointed out in this chapter, some interesting examples of the opposite can be found in the case studies.

In addition, anecdotal evidence suggests that large companies deliberately influence national regulatory arrangements (their content, timing and form) and use self-regulation with an eye to foreign competitors. This can serve to protect the domestic market or obtain an advantage when operating abroad; it can occur in both shrinking or growing industries. The type of market and sector is important for the relationship with domestic legislators. The specific configuration may influence the response depending on whether there is a monopoly or not, whether there is a large or small number of companies, whether these are large or small, whether domestic and/or foreign companies are involved, and how important specific sectors or companies are to the domestic economy.

Whatever the companies' reasons for adopting or further integrating environmental considerations, once this process has begun within core companies, it may at some point in time come to have implications for other companies within their network. With increasing experience of incorporating environmental issues, both by regulators and managers, the likelihood increases that end-of-pipe measures are replaced by improvements in processes or even products. Changes in the entire product life cycle fundamentally affect production, supply and distribution as well, and will continue to do so in case of restructuring. Whether in the context of expansion or decline, the explicit promotion of strategic environmental management within core companies seems to be a way to further green industrial restructuring.

References

Algemene Rekenkamer (1995): Convenanten van het Rijk met bedrijven en instellingen. The Hague: Tweede Kamer, 24480, 1–2.

Andrews, C. (1994): Policies to Encourage Clean Technology. In: Socolow, R., Andrews, C., Berkhout, F., Thomas, V. (eds): Industrial Ecology and Global Change. Cambridge: Cambridge University Press, pp. 405–422.

Blodgett, J.E. (1997): Environmental protection: how much it costs and who pays. Washington: Congressional Research Service.

Dutch Government (1997): Nota Milieu en Economie. Op weg naar een duurzame economie. The Hague, No. 25405.1, 18 June.

Hikino, T., Harada, T., Tokuhisa, Y., Yoshida, J.A. (1998): The Japanese puzzle: rapid catch-up and long struggle. In: Arora, A., Landau, R., Rosenberg, N. (eds): Chemicals and long-term economic growth. Insights from the chemical industry. New York: John Wiley, pp. 103–135.

Jaffe, A.B., Peterson, S.R., Portney, P.R., Stavins, R. (1995): Environmental regulation and the competitiveness of US manufacturing. Journal of Economic Literature 33 (1), pp. 132–163.

Kolk, A. (1996): The limited returns of Dutch rainforest policy. International Environmental Affairs 8 (1), pp. 41–49.

Kolk, A. (1999): Evaluating corporate environmental reporting. Business Strategy and the Environment 8.

Kolk, A. (2000): The economics of environmental management. London: Financial Times Pitman Publishing.

Michalet, C. (1997): France. In: Dunning, J. (ed.): Governments, globalization, and international business. Oxford: Oxford University Press, pp. 313–334.

Mintzberg, H., Quinn, J.B., Ghoshal, S. (1998): The strategy process. Revised European edition. London: Prentice Hall.

OECD (1996): Pollution abatement and control expenditures in OECD countries. Paris, report No. OCDE/GD(96)50.

Porter, M.E. and Van der Linde, C. (1995): Green and Competitive: Ending the Stalemate. Harvard Business Review, September-October, 120–134.

Rennings, K., Brockmann, K.L., Bergmann, H. (1997): Voluntary agreements in environmental protection: experiences in Germany and future perspectives. Business Strategy and the Environment, 6 (5), 245–263.

Rondinelli, D.A. and Vastag; G. (1996): International environmental standards: an integrative framework. California Management Review 39 (1), 106–122.

Rugman, A. and D'Cruz, J. (1997): The theory of the flagship firm. European Management Journal 15 (4), 403–412.

Ruigrok, W. and Van Tulder, R. (1995): The logic of international restructuring. London and New York: Routledge.

Schlegelmilch, K. (1998): Energy taxation in the EU and some member states: looking for opportunities ahead. Wuppertal Institute for Climate, Environment and Energy.

Steger, U. (1993): The greening of the board room: how German companies are dealing with environmental issues. In: Fischer, K. and Schot, J. (eds): Environmental Strategies for Industry. International Perspectives on Research Needs and Policy Implications. Washington: Island Press, pp. 147–166.

Van Tulder, R. (1998): Governing with multinationals. Dutch smartness in coping with home-based and host multinationals. Administrative Studies 3, 160–172.

Wallace, D. (1995): Environmental Policy and Industrial Innovation. Strategies in Europe, the US and Japan. London: RIIA/Earthscan.

Opportunities for a Green Industrial Policy from an Evolutionary Technology Perspective

René Kemp
Maastricht University, MERIT, P.O. Box 616, 6200 MD Maastricht, The Netherlands

1 Environmental Restructuring and Modernisation of Industry

This chapter examines the opportunities for a green industrial policy aimed at making industrial sectors more environmentally friendly through a process of restructuring. The chapter is about "eco-restructuring", the restructuring of sectors and product chains in environmentally beneficial ways. Eco-restructuring focuses on issues of industrial decline and the process of creative destruction – issues neglected in the literature on environmental policy and environmental management, which tends to deal with changes in production processes and organisational structures within companies rather than structural change within product chains.

Eco-restructuring is a broader term than environmental modernisation – it describes sector-wide upgrading through the use of science and technology (Huber and Mol 1995). It involves downsizing environmentally harmful activities and creating new types of economic activity to compensate for the decline of these activities. Like environmental modernisation, eco-restructuring may or may not be motivated by environmental objectives. In fact, eco-restructuring in old polluting sectors (steel, coal) is often not motivated by environmental considerations, but the result of a decline in demand from saturated markets, or is caused by increased competition from developing countries rich in mineral resources which they process themselves. Environmental regulation may contribute to the contraction of an environmentally intensive sector but – as the case studies show – is not usually the prime cause of decline or restructuring.

Environmental modernisation, as commonly understood in political discourse and business, is usually confined to eco-efficiency options, win-win solutions that combine environmental and economic benefits. Examples are energy-efficient processes, recycling systems or low-solvent paints and coatings. It is promoted by environmental policy, not industrial policy.

Little experience has been gained in green industrial policy as yet. Industrial policy in the way in which it has been typically implemented aims to protect existing industries or support new high-tech ones – sectors that are viewed as a source of growth or as vital to the economy as a whole. State environmental policy and company environmental management is generally aimed at environmental modernisation, at picking the low-hanging fruit. There have been few attempts to achieve an eco-restructuring of old polluting sectors or to develop new technology systems which offer improvements in environmental performance. Most environmental policy has been geared towards upgrading existing processes and products with regard to their environmental impact. One might say that these policies were aimed at *optimising* rather than renewing systems. System optimisation means using options which do not entail any fundamental change within a product chain. In the case of environmental protection, this consists of changes at the component level, such as adopting end-of-pipe technologies or redesigning products (e.g. to use light-weight materials). System renewal involves wider change, going beyond the level of components, entailing the use of new technology, new markets and major organisational change.[1]

One example of system renewal is the shift towards renewables for electricity generation at the local level, and a shift towards demand side management, whereby the current logic of accommodating electricity demand through centralised electricity generation is replaced by one of providing energy services, rather than supplying electricity, in an environmentally sound way. Such changes involve a change in the basic rules and logic by which industry operates.[2] A second example of a shift in the regime is where people engage in intermodal travel as the dominant mode of transport, to satisfy their mobility needs. Regime shifts are accompanied by social and organisational, alongside technical innovation. They include, in the case of the intermodal travel regime, the collective ownership and use of cars (car-sharing and car pools), the creation of so-called mobility agencies to provide intermodal transport services,[3] the integration of collective transport schemes by transport companies (particularly railway and bus companies, but also car rental companies), and the introduction of traffic management systems in companies[4].

An example of system renewal by industry is industrial ecology or symbiosis – closing material flows through the use of waste outputs from one company by another. A well–publicised example of industrial symbiosis is found in Kalundborg, an industrial area on the Danish coast, where companies exchange material

---

[1] One may also define a third category, of partial system redesign, where there is a partial change to a system (Weterings et al. 1997).

[2] The words system and regime are often used interchangeably but are not identical. When we talk about a technological system we mean the socio-technical components and relationships between such components (actors, actions and artefacts). The term regime refers to the rule-set or grammar embedded in the complex of engineering practices, production process technologies, product characteristics, skills and procedures, infrastructures and institutions that make up the whole of a technology (see Kemp 1994; Kemp et al. 1998; Rip and Kemp 1998).

[3] These agencies provide the customer with a personalised travel plan involving the use of different transport modes (bus, train, taxi, rental car with or without chauffeur). They also take care of the financial and administrative details. Today the main customers are business users, but once the services become cheaper, private users are likely to follow suit.

[4] Company travel management systems are informational and financial schemes that encourage employees to share vehicles, and to use bicycles and collective forms of transport instead of automobiles. Such schemes help to reduce company costs for parking space and travel allowances. Apart from the companies' self interest there often is an important social motivation behind the introduction of travel management schemes in companies.

and energy outputs. Sulphur by-products from a refinery are passed to a local sulphuric acid manufacturer and hot water to local greenhouses. Waste heat and steam from the local power station are used by a company producing insulin and industrial enzymes (Cohen-Rosenthal and McGalliard 1998). A second example of industrial transformation is the provision by companies of product-service systems, as opposed to their manufacturing and selling products, such as leasing carpets and organised car sharing, where the product is not purchased and owned, but hired. The goal for companies is to fulfil a function for users rather than sell a product (Goedkoop et al. 1999).

The distinction between optimising and renewing a regime is important for (the management of) responses to environmental technology. Environmental benefits may be achieved through both types of change. Some writers – for example Clayton et al. 1999 – have argued that significant environmental benefits may be achieved through incremental change over an extended period. Whilst this is undoubtedly true, such changes will run into increasing marginal costs per unit of improvement. If we want to achieve a 10- to 50-fold improvement in resource productivity – which some claim will be necessary within the next 50 years (Weaver et al. 1999) – we need shifts in the regime (involving structural change) in addition to the optimisation of existing systems or product chains. As argued in Jänicke (1985), Ayres and Simonis (1994) and Kemp (1994), environmental sustainability requires change beyond the eco-efficiency options of environmental modernisation. In the Dutch DTO programme, several sustainable technologies (technology systems) offering quantitative improvements have been identified. These include: new protein foods (factor 10–30 improvement), precision agriculture (up to factor 50 improvement), decentralised electricity production using renewables, and underground pipe line transport of commodities (factor 10 improvement in energy efficiency), and industrial ecology.[5]

It should be noted that shifts in technological regimes do not mean the diffusion of so-called environmental technologies but changes in the system, which produce environmental benefits because the new trajectories are *inherently* more environmentally benign.

This chapter will not go into the topic of achieving technological regime shifts towards environmental sustainability but deal with the issue of eco-restructuring: upgrading existing sectors through structural change. In doing so, I will make use of insights from technology studies, especially evolutionary approaches based on the notion of systems.

[5] The factors given indicate the magnitude of the possible improvement. There may be boomerang effects which reduce the level of the improvement; different types of environmental effects are lumped together in one index, some types of environmental damage and risks may been assigned too high a score, while others may be attributed too low a weight.

2 The Production Chain as the Context for a Green Industrial Policy

One of the key insights of evolutionary theory is that innovation by companies should not be understood purely in terms of independent decision-making at company level. The development of specialisation, accompanied by inter- and intra-industry flows of technology, implies viewing the technological and economic structure of an economy not as an agglomeration of independent micro-level decisions, but as an integrated system shaped partly by the input-output relationships between companies, and partly by intra-company specialisation (Kemp et al. 1999). The supra-organisational context consists of the product chain, which denotes the input-output relations, and the set of conditions governing such relations; these conditions consist of product-specific supply and demand (willingness to pay for product services, usage of process technologies, skills and capabilities) and framework conditions such as (cf. Fig. 1):

- the overall state and development of the economy, including price stability, exchange rates or the situation in financial and labour markets
- the availability of an efficient and complete tangible and intangible infrastructure
- regulations within collective agreements and labour law
- political determinants, such as command-and-control policies, economic policy,

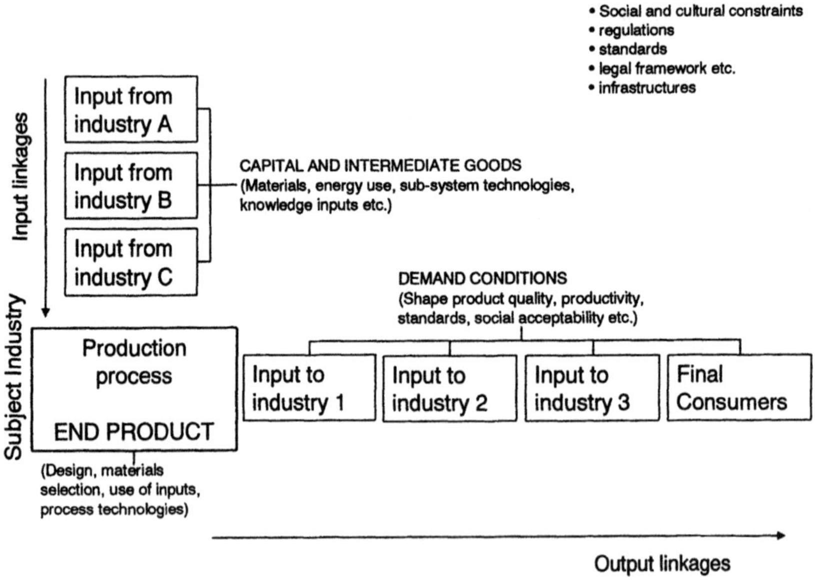

Fig. 1. The production chain

financial policy or the vast range of regulations, for instance in consumer protection, insurance, banking, the transport sector, the energy industry or in social regulation

- influences emanating from society and impacting on the economic players, such as social stability, a society's openness to technological innovation and economic growth in general, willingness to put up with negative environmental impact, etc. (Kemp et al. 1999)

The production chain is a useful starting point for industrial environmental policy. It reveals the ways in which companies and sectors are linked, economically and technologically. It helps in pointing up the inter-industrial and wider systemic aspects of environmental benefits. Such aspects differ according to the type of environmental technology. Pollution control and clean-up technologies are usually provided by EGS firms who have specialised knowledge of pollution and how it may be handled. Prevention usually requires environmental management and auditing systems, which may or may not be absent, depending on whether the company or sector ascribes great importance to environmental issues – which in turn depends in part on the kinds of market (consumer or industrial). Downstream, there are customers who may or may not be willing to pay extra for environmentally superior products. In addition, there are the waste sector and the framework conditions governing waste management practices – such as requirements to treat waste before disposal, restrictions on waste transport, availability of landfill, and so on. The waste business has a chain of its own, consisting of activities such as collection, transport, separation, treatment, recycling and controlled disposal. The costs and demand within this chain, and the regulations governing waste management practices within the chain have implications not just for what is done with waste in the waste sector (whether waste is disposed, burned, or upgraded for re-use) but also for companies upstream, in that they influence choices for manufacturing technology and materials.

The product chain not only reveals the interdependencies and "strategic knots" (*Jacob's* chapter in this book) but also the fact that environmental improvements may be made at different points along the production chain. What it does not show is that the economic opportunities and responsibilities for changing technological parameters differ widely along the chain. In general, the window of opportunity for making integrated process change is limited; it is very much restricted to periods of new investment (replacing old plant or creating new production capacity), and even then it is difficult to factor in environmental considerations. The same is true for product design, where many of the design parameters have already been set, especially for standardised products; changing these entails considerable costs and potential risks to the company, which they may be unwilling to accept. Environmentally improved products must meet important user requirements. It is difficult to sell a product on the basis of environmental advantages alone. The success of the Body Shop chain is misleading, because the products are bought for health rather than environmental reasons. The market for cosmetics with natural ingredients in which the Body Shop is so successful is still a small one, as is the market for organic food products (1% in the Netherlands). Environmental products often suffer from disadvantages of scale and a poor distribution system. Again, this shows the importance of the *filière* for innovation. Environmental innovation requires changes at different points on the *filière*. For example, recycling depends on collection and separation, on designing for dismantling and re-use, as well as on methods to identify impurities. It requires coordinated change along an entire chain of activities, which is difficult to manage. It is only the larger companies such as the Dutch electronics company Philips and paints manufacturers who are active in environmental chain management, by placing environmental demands on their suppliers and educating customers.

The relationships and conditions within a production chain are important contextual factors for any attempt by business to achieve change, but also for governments who are unwilling to cause disruption. Government policy (especially environmental policy) is not formulated independently of the production chain or the available solutions to perceived problems. There are overlapping systems in which actors from different worlds interact with one another (actor networks). With respect to the environmental, three important networks are: 1) the economic network of suppliers and users within a production chain, 2) the policy network of regulators, business, epistemological communities and societal groups, and 3) the innovation or knowledge network of universities, research institutes, consulting firms and of course industry (Fig. 2). Industry is thus a player in all three networks – as manufacturer, as problem, as technology user and as political actor. The networks are the historical outcome of interaction which has given rise to patterns of innovation, policy style (confrontational or co-operative) and outcome.

Actors in the networks influence each other mutually. The networks incorporate a system of interaction, a set of values, norms and belief systems, and balances of power. It would be wrong to see government actors at the top of the hierarchy, so to speak in the driver's seat, or to see industry in control of everything (as in regulatory capture theories and Marxist views, where both workers and consumers are ascribed minor influence). Change at one point in the network, such as a manufacturing company or a sector, requires change in knowledge, process technology, values, regulations, etc. at other points. A systems perspective helps to appreciate this aspect but must be integrated into a dynamic framework. This is accomplished in an evolutionary approach to technology systems developed by Freeman, Nelson and Winter, Dosi, Malerba, Schot, Soete and Arundel, Kemp and Rip. A description of this perspective is given in this book in the chapters by *Nill and Petschow*, and *Joest*.

3 Incremental Versus Radical Change

An evolutionary perspective on technology systems is useful because it helps grasp not only the relevant actors and relationships but also the technological

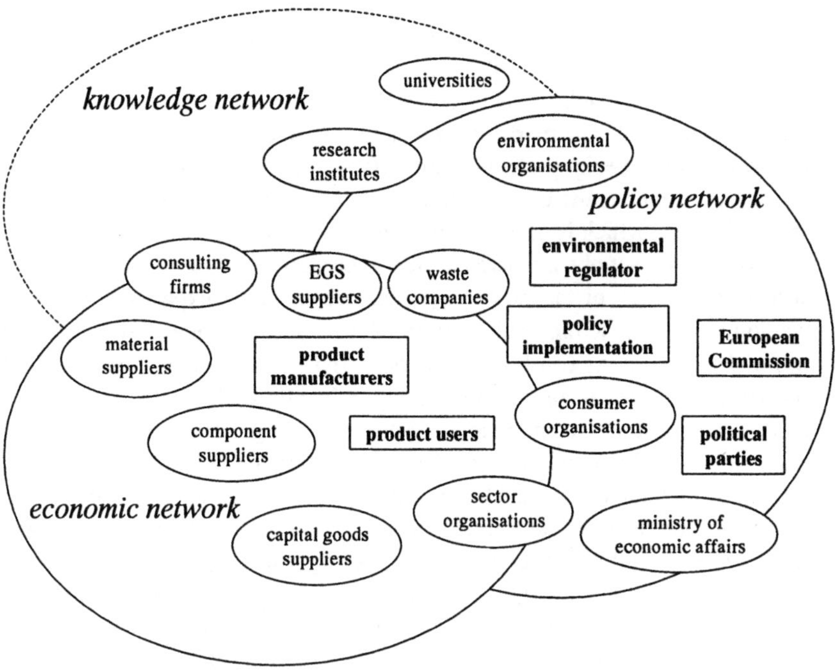

Fig. 2. Overlapping systems of interaction with respect to environmental issues

basis, the sunk costs, economies of scale, complementary technologies, skills and so on – how these create barriers to new technologies and market entry – and the technological possibilities for change. Technology is not perfectly malleable, it is "lumpy", involves knowledge and skills; moreover, as shown by studies of technological and industrial development (Abernathy and Clark 1985; Abernathy and Utterback 1978; Rosenberg 1976, 1982), advances are not random, but tend to occur along certain trajectories, in which a certain design is used as the basis for development decisions. Some technological possibilities are locked out or disadvantaged. This is especially true for more radical solutions which require change at different levels: in skills, plant, production routines, as well as ideas and institutional arrangements. The latter include dissolving existing regulations or developing new ones. All this means that there are switching costs involved in system change; such costs (of which financial costs are only a part) constitute a barrier to systemic technologies: technologies which depend for their use on complementary technologies (co-assets) and changes in the institutional environment.

Radically new technologies may be actively supported by an innovator or consortium, but there may also be limits to their willingness to invest in the new technology. For a dynamic process of radical change to occur, whereby an old system is overhauled by a new one, it is important that the radical technology find some domain of application, a niche. Without a market for its use, industrial interest in developing a new technology will be marginal. Put simply, new technologies need a champion and the champion needs a market. Given the imperfection of a new technology at the beginning of its life cycle and competition from entrenched mature technologies which have benefited from economies of scale and learning, this market will initially be small. The market may be a localised market or a market of idiosyncratic users with a special preference for a new product because of unique features. Such markets (niche markets) are important in two ways: it provides vendors with a source of income, which is necessary for them to continue, and it facilitates learning and adjustment on supply and demand sides, necessary for the further development of a new technology or technology system. For example, experience with a new technology in the niche market help promote user acceptance and alter established views and expectations (on both the supply and demand sides). Feedback from users (about their needs and the technology's functions) helps to inform companies' research, production and marketing policies, bring cost savings in production and use, promote the development of complementary assets, and foster building a constituency behind the product. The latter is important in exercising political influence – gaining political clout or access to the political arena, and fighting vested interests – and for programming or pooling research, or introducing quality assurance schemes. Niches thus provide an impetus to learning, investment and alignment. Real use of a new technology is crucial, as some things can only be learnt from experience. Real-life experience is often necessary for institutional adaptation.

New technologies or economic activities initially need some form of protection. This protection may be offered by an industry willing to invest money, time and effort in a certain endeavour, or by other actors, such as users or governments, interested in creating new business and employment, environmental authorities interested in new environmental solutions or governments, for both industrial and environmental reasons.

The multiple dependencies and influence structures, and the multiplicity of technological options for achieving environmental benefits, with varying costs and benefits for economic and social actors, imply that there is no clear recipe for action, at least no single recipe. What is needed is exploration of the options, before any attempt is made to support an environmentally beneficial project. Policy, especially green industrial policy, should be aimed at identifying windows of opportunity and acting on any possibilities identified. This is a collective task, not just a task for the polluting industry or government.

Windows of opportunity may be divided into two types: *economic* windows of opportunity for a new technology or new activity, and *political* windows of opportunity. The latter consists of policies which are not only conceivable but *possible*, in view of the local problem, the political agenda at a particular moment in time, and the available resources. These resources consist of financial resources (the funding available for programmes), organisational resources – organisations capable of implementing the programmes – and the presence of a political entrepreneur who is able to build support for a policy and manage the process from the idea through to its political implementation.[6]

Looking at the case studies of environmentally intensive sectors we see many economic opportunities for achieving environmental benefit. The reason is that these sectors are resource-intensive. Material resources and energy constitute a major cost category. A higher resource productivity helps to save costs and at the same time yield environmental benefits, because pollution is often associated with use of resources. Although there are powerful incentives for processing industries to reduce their use of materials and energy, this is not automatic. It depends upon company management and auditing systems, such as energy management systems, identifying the economic opportunities in saving inputs. Most companies in processing industries use such systems and exist in culture which ascribes a high value to resource productivity (Clayton et al., 1999). This is supported by the evidence from the case studies, where material savings and energy savings per unit of output are reported. As an example, Ålborg Portland achieved a reduction of 33% in its energy consumption between 1979 and 1992. This was achieved through material substitution and re-use, such as the use of fly-gas, rubber tires and other types fuel in cement kilns (see *Klemmensen's* chapter in this book). There are few examples of major change in products and processes. Two examples are Esso's flexicooker refinery process and the development of CFC substitutes. In the first case, the development of the market – greater demand for light fuels, especially for transport – fuelled the development and use of a new technology that was also attractive from an environmental point of view; in the second case, there were important non-market pressures to come up with a substitute for CFCs, which were destroying the ozone layer that protects the Earth against UV radiation. It was not regulation, however, which brought about the innovation, it was the *threat* of regulation and the increasing lack of acceptability for CFCs, as evidence abounded that CFCs were indeed destroying the ozone layer, which prompted CFC producers to search for alternatives and introduce them onto the market. In both cases, market considerations played an important part (see Chaps. by *Van der Straaten* and *Jacob* in this book; cf. Gabel 1995; Rip and Kemp 1998).

The case studies describe political interaction in controlling environmentally harmful activities. From the descriptions of policy it is unclear how political windows of opportunity may be created. One thing that they show is that it is hard to make industry adopt environmental measures which entail financial losses for the company. In general, regulation is a lengthy, conflict-ridden process. The case studies also show that policy is mostly reactive; it tries to limit environmentally harmful emissions from existing processes. Environmental solutions sometimes create new problems, such as the incineration of plastics and waste by cement manufacturers to generate heat. Innovations are not actively sought. Environmental authorities rely on industry to come up with innovative solutions. Until recently, environmental policy was pollution oriented rather than process or product oriented, although this is now changing.

---

[6] This may be accomplished by more than one actor. The same actor need not be involved at every stage.

Some companies have diversified their activities and created new markets for environmentally superior products. Government policy has played a very minor role here. One exception is lead-free petrol and CFC substitutes. Perhaps more could have been achieved through a green industrial policy. Industrial policy and environmental policy are distinct. There is room for integrating the two but no basic theory of green industrial policy to guide such an integration. At the same time, we should point out that green industrial policy is far from simple. It throws up thorny problems regarding government's ability to pick winners, the amount of protection to offer, the nature of the protective measures, and how to end supportive measures on which industry has become reliant. There is also the problem of opposition by vested interests and managing the process of creative destruction, which poses an additional problem for public policy-makers.

Suggestions for innovation-friendly regulation are offered by Blazejczak et al. (1999). They identify innovation-friendly instruments, policy styles and configurations of actors.[7] The issue of environmental policy instruments promoting innovation is also discussed in Kemp (1997), where the results of empirical studies of the innovation impact of environmental policies are studied and recommendations are made for using environmental policy instruments in different contexts. This chapter looks at a method for *exploring* the potential of environment-friendly innovations and managing their introduction. This method is called *strategic niche management* (SNM). SNM is a probe-and-learn strategy for innovations which relies on the real-life use of new technologies by selected users. Although it concerns the introduction to society of a new technology rather than the creation of new business, it provides a useful framework for this and for green industrial policy, through its focus on issues such as the balance between protective measures and selective pressure, the up-scaling of projects, appropriate roles for private and public decision-makers, and decision-making where there is uncertainty and conflicting views and interests.

4 Strategic Niche Management. A Possible Model for Green Industrial Policy

New technologies, like new economic activities, are surrounded by uncertainty – about their acceptance, or the best way to design, produce and market them. Their feasibility remains to be proven to potential users but also to the company who may champion it. One way to find out about their potential, and how this potential may be tapped, is through the use of experiments, involving especially the experiences of real users. The experiment facilitates learning, provided that it is well-conducted. This raises the question how experiments should be conducted and

[7] According to the writers, innovation-friendly policy instruments should rely on the use of economic incentives, act in combination, be based on strategic planning and formulation of goals, support innovation and take account of its different phases. Innovation-friendly policy styles are based on dialogue and consensus; they are decisive, proactive and ambitious, open, flexible and knowledge-oriented. Innovation policy should include network management.

what might be learned from them. These issues have been examined in a project for the Environment and Climate programme of DG–XII, "Strategic Niche Management as a Tool for Transition to a Sustainable Transport System", in which I was involved.

Strategic niche management (SNM) is a new approach, first suggested by Rip and further developed by Schot et al. (1994), Kemp et al. (1998) and Weber et al. (1998). It is a kind of probe-and-learn approach aimed at learning and adaptive decision-making. This is accomplished through managing protected spaces for the use of new technology. In these spaces, the technology is partly protected from the normal selective pressures of business. This protection should be temporary and partial. Apart from protection, there should also be some selective pressure. SNM is about finding a balance between the two. Some protection may be warranted even when the technology is economic to use. This is especially true for environmental technologies whose the social benefits are undervalued in the market place.

The spaces should be carefully selected. Important considerations are firstly the attractiveness of the technology in the local context, and secondly the extent to which the setting allows for learning (about the technology's market and the best design and policies to support it). The heterogeneity of the selective environment implies that there will always be spaces (application domains) where the technology is attractive.

SNM thus consists of the real use of technologies, in selected settings. This actual use of a new technology is important if articulation is to take place, and to learn about the viability of the new technology and build a network around the product. SNM helps to inform development decisions by the innovating company and other companies or government officials interested in the development of a new technology.

SNM is a concentrated effort to develop protected spaces for certain applications of a new technology.[8] The protection may be provided by private or public decisions: by suppliers, development and regulatory agencies, local authorities and/or users.

SNM may be used for different types of innovation, but is especially suited for sustainable technology systems and products offering environmental and social benefits which are undervalued in the market place.

The idea of SNM for new technologies comes from the literature on sequential decision-making to deal with complexity (Simon), trial-and-error modes of learning, evolutionary theories of technical change (Nelson and Winter 1982; Ayres;

[8] In the EU project "SNM as a tool for transition to a sustainable transport system", SNM is defined as the creation and management of protected spaces (niches) for promising technologies by means of experimentation, with the aim of learning about the performance, effects, economic viability and social desirability of the technology, and of using this knowledge to inform private and public (support and control) policies needed for the further development and rate of application of new technologies and technology systems.
(see http://www.jrc.es/snm/)

Rip and Kemp 1998), and studies on the critical role of users in innovation (von Hippel 1988).

Arguments in favour of technology experiments involving real users:

1. Under-utilisation of technological possibilities: many ideas and prototypes never reach the stage of market introduction, which means that their potential and viability is not adequately investigated and abundantly tested; this applies especially to radical, systemic technologies and technologies offering social benefits as opposed to benefits for individual users.
2. Some things can only be learnt from real-life use of new technologies: market acceptance, their actual performance in real situations, the best design configuration for different groups of users etc.
3. Users are an important source of knowledge; involving users helps deal better with their needs. Experience with new technologies helps users to develop a better understanding of them and makes them learn about their own behaviour and mobility needs.
4. Experiments help to set into motion or facilitate institutional embedding, and by doing so they act as a stepping stone or engine for change.

SNM is more than a learning strategy involving user experiments. It helps create a platform for interaction and builds a constituency for the product among private and public decision makers. The latter is especially important for escaping the gridlock of dominant practices and vested interests or belief systems. SNM thus not only helps deal with problems of uncertainty and fragmented or disparate knowledge and organisational resources – common obstacles to change – but also helps deal with opposition from vested interests and with the conservatism in established views, beliefs that the innovation is not economically or technologically viable, or inferior to alternative solutions within the existing technological regime. SNM may thus help to set into motion processes of alignment and institutional embedding necessary for the development and wider use of a new technology. Institutional embedding involves forming networks, coordinating the policies of private and public actors, establishing practices for production and use, standardising a technology, and developing shared belief systems, roles and expectations.

After having talked about the "what", we now turn to the "how" of SNM. What does it imply in practical terms? In the SNM project mentioned above, we developed a five step plan, delineating key elements of SNM.[9] The five steps are: 1) choosing a technology, 2) selecting an experiment, 3) balancing protective measures and selective pressure, 4) scaling the experiment up, 5) dismantling the protection. While a stepwise plan may seem to be a deterministic exercise, obfuscat-

[9] See also Weber and Dorda (1999) for a discussion of SNM focussing on dilemmas and trade-offs that occur at different stages of an SNM process, and how these dilemmas may be dealt with.

ing the feedback loops and reflexive nature of SNM, it does bring out the key features. The steps are now elaborated in detail.

- *Choosing a technology:* A decision must be made on which type of technology to experiment with. From a business point of view, the technology should be economically attractive to sell and use. For environmental authorities, the environmental benefit is an important consideration. Although all kind of technologies should be experimented with, there is a special need to support technologies outside the existing regime or paradigm. Of course, it is exactly these kind of solutions which users often find unattractive to use. On the other hand, there may be niches in which the costs do not bear so much weight. If such niches exist, we may use SNM. The choice of technology should also be informed by the possibilities for improvement and branching. Attractive technologies eligible for support are technologies with a large development potential and a synergy with parallel developments (such as the evolution of user preferences and society's values, political development, areas of rapid technical advance).
  The selection of the technology can be accomplished in a consultative co-operative mode, together with stakeholders, or in a more competitive mode through the use of tenders, where applicants are asked to come up with proposals. Design choices should not be fixed from the outset, to prevent lock-in at suboptimal designs.
- *Selecting an experiment or niche:* After choosing an eligible technology, an appropriate setting for the new technology needs to be found. This may be the company in which the technology was developed, another company, or private households. The choice should be based on the relative attractiveness of the technology for the user and the potential for learning something about the technology and its market from the users. The heterogeneity of the selective environment means that there are almost always areas and types of application for which the new technology is attractive, where the disadvantages weigh less heavily and the advantages are more highly valued. Electric vehicles which emit no pollutants at the point of use are attractive for use in cities with high levels of pollution. Their small size makes them well-suited for tourist resorts with narrow roads. They are also attractive to use in fleets because of the low operating costs. These are just three applications for EVs. Users should be critical and able to communicate their requirements, so as to push suppliers to constantly improve the product.
- *Balancing protection with selection:* When conducting the user experiments, a balance must be struck between protection and selective pressure. This is perhaps the most difficult step. Protective measures should be oriented around the barriers to the use and diffusion of the new technology. These barriers may be economic, if the new technology is unable to compete with conventional technologies within the prevailing cost structure. They may be technical, in the form of a lack of complementary technologies, new infrastructure or appropriate skills. And they may be social and institutional, concerned with existing

laws, practices, perceptions, habits and interests. An integrated and coordinated support policy is required to deal successfully with these barriers. The possible elements of such a policy are formulating long-term goals, creating an actor network and the use of taxes, subsidies, public procurement and standards. However, the policy should not attempt to lift all barriers, as this would remove all the selective pressures which prevent the emergence of "white elephants". Industry should bear a significant part of the cost and risks.

SNM is not just something for governments: industry and NGOs are well-placed to initiate and run niche projects. As a rule, government should assume those roles where they operate better than others; they should not assume responsibility for running the experiment, as this is probably best done by professionals with their own social networks. As noted, governments have a special role as facilitator, to make sure that something happens, and that the project yields satisfactory results (which requires monitoring, evaluating outcomes and policies and, in the case of undesirable outcomes, the judicious exertion of pressure to correct adverse activities and policies). As for the role of different levels of government, local governments are best placed to engage in local affairs such as network management, while regional and state governments may act as cosponsor for projects that may be used on a larger scale. They could also help in scaling successful experiments up, through sponsorship or macro-policy (like changing the regulatory framework and the use of fiscal incentives). National and regional governments also have a special responsibility for making sure that there is broad social learning. This could be accomplished by supporting a portfolio of niche projects, institutional technology appraisals and social discourse (in which the technologies are evaluated over a wide range of dimensions), and disseminating the knowledge gained from the projects.

- *Scaling the (niche) experiment up:* The next step concerns scaling up the experiment or niche through policy measures. Even a highly successful experiment may require some form of support from public policy-makers or industry. This applies especially to environmentally benign technologies whose benefits are undervalued in the market place. Again this raises the question of how far governments should go in support of a particular technology. It also raises the issue of the nature of the policies. The above provides some suggestions. For a further exploration see Schot et al. (1994) and Schot and Rip (1997).
- *Dismantling protection:* The final step is the phased dismantling of protection. Support for the new technology may no longer be necessary or desirable if results are disappointing and prospects bleak. Dismantling the protection is also necessary to create additional selective pressure and reduce windfall profits. The dismantling should best be controlled and phased, in order not to cause too severe problems for the companies involved.

SNM may be described as learning by doing or a probe-and-learn policy. It is not technology push although there are "push" elements. The goals of SNM are learning and institutional alignment or embedding, two key elements in coevolution.

Learning and institutional alignment are important in situations of "interactive complexity" (Perrow 1984), when it is hard to design the "best" system a priori (Karnoe and Garud 1999). Probe-and-learn strategies are particularly useful in situations of uncertainty and opposition. SNM helps actors to negotiate and explore various interpretations of the usefulness of specific technological options and the conditions for their application. SNM highlights choices and options and makes the introduction process more transparent and doable for all parties involved, including producers, users and policy-makers (Weber et al. 1998). Through management of niches, a new type of economic activity and development path may be created in a bottom-up, gradual, non-distorting manner, keeping adjustment costs to a minimum. It provides a model for industrial policy and legitimises government support for experimentation.

5 Concluding Remarks

This final section draws some conclusions about eco-restructuring. Eco-restructuring is a multicoloured creature. There are different types of greening and different shades of green. The dirty sectors described in the book have all undergone a process of eco-restructuring best described as "pale greening". None of the sectors has really become green. The sectors are still environmentally intensive, as major consumers of energy, emitters of pollutants, and producers of environmentally harmful waste material. Governments have not for the most part actively pursued eco-restructuring. They relied very much on the technological responses favoured by industry. They helped industry go through the restructuring process in a relatively painless manner, through provision of subsidies and the use of soft, time-phased regulations. They did not pursue a policy of creative destruction: the creation of less environmentally harmful business activities to substitute unsustainable types of production.

This raises some questions: what possibilities do government authorities have for promoting creative destruction? Is it possible? Is it desirable? Do governments have a role in creating business? What does this role consist of? And how to deal with obstruction and uncertainty?

Is creative destruction desirable? The answer is that it is desirable from an environmental point of view and from the perspective of regional development and employment, but not necessary desirable for the industry itself, because creative destruction means that a part of previous activities is closed down or phased out. The creation of business may take place not in the existing industry but in new companies, exploiting some of the skills of workers previously employed in the old industry. On the other hand, an old industry may find new areas of economic activity, such as waste management or environmental consulting, as in the case of the chemicals company DSM in the Netherlands.

What is the role of government? There are many conceivable roles, but one important role is to facilitate learning aimed at identifying new business opportunities, maybe in eco-business or some other line which is relatively harmless to the environment. This can be accomplished in two ways: through the use of a consensual approach, aimed at bringing knowledge holders together, encouraging them to do something innovative that is environmentally beneficial, or through a more confrontational approach, setting constraints on processes and products to promote their replacement by more environmentally friendly ones. The latter is only possible if the activities and products to be controlled pose an urgent and serious problem.

One example of a consensual approach is network management and the use of measures to support innovation. The role of government is that of facilitator. When using a confrontational approach, government authorities create conflict, for example through the threat of regulation. Here it tries to exploit differences in interests by intervening in the strategic games between companies: industrial producers and the suppliers of innovative solutions. A mixed strategy is the organisation of a discourse of sustainable futures, where industrialists meet with environmentalists, technology experts and policy makers. This was the case in Germany in respect of sustainable chemistry. On a smaller scale, the same has been done by individual companies such as Monsanto and Hoechst, adopting a stakeholder view (or enlightened shareholder view, depending on how you look at it). Governments could promote such exploration and learning partnerships. They could also promote the use of experiments with new technologies. As new technologies involve uncertainty, there is a need to learn more about them. One way of learning more about a new technology is through the use of technology experiments in selected settings. The rationale for conducting experiments is that new activities require some kind of protection and need to go through a learning process. They need to prove themselves to the outside world: bankers and users. They also need to prove themselves to the people within the company who are sceptical or critical about a new endeavour. In this chapter we have argued for more experimentation, and promoting broad learning about green technological options, especially radical options. The issues involved in such endeavours are discussed under the label of SNM, which, by highlighting trade-offs and dilemmas, offers a guide for private and public decision-makers in managing the process of structural change inevitable in creative destruction.

References

Abernathy, W.J. and Clark, K.B. (1985): Innovation: Mapping the winds of creative destruction Research Policy 14, 3–22.

Abernathy, W.J. and Utterback, J.M. (1978): Patterns of Industrial Innovation. Technology Review. 50 (7), 41–47.

Arthur, W.B. (1988a): Competing Technologies: An Overview. In: Dosi et al. (1988), pp. 590–607.

Arthur, W.B. (1988b): Self-Reinforcing Mechanisms in Economics. In: Anderson, P. et al. (eds.): The Economy as an Evolving Complex System. Reading, MA: Addison-Wesley.

Ayres, R.U. and Simonis, U.E. (eds.) (1994): Industrial Metabolism. Tokyo: United Nations University Press.

Blazejczak, J., Edler, D., Hemmelskamp, J., Jänicke, M. (1999): Environmental Policy and Innovation: An International Comparison of Policy Patterns and Innovative Impacts. In: Klemmer, P. (ed.): Innovation and the Environment. Berlin: Analytica: pp. 9–30.

Clayton, A., Spinardi, G., Williams, R. (1999): Policies for Cleaner Technology. A New Agenda for Government and Industry. London: Earthscan.

Cohen-Rosenthal, E. and McGalliard, T.N. (1998): Eco-Industrial Development: The Case of the United States. IPTS Report no. 27, Sept 1998, 23–34.

Dosi, G. (1982): Technological Paradigms and Technological Trajectories: A Suggested Interpretation of the Determinants and Directions of Technical Change. Research Policy 6, 147–162.

Dosi, G. (1988): The Nature of the Innovation Process. In: Dosi et al. (1988).

Dosi, G.. Freeman, C., Nelson, R., Silverberg, G., Soete, L. (eds.) (1988): Technical Change and Economic Theory. London: Pinter Publishers.

Edquist, C. (ed.) (1997): Systems of Innovation. Technologies, Institutions and Organizations. London and Washington: Pinter.

Van Est, R. (1997): Winds of Change. A Comparative Study of the Politics of Wind Energy Innovation in California and Denmark. Ph.D. manuscript, Draft, University of Amsterdam.

Freeman, C. and Perez, C. (1988): Structural Crises of Adjustment, Business Cycles and Investment Behaviour. In: Dosi et al. (1988), pp. 38–66.

Gabel, L. (1995). In: Folmer, H., Gabel, L., Opschoor, H. (eds.): Principles of Environmental and Resource Economics. A Guide for Students and Business Makers. Aldershot: Edward Elgar.

Goedkoop, M.J., Van Halen, C.J.G., te Riele, H.R.M., Rommens, P.J.M. (1999): Product Service Systems, Ecological and Economic Basics. Den Haag: PriceWaterhouseCoopers.

Green, K. (1992): Creating Demand for Biotechnology: Shaping Technologies and Markets. In: Coombs, R., Saviotti, P., Walsh, V. (eds.): Technological Change and Company Strategies. London: Academic Press, pp. 164–184.

Green, K., McMeekin, A., Irwin, A. (1994): Technological Trajectories and R&D for Environmental Innovation in UK Firms. Futures 26 (10), 1047–1059.

Harms, S. and Truffer, B. (1998): The Emergence and Professionalisation of two Carsharing Co-operatives in Switzerland. Final report for SNM project.

Hughes, T.P. (1979): The Electricification of America: The Role of System Builders. Technology and Culture 20, 124–161.

Hughes, T.P. (1983): Networks of Power. Electrification in Western Society, 1880–1930. Baltimore: The John Hopkins University Press.

Hughes, T.P. (1989): The Evolution of Large Technological Systems. In: Bijker, W.E., Hughes, T.P., Pinch, T.J. (eds.): The Social Construction of Technological Systems: New Directions in the Sociology and History of Technology. Cambridge MA: MIT Press, pp. 51–82.

Jänicke, M. (1985): Preventive Environmental Policy as Ecological Modernisation and Structural Policy. IIUG dp 85–2, Berlin, IIUG.

Karnoe, P. and Garud, R. (1999): Path Creation and Dependence in the Danish Wind Turbine Field. In: Porac, J. and Ventresca, M. (eds.): Social Construction of Industries and Markets. Pergemon Press (forthcoming).

Kemp, R. (1994): Technology and the Transition to Environmental Sustainability. The Problem of Technological Regime Shifts. Futures 26 (10), 1023–46.

Kemp, R. (1996): The Transition from Hydrocarbons. The Issues for Policy. In: Faucheux, S., Pearce, D., Proops, J.L.R. (eds.): Models of Sustainable Development. Cheltenham: Edward Elgar, pp. 151–175.

Kemp, R. (1997): Environmental Policy and Technical Change. A Comparison of the Technological Impact of Policy Instruments. Cheltenham: Edward Elgar.

Kemp, R. and Soete, L. (1992): The Greening of Technological Progress: An Evolutionary Perspective. Futures 24 (5), 437–457.

Kemp, R., Miles, I., Smith, K. et al. (1994b): Technology and the Transition to Environmental Stability. Continuity and Change in Complex Technology Systems. Final report for SEER research program of the Commission of the European Communities (DG XII).

Kemp, R., Schot, J., Hoogma, R. (1998): Regime Shifts to Sustainability through Processes of Niche Formation. The Approach of Strategic Niche Management. Technology Analysis and Strategic Management 10(2), 175–195.

Kemp, R., Truffer, B., Harms, S. (2000): Strategic Niche Management for Sustainable Mobility. In: Rennings, K., Hohmeier, O., Ottinger, R.L. (eds.): Social Costs and Sustainable Mobility – Strategies and Experiences in Europe and the United States. Physica Verlag (Springer), Heidelberg, New York (forthcoming).

Kemp, R., Smith, K., Becher, G. (1999): How Should We Study the Relationship between Environmental Regulation and Innovation? Paper for International Conference "Innovation-Oriented Environmental Regulation: Theoretical Approaches and Empirical Analysis". Potsdam (Germany), May 27–29, 1999.

Kooiman, J. (ed.) (1993): Modern Governance. New Government-Society Interactions. London: Sage.

Metcalfe J.S. (1998), Evolutionary Economics and Creative Destruction. Routledge: London

Mol, A. (1995): The Refinement of Production. Ecological Modernisation Theory and the Chemical Industry. The Hague.

Molina, A.H. (1993): In Search of Insights into the Generation of Techno-economic Trends: Micro- and Macro-constituencies in the Microprocessor Industry. Research Policy 22, 473–506.

Nelson, R.R. (1994): The Co-Evolution of Technology, Industrial Structure, and Supporting Institutions. Industrial and Corporate Change 3 (1), 47–63.

Nelson, R.R. and Winter, S.G. (1977): In Search of Useful Theory of Innovation, Research Policy 6, 36–76.

Nelson, R.R. and Winter, S.G. (1982): An Evolutionary Theory of Economic Change. Cambridge MA: Bellknap Press.

Perez, C. (1983): Structural Change and Assimilation of New Technologies in the Economic and Social System. Futures 15 (4): 357–375.

Perrow, C. (1984): Normal Accidents. Living with High-Risk Technologies. New York: Basic Books.

Rip, A. and Kemp, R. (1998): Technological Change. In: Rayner, S. and Malone, E.L.: Human Choice and Climate Change. An International Assessment. Vol. 2. Batelle Press, Washington D.C., pp. 327–400.

Rip, A., Misa, T. Schot, J. (eds.) (1995): Managing Technology in Society. New Forms for the Control of Technology. London: Pinter Publishers.

Rosenberg, N. (1976): The Direction of Technological Change: Inducement Mechanisms and Focussing Devices. In Rosenberg, N.: Perspectives on Technology. Cambridge: Cambridge University Press, pp. 108–125.

Rosenberg, N. (1982): Inside the Black Box. Technology and Economics. Cambridge: Cambridge University Press.

Schot, J.W. (1992): Constructive Technology Assessment and Technology Dynamics. The Case of Clean Technologies. Science, Technology and Human Values 17 (1), 36–57.

Schot, J., Elzen, B., Hoogma, R. (1994): Strategies for Shifting Technological Systems. The Case of the Automobile System. Futures 26 (10), 1060–1076.

Schot, J. and Rip, A. (1997): The Past and Future of Constructive Technology Assessment. Technological Forecasting and Social Change 54, 251–268.

Soete, L. and Arundel, A. (eds.) (1993): An Integrated Approach to European Innovation and Technology Diffusion Policy. A Maastricht Memorandum. Brussels: Commission of the European Communities.

Sonntag, V. (1998): Changing the Direction of Economic Advance: The Economics of Flexible Production Systems, Market Conditions and Sustainable Consumption. MERIT/INTECH-UNU dissertation proposal, Maastricht.

Stahel, W.R. (1994): The Utilization-Focused Service Economy: Resource Efficiency and Product-Life Extension. In: Allenby, B.R. and Richards, D.J. (eds.): The Greening of Industrial Ecosystems. Washington D.C.: National Academy Press, pp. 178–190.

von Hippel (1988): The Sources of Innovation. Oxford: Oxford University Press.

Weaver, P. et al. (1999): Sustainable Technology Development. Sheffield: Greenleaf Publishing.

Weber, M., Hoogma, R., Lane, B., Schot, J. (1998): Experimenting with Sustainable transport Innovations. A Workbook for Strategic Niche Management. Seville/Enschede.

Weber, M. and Dorda, A. (1999): Strategic Niche Management: A Tool for the Market Introduction of New Transport Concepts and Technologies. IPTS Report, Febr 1999, 20–27.

Weterings et al. (1997): 81 Mogelijkheden: Technologie voor duurzame ontwikkeling. Final Report. Environmental Technology Exploration for the Ministry of Housing, Spatial Planning and Environment (VROM) in the Netherlands.

Global Change and the Economic Charm of the Environment

Franz Lehner
Institute of Work and Technology, Munscheidstraße 14, 45886 Gelsenkirchen, Germany

Globalisation presents a severe challenge to industrial competitiveness in Europe. Thus far, Europe has not been very successful in meeting this challenge, as indicated by high and persistent unemployment, and stagnating real incomes in many countries. This is revealed clearly in the Competitiveness Report of the European Commission (Commission of the European Union 1997).

Whilst the diagnosis is widely accepted, the reasons for this poor performance are disputed. There is a widely held view that high labour costs, over-burdened welfare systems and rigid labour regulations are the major causes of both high unemployment and the poor competitiveness of European industry. In some countries it is argued that the problems are compounded by excessively strict environmental regulations and standards. From this perspective, the solution to problems of competitiveness and unemployment in the advanced economies of Europe is to cut labour costs and welfare payments, deregulate employment and moderate environmental policy. In general, business leaders and politicians recognise the attraction of the environment if it can be used to demonstrate a commitment to the general good and concern for the natural world, providing that it does not involve too much cost or effort.

The approach adopted here is radically different. It begins with an argument elaborated by the OECD (1994, 1996a) (and many other studies), namely, that Europe's current employment and competition problems (stagnating real incomes and increasing poverty in particular) are primarily the result of delayed structural change. The logical conclusion to be drawn from this argument is that these problems cannot be solved without more rapid and radical structural change.

However, there is no reason to moderate environmental policy. On the contrary, since delayed structural change is also a major cause of the current environmental problems, Europe needs an environmental policy which drives innovation and structural change towards a knowledge-based and service-oriented economy. The key to such a policy is to increase resource productivity and dematerialise production.[1] The economic charm of the environment is that it can serve as an effective vehicle to advance structural change, enhance growth and create new employment.

[1] This paper is based on Lehner and Schmidt-Bleek 1999.

1 Industrial Competitiveness in Europe[2]

In its Competitiveness Report (1997), the European Commission describes a competitiveness pyramid where standard of living is the top concern and where productivity and employment are the secondary goals. A competitive economy is described as an efficient economy which creates jobs and raises the standard of living (Commission of the European Union 1997, p. 7). A similar definition is suggested by the American Competitiveness Policy Council which defines competitiveness as the capacity of an economy to maintain or increase the living standards of a society in a productive way (Competitive Policy Council 1992).

According to the European Competitiveness Report, the focus of public (and private) activities on competitiveness should be to improve the performance of enterprises, since "growth and jobs are produced by enterprises operating in a competitive market" (Commission of the European Union 1997, p. 8): a competitive economy is one in which the overwhelming majority of companies are competitive. With respect to individual companies, competitiveness can be defined as an enterprise's capacity to withstand competition and secure profitability. Companies are competitive if their products and prices meet the requirements of the relevant markets, and if they can profitably remain in business under "free and fair" market conditions.

A cursory examination of EU companies' capacity to withstand international competition provides no cause for concern. However, there are warning signs. In recent years, the countries of the European Union have experienced only modest growth in exports, and growth has stemmed primarily from a general increase in demand rather than from increased market share or export diversification, i.e. from the development of new markets, businesses or technologies. As the EU Commission points out, "export decline is not due to a lack of performance of manufacturing industry itself, but to poor initial specialisation and adaptation of European Union exports to changes in world trade" (Commission of the European Union 1997, p. 35). Industry in Europe is globally competitive primarily in medium technology exports, whereas its export performance in the high technology industries is poor. In contrast, the USA and Japan rank higher in high technology industries and have improved their position during the 1980s and 1990s. This is not only an indication of poor specialisation, but also of delayed structural change in Europe. As the OECD argues, industry in Europe does not respond to global change with sufficiently radical and rapid structural change (OECD 1996a). Compared to the USA and Japan, industry in most EU countries is weak in product innovation and product development. Even worse, product innovation in Europe is concentrated in established, stagnating or even declining markets, whereas in the USA and Japan, it is directed more towards new and growing markets.

---

[2] This section summarizes an argument which is developed in Charles and Lehner 1998 and is elaborated in Lehner et al. 1999.

Similar innovation problems also exist with respect to productivity. During the past three decades, industry in Europe has made determined and successful attempts to catch up with the USA in productivity. As a result, total labour productivity increased from roughly 40% of that in the USA in 1960 to about 80% by 1990. Since then, however, European industry has been unable to close the gap further. One major reason is that the productivity strategies of many companies in Europe has concentrated on cutting labour costs and emplyoment, using traditional means of rationalisation. Only a very small section of European industry considers productivity across the entire organisation and business processes, engaging in systematic re-engineering, which would do more to increase competitiveness than the traditional tool of rationalisation (for Germany, this is demonstrated by Naschold 1997).

The European Competitiveness Report describes a successful competition strategy as one based on innovation, product differentiation and productivity rather than on reducing costs. In contrast, much of industry in Europe is structurally conservative. Many enterprises remain in markets where they have been successful in the past and concentrate on established core business rather than on developing new business. Export growth is created by demand growth rather than by increasing market share and export diversification. Driven by global competition, many companies are making determined efforts to reduce costs, but are failing to pursue organisational or process innovation. This also applies to R&D, where input-output ratios are often unfavourable. Last but not least, although Europeans tend to be proud of their well-educated and skilled labour force, human capital accumulation in many European countries is rather modest in comparison with that in the USA. Whilst these problems have as yet had no negative impact on business competitiveness, and will only exert their effects in the future, they have already severely damaged the competitiveness of European economies as a whole. They are a major cause of the high unemployment and comparatively modest economic performance of EU economies.

Both the Competitiveness Report of the EU Commission and the American Competitiveness Policy Council use living standards as the ultimate yardstick for the competitiveness of economies, and employment as an essential determinant of living standards. If we accept this yardstick, European economies do not seem to perform well. The standard of living, measured by per capita GDP, is still lower in the European Union than in either the USA or Japan. In the 1970s and 1980s, Europe and Japan narrowed the gap with the USA, but convergence came to a halt in the 1990s. Up to the 1980s, living standards in the European Union were higher than in Japan, but in the meantime Japan has overtaken the European Union. However, it should be noted that per capita GDP provides only a very limited picture of living conditions within a country. If we look at non-monetary indicators of the standard of living, we find that the competitiveness of the European economy may be somewhat better than is suggested by per capita GDP.

The central competitiveness problem of EU economies is employment. Here, the European Union is clearly under-performing relative to both the USA and

Japan. This is indicated by higher unemployment rates in the European Union and a much larger level of long-term unemployment which, in many European countries, is almost four times higher than in the USA. A far better indicator is job creation. Here, the European Union economies under-perform both the American and the Japanese economies. Between 1960 and 1995, the European Union created jobs at a rate of only 8%, whereas Japan achieved 44% and the USA 84%. Between 1960 and 1994, total employment in manufacturing and services increased by about 20% in Europe, whereas it doubled in the USA and Japan. Moreover, in the USA, growth is translated into employment more effectively than in Europe. In the USA, 1% growth in GDP leads to almost 1% increased employment, compared to less than a 0.5% in Europe. At the same time, the working population has been growing far more slowly in Europe than in the USA. The Competitiveness Report summarises the low performance of the European Union thus, "...in contrast with the USA where job creation has kept pace with the increased supply of labour, the European Union is facing relative employment stagnation despite less demographic pressure" (Commission of the European Union 1997, p.15).

There is a widely held assumption that the better employment performance of the USA is largely explained by the increase in cheap and low-skilled service jobs. However, this assumption is false. In the 1980s, many low-wage, low-skilled jobs were indeed created – particularly in services. However, many high-wage, high-skilled jobs were also created in the 1980s, and even more so in the 1990s. This applies to manufacturing as well as to services, where skilled employment grew in the finance, maintenance and health sectors. In this context, it is interesting to note that, in the USA, employment increased far less in those sectors characterised by a high proportion of low-wage, low-skilled jobs than in sectors with many high-skilled, high-wage jobs. In the service sectors such as transport and retail, low and even declining wages correlate with slow growth in employment. In contrast, other sectors, such as computing, combined increased wages with increased employment. This demonstrates that the widespread assumption that labour costs are a major determinant of job creation is misleading (Gaß and Schmid 1994).

The USA's positive employment performance can be ascribed to two major factors. Firstly, increasingly flexible working conditions allow more efficient organisation of labour and higher wage differentials. The result is that more low-skilled, low-wage jobs are retained, whilst high-skilled work which enhances value added is rewarded disproportionately. Secondly, the pursuit of "radical" innovation leads to new products and processes for new markets, enhancing structural shifts towards services and new industries. The second strategy has been particularly effective in the Clinton era, whereas the reduction of labour costs and deregulation of labour markets played a more important role in the Reagan era. In contrast to Europe, where public policy promotes innovation primarily via technology programmes, innovation policy in the USA has combined instruments for technology push and market pull (Lehner and Nordhause-Janz 1998).

In view of rising unemployment, there is an ongoing debate about the future of work itself. As Manuel Castells (1996, p. 255) points out, "the prophets of massive unemployment, led by the honourable Club of Rome, argue that ...our societies will have to choose between massive unemployment (...) or else a redefinition of work and employment". This is indeed the choice described in the report to the Club of Rome by Giarini and Liedtke (1998). But this is not the only choice we have. If the European economies had been as successful in generating jobs during the last two decades as the USA and Japan, there would be zero unemployment in the EU.

We have a better option in Europe: radical innovation and structural change. If innovation and structural change could be accelerated in Europe, if European industry were to be more successful in developing new markets and if the transformation of the European economies were to be driven more consistently and radically towards a knowledge-based and service-oriented economy, there would be no danger of a jobless future.

2 Trends and Directions in Global Change

Predicting social and economic change is difficult. All that can be done is to observe current trends, apply our theoretical knowledge to these observations and attempt to develop a reasonably coherent picture of what might emerge in the coming years. At present, we can observe two trends which may extend well into the 21st century: the knowledge-based economy, and the new highly customer-oriented service economy.

In recent studies, including that by the OECD, it is argued that the advanced economies are increasingly becoming knowledge-based economies (OECD 1996b; Neef 1998; Neef et al. 1998). This does not seem to be a new development – knowledge and information have always played an important role in the production of goods and services, and their importance has been growing for many years. What is new however, is that knowledge has become the principal factor of production. As Castells (1996, p. 17) argues, knowledge has finally become "the main source of productivity". Rather than being merely an auxiliary function, generating and applying knowledge is increasingly becoming the core of the production process. It is no longer simply a means towards efficient use of production factors, but has become the most important production factor itself. Knowledge constitutes the value added by products and services, whereas other production factors play a minor role. For example, biotechnologists are in the process of isolating those proteins which spiders use to construct their webs. In a few years, these proteins will be used to develop a material which may replace steel and other materials. The production of this new material does not require significant natural resources, nor does it involve much work. What it does require is considerable knowledge of the proteins concerned, the process to reproduce these proteins artificially and the application of the new material. Even in very traditional product lines, such as machine-tools, the knowledge incorporated in technology, design and software constitutes 80% of the total value. According to an OECD study, knowledge-based products already constitute more than 50% of GDP by the advanced economies (OECD 1996b, p. 9).

What is referred to as the knowledge-based economy is simply the result of a huge increase in knowledge production in the 20th century. If we look at the number of scientific publications, we can conclude that our knowledge base has doubled every seven to ten years. Similarly, the number of researchers has increased dramatically. Of all the researchers active during the history of humankind, more than half are working at present. Moreover, knowledge has been used to create modern information technology, which in turn helps to increase knowledge productivity at an even faster pace. As a result, we have far more knowledge, which can be distributed far more easily through the world than ever before, and can be used to generate new goods and services which we could not even have dreamed of a few decades ago.

Paradoxically, whilst knowledge has become the central production factor, it is not a scarce resource: it is in fact in ample supply. What is scarce is the ability to exploit knowledge for profit. This includes discerning market applications for new technological or organisational solutions, rapidly commercialising new products and getting them into regular production speedily. In the knowledge-based economy, commercial success (and with it economic growth), depends as much on intelligent marketing and organisation as on technological progress.

This illustrates a major problem in European economies. Europe has a high level of competence in knowledge and technology, but its industry is weak in developing new markets and organisational restructuring. This is not only a major source of unemployment, as noted above, but also constitutes a significant impediment to long-term industrial competitiveness in Europe.

The development of a knowledge-based economy is increasingly linked with the development of the service economy – another long-standing development. The approach adopted in the mainstream economic and sociological literature is that global change, industrialisation and a change in the division of labour in particular, force advanced economies to transform into service societies in order to maintain productivity (or to use the modern term: competitiveness). The argument is not merely that the share of manufacturing in total production and employment is declining whilst that of services is increasing. Rather, it is suggested that services are replacing manufacturing as the economic basis of society and that social structures and values will change accordingly.

For several decades, the share of services in both GDP and employment has increased in the advanced economies, whereas employment from manufacturing has stagnated or even declined in the last two decades. This is particularly true for the USA, where an overwhelming share of new jobs are service jobs – as mentioned above, these are high-wage, high-skills jobs. It should be noted, however, that many of these service jobs have been created in services to production, such as software and R&D, and in manufacturing industry itself. In many of the advanced economies, about half of all commercially produced services are production-oriented.

The important role of production services in the creation of new service jobs, particularly of high-wage, high-skills jobs, indicates that the established theory of a service society – namely that there is a simple transformation from industry to services – is misleading. Rather, what is observed is that manufacturing and services are becoming hard to distinguish: the traditional demarcation between manufacturing and services has become obsolete. In manufacturing, the service content of goods is rising, and in many cases outweighs the material content in terms of value added. In the service sectors, on the other hand, technical content is often high, and many services are produced with or for manufactured technologies and products. In addition, organisational forms and processes (such as lean production) developed in manufacturing are increasingly applied to the production of services. Even more important, in many industries "the product" is no longer purely manufactured, but is rather a bundle of goods and non-material services.

The convergence of manufacturing and services is also evident if we consider the characteristics which traditionally distinguished manufacturing from delivering a service. Apart from the distinction between manufactured goods as material and services as non-material, the major distinction was (and often still is) that services are personalised. They involve the customer as a direct participant in the production process. Currently however, many services, e.g. insurance and banking, are not in fact personalised, but are at best a mass product adjusted to meet customer requirements. A similar or even higher level of service content is to be found in many segments of manufacturing. Indeed, flexible production systems have greatly advanced the capability of manufacturing industries to customise their products. Beyond that, in a number of industries, new production concepts have been developed which attempt to achieve the full personalisation of products, including mass products. On the demand side, customisation and personalisation are, as Lehner et al. (1999) argue, increasingly supported by changing values and social structures.

With powerful trends towards customisation embedded even in mass production, as "mass customisation", a new service economy is emerging which can no longer be described in terms of the traditional demarcation between services and manufacturing. In the new service economy, customer utility will be the primary yardstick for the design of products – whether "services" or "manufactured goods" in the traditional sense.

Interaction between the two trends discussed here – the development of a knowledge-based economy and the development of a new service economy – may lead to a new type of economy, which Lehner and Schmidt-Bleek (1999) term a tailored economy. It is tailored in two respects: it is tailored firstly to the individual customer's needs and secondly to an efficient use of resources.

The economy which developed in the industrial era is a throughput economy, an economy in which the guiding principle is efficient production of large quantities of goods at low prices for mass markets. Growth results from volume and speed. The more and faster materials are transformed into manufactured products, the more and faster the economy grows.

The throughput economy was economically and socially (although not environmentally) adequate, as long as the material wealth of large parts of the population was modest and where large numbers of people did not own consumer goods such as cars and washing machines. Highly standardised mass production of somewhat modest technical and functional quality but great economies of scale, introduced to the automobile industry by Henry Ford, was highly successful in many industries until the 1970s, because it made the automobile and many other products genuine mass products. In advanced economies, this brought unprecedented prosperity.

As a result of this prosperity, the values, needs and demand itself in the advanced economies have changed over the past two decades. Technical and functional quality as well as non-material requirements have replaced simple material considerations. Consequently, mass production has lost impetus since the 1970s. The alternative seemed to be quality production, customised production of goods (and services) with high technical and functional quality. This, in a nutshell, was the argument of Piore and Sabel in their well-known book "The Second Industrial Divide" (Piore and Sabel 1984). They predicted a shift towards customised quality production in the advanced economies as a response to the crisis in traditional mass production.

Piore and Sabel certainly had a clear understanding of the dynamics of needs and demand, and of its implications for quality production. Nevertheless, their conclusion proved false. Instead of a dichotomy between mass production and customisation, new organisational forms emerged, and what might be termed "anthropocentric production systems" enabled the transfer of traditional mass production into new directions. These systems facilitated a highly flexible mass production which succeeded in combining the economies of scale and productivity of mass production with the advantages of quality production, namely high functional and technical quality and customisation. The result was a form of production which could be described as "mass customisation".

Whilst the development of mass customisation is still at a rather early stage, it nevertheless signals a way out of the throughput economy towards the highly customised economy described above. This is promoted, as Lehner and Schmidt-Bleek illustrate, by a number of economic, technical and organisational conditions which act as motors towards a customisation-based economy (Lehner and Schmidt-Bleek 1999, pp. 109–112).

The shift from a throughput economy to a customisation-based economy is of crucial importance for the environment. It fundamentally changes the relationship between the environment, growth and employment. In the throughput economy, growth was created by increasing the volume of production, which was associated with an increasing exploitation of nature. Environmental restrictions have therefore always limited growth. In a knowledge-based economy, this is no longer the case. Growth derives from the accumulation and exploitation of knowledge, and does not necessarily imply increasing the volume of production. Moreover, the use of natural resources can be substituted by knowledge then used to create artificial materials. Environmental restrictions create an incentive to accumulate and exploit more knowledge, and hence may act as a motor of growth and employment.

Before we discuss the economic opportunities opened up by an environmental policy systematically designed to drive structural change towards a knowledge-based, service-oriented economy, we turn to the case studies.

3 Environmental Policy in Europe: Missed Opportunities

In view of the argument outlined so far, three questions are particularly important in the case studies:

1. What is the impact of environmental policy on production concepts and company strategies in heavy industries and their customer industries?
2. Which kind of innovation pattern does environmental policy promote in heavy industries and their customer industries?
3. How does this affect structural change in the advanced economies?

As far as the first question is concerned, the case studies show clearly that environmental policy has an impact on technology, production concepts and marketing strategies in heavy industries. However, they show equally clearly that environmental policy (with the notable exception of chlorine production), is not a real driving force for new technology and organisational forms. Rather, the real driving forces have been broad changes in demand, competition and technological development, which occur independently of environmental regulation. In a few cases, in particular the Danish cement industry and fertiliser production in Austria, environmental regulation has significantly enhanced these driving forces and the resulting modernisation of the industries concerned. In most other cases, the impact of environmental policy on industrial modernisation was supportive but by no means decisive.

With respect to the second question, the answer is very simple: the types of innovation which can be observed in all cases are process innovations and incremental product innovations. As heavy industries operate in markets where there is over-capacity and increasing global competition, the major strategic goal is to secure competitiveness by means of rationalisation and cost-cutting. In a number of cases, cement and steel for example, the entire industry has shifted to higher quality products and niche markets, but innovation is still being developed along established product lines and has not been used to develop new businesses and markets. The type of environmental policy described in the case studies fits this pattern of innovation very well and may even reinforce it. Since regulation and other measures are moderate, they stimulate only incremental innovation in heavy industries and in their customer industries, not innovation which leads to new products and markets. Instead, some of the industries concerned are looking to foreign markets for the same products, in order to adjust to the increasing costs resulting from environmental regulation. This is currently a dominant trend – the products of heavy industry are globally traded commodities.

The other side of this coin is that environmental policy affects the competitiveness of heavy industries much less than is often feared. As the case studies generally indicate, environmental policy is not an important negative factor in the competitiveness of heavy industries.

The answer to the third question is even simpler. The type of environmental policy described in the case studies has no effect on structural change – it may not hinder change, but certainly does nothing to advance it. It is a type of policy which is integrated into dominant industrial structures and fits the incremental pattern of innovation and change predominant in most European industry. Contrary to the optimistic assumptions of Jänicke and others (in this volume), this condition does not lead to significant dematerialisation and certainly does not advance an economy in Europe which is environmentally, socially and economically sustainable.

We can see from the case studies that the concepts and behavioural patterns display striking similarities across sectors and regions. Although there is some national variation in policy measures and standards, environmental policies in Europe are still dominated powerfully by considerations of waste, emissions, pollution and toxicity, whereas more recent ideas on the overall use of natural resources and resource productivity play little or no part. The belief underpinning all these policies is that the problems of employment and competitiveness are so serious in Europe (and elsewhere) that environmental policy must be very moderate and cautiously designed. This is reflected in the behaviour of industries, where environmental policy is accepted and sometimes even supported as long as it is moderate.

In this way, European environmental policy becomes an integral part of the structural conservatism which dominates most European economic development. It is worth remembering that, in the 1960s and 1970s, environmental policy in a number of European countries was more radical, and forced a number of industries to significantly change their products and processes. Thus, environmental policy was able to induce considerable technological innovation and structural change.[3] The 1960s and 1970s was a period of growth and prosperity, where industry and the public could afford costly environmental policy and where changing values and the rise of green political movements supported such policy. Now, with the European economies performing less effectively, particularly with respect to employment, and with fiercer competition, environmental policies (and business strategies) are becoming more conservative.

There seems to be no other choice. The case studies, at least, say little or nothing about strategic choice, about what would constitute a profitable, environmentally sustainable alternative to current activities and businesses, or about what the role of environmental policy could be in these strategic alternatives. In the con-

[3] This is very well demonstrated for North Rhine-Westphalia, which is one of the oldest industrial areas, in Nordhause-Janz and Rehfeld 1995.

struction industry, for example, there is an interesting option: a more radical environmental policy could be used to develop new and more advanced technologies and production concepts, which would lead to customised and even personalised mass production. Such production would cut costs considerably whilst increasing quality. Above all, it would significantly improve the environmental quality of the construction industry – at present the most environmentally damaging sector.

This is not simply a deficiency in the case studies, but represents the widespread structural conservatism in the economies of the European Union. Structural conservatism is, as previously argued, the root cause of industrial employment and competitiveness problems in Europe. It also characterises the widely shared view of the relationship between environment and economy among business people, politicians and union representatives as well as among environmentalists: that there are environmental limits to growth and employment which create a conflict, or at any rate a necessary trade-off, between environmental priorities and economic imperatives. As previously discussed, we believe that this assumption is no longer valid for a knowledge-based economy.

4 The Environment as an Engine of Structural Change

The cases studies in this volume clearly demonstrate that, given the dominant environmental policy in EU member states, we will remain within economic structures which effectively damage employment and the competitiveness in the advanced European economies. What we need is an economically feasible alternative – "Factor 10".

When Frank Bosshardt prepared the World Business Council for Sustainable Development's statement to UNCED in 1992 at Rio de Janeiro, he developed the concept of eco-efficiency. Eco-efficiency means that goods and services are produced at competitive prices which satisfy the needs of people and improve their quality of life, while simultaneously reducing their environmental impact and the use of natural resources throughout their life-span. About a year later, Friedrich Schmidt-Bleek developed Factor–10, which takes this idea much further. Factor–10 basically looks for a ten-fold increase in resource productivity, or put simply: the same amount of goods and services to be produced using a tenth of the natural resources (Schmidt-Bleek 1994).

What makes the Factor–10 concept particularly promising for both environment and economy is firstly, that it introduces the concept of an ecological rucksack for each product, and secondly that it advances the MIPS-concept. The idea of the ecological rucksack simply points to the fact that many of the resources which we extract from nature are not actually used in final products, but are simply wasted in the process of production. In order to get one ton of copper, some 500 tonnes of natural resources are used, whereas some 70 tonnes are exploited for a tonne of steel. This points to a way for massively reducing environmental damage without sacrificing economic utility – we need to produce goods from materials with small rucksacks, and develop technologies to keep all rucksacks as small as possible. As many promising examples demonstrate, rucksacks can often be dramatically reduced through advanced technology and innovative design.

MIPS stands for Material Input Per Service Unit, the material input per unit of utility. The message contained in the MIPS-concept is simple: try to extract from every unit of natural resources the maximum utility for each customer. This simple guideline shows how we can improve resource productivity not only through reducing material inputs to production, but also by increasing the utility derived from this material input. A classic example is the photocopier, which is built so that the technologically sensitive parts and those parts which have heavy wear and tear can be replaced easily, while the rest of the machine can function over a considerable period. With this device, material input for photocopying is drastically reduced and utility is increased, as the machines can easily be modernised. A modern aeroplane is built in a similar way, and has a product life of more than thirty years, while it can be continuously upgraded to current technical standards.

There are many practical examples from a variety of industries which demonstrate that Factor–10 is economically feasible (see for example Fussler 1996, Lehner and Schmidt-Bleek 1999). This may often be radical, in the sense that it cannot be implemented with the current state of technology and organisation, but it offers a solution to the problems of competitiveness and employment in the advanced economies, because dematerialisation in the economy by a factor of 10 (or more) goes hand in hand with the development of a knowledge-based economy and a new service economy. It calls for replacing material consumption with knowledge and better service. There is a largely unexploited synergy between dematerialisation, new employment and the knowledge-based economy.

With innovative strategies, the Factor–10 concept offers a promising avenue for the advancement of the knowledge-based economy and the new service economy, and thus for sustained growth and full employment in the advanced economies. This also provides reasonable opportunities for poorer nations and newly industrialising countries to industrialise and develop – and thus sets a reasonable trajectory for globalisation.

References

Castells, M. (1996): The Rise of the Network Society. The Information Age: Economy, Society and Culture, Vol. 1. Malden-Oxford: Blackwell.

Charles, T. and Lehner, F. (1998): Competitiveness and Employment: A Strategic Dilemma of Economic Policy. Competition and Change 3, 207–236.

Commission of the European Union (1997): The Competitiveness of European Industry. Luxembourg: Office for Official Publications of the European Communities.

Competitiveness Policy Council (1992): Building A Competitive America. First Annual Report to the President and Congress. Washington.

Fussler, C. (1996): Driving EcoInnovation. A Breakthrough Discipline for Innovation and Sustainability. London, Hongkong, Johannesburg, Melbourne, Singapore, Washington: Pitman Publishers.
Gaß, G. and Schmid, A. (1994): Lohn- und Arbeitsmarktentwicklung in den U.S.A. WSI Mitteilungen, 290–299.
Giarini, O. and Liedtke, P. M. (1998): Wie wir arbeiten werden. Der neue Bericht an den Club of Rome. Hamburg.
Lehner, F., Bierter, W., Charles, T. (1999): Resource productivity, competitiveness and employment. A Factor 10 Club paper. Gelsenkirchen: Institute of Work and Technology.
Lehner, F. and Nordhause-Janz, J. (1998): Beschäftigung durch Innovation: Strategische Optionen im globalen Strukturwandel. In: Lehner, F., Baethge, M., Kühl, J., Stille, F. (ed.): Beschäftigung durch Innovation. München-Mering: Hampp, pp. 59–97.
Lehner, F. and Schmidt-Bleek, F. (1999): Die Wachstumsmaschine. Der ökonomische Charme der Ökologie. München: Droemer.
Naschold, F. (1997): Ökonomische Leistungsfähigkeit und institutionelle Innovation – Das deutsche Produktions- und Politikregime im globalen Wettbewerb. In: Naschold, F., Soskice, D., Hancké, B., Jürgens U. (ed.): Ökonomische Leistungsfähigkeit und institutionelle Innovation: das deutsche Produktions- und Politikregime im globalen Wettbewerb. Wissenschaftszentrum Berlin Jahrbuch 1997. Berlin: Edition Sigma.
Neef, D. (ed.) (1998): The Knowledge Economy. Boston, Oxford, Johannesburg, Melbourne, New Dehli, Singapore: Butterworth-Heinemann.
Neef, D., Siesfield, G.A., Cefola, J. (eds.) (1998): The Economic Impact of Knowledge. Boston, Oxford, Johannesburg, Melbourne, New Dehli, Singapore: Butterworth-Heinemann.
Nordhause-Janz, J. and Rehfeld, D. (1995): Umweltschutz "Made in NRW". Eine empirische Untersuchung der Umweltschutzwirtschaft in Nordrhein-Westfalen. München: Hampp.
OECD (1994): The OECD Jobs Study: Evidence and Explanations; Part I: Labour Market Trends and Underlying Forces of Change. Paris: OECD.
OECD (1996a): Technology, Productivity and Job Creation. Analytical Report. Paris: OECD.
OECD (1996b): The Knowledge-based Economy. Paris: OECD.
Piore, M.J. and Sabel, C. (1984): The Second Industrial Divide. Possibilities for Prosperity. New York: Basic Books.
Schmidt-Bleek, F. (1994): Wieviel Umwelt braucht der Mensch? Berlin et al.: Birkhäuser.

B. The Case Studies

Chlorine Production in Germany

Klaus Jacob
Free University of Berlin, FFU, Ihnestr. 22, 14195 Berlin, Germany

1 The Development of German Chlorine Production

1.1 General Features of the German Chemicals Industry

Next to automobile manufacture and mechanical engineering, the chemicals industry is Germany's third largest industrial sector. In 1995, its most significant lines in terms of value produced were basic chemicals (DM 74.7 billion) including plastics (DM 28.2 billion) and pharmaceutical products (DM 32.5 billion) (VCI 1996a). In 1995, total production was DM 169 billion (ibid.), occupying about 536 000 employees. Since 1991, when 716 000 people were still working in the chemicals branch, the number of employees has therefore decreased significantly. This loss of 180 000 jobs is due in part to dramatic cuts in employment in the former East Germany. The chemicals industry in Germany consists of around 4 000 companies, of which 100 are large concerns, employing more than 1 000 and accounting for about 2/3 of total turnover.

In 1994 the chemicals industry accounted for 10.3% of total turnover by processing industries. However, it employed only 8.3% of the total workforce, paying out 10.1% of total wages, exporting 13.9% and importing 9.2% of total trade within the processing industries (VCI annual report 1995, own calculations), indicating high labour productivity and export orientation. In 1994, 54% of total turnover was exported, compared with 40% of the turnover by the processing industries overall (VCI annual report 1995).

Until the end of the 1980s, the chemicals industry in post-war Germany grew faster than other branches: whereas its share of GNP by the processing industries was 3.6% in 1936, today it is over 10% (Amecke 1987, p. 28). In the recession year of 1993, turnover fell off dramatically. Reaction varied to the sometimes considerable structural change within the branch which accompanied this decline. While some companies (e.g. Hoechst) sought to expand their "high-chem" sector, i.e. pharmaceuticals or other specialist products, other companies (e.g. BASF, Hüls, Dow) concentrated on their core business of producing and processing basic materials. Both strategies were accompanied by significant reductions in the workforce (FAZ 10.4.95, 21).

Until about 1970, investment was 10% of turnover, after which it fell to 5% by the mid-1980s. From the 1990s on, it rose again slightly to 6–7% by 1992, but had fallen again to only 5% in 1993. Expenditure on research and development (R&D)

Table 1. Sales of chemical products

Customer sector	share
chemicals	21.5%
health	10.6%
transport	8.8%
construction	8.3%
private consumption	8.0%
textiles	6.2%
agriculture	5.8%
packaging	5.4%
electrical goods	4.4%
printing	2.1%
paper and paper products	2.0%
iron and steel	1.8%
food, beverages and tobacco	1.5%
machinery	1.4%
petroleum products	1.3%
other	8.9%

Pohle 1991, p. 9

has remained constant at about DM 10 billion p.a. in recent years, its share of turnover has now fallen to 5.8% (VCI annual reports of several years).

The chemicals industry supplies almost every other branch of industry, but the most important customer for chemical products is the chemicals industry itself (Table 1). The comparatively low level of private consumption, 8% of sales, qualifies the chemicals industry as a primary industry, involved with early stages of production; the broad range of customer sectors qualifies it as strategically important.

The three largest companies in the German chemicals industry – Bayer, Hoechst and BASF – account for about 30% of total turnover in the chemicals branch (Pohle 1991, p. 41) and at the same time rank among the largest industrial concerns in Germany (they also have the highest turnover world-wide with respect to the chemicals industry). Chlorine production is a highly capital-intensive task, and capacity is therefore concentrated in the sector's largest companies (Table 2).

1.2 The Significance of Chlorine for the Chemicals Industry

Alongside oil products, which find their way into the chemicals branch as naphtha, chlorine is the most significant primary substance for the chemicals industry. According to the VCI, the production of around 60% of chemical products (in terms of both quantity and value produced) involves the use of chlorine (Meerkamp v. Embden 1992, p. 6). Nader (1996) points to a figure of 300 basic chemical substances and 30 000 "refined" products made with chlorine. In many product lines, chlorine is used only as an auxiliary agent and forms no part of the end product. Except for the relatively insignificant application (in terms of consumption) of chlorinating drinking water, pure chlorine is not considered an end product, because it is rather used by the branch itself. Its share of turnover, including products made with auxiliary chlorine, is DM 100 billion. Taking an average price of DM 320 per tonne of chlorine (Nolte and Joas 1992, p. 45), however, direct production value amounts to just DM 1 billion[1].

[1] The value of NaOH (also about DM 1 billion) should be added since both chemicals are produced in the same process.

Table 2. Turnover by the major chemicals enterprises and chlorine manufacturers

company	turnover	owner	capacity[a] (kt/a)
BASF	DM 48.8 bn. (1996, world-wide)	public shareholders (ca. 375 000)	360
Bayer	DM 48.6 bn. (1996, world-wide)	public shareholders (ca. 295 000)	938
BSL Olefinverbund	DM 760 mil. (1992)	80% Dow Deutschland Inc. 20% BvS	200
Chemie GmbH Bitterfeld-Wolfen	DM 170 mil. (1995)	100% BVS[b]	118
Dow	US$ 20.0 bn. (1996, world-wide) DM 3.7 bn. (Germany)	Dow Inc. (USA)	1023
Elektro-Chemie Ibbenbüren	DM 148 mil. (1993)	Preussag AG (50%) Akzo (NL, 20%), ARBEGO (NL, 20%), Ribaja (NL 10%)	120
Hoechst	DM 50.9 bn. (1996, world-wide)	Kuwait Petroleum Corporation (<25%), Rest public shareholders (330 000)	471
Hüls	DM 10.3 bn. (1996, world-wide)	VEBA AG (public shareholders)	180
ICI	£10.5 bn. (1996, world-wide) (DM 2.58 bn.. 1992, Germany[c])	ICI GB	130
Solvay	DM 13.7 bn. (1996, world-wide) DM 3.7 bn. (1996, Germany)	Solvay Belgien	215
Wacker	4.1 bn. DM (1996, world-wide)	Hoechst AG (50%) Family (50%)	157

The companies mentioned account for most of the chlorine capacity in Germany.
[a] chlorine capacity in Germany
[b] in January 1997 the Elektro-Chemie GmbH, Ibbenbüren was about to take over the company (Chemische Rundschau No. 1 (10.1.1997))
[c] PVC production was sold off in 1994.

Sources: SRI 1996; Schmacke and Jaeckel 1996; Hoppenstedt 1997; Kompass 1994; FAZ 22.6.1993, 27.6.97; annual reports of Bayer, Hüls, Hoechst, BASF, Solvay, ICI, Dow.

In Germany, 92% of chlorine is processed directly where it is produced. Foreign trade in chlorine is scarcely worth mentioning, nor is its share in the production which is meant for sale (Table 3). The most important countries exporting to Germany are Austria and the Benelux countries; exports from Germany are almost exclusively to Belgium and the Netherlands.

The proportion of chlorine processed internally is above average in international terms, as ca. 20% of the produced chlorine is generally traded world-wide (Seidel 1993). In the U.S., for example, 25% of the estimated production capacity of 13 billion tonnes p.a. is used outside the chemicals industry (Hileman et al. 1994). This comparatively high proportion can be explained by a higher level of drinking water chlorination and by exports of chlorine to countries without a chlorine industry of their own. But the German industry hardly participates in this trade (Schlegel 1993), and it is rather intermediate and end products of the chlorine industry which provide foreign trade from Germany.

Table 3. Foreign trade in primary chlorine

year	import (t)	= of total[a]	export (t)	= of total[a]
1987	119 075	3.45%	48 096	1.39%
1988	145 687	4.16%	75 910	2.17%
1989	114 041	3.31%	73 837	2.14%
1990	145 603	4.47%	70 245	2.16%
1991	115 546	3.81%	58 028	1.91%
1992	117 955	4.42%	45 765	1.71%
1993	87 615	3.07%	36 600	1.28%
1994	118 359	3.77%	56 458	1.80%
1995	95 872	2.92%	57 215	1.74%

[a] total chlorine production in Germany; Source: StaBu FS 7, Reihe 2, several volumes.

1.3 International Change in Chlorine Production

As shown in Table 4, the per capita production by chlorine-producing countries world-wide varies widely. These countries can be divided into three groups. Firstly, there are countries whose more or less stable per capita production is less than 10 kg/a (e.g. Australia, Denmark, Greece, Turkey). These countries usually have a single plant of below-average capacity. Some of them (like Australia and Denmark) have already ceased chlorine production. Complete data on imports of chlorine or its immediate secondary products are not publicly available. The second group of countries (e.g. UK, Spain, Italy, the Czech Republic) has a per capita production of 10 to 20 kg/a. Production here expanded considerably in the late 1960s and early 1970s, but has since stagnated. In these countries chlorine is most often produced in several plants of average capacity. Lastly there is the group (e.g. the USA, Canada, Germany) of major chlorine producers (20 to 60 kg/a), which experienced rapid growth in production up to the beginning of the 1980s. Most of the dynamism in the chlorine industry at the end of the 1980s and the beginning of the 1990s can be found within this group. Production in the Scandinavian countries has declined to such an extent that a large number of plants have had to be closed down. There are only a few countries which fail to fit into this scheme: at the end of the 1970s, Norway rapidly grew to one of the major producers of chlorine, while in the mid to late 1980s, production fell by a comparable amount in Bulgaria and Romania.

Almost all the major chlorine-producing countries faced a decline in production following the prohibition of CFCs. Major investment in the chlorine processing industry became necessary, for example building high temperature incinerators for the chlorinated waste which had been processed by chlorolysis. The Scandinavian countries seem to have phased out chlorine production completely, while growth in production has been observed in Germany since then (due mainly to a rise in PVC and polyurethane consumption, see below).

Table 4. Chlorine production per capita in industrial countries 1960–93

data in italics: based on time-series with missing values	kg/cap annually 60-72	73-86	87-93	annual growth rate 60-72	73-86	87-93
Canada 1960–89	27.6	47.7	*62.8*	7.4%	4.3%	*0.1%*
Norway 1960–92	10.8	32.6	*51.0*	8.3%	9.8%	*-1.9%*
Belgium 1986–93		*43.9*	45.6			*-0.4%*
USA	33.2	42.1	43.1	5.3%	-0.2%	1.1%
Germany 1960–64, 1970–93	*20.3*	*44.4*	43.0	*8.2%*	*4.1%*	-2.2%
Sweden 1960–92	30.5	42.6	*34.7*	5.5%	0.8%	*-12.1%*
Finland	22.5	40.8	25.3	9.8%	2.1%	-23.6%
France	20.0	31.5	24.2	9.9%	-0.6%	-0.8%
Ex-Czechoslovakia 1966–91	*4.2*	17.7	*19.2*	*19.7%*	7.7%	*-2.4%*
United Kingdom 1974–91		*16.6*	17.3		*3.4%*	-2.2%
Italy	11.7	16.2	17.0	10.9%	-0.1%	-1.1%
Romania 1960–80, 1986–93	6.0	*18.9*	16.2	26.7%	*5.9%*	-15.6%
Spain	2.8	11.6	15.1	22.9%	6.1%	-0.6%
Poland	4.0	8.5	8.6	11.0%	3.5%	-4.8%
Portugal 1970–92	0.8	4.6	*7.4*	112.8%	18.0%	*-0.3%*
Japan	2.8	6.2	7.9	8.6%	4.2%	0.1%
Bulgaria 1985–93		*19.4*	6.5		*-10.5%*	-12.8%
Ex-Yugoslavia Sfr 1970–93	*2.4*	3.8	4.7	*8.7%*	8.6%	-9.4%
Greece 1968–92	*2.4*	3.1	*3.3*	*-5.2%*	-2.3%	*4.6%*
Hungary 1960–91	1.4	1.8	*2.3*	46.1%	3.1%	*13.2%*
Denmark 1966–84, 92	1.5	2.7	0.6	9.9%	7.8%	
Turkey 1970–93	*0.3*	0.5	0.7	*31.8%*	183.6%	29.8%
Australia 1960–79	5.0	*7.9*		9.9%	*0.3%*	

Source: UN, StaBu, own calculations

1.4 Structural Change in Germany's Chlorine Production and Use

After having grown steadily until the mid-1970s, German chlorine production fell, due to the general recession, but recovered again by the mid-1980s (Fig. 1). Between 1983 and 1989, the production rate stabilised at a level of 3.5 Mt/a., falling again to 2.67 Mt by 1992, although since 1991, production in the former GDR is also taken into consideration (about 600 kt/a up to 1989). The decline in the former GDR was more drastic because of the shutdown of several chlorine plants, but there is one new plant under construction. Since the beginning of the 1990s, consumption of chlorine has been rising again, but without regaining its former levels.

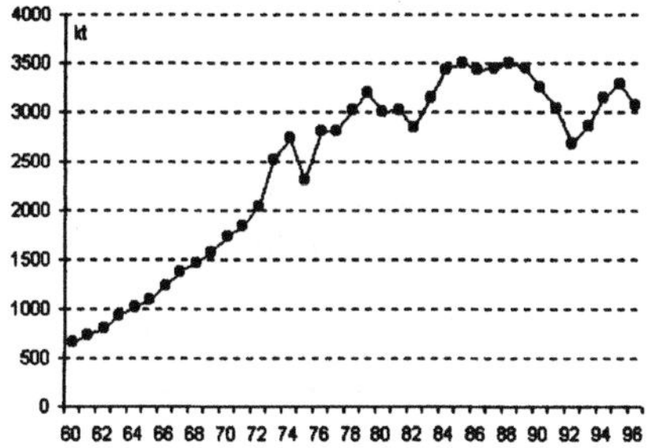

Sources: UN; StaBu: Produktionsstatistik

Fig. 1. Chlorine production in Germany 1960–96

Chlorine's high reactivity makes oxidation processes easier. Used as an auxiliary agent, chlorine is most often emitted in the form of common salt. In other processes, chlorine is used within closed material flow systems, so that only a small quantity of chlorine is emitted into the environment (Table 5). The majority of chlorine is used in processes which are open to the environment, but the quantity of chlorine in end products which can get into the environment has declined significantly – with the exception of PVCs (Table 6).

PVC, with its associated commodity chemicals ethylendichloride and vinylchloride, is the segment which uses the most chlorine. In second place are the pre-products of polyurethane, which return the used chlorine as salts (NaCl/KaCl) or HCl. In 1992, third place was taken by CFC pre-products, i.e. TETRA and, to a lesser extent, components in chloroform and perchloroethylene production. These uses of chlorine for CFC production have now been completely discontinued. The chlorinated solvents segment was and still is another significant user of chlorine (see perchloroethylene and dichloromethane in Table 6).

Between 1987 and 1992, there was a significant decline in PVC, CFC and chlorinated solvents production, whereas slight growth could be seen in PUR production. The decline in PVC production has since been largely compensated, due to growing demand from the building and construction industry; short-lived PVC applications, however, such as packaging materials, fell back tremendously.

Chlorine is increasingly being used as an auxiliary agent: With the exception of one propylenoxide producer – who has ceased production on the grounds that they wanted to avoid salt emissions (Kleinloh 1994) – all other producers of these product lines argue that this form of pollution is ecologically negligible in comparison to other emissions of salt or its natural incidence in the ocean. They reject the idea of reprocessing these salts and the possibility of feeding them back into chlorine-alkali electrolysis because of the high associated energy requirement (Bayer AG 1995, p. 24).

This modernisation of the chemicals industry can also be demonstrated by analysing the substance reports which are published on existing chemicals up to the: use of halogenated organic substances where they can enter the environment is

Table 5. Types of application for chlorine and chlorinated chemicals

	open to environment	closed systems
chlorinated chemicals as auxiliary agents	emissions most often as NaCl	certain processes, where most often HCl, in some cases also Cl is used[a]
Cl as part of final products	e.g. PVC, dichloromethan as paint stripper, chlorinated solvent in *old* degreasing processe	e.g. chloroform as extraction agent in pharmaceutical production, chlorinate solvents in *new* degreasing processes

[a] For example in a new technology at Hüls AG for producing silicone (Kleinloh 1994, p. 63), a technology for producing Bisphenol A, developed by Bayer AG or chlorine-free technologies for producing polycarbonate (Zlokarnik 1996, p. 469).

more often in decline than the use of chlorine-free products. Table 7 presents a summarised evaluation of the 182 substance reports published by the BUA[2] up to the end of 1996.

Table 6. Consumption of chlorine in final products and production segments, 1987–92

	kt[a] 1992	% of total	change 87–92	residual Cl	most important final product
ethylen-dichloride	1 249	32.6	-11.0%	mostly in final product; little as waste	PVC
propylen-oxide	758	19.8	+4.3%	as NaCl or KaCl	polyurethane
phosgene	286	7.5	+8.3%	as HCl; as salt	polyurethane, polycarbonate
tetrachloro-methane	219	5.7	-36.0%	mostly in final product; partially as HCl and waste	CFCs
miscella-neous[b]	128	3.4	-41.0%		
allyl-chloride	120	3.1	+5.3%	in final product and as HCl (65% of the chlorine used)	epichlorohydrine, epoxy resins (Cl-free)
perchloro-ethylene	104	2.7	-57.4%	mostly in final product; partially as HCl and waste	mostly used as final product (solvent)
dichloro-methane	88	2.3	-56.7%	mostly in final product; partially as HCl and waste	final product: solvent, paint stripper, etc.
chloroform	77	2.0	-12.5%	partially as HCl and waste	final product: solvent in pharmaceutical industry
chloro-benzol	72	1.9	-6.5%	partially as HCl and waste; about 50% in final product	various, also chlorine-free products in pharmaceuticals, colours, pesticides, etc.
epichloro-hydrine	71	1.9	0.0%	as salt (65% of the chlorine used)	epoxy resins (Cl-free)
methyl-chloride	55	1.4	-6.8%	in final product	intermediate for silicon production (chlorine-free)
vinyl-chloride	38	1.0	-67.2%	partially in product and as HCl	PVC
R113/R114	3	0.08	-40.0%	as HCl (50%) and in final product	CFCs
Total	3 268	85.2	-11.1%		
Overall	3 836				

[a] measured as chlorine equivalents: the indicator includes recycled chlorine (mostly HCl) and is therefore higher than the total production volume

[b] monochloroacetic acid, dichlorobutene, chlorotoluols, chlorinated paraffin, bleaches

Source: Nolte and Joas 1995, own calculations

[2] Beratergremium Umweltrelevante Altstoffe: A council whose members represent the government, science and industry. They collect data and publish reports on so-called "existing" chemicals in order to advise regulators in drafting further regulations. These chemicals were on the market before the chemicals acts were enacted at the start of the 1980s.

- Although the chemicals industry claims that 60% of turnover and production volume depends on chlorine, the halogenated substances appear to be slightly *underrepresented* (52%) in the substance reports published to date. This figure represents only the number of substances reported and not the value or volume of the chemicals produced.
- Substances containing halogens are used significantly more often as a purely intermediate product. The analysis cannot say whether the end products still contain chlorine or other halogens, and whether these substances are emitted during further processing.
- Chemicals containing chlorine are more frequently proven or suspected to be carcinogenic than chlorine-free substances. However, the possibility cannot be ruled out that carcinogenic properties have been more thoroughly investigated in this substance group.
- The use of particular chlorine-based substances has, in the past, declined significantly more often than that of other substances, especially when used in open situations.
- The use of chemicals containing chlorine is more often regulated than that of comparable groups. Different forms of regulation are considered here, ranging

Table 7. Chlorinated chemicals and processes listed in BUA reports

	n	declining[a]		explicitly regulated[b]		carcino-genic[c]		used in open environment[d]		*of which: declining*	
		%	n	%	n	%	n	%	n	*%*	*n*
halogenated chemicals	70	50%	35	31%	22	44%	31	34%	24	*75%*	*18*
Cl as auxiliary agent	25	24%	6	16%	4	12%	3	60%	15	*27%*	*4*
halogen-free chemicals	87	28%	24	25%	22	28%	24	63%	55	*35%*	*19*
all chemicals	182	36%	65	26%	48	32%	58	52%	94	*44%*	*41*

[a] Consumption data for a time series is not always included in the BUA reports. A chemical has been included under this category if there was a decline in consumption (or production, where there was no indication of a substitution by imports) of >10% in the past ten years. In a few cases, qualitative statements like "sharp decline in consumption" in the reports have been considered equivalent evidence.

[b] Included are any explicit regulation in one or more decrees based on any chemicals-related law in Germany and formal voluntary agreements. Not included are classifications in lists drawn up on the basis of these laws, since such classifications most often do not influence production volumes.

[c] All possible classifications (whether of proven or suspected carcinogenicity) were counted together.

[d] Included are also those chemicals with only a small share of production which are used in an open environment.

Sources: BUA reports 1985–96 (1–186), author's analysis.

from the restriction of certain uses under food hygiene laws[3] to complete bans on production or use[4].

The high incidence of regulation among these substances, and the frequency of declines in their use suggest a causal relationship. Case studies of especially strictly regulated substances, however, have shown that this assumption only holds in exceptional cases (Jacob 1999).

1.5 Social and Economic Consequences of Falling Production

The loss of jobs, turnover and capital or R&D investment for producers affected by declining chlorine consumption is hard to estimate. Relatively few jobs in production and the early stages of chlorine processing have been affected by the potential closure of plants. As Haas (1990, p. 83) has calculated, the distribution of costs in a typical chlorine production plant is only 4% on labour, 35–50% on energy and 13–15% capital investment. Nor are chlorolysis plants for the production of PVC and PUR very labour-intensive. Most of the jobs affected by the closure of chlorolysis plants have been replaced in other sectors (e.g. Hoechst AG in TAZ, 29.6.1993, p. 7).

Since the second half of the 1980s, all the companies active in chlorochemistry have been investing heavily in the construction of incineration plants, making chlorolysis (with its by-products of CFC and chlorinated solvents) unnecessary. The considerable capital and R&D investment in finding a substitute for CFC – HCFC – do not seem to have paid off, as the technology for this has since been sold on.

PVC and polyurethane are also *produced* with low labour intensity, but in capital-intensive plants. The total number of jobs in PVC production is estimated at 6 000 (Plinke et al. 1994), whereas the number of jobs in PVC *processing* is ten times higher. In the processing industries, however, the number of jobs does not generally depend on which material is used – it is even estimated that substituting PVC with wood would bring positive net effects (Plinke et al. 1994).

2 Environmental Effects of the Production and Use of Chlorine

Since the end of the 1970s, the production and use of chlorine has been an issue in environmental debates. At first, substances which are highly toxic to humans (DDT, PCBs, dioxins) were the centre of attention, as well as those – such as CFCs – whose extreme persistence in nature represents an ecological threat. In the 1980s, the debate spread to cover other mass products, for instance:

---

[3] e.g. a limit of DCM in cosmetics or the prohibition of vinylchloride as a propellant: In these regulations the major application is not taken into account.

[4] e.g. PCP or CFCs

- chlorinated solvents
- PVC and its additives
- chlorine bleaches for processing paper and textiles

This debate embraces a number of different goals and environmental media: while their high toxicity to humans was central in the discussion of dioxins, the toxicity of CFCs is negligible[5]. Despite this broad range of possible effects, there is nonetheless a number of reservations about the use of chlorine which cast a similar pall of suspicion over the entire substance group.

2.1 The Risks of the Products

Chlorinated organic chemicals do not generally occur in nature (e.g. AKN 1995, p. 9), with the exception of methyl chloride, which is a metabolic product of certain algae, and the natural occurrence of which is thought to exceed industrial production. Particularly problematic is the fact that there are hardly any organisms capable of decomposing these substances (AKN 1995, p. 10). Even today, we do not fully understand the mechanisms by which microorganisms decompose such chemicals: in the case of dichloromethane (DCM), a frequently used chlorinated solvent, for instance, there are possibilities for aerobic decomposition. It is assumed, however, that extensive adaptation to this substance was required on the part of the organisms, which was only made feasible by virtue of its industrial use (cf. Leisewitz and Schwarz 1994, p. 242). Another factor in blanket suspicion of this group is the possibility of the emission of dioxins or furanes when waste is burnt or when chlorinated organic substances catch fire. The chemicals industry counters that natural occurrence and formation of chlorinated organic compounds have not yet been sufficiently explained. It is assumed that the presence of sodium chloride (common salt) during incineration is enough to cause the emergence of dioxins. This would explain why dioxins can even be shown in sediments dating from pre-industrial times (Bayer AG 1995, p. 18). Of the 2 500 to 3 000 naturally occurring substances discovered every year, mostly in pharmaceutical research, 30 to 40 are chlorinated organic compounds (Bayer AG 1995, p. 17).

Chlorinated organic compounds can often be found on lists of substances which are harmful to the environment (e.g. European directive 76/795/EEC and its list of 129 chemicals, the priority list of the BUA, substance lists from the Conference for the Protection of the North Sea). It is very frequently assumed that these substances are carcinogenic (Henschler 1994).

2.2 Risks from Production

Another long-term threat to the environment is the high consumption of electricity during the production of chlorine. In 1987, production of 3.45 million tonnes of

[5] The Federal Health authorities (BGA) stated in 1988 that the possibility of carcinogenic properties in the CFC R-22 cannot be ruled out (Broecker 1988, p. 11)

chlorine required 10 million MWh, 2.8% of total electricity consumption in Germany (Nolte and Joas 1992, p. 2). The relatively low electricity prices for chlorine producers are sometimes suspected of being hidden subsidies (e.g. Gremler 1991, p. 208). On the other hand, chemicals companies are increasingly becoming energy suppliers themselves.

The high reactivity of the intermediate products in chlorine chemistry, often linked to acute and chronic toxic properties for humans, leads to a heavy risk to humans and the environment from production. The severity of many accidents in the past (Seveso, Boehringer Hamburg, Bophal) can be linked to chlorine and its by-products (Winteler and Ahrens 1991, p. 8). The EU Commission's Major Accident Hazards Bureau (MAHB) states that, in about 10% of accidents recorded, the element chlorine was emitted from stationary plant (Hendriks and Papameltiou 1996, p. 11). The potential threat from individual by-products such as phosgene could be even more problematic.

2.3 Industry's View: a Positive Net Impact

Representatives of the chemicals industry cast doubt upon the existence of common features for chlorinated organic compounds and – pointing to the wide variance in acute toxicity – demand that each substance be analysed individually (e.g. Meerkamp v. Embden 1991, p. 50; Nader 1996). They also deny blanket ascriptions of slow decomposition rates (e.g. Fricke et al. 1995, p. 279).

On the contrary, they favour the large-scale use of chlorine, basing their arguments on what amounts to blanket praise for the substance group:

- The high reactivity of chlorine makes selective production possible, helping to minimise the accumulation of by-products.
- In spite of their high energy requirement, chlorination processes are usually exothermic, and thus contribute to what may be a net energy gain.
- Considering the fact that reserves of the raw materials for chlorine (NaCl in particular) are virtually inexhaustible – in contrast to the raw materials for petrochemistry – substitution of PVC by purely petrochemical synthetic substances (e.g. polyethylene), with their heavy oil consumption, cannot be justified. In many cases, more energy is required to produce equivalent quantities of possible PVC substitutes.
- In contrast to many primary materials in the petrochemicals industry, chlorine is not explosive.
- Chlorine was not involved in a great many of the serious chemicals accidents in the past (e.g. Bophal[6]).
- In general, production costs are lower than those for the proposed substitutes[7] (e.g. Deutsch 1993, p. 32; Bayer AG 1995; Budzinski 1993, p. 13; Schlegel 1993; Nader 1996).

---

[6] In fact, methylisocyanate was emitted, which is part of the production of polyurethane and is produced by, among others, phosgene (CO-Cl).

3 Political Processes and Possibilities for Intervention

3.1 Governmental Actors: Policy by Research

In contrast to other industrial branches, the chemicals industry receives extremely little in the way of direct subsidy. The level of regulation, however, is comparatively high: plant and processes, products, waste and emissions are all subject to more or less complete government licensing and control.

Chlorine came onto the political agenda when several studies of the possible health and environmental effects of several chlorinated chemicals were published, starting at the end of the 1970s. Substances like DDT, PCB, PCP were discussed in public, but it was the Federal Environmental Agency (UBA), along with some environmentalist associations, who extended the discussion to chlorine itself at the beginning of the 80s. Further studies were commissioned and preliminary policy goals were formulated. These studies and policy goals have had a considerable effect on the sector, as is shown in the following sections – and in more depth in other studies (e.g. Jacob 1999).

The debate at the level of a basic chemical like chlorine was also motivated by the enactment of the Chemicals Act in 1980. There was a large number of different industrial chemicals with unknown toxicological properties, and these were perceived as an environmental and regulatory problem for the enforcement of chemicals law. It was estimated that between 40 000 and 100 000 chemicals were on the market, most of them without any scientific data on the environmental risks. The large number of chemicals suggested clustering as a method for simplifying the risk assessment. In 1980 the former head of the Federal Environmental Agency, von Lersner, proposed starting with an examination of the potential risk from those chemicals containing chlorine. However, a chlorine atom alone cannot justify blanket prohibition of all chemicals of this kind (von Lersner 1980).

Public debate was promoted by several publicly financed studies. From the end of the 1970s, several studies on different aspects of chlorine chemistry were commissioned by the Federal Environmental Agency, e.g.:

- representation of the emissions of halogenised hydrocarbons (Oberbacher et al. 1977)
- possibilities for avoidance or recycling of chlorinated waste (Schulze and Weiser 1985)
- handbook of chlorine chemistry (Nolte and Joas 1992)
- follow-up studies on the handbook chlorine chemistry (Nolte and Joas 1995)

Other agencies also became interested in research into the environmental effects of chlorine. In 1982 the Ministry of the Interior and the Federal Environmental

[7] Besides citing these advantages of chlorine-based technologies, the industry emphasises that there is no ideological motivation for producing chlorine, but that these technologies must continuously prove their economic advantages over alternative technologies (e.g. Nader 1996).

Agency commissioned a study on the sources and final destinations of halogenised chemicals in water bodies, at the German association of chemists (GdCH) (Ballschmitter et al. 1987). Another study, a technology assessment of chlorine chemistry, was commissioned by the ministry for research (BMBF) at Prognos (Wolff et al. 1993). There was no official focus on chlorine chemistry on the part of of the BMBF, but in isolated segments (CFC, chlorinated solvents, production of paper and pyrolysis of PVC) there were considerable efforts on research into possible substitutes (Schött 1993, p. 100). The fact that the study was funded by the Ministry of Research had a considerable positive impact on the implementation of respective regulations (Angerer et al. 1996).

Since 1974, the German Council of Environmental Advisors (SRU) has criticised the environmental impact of chlorine chemistry in its annual reports. This criticism is most pronounced in the special report on waste management:

> Chemists and other engineers in industry and universities are increasingly of the opinion that the rapid expansion of chlorine chemistry in the 1950s and 1960s was a serious error in industrial development. It would not have happened if the present state of knowledge on the risks for the environment and human health from chlorinated chemicals had been available. A large part of the present activities in environmental and health policy is aimed at compensating for problems caused by chlorine chemistry, inasmuch as their full extent is already known. (SRU 1990, p. 219, author's translation)

The report recommended that further expansion in chlorine chemistry should be prevented, and its dominant position should be eroded step by step. The priority should be in final products containing chlorine, but chlorine-based processes should also be substituted in the long term (SRU 1990, no. 752–753).

Various areas of chlorine chemistry (PVC, chlorinated solvents, propylenoxide), and their associated environmental, economic and social risks (and opportunities) were on the agenda of the German government's commission of enquiry "Schutz des Menschen und der Umwelt" (Enquete-Kommission 1993, 1994, p. 158). While there was agreement on the need to reduce the use of toxic chlorinated substances in an open environment, the issue of PVCs remains a disputed one, and no policy goals have been developed for chlorine-based *processes*.

At the level of the *Länder*, chlorine chemistry has been discussed in several coalition agreements between the Green party and the Social Democrats (e.g. in North Rhine-Westphalia, Sachsen-Anhalt, Hessen). The Hessian environment ministry commissioned two studies[8] on the possibilities for converting the chlorine industry (Wolff and Meckel 1987; Plinke et al. 1994). The Green party views chlorine chemistry as an example of "hard" chemistry, which should be replaced by – more natural – "soft" chemistry. Their preference is for using naturally occurring substances in the production and processing of chemical goods, and most chlorinated hydrocarbons do not occur in nature (von Gleich 1991, p. 4).

Around 200 local communities and several *Länder* have banned the use of PVCs in building projects which they support or finance, or have at least required

[8] The first study includes other products and emissions beside those from chlorine chemistry, e.g. heavy metals. However, chlorinated products remain the key focus of the report.

the use of a high proportion of recycled PVC. This has not had any marked effect on private house building. The first regulations of this type (in Bielefeld, Nürnberg etc.) were attacked vigorously by the chemicals industry. Since the 1990s, there have been hardly any new total bans on the use of PVCs, but quotas have been introduced for the use of recycled material and the municipalities require take-back guarantees for the materials used. Today, these demands can be met by the PVC industry and the days of heated disputes over regulations appear to be at an end.

Emissions by chlorine production plants are subject to specific EU guidelines (European directive 82/176/EEC). Further regulations apply to by-products containing chlorine, which directly or indirectly affect all the important subsequent products. The use of CFCs is by now almost completely prohibited, and the production of PVC is subject to special requirements (regarding VC emissions, for example) as is the use of additives (such as cadmium). The use of chlorinated solvents is also covered by detailed regulations.

As regards chlorine-free end products where chlorine is used as an auxiliary agent, regulations which refer explicitly to products are rare (see also Table 7). Because of the frequently high potential for risks from the intermediate products, regulations tend to apply more to production technology and to emissions: phosgene and propylenoxide, for instance, come under explicit state supervision, and both production and emissions require official authorisation.

3.2 Societal Actors

The demand for closing down the chlorine industry has been an important issue in "chemicals policy" as formulated by the major environmentalist association BUND since the middle of the 1980s (BUND 1984). This demand was later reformulated as "conversion" of chlorine processes. Several environmentalist associations joined in common campaigns on the prohibition of CFCs or ending the incineration of chlorinated waste on the North Sea. Greenpeace mounted its own international campaign on chlorine chemistry. AKN and ÖKOPOL also were engaged in issues regarding chlorine chemistry. Several environmentalist associations joined together in a common initiative "Chlorine has no future". There has been no fragmentation wother of mention within this subtype of actors.

The proposals for a chemicals policy as formulated by BUND were taken up in parliament by the Social Democrats around 1985 (Grant et al. 1988). They organised hearings and inquiries on single chemicals[9]. However, the positions were divided on some issues: while Social Democratic members of the Enquete-Commission argued for regulation of PVC products (Enquete-Kommission 1994, p. 174), the former head of the chemicals industry union and SPD MP, Rappe, refused to adopt any policy of this kind (Anonymous 1993).

[9] The Social Democrats raised the issue because they were seeking a profile in environmental policy independent of the Green Party which was elected for the first time.

Further societal actors have been engaged in various issues in chlorine chemistry. For example, the biggest German union, the IG Metall, campaigned against the use of chlorinated solvents at workplaces in the metallic industries. On the other hand, the union for the chemicals industry (IG Chemie) argued that there is no need for a generalised end to the use of chlorinated chemicals, which would bring heavy job losses and a decline in turnover. There are close links and a great many shared positions between IG Chemie and the industry's employers association (VCI) on most issues.

Another actor who should be mentioned is the Protestant Church, which started a dialogue between the main actors (industry, government and environmentalists) in two of its academies (Tutzing and Bad Boll) and a round table discussion (Oeser-Kreis) with representatives from industry, government, environmental action groups and scientists (Held 1988, 1991; Evangelische Akademie Bad Boll 1988). Similar meetings were organised by the municipality of Frankfurt (Ueberhorst et al. 1990, Ueberhorst and de Man 1992).

3.3 Enterprises and Their Industrial Associations

Chlorine producers are members of an industrial association (VCI), which became active in response to several questions about chlorine production and chlorinated products. Several working groups were set up on chlorinated products and processes. Since 1986, a number of voluntary agreements has been declared on the phasing out or restriction of products such as CFCs or chlorinated solvents. Producers and manufacturers of PVC formed an association concerned with PVC environmental issues (Arbeitsgemeinschaft PVC und Umwelt) in 1988.

European producers have an industrial association, "Eurochlor", and there is also co-operation world-wide with other industrial associations in the chlorine industry: there have been several symposia involving representatives of the chlorine industry from all over the world in the past. The 12 top ranking European PVC producers joined in the "European Council of Vinyl Manufacturers" (ECVM) in 1988.

The VCI and individual producers claim that they halted the use of highly toxic chlorinated chemicals in an open environment very early on, as soon as there was evidence of harm. They claim that there may have been mistakes in categorising individual chemicals in the past, but that today, as soon as the use of certain substances such as PCP, PCBs, DDT, CFCs and some applications of PVC or chlorinated solvents finally ceases, there will be no more major environmental risks: the chemicals industry's product portfolio is thus now free from undue environmental risk, and use of the remaining chlorinated products, especially PVC in the construction sector, is well-justified (VCI 1996b, pp. 3, 6).

Other industrial associations and enterprises are active *against* single chlorinated products. Many "downstream" branches supported the substitution of PVC for disposable applications, and a number of firms have explicitly abandoned the use of PVC in packaging (Herlitz, Henkel) or as a building material for furniture

(IKEA). This abandonment of PVC was to some extent a question of economic self-interest: automobile scrap merchants, for example, asked (unsuccessfully) for an end to the use of PVC in car manufacture because of the considerable costs incurred for decontamination. Following this, and fearing government regulations requiring the recall of older models, some car producers discontinued PVC-based applications, e.g. for under-body sealants (Claus et al. 1990).

3.4 International Agreements – the Discussion in Other Industrialised Countries

There has been no explicit discussion of chlorine chemistry as a whole among governments. However, international organisations and agreements have become very important for individual segments of the chlorine chemistry:

- a variety of agreements on CFC use
- the meetings on the protection of North Sea, especially in 1990 and 1995, where a sharp reduction in emissions of chlorinated substances and mercury during the amalgam process in chlorine production was agreed
- the EC/EU and their regulations on both products and processes in chlorine chemistry – the EU issued regulations on water emissions as early as 1974 (RL 76/464/EWG[10]) – which were important for several chlorinated chemicals but which were only partially implemented at national level; as a specification of the aforementioned guideline a regulation of mercury emissions by amalgam plants was also issued in 1982; the EU policy on VOC emissions[11] might have some relevance too; the Commission has constituted a VOC task force and made several proposals for a regulation on reducing VOC emissions
- the efforts of the OECD, especially in the 1970s, on a common end to the use of PCBs (Nichols and Crawford 1983)

There has also been a debate in other western industrialised countries on the industrial use of chlorine:

There has been discussion between the industry, environmentalists and regulators for a considerable time in *Austria*. This dialogue was initiated under the auspices of the Austrian society for technology and the environment (Österreichische Gesellschaft für Umwelt und Technik, ÖGUT). This is perceived as a model for green corporatism between industry, authorities and environmentalists. In a joint paper, all participating groups underlined the necessity of a stop to the use of chlorinated organic chemicals. The central issue of production and consumption of

[10] The list of 129 chemicals which should not be released in waters before prior allowance by the authorities is by 2/3 chlorinated organic chemicals.

[11] The norm-setting and other activities like the VOC task force of the EU referring most often to emissions of chemicals other than chlorinated substances. Because of the rather small potential for ozone formation (which was believed to be completely different in the US at the end of the 60s) most of the producing firms reprehend that a substitution of chlorinated solvents by non chlorinated is not necessarily desirable (Deutsch 1993, pp. 35–36).

PVC remains disputed. The use of chlorine in drinking water has not been discussed and the use of hydrochloric acid and ferro(III)chloride[12] has been approved (Leutgeb 1993, pp. 13–14).

In *Sweden*, a substance flow analysis has been published for chlorine (Swedish National Chemicals Inspectorate 1995). The aim is to phase out the use of chlorinated solvents, their sale as a component in final products for private households has been forbidden since 1993, and commercial use – with the exception of perchloroethylene (PER) – has been forbidden since 1996. With regard to PER, the environmental authorities have proposed a tax, in order to accelerate the development of substitutes and to provide further incentives for reducing emissions. In the long run, a phasing out of PER is also an aim (ibid., p. 106), and a phasing out of the production and consumption of chlorinated paraffins is planned. However, officials do not generally see any greater risk from chlorinated chemicals compared to non-chlorinated ones. There is still a need for a risk assessment of each chemical or at least chemicals group (ibid., p. 101). A general phasing out of PVC has been rejected by the Swedish environmental authorities (Anonym 1995).

In *Norway* there has been a joint research project with MIT in the USA, the Norwegian industry, authorities and environmentalists, covering the economic, social and ecological consequences of a complete ban on the use of chlorine. The study has not as yet resulted in any regulatory action (ibid., p. 76).

In *Denmark*, the debate on chlorine has focused on the use of PVC. In 1987, the environment minister proposed a phasing out by 1993, triggering a fierce public debate on the issue. Consumption of PVC as a coating material declined from 10.5 kt in 1987 to about 5.7 kt in 1992 (Holm 1995, p. 304; Greenpeace 1992).

In the *Netherlands*, the government concluded on the basis of a material flow analysis for chlorine that the risks are manageable and a new policy on chlorine is unnecessary. (Hendriks and Papameltiou 1996, p. 7). In the old National Environmental Policy Plan (NEPP), the chemicals industry was asked to look for possible alternatives to the use of chlorine as a basic material or – where its use is essential – to reduce the risks as far as possible (Swedish National Chemicals Inspectorate 1995, p. 75).

In *North America*, the US-Canadian International Joint Commission (IJC) on the protection of the Great Lakes suggested in 1992 a plan for reducting and phasing out chlorinated organic chemicals. Both the US and Canadian governments refused to adopt the plan. However, they began to tighten up regulations, e.g. on CFCs, or the emissions standards for the paper industry. In 1994, the IJC repeated its demands but suggested considering social and economic criteria in the phasing out (Hileman et al. 1994). During the proceedings surrounding the re-enactment of the Clean Water Act, the EPA was asked by the Clinton administration to consider the possibilities of reducting or phasing out the use of chlorinated organic chemicals (Swedish National Chemicals Inspectorate 1995, p. 71).

[12] This chemical is used frequently for the sewage-cleaning.

3. The chlorine debate: actors and issues

	DDT/PCB	PCP	CFCs	CKW	PVC	chlorine based processes	waste incineration on the North Sea
- its	phase out	phase out	phase out	phase out	phase out	reduction to essential applications only	phase out
mental ies	phase out	phase out / health authorities: no danger to public health	phase out / doubt of scientific proof of harmfulness	reduction to closed cycle applications	reduction of dangerous applications	reduction of chlorinated waste, in general: approval	phase out
ls and conomic ies	phase out when substitutes are available	no danger for public health	no proof of danger (when substitutes were developed: approval of phase out)	not dangerous if used in a responsible way	not dangerous, even sustainable	most efficient processes	most efficient / no dangers for health or environment
dustrial			phase out, because of image	some sectors: phase out because of image	few sectors: phasing out of PVC based coatings; few companies: phase out of PVC based construction materials	no opinion	no opinion
n other		rapid phase out	no opinion	phase out when substitutes are available	some unions: phase out because possible substitutes are more labour intensive some states: phase out	no opinion	no opinion
untries	early phase out (USA)	n.a.	early phase out (USA)	more radical phase out (Scandinavia)	more radical policy regarding short term applications (e.g. Denmark)	n.a.	earlier stop of incineration (Netherlands)

3.5 Summary: Issues and Actors

With regard to the (temporary) decline in chlorine consumption, the most important political intervention and regulations did not take place at the level of chlorine production itself, but of certain subsequent products. Table 8 provides an overview of the most important issues, actors and their respective policy targets in the chlorine debate. Of course the table cannot do justice to the complexity of the actors' strategies. It highlights their positions at the climax of the conflict. Frequently, learning or tactical changes of position cannot be considered in the table.

There is a powerful coalition between the chemicals industry, its union and the ministry of economic affairs. On the opposing side, environmentalists are the only actors who promote a strict policy against chlorine in all areas. Environmental and health authorities are frequently divided internally on specific issues, which makes targets difficult to identify. The environmental administration most often has to find targets and policy measures in conjunction with the ministry of economic affairs, which is naturally hard to achieve. National government actors therefore provide a rather weak driving force in most of the issues.

Other possible actors, such as the *Länder*, unions from other sectors or other industrial sectors themselves, are most often not involved in the political process. However, if they join the process, and if there is a coalition between e.g. environmentalists, unions and authorities, they are very often successful in enforcing their targets. It is also worth mentioning that if only one member of the coalition between the chemicals industry, its union and the economic authorities falls apart, the coalition becomes very weak. Each member of this three-way coalition is necessary to legitimate the positions of their partners. If one is missing, strong pressure for change is created. This was the case with PCP, for example, where the chemicals union was internally divided on the issue, and sections of the union (those which represented workers in sectors like the paper industry, which uses rather than produces PCP) put heavy pressure on the industry to substitute the chemical. This also happens when a single company renounces its former antagonistic position, for example because it is able to provide a technological superior substitute, and the entire sector comes under considerable pressure. Other companies are no longer able to justify their environmental harmful products.

4 Technological Possibilities and Obstacles to Environmental Improvement

4.1 Chlorine Production

Since the turn of the 19th century, industrial production of chlorine has used electrolysis of a common salt solution. The products are sodium hydroxide (NaOH), chlorine and – in small quantities – hydrogen. These substances must be separated, either with a diaphragm of asbestos, mercury (the amalgam method) or the membrane process, which was developed by ICI only in the 1960s, (Dibble and Argust 1986, p. 89; Eklington and Burke 1989, p. 115), and which avoids the use of asbestos or mercury emissions. There are also significant differences between the specific electricity requirements of the processes: to produce one tonne of Cl, the amalgam method requires 3 300 kWh, the diaphragm method 2 800 kWh, the membrane method 2 200 kWh (Dibble and Argust 1986: p. 89). The balance of the energy savings remains disputed, however, since the sodium hydroxide (NaOH) produced with the membrane or diaphragm methods must then be concentrated. Furthermore, the necessary prior investment makes the membrane method significantly more expensive (Müller 1993, p. 48). This new membrane technology is used in 10% of global production capacity and predictions were that this figure would have risen to 50% by the year 2000 (Seidel 1993, p. 26). Although the technology is available in Germany – the German plant manufacturer Uhde has set up membrane-based facilities all over the world (FAZ 9.9.1994, p. 21) – the German chlorine industry as a whole is not investing in this energy-efficient technology. There are two possible reasons for this:

- Since consumption of chlorine is declining, it is uncertain how chlorine production capacity will develop, even in the near future. Planning for capacity seems impossible in the light of the heavy criticisms directed at products and processes which use chlorine. Furthermore, there is no urgent need to replace old plants. Although these plants are sometimes several decades old, they have been well maintained and run at low capital costs.
- While energy-efficient membrane plants make financial sense in countries like Taiwan or India, where energy is estimated to be more expensive, similar incentives are lacking in Germany (Bergner 1993, p. 68), although with an eye to proposed energy taxes, representatives of the chemicals industry do complain about comparatively high energy costs in Germany (e.g. Amerschläger 1996).

The members of the Conference for the Protection of the North Sea have agreed to considerably reduce mercury emissions into the North Sea by the year 2010. This will make the use of mercury-free methods in Germany unavoidable (Darimont 1995, p. 286, Nolte and Joas 1995, p. 41). The ordinance prohibiting the use of asbestos (ChemikalienVerbotsV of 1993) outlaws the diaphragm method as of 31st December 2010. Furthermore, government licensing of new planned chlorine production sites will be granted only if they use membrane technology.

The chemicals industry has refused to modernise its amalgam plants by 2010, as this would incur costs of about DM 100 million per 100 kt/a production capacity. They point instead to the fact that mercury emissions from existing plants are far below current legal limits, and that complete substitution with membrane technology would cost DM 3 billion, ca. DM 1 million for every kilogram of mercury emissions prevented annually in future (Schmidt 1993, p. 49). Moreover, other industrial sources, such as coal-fired plants, emit far more mercury in terms of volumes produced (Bayer AG 1995). At first, only the chlorine producer Dow accepted that the membrane method is the best available technology for any new plant (Schmidt 1993, p. 49). However, in a recent publication, Bayer has also acknowledged the fact, albeit stressing the fact that an obligatory modernisation of all plants by 2010 would be unreasonable (Bayer 1995, p. 30). To date, Dow – or rather their subsidiary BSL (formerly Buna) – has been the only producer to build a membrane plant, with a capacity of 200 kt/a. The chlorine produced there will be used for production of PVC and Propylenoxide.

Environmental groups have also warned against upgrading the plants, as this would lock capital into large-scale chlorine production capacity for a long period of time, and thereby become a disincentive to reviewing the products or the chlorine processing methods themselves (Ahrens 1993, p. 57).

4.2 By-Products

Substances which require considerable quantities of chlorine – CFCs (1987 ca. 6% of total chlorine consumption) and chlorinated solvents[13] (in the 1970s ca. 20% of total chlorine consumption, in 1992 ca. 10%) – have substitutes which have already found acceptance in the relevant sectors. These product lines were important for reprocessing chlorinated chemical residues from other processes. The chlorinated solvents TRI and PER were produced by means of chlorolysis, as was the most important pre-product of CFCs, TETRA. Incinerating them in the North Sea was an alternative method of disposing of these chlorinated residues. However, both the product lines, chlorinated solvents and CFCs, and this method of waste disposal came under considerable pressure from environmental interests in the mid-1980s. The CFCs came under international regulation in the Vienna protocol in 1985 and the Montreal protocol in 1987. In response, chemical enterprises built land-based high temperature incineration plants, which made chlorolysis and marine incineration superfluous. All but one producer has abandoned chlorolysis and with it the production of CFCs and chlorinated solvents. Land-based incineration yields hydrochloric acid, which can substitute chlorine in certain processes, thus making an additional contribution to reducing chlorine consumption.

PVC has been, and still is, the subject of heated debate, but this has not led to plants being decommissioned or a large-scale reduction in the quantities produced. Instead, the industry seems to want to retain this product line, because they are building new plants and modernising existing ones (this includes BSL and Hüls). Disposable products – especially packaging – are no longer seen as a sensible application for PVC. The declining demand for these types of product has been significant. But in the construction sector (especially for windows, pipes and floor coverings), the industry is further developing this material, eliminating particularly problematic substances such as cadmium (a stabiliser) or certain flame retardants, which can lead to the formation of dioxins during incineration. At the same time, more or less effective recycling facilities are being built.

[13] Chlorinated hydrocarbons, used as solvents in metals processing, dry cleaning and (a small share of consumption) as solvents in coatings, adhesives, etc.

Although the industry refuses to recycle PVC on cost grounds (ca. DM 3 500 per tonne, compared with DM 1 000 per tonne for new PVC), some pilot plants have been built, using a variety of technologies. But incineration is still the preferred disposal method for plastics and in the case of PVC, mono-incineration, which produces HCl as a by-product. The HCl can, however, be fed back into the oxichlorination process (and thus re-used in PVC production), or into other processes requiring hydrochloric acid. Environmentalists do, however, reject the idea of a closed salt system, in which the hydrochloric acid produced in common incineration plants is neutralised and then fed back into electrolysis, because of pollutants in the sodium chloride (Winteler and Ahrens 1991; BUND 1994).

As regards the third major area of chlorine consumption, production of chlorine-free polyurethane, economically viable chlorine-free options are also available, but they play no part in the German chemicals industry. In other countries, hydroperoxide processes using isobutane or ethylbenzene have already been implemented on a large scale, yielding either methyl-tert-butyl-ether (MTBE) or styrene as by-products. These yields outweigh the propylenoxide (PO) output by a ratio of between 2.5 and 3 to one. Styrene is a pre-product of polystyrene, whereas MTBE is used as an additive in "Super Plus" fuel, to increase its octane value and as a substitute for carcinogenic benzene. The quantity of MTBE produced in Germany was 153 000 tonnes in 1992, but imports are estimated to be 500 000 tonnes per year (Plinke et al. 1994, p. I-52).

The costs of producing PO in chlorine-free processes are 40–50% lower than those of conventional chlorine-based technology: converting to these processes would therefore seem economically well-founded. According to a study by Prognos, the time-scale for such a conversion would depend on the time required for planning, authorisation and construction, but lie somewhere between 5 to 10 years (Plinke et al. 1994, p. I-55). Nolte and Joas (1995, p. 81), however, do not favour substitution of conventional technology, neither in the short nor the long term. Major arguments against such a substitution are the highly selective production (ignoring emissions of NaCl), the lack of a market for the by-products and the concomitant investment required in the plants themselves (additional refineries).

About half of the PO produced world-wide is produced with chlorine-free processes (Plinke et al. 1994, p. I-26). In the US, demand for the by-product MTBE increased after restrictions were placed on benzene as a fuel additive (Anonym 1991), placing higher costs on refinement processes. In Germany, these extra costs were only levied on "Super Plus" fuels (FAZ 8.8.1995, p. T3) and, although this technology was developed in Germany, the global shift in production left Germany untouched (Rohe 1990).

In view of these options for reducing the quantities of unused by-products (recycling HCl in incineration plants and in PVC reprocessing, economically viable chlorine-free processes in the PUR product line), as well as the tentativeness in building new chlorine production plants (explained by uncertainty as to future demand for chlorine) it can be inferred that – in spite of a short-term increase over recent years – chlorine consumption will decline further in the long run. Instead, there will be growth in the use of processes using HCl, and it could be said that chlorine chemistry is evolving into hydrochloric acid chemistry. There are also technologically and economically feasible ways[14] of satisfying the growing demand for sodium hydroxide, a by-product of chlorine production, for instance from naturally occurring sodium carbonate (Nolte and Joas 1993). However, since these technologies require considerable investment and the chlorine/sodium hydroxide plants are up and running already, there is a strong constituency in favour of the old technology.

There is no single substitute for chlorine, but a wide range of substitutes for chlorine products and alternative chlorine-free technologies. The economic and environmental consequences of substitution have been evaluated in a number of studies, with the overall potential for substitution estimated at between 50% (Winteler and Ahrens 1991), 70% (BUND 1994), and 65–75% (Plinke et al. 1994[15])

4.3 Future Developments

The chemicals industry claims that the environmentally relevant modernisations in its product lines are now complete. The use of PVC in durable goods which are easy to recycle is environmentally reasonable, especially when critical additives such as bromide or cadmium are eliminated. Environmental activists are not satisfied. Only recently, Greenpeace has started a campaign in several countries to attack PVC use in children's toys. Several enterprises in the sector have promised to phase out PVC use, and some governments, along with the EU Commission, have discussed possible regulations. A negative impact on the key construction sector applications is not yet in sight.

It is generally beyond dispute – even among representatives in industry – that further reductions in consumption for non-durable goods are possible and desirable. This may entail further slight decreases in chlorine production. Growth in the main downstream sector – construction – explains the resurgence of chlorine production in the recent past. With the expected flattening of the economic cycle, reduced investment rates and the stabilisation of building activity in the *New Länder*, there will be a further decrease in chlorine production. More reduction is possible if large-scale incineration of PVC and recycling of hydrochloric acid begin. Hydrochloric acid can substitute chlorine in many processes, for example in PVC production.

Beyond that, environmentally and economically sensible opportunities for modernising chlorine-free final products, especially propylenoxide, are not being

[14] Representatives of industry (e.g. Höllbacher 1993, p. 38; Schlegel 1993) dispute the environmental benefit as well as the economic feasibility of alternative technologies. A considerable amount of NaOH is used in the chemicals industry itself as a neutralisation reagent, for example in the chlorine-based technologies for producing propylene oxides. These technologies could be substituted.

[15] The study quoted is examining the four most important fields of chlorine consumption: production of propylene oxide, PVC, phosgene and epoxy resin.

taken, or only very slowly. Although some interesting technological solutions are available, political pressure is weak compared to that in other segments, such as CFCs, chlorinated solvents, waste incineration on the North sea or PVC. There is therefore hardly any need for the chemicals industry to adopt these technologies. There is a lack of environmental targets set by the parliamentary Enquete-Commission or the environmental authorities, which have discussed these products of the chlorine industry at length. Environmental action groups have been concentrating on the continued use of PVC, while the use of chlorine as a processing agent rarely reaches the agenda.

5 Conclusions

In spite of the intense public debate on possible risks, industrial production and use of chlorine is basically sanctioned by current law. A clear policy for phasing out or substantially reducing chlorine consumption is frequently called for (especially by environmentalists) but has not been implemented, nor is it to be expected.

Regulations have affected only downstream products. But even here, it should be mentioned that these regulations often came into effect only after the decline had set in, or in some cases even when production and consumption had already ceased. In general – with the exception of PVC – there is a tendential decline in the use of chlorinated final products in an open environment, while for applications in which chlorine is used as an auxilliary reagent only, consumption is rising. The structural change in the chlorochemicals industry can be described as the abandonment of chlorolysis technology, and subsequently of CFCs and chlorinated solvents, product lines which were dependent on this chemical process. At the same time, chlorolysis had helped reprocess chlorochemical residues into marketable products. Due to pressure on CFCs and to a lesser extent, chlorinated solvents, from the middle of the 1980s, and with the abandonment of high seas incineration, this technology has become unnecessary. Instead of reprocessing waste, land-based high temperature incineration plants, yielding hydrochloric acid, have been built. This has led to a decline in demand for chlorine, not only because chlorolysis has been abandoned but also because HCl can substitute for pure chlorine in some processes. For this reason we can speak of a partial transformation from chlorochemistry to hydrochloric acid chemistry. The construction of PVC recycling facilities will also promote this transformation, if mono-incineration becomes the major recycling technology for this material. Sizeable quantities of hydrochloric acid are produced here also, which is liable to lead to an even greater reduction in the use of primary chlorine. Alongside this technological change, production of PVC was sharply reduced at the end of the 1980s and the start of the 1990s, mostly for economic reasons. Furthermore, global over-capacity led to heavy pressure on domestic producers from imports, although this has since been balanced out.

In recent years, the use of PVC has been increasingly restricted to the construction sector. Disposable uses, such as packaging, have come under particular pressure. Thus a decline in domestic use of PVC can be expected to accompany the downward trends in the construction industry itself.

There is insufficient pressure in Germany for change to polyurethane production technologies, even though the economic advantages of chlorine-free processes have been rewarded with a 50% share of global production capacity. Despite current statements by industry representatives, conversion of production processes is to be expected in the medium term; one third of chlorine production in this sector could be substituted.

At present, (somewhat) more environmentally friendly chlorine production plants (lower emissions and energy consumption) are not being built, because of uncertainties about medium-term demand for chlorine. There is a chance for further structural change if clear targets are set for the remaining key product lines like PVC or polyurethane. At present, these targets are being discussed by the different actors, but if they are clearly defined there is a good chance that the industry will adopt chlorine free technologies even without formal regulation.

To sum up, reductions in chlorine production can be put down to a complex interplay of economic, technological, structural and environmental factors. Whether the results of this case study, given the complexity of chlorine's material flows and those of its downstream products, can be transfered to other basic industries is questionable, as these most often deal with a single product and a small number of technologies.

The debate on chlorine started with substances which were exiguous with respect to the quantities produced, but which produced significant environmental harm, such as chlorinated pesticides (DDT) or PCBs. Phasing out these substances and regulatory measures had virtually no effect on overall consumption of chlorine.

In a second phase of the public debate, the environmentally harmful substances produced by chlorolysis (CFCs and chlorinated solvents) came onto the agenda. Together with the halt to high sea incineration on the North Sea and the nearly total phasing out of chlorolysis, a "strategic knot" was encountered in the substance flow. We can define a strategic knot as a process which is central to the substance flow: several different inputs are processed and a variety of products are produced. As chlorolysis was phased out in all but one of the companies, several products were hit (most of the chlorinated solvents, as well as CFCs), but more importantly, considerable waste streams from other processes were exposed. There was an urgent need for special high temperature incineration plants to process this waste in order to be able to give up on CFCs. The death of this particular process can explain most of the temporary declines in chlorine consumption in the past. The short term crisis in the market for PVC, caused by over-capacity on the European market and temporarily rising imports, was the other important factor in explaining declining chlorine production. However, the crisis was only temporary and production of PVC has regained its the former levels.

Lessons can be drawn for other basic materials. Regulators (or other interested organisations) should start with a careful material flow analysis, from the extraction of raw materials through to final consumption, in order to identify strategic knots. These key products/processes should be placed at the centre of any efforts. It will not be generally possible to put in question an important basic industry as a whole from the outset. Central actors simply cannot conceive of a total phasing out. At first, the environmental modernisation of single processes, product lines or products will be in the foreground.

Clear targets should be formulated to promote green modernisation of the critical products or processes. These targets should not impede long term structural change. For example, optimisation by EOP measures may have negative effects on structural change, because the often huge investments represent sunk costs. The same might hold for modifying production processes while leaving the final products unchanged.

Industry will presumably start to examine closely related products of itself, if sufficiently abstract targets are set (e.g. phasing out chlorinated solvents in five years). This screening may lead to environmentally benign substitutes. Of course this possibility should not lead to complete renunciation on the part of environmental policy, because innovation by companies is not in itself sufficient guarantee of safe products. In particular where there is immediate danger to the environment and/or human health, regulators cannot renounce command-and-control completely.

In order to implement environmental targets as rapidly as possible, regulators should try to identify possible innovators in a branch which could offer substitute products or processes. If there is an innovative firm in a sector, this puts heavy pressure on its competitors to adopt this technology as well. It is most sensible to conduct the environmental discourse not only with the industrial associations, but also with individual companies.

In the case of the chlorine industry, social problems were avoided because the structural change took place at the peak of the industry's business cycle. The (relatively few) jobs affected by the closure of single plants were recovered in the companies' other areas of activity, and employees were transferred to other divisions. It seems more difficult to manage structural change if the future of the entire sector is in doubt, rather than a single (albeit important) product in an otherwise highly integrated branch.

References

Ahrens, A. (ÖKOPOL) (1993): Beitrag zur Expertenrunde. In: IG Chemie – Papier – Keramik 1993, p. 56.

Aktionskonferenz Nordsee e.V. (AKN) (1995): Chlorchemie. Materialien für Schule, Fortbildung und Öffentlichkeitsarbeit. Bremen.

Amecke, H.-B. (1987): Chemiewirtschaft im Überblick. Produkte, Märkte, Strategien. Weinheim, VCH Verlagsgesellschaft.

Amerschläger, R. (1996): Interview. Bewegung in der Landschaft. Der Chemie-Gigant Hoechst laesst sich in Zukunft vom Strom-Riesen RWE mit Energie versorgen. Energie und Management, No. 10, 1–2.

Angerer, G. et al. (eds.) (1996): Einflüsse der Forschungsförderung auf Gesetzgebung und Normenbildung im Umweltschutz. Physica Verlag.

Anonym (1989): Henkel Initiative'89: Phosphatfrei und PE-verpackt. Chemische Industrie 3/89, 66.

Anonym (1991): Clean Air Act. Auftrieb für MTBE. Chemische Industrie 7/91, 10.

Anonym (1993): Rappe: PVC Verbot ist unseriös. Chemiereport 7/93.

Anonym (1995): Schwedische Umweltschutzbehörde gegen einen Ausstieg aus der PVC Herstellung. Umwelt Contact, No. 2 April/Mai, 13.

Ballschmitter, K., Haltrich, W., Kühn, W., Niemitz, W. et al. (1987): HOV-Studie. Halogenorganische Verbindungen in Wässern. Eine wissenschaftlich-technische Studie herausgegeben von der Fachgruppe Wasserchemie in der GdCh i.A. des UBA. Berlin. FB 89/022.

Bayer AG (1995): Chemie mit Chlor. Chancen – Risiken – Perspektiven. Leverkusen.

Bergner, D. (Hoechst AG) (1993): Beitrag zur Expertenrunde. In: IG Chemie – Papier – Keramik 1993, p. 68.

Broecker, B. (1988): Gefährdungsabschätzung und Risikobegrenzung bei Produktion, Anwendung und Entsorgung von Chemikalien. In: Evangelische Akademie Bad Boll 1988.

Budzinski, A. (1993): Kurs aufeinander zu. Chemische Industrie 10/1993, p. 13.

BUA (Beratergremium Umweltrelevante Altstoffe in der Gesellschaft deutscher Chemiker) (ed.) (1985–1996): Stoffberichte 1–186.

BUND (Bund für Umwelt und Naturschutz Deutschland e.V.) (1984): Chemiepolitik, BUND fordert einen neuen Politikbereich, BUND Position 10.

BUND (1994): Chlorchemie – Eine Ära geht zu Ende. Bund-Positionen 24.

Claus, F., Friege, H., Gremler, D. (1990): Es geht auch ohne PVC. Einsatz – Entsorgung – Ersatz. Ein Ratgeber. Hamburg: Rasch und Röhring Verlag.

Darimont, T. (1995): Konversion Chlorchemie. Perspektiven eines Umbaus. In: Umweltwissenschaften und Schadstoff-Forschung. Zeitschrift für Umweltchemie und Ökotoxikologie 7 (5) November, 280–286.

Deutsch, E. (Dow) (1993): Die Zukunft der Chlorchemie aus der Sicht der chemischen Industrie. In: IG Chemie – Papier – Keramik 1993, p. 31.

Dibble, G., Argust, J.P. (1986): Membran-Chloralkali-Elekrolyse reduziert Umweltverschmutzung und Energieverbrauch. Chemische Industrie 2/86, 89.

Eklington, J. and Burke, T. (1989): Umweltkrise als Chance. Ökologische Herausforderung für die Industrie. Zürich, Wiesbaden: Orell Füssli Verlag.

Enquete-Kommission (1993): Verantwortung für die Zukunft. Wege zum nachhaltigen Umgang mit Stoff- und Materialströmen. Zwischenbericht der Enquete-Kommission "Schutz des Menschen und der Umwelt – Bewertungskriterien und Perspektiven für umweltverträgliche Stoffkreisläufe in der Industriegesellschaft" des 12. Deutschen Bundestages. Bonn: Economica Verlag.

Enquete-Kommission (1994): Die Industriegesellschaft gestalten – Perspektiven für einen nachhaltigen Umgang mit Stoff- und Materialströmen. Bericht der Enquete-Kommission "Schutz des Menschen und der Umwelt – Bewertungskriterien für umweltverträgliche Stoffkreisläufe in der Industriegesellschaft". Bonn: Drs. 12/8260.

Evangelische Akademie Bad Boll (1988): Naturwissenschaft und Theologie im Gespräch (5): Chemie – Im Dienst des Lebens? Mitweltfreundliche Produktion, Anwendung und Entsorgung von Chemikalien als Herausforderung an Gesetzgeber, Industrie und Verbraucher. Tagung vom 29. bis 31. Januar 1988 in Bad Boll. Protokolldienst 7/88.

Fricke, G., Hellhammer, J., Klamt, A., Meerkamp van Embden, J., Rose, M., Sewekow, B., Wittlinger, R. (1995): Zum indirekten Phtoabbau chlororganischer Verbindungen durch OH-Radikale in der Atmosphäre. In: Umweltwissenschaften und Schadstoff-Forschung. Zeitschrift für Umweltchemie und Ökotoxikologie 7 (5) November, pp. 275–279.

Glatz, H. (ed.): Perspektiven der Chemiepolitik III: Chlorchemie. "Setzen wir der Chemie Grenzen?" Fachtagung am 14. und 15. Jänner 1993 Veranstaltet von der Kammer für Arbeiter und Angestellte für Wien, Gewerkschaft der Chemiearbeiter und der Gewerkschaft der Privatangestellten, Sektion Industrie. Wien.

von Gleich, A. (1991): Über den Umgang mit Natur – Sanfte Chemie als wissenschaftliches, chemiepolitisches und regionalwirtschaftliches Konzept. In: Wechselwirkung No. 48.

Grant, W., Paterson, W., Whitston, C. (1988): Government and the Chemical Industry. A Comparative Study of Britain and West Germany. Oxford: Clarendon Press.

Greenpeace International (1992): PVC: Toxic Waste in Disguise (GP). http://www.greenpeace.org/ftp/pub/campaings/cdromgp/toxics/pcv.txt

Gremler, D. (1991): Reduzierung der Chlororganika. In: Held 1991.

Haas, U. (1990): Standortperspektiven stromintensiver Produktionen in der Bundesrepublik Deutschland. Mögliche Konsequenzen umweltinduzierter Strompreise bei Aluminium und Chlor. Göttingen: Vandenhoeck & Ruprecht.

Held, M. (ed.) (1988): Chemiepolitik: Gespräch über eine neue Kontroverse. Beiträge und Ergebnisse einer Tagung der Evangelischen Akademie Tutzing. Weinheim: VCH-Verlagsgesellschaft.

Held, M. (ed.) (1991): Leitbilder der Chemiepolitik. Stoffökologische Perspektiven der Industriegesellschaft. Frankfurt a.M., New York: Campus Verlag.

Hendriks, C. and Papameltiou, D. (1996): Strategien zur Zukunft der Chlorindustrie. The IPTS Report 05 1996.

Henschler, D. (1994): Toxikologie chlororganischer Verbindungen. Der Einfluß von Chlorsubstituenten auf die Toxizität organischer Moleküle. VCH, Weinheim.

Herlitz AG (1995): Umweltbericht. http://www.herlitz.de/umwelt/umw06.html

Hileman, B., Long, J.R., Kirschner, E. (1994): Chlorine Industry Running Flat Out Despite Persitent Health Fears. Industry, government, and environmental groups still lack enough common ground to develop sound regulatory policies that won't cripple industry. Chemical and engineering news, No. 72 (47), pp. 12–26.

Höllbacher, G. (1993): Die Zukunft der Chlorchemie: Umstieg statt Austieg. In: Glatz 1993, p. 31.

Holm, J. (1995): Substitution of PVC in Danish Plastic Packaging. In: Jänicke, M. and Weidner, H. (eds): Successful Environmental Policy. A Critical Evaluation of 24 Cases. Berlin: Edition Sigma.

Hoppenstedt (1997): Firmen der neuen Bundesländer. Darmstadt et al.

IG Chemie – Papier – Keramik (ed.) (1993): Auf dem Weg zu einer umweltverträglichen Industriegesellschaft. Workshop-Reihe der IGCPK, Vorstandsbereich Bildung-Umweltschutz. Workshop III: Chlorchemie. 23.–24.2.1993 in Bad Münder.

Jacob, K. (1999): Innovationsorientierte Chemikalienpolitik. Politische, soziale und ökonomische Faktoren des verminderten Gebrauchs gefährlicher Stoffe. München.

Kleinloh, W. (1994): Die Chlorchemie als Kreislaufwirtschaft. In: Lucas, R. (ed.): Strukturwandel der chemischen Industrie. Schriftenreihe des IÖW 73/94, p. 62.
Kompass (1994): Jahrbuch der Deutschen Wirtschaft, 22nd edn. 1994/95.
Leisewitz, A. and Schwarz, W. (1994): Metalloberflächenreinigung mit CKW, KW und wäßrigen Reinigern. Stoffstromanalyse und Umweltbelastungsvergleich. UBA FB 94–144. Berlin.
von Lersner, H. Freiherr (1980): Vortrag vor dem Kongreß "Chemie – Mensch – Umwelt" 13.-15.10.1980 im Gottlieb Duttweiler Institut, Rüschlikon (CH).
Leutgeb, Franz 1993: Chemiepolitisches Exempel Chlorchemie: Vom richtigen Umgang mit Kuppelprodukten. In: Glatz 1993, p. 2.
Meerkamp van Embden, I. (1992): Pauschalausstieg wenig hilfreich. Chemische Industrie 3/92, 6.
Meerkamp van Embden, I. (1991): Thesen zu Chlororganica. Überlegungen aus der Sicht der Chemischen Industrie. In Steger, Ulrich (ed.): Chemie und Umwelt. Berlin: Schmidt, pp. 49–56.
Müller, W.D. (Bayer AG; VCI) (1993): Beitrag zur Expertenrunde. In: IG Chemie – Papier – Keramik 1993.
Nader, F. (1996): Perspektiven der Chlorchemie. Chemie Technik 25 (8), 66.
Nichols, J.K. and Crawford, P.J. (1983): Managing Chemicals in the 1980s. Paris: OECD.
Nolte, R. and Joas, R. (1992): Handbuch Chlorchemie I. Gesamtstofffluß und Bilanz. Handbuch Chlorchemie II: Ausgewählte Produktlinien. Berlin: UBA Texte 55/91, 42/92.
Nolte, R. and Joas, R. (1993): Kuppelproduktion von Chlor und Natronlauge – Alternativen und deren Folgen. Berlin: UBA Texte 61/93.
Nolte, R. and Joas, R. (1995): Handbuch Chlorchemie I – 1992. Produkte, Unternehmen, Stoffflüsse. Berlin: UBA Texte 25/95.
Oberbacher, B., Deibig, H., Eggersdorfer, R. (1977): Darstellung der Emissionssituation der halogenierten Kohlenwasserstoffe aus dem Lösemittelsektor. 1. Stufe: Schätzung der Emissionsmenge an Hand des Verbrauchsstrukturbaums. UBA FB 77–071.
Plinke, E., Schüssler, R., Kämpf, K. (Prognos AG) (1994): Endbericht Konversion Chlorchemie. Studie im Auftrag des Hessischen Ministeriums für Umwelt, Energie und Bundesangelegenheiten. Wiesbaden.
Pohle, H. (1991): Chemische Industrie. Umweltschutz, Arbeitsschutz, Anlagensicherheit. Rechtliche und Technische Normen. Umsetzung in der Praxis. Weinheim: VCH Verlagsgesellschaft.
Rohe, D. (1990): Methyl-tert-butylether. Super für Super. Chemische Industrie 4/90, 60.
Schlegel, W. (SRI) (1993): Rahmenbedingungen und Entwicklungstendenzen der Chlorchemie In: IG Chemie – Papier – Keramik 1993.
Schmacke, E. and Jaeckel, R. (Ed.) (1996): Die großen 500. Neuwied et al.: Luchterhand.
Schmidt, L. (Bayer AG) (1993): Beitrag zur Expertenrunde In: IG Chemie – Papier – Keramik 1993.
Schött (BMFT) (1993): Welche Forschungsmittel werden für die Entwicklung neuer Technologien zur Ablösung chlorchemischer Verfahren in der Bundesrepublik durch das BMFT bereitgestellt und welche Ergebnisse liegen vor? In: IG Chemie – Papier – Keramik 1993, p. 100.
Schulze, J. and Weiser, M. (1985): Vermeidungs- und Verwertungsmöglichkeiten von Rückständen bei der Herstellung chlororganischer Produkte. Berlin: UBA Texte 5/85.
Seidel, H. (1993): Unersetzlich, aber stagnierend. Chemische Industrie 1/93, 26–27.

SRI (Stanford Research International) (1996): 1996/97 Directory of Chemical Producers. Europe Vol. 2. Menlo Park.

SRU (Rat von Sachverständigen für Umweltfragen) (1990): Sondergutachten Abfallwirtschaft. Deutscher Bundestag (Ed.), BT Drs. 11/1568.

StaBu (Statistisches Bundesamt): Fachserie 7 Außenhandel. Reihe 2 – Außenhandel nach Waren und Ländern. Various issues.

StaBu: Produktionsstatistik: Fachserie 4 Produzierendes Gewerbe. Reihe 3.1 Produktion im Produziernden Gewerbe. Various issues.

StaBu: Statistisches Jahrbuch der Bundesrepublik Deutschland. Various issues

Swedish National Chemicals Inspectorate, The (1995): Chlorine and Chlorine Compounds. Use, occurence and risks – the need for action. Report on a Government Assignment. KEMI Report 1/95.

Ueberhorst, R. and de Man, R. (1992): Zweite Frankfurter Studie zur Förderung chemiepolitischer Verständigungsprozesse. Teil I: Zur Initiierung und Fortsetzung erster kooperativer Arbeiten mit gewerkschaftlichen, industriellen und ökologischen Partnern. Frankfurt/Elmshorn, Stadt Frankfurt a.M., Umweltdezernat.

Ueberhorst, R., de Man, R. et al. (1990): Erste Frankfurter Studie zur Förderung chemiepolitischer Verständigungsprozesse: Planungsstudie und diskursanalytische Vorstudien. Herausgegeben von: Stadt Frankfurt a.M., Umweltdezernat. Frankfurt a.M.

UN (United Nations): Industrial Statistics Yearbook. New York. Various issues.

VCI (Verband der Chemischen Industrie) (1996a): Chemiewirtschaft in Zahlen. Frankfurt/Main. http://www.vci.de/a-f-z.htm.

VCI (1996b): Positionen zur Chemie mit Chlor. Stand: April 1996. Frankfurt/Main.

VCI: Annual Report. Frankfurt/Main. Various Volumes.

Winteler, S., Ahrens, A. (1991): Chlor hat keine Zukunft. Kurzfassung im Auftrag der Aktionskonferenz Nordsee e.V. Bremen.

Wolff, H. and Meckel, H. (1987): Chemiestandort Hessen. Eine Umwelttechnische Herausforderung. Untersuchung der Prognos AG im Auftrag der hessischen Landesregierung. Basel.

Wolff, H., Alwast, H., Buttgereit, R. (1993): Technikfolgen Chlor. Szenarien für die ökonomischen und ökologischen Folgen technischer Alternativen. Prognos AG (Hrsg.). Stuttgart, Schäffer Poeschel Verlag.

Zlokarnik, M. (1996): Produktionsintegrierter Umweltschutz in der chemischen Industrie. In: Brauer, H. (Ed.): Handbuch des Umweltschutzes und der Umweltschutztechnik. Band 2: Produktions- und produktintegrierter Umweltschutz. Berlin et al.: Springer Verlag, p. 444.

Fertiliser Production in Austria

Harald Payer[1]
CULINAR, Mayerhofgasse 4/11, 1040 Vienna, Austria

In contrast to all other annual bulk material flows through the Austrian economy, the use of mineral fertilisers has experienced notable long-term falls over recent decades (Jänicke et al. 1992; Payer et al. 1997). Alongside fertiliser production in some other European countries, the Austrian mineral fertiliser industry therefore presents itself as a highly interesting field for research into structural change and environmental restructuring. This chapter starts with an overview of international trends in fertiliser supply and demand. The second part focuses on the current situation in the Austrian fertiliser market, followed by an analysis of the structural change in Austria's fertiliser industry since the early 1970s.

The rest of the study deals with the process of political decision-making in relation to the structural change in this sector during the last three decades. Special emphasis is placed on the environmental aspects of this structural change. One of the main ideas of this study is to identify the key actors in the decision-making process in the fertiliser sector, their missions, aims and strategies, and to explore the possibilities for intervention by a more environmental industrial policy. Because of the serious lack of relevant literature on the fertiliser sector in Austria, the second part of the case study is based mainly on a set of structured interviews with selected experts in the fertiliser sector. From June to July 1998, we conducted eight interviews, each about one and two hours long. The range of interviewees was as wide as possible, to guarantee a broad variety of insights from different core actors: apart from representatives of three production firms, we also interviewed representatives of three wholesalers and two representatives from public authorities (the references include the list of interviewees). The interviews took place under a guarantee of anonymity, so that all parties were happy to participate and to provide confidential information, where necessary. All the interviews yielded satisfactory responses to the questionnaire. By and large, the results portray the current situation as well as the recent past. A deeper analysis of the long-term past development of the fertiliser market was not possible, since none of the selected persons was employed in the industry before the mid-1980s. Nevertheless, the selected sample offers a broad impression of developments over the last ten years. Deeper analysis of the issues requires more than interviews alone.

---

[1] Research for this chapter was carried out with the IFF, Schottenfeldgasse 29, 1070 Vienna, Austria. Special thanks to Karin Baumgardinger for her help with the data.

1 General Trends in International Fertiliser Supply and Demand

The fertiliser industry supplies agriculture with mineral fertilisers containing one or more of three main nutrients: nitrogen (N), phosphate (P) and potash (K). For several years, the international fertiliser market has been characterised by a fall in overall demand, caused mainly by reductions in Europe and the Commonwealth of Independent States (CIS). The roots of this development are political and economic change in the CIS and Central and Eastern European countries (CEE), as well as stricter environmental legislation and an intensified set-aside policy, imposed within the EU in response to general structural change in agriculture and the reform of the Common Agricultural Policy (CAP) in particular. Unlike in Europe, aggregate demand in developing countries has been growing rapidly since the 1980s, while demand in North America has remained relatively constant. The combined share of Asia, Latin America and Africa in world consumption has now risen to about 60% (Fig. 1).

Following developments on the demand side, the fertiliser industry in the EU experienced a drop in its share of world nitrogen and phosphate output, which now stands at about 10%. The EU fertiliser industry, which occupied a dominant position in world mineral fertiliser production (40%) in the 1950s, has adapted, primarily to meet the demands of EU agriculture. During the heavy restructuring of the industry in the early 1990s, the EU fertiliser industry reduced nitrogen fertil-

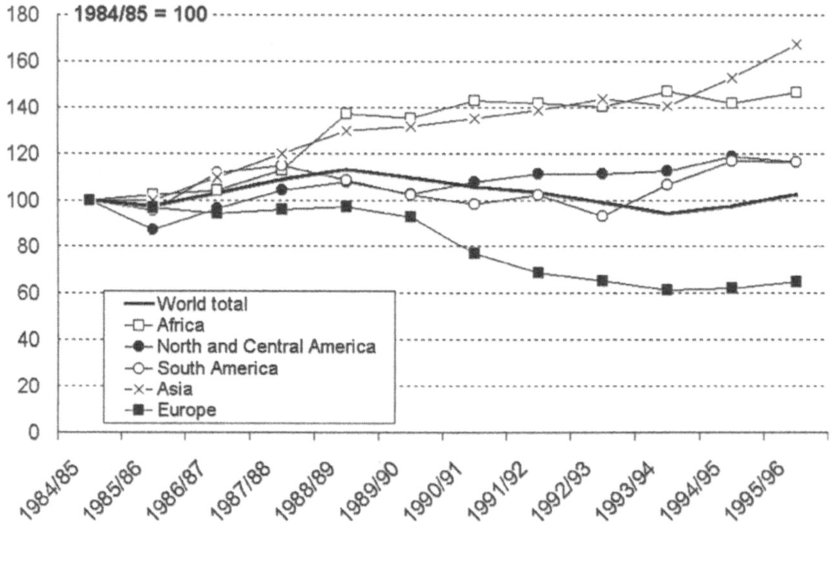

Source: FAO

Fig. 1. World fertiliser production 1984–96

iser capacity by 25% and phosphate fertiliser capacity by 33% (cf. Fig. 2). Some 66 plants closed down permanently and the workforce was halved. The plants which remain have been modernised, and logistics and sales units have been streamlined in order to improve competitiveness and to adapt to the changing supply and demand conditions which followed reform of the CAP and the introduction of a more liberal import regime. The loss of surplus capacity has led to an improvement in capacity utilisation and a reduction in unit costs (EFMA 1997a).

The main producers of mineral fertilisers in Western Europe are Norsk Hydro (N), Kemira Oy (SF), BASF (D), Grande Paroisse (F), DSM (NL) and Fertberia (E), which has been taken over by the US Freeport McMoran Resource Partners. Today, the EU fertiliser industry manufactures mainly nitrogen products (about three quarters of its total production volume). In comparison to foreign competitors, especially from CEE countries and the former Soviet Union, however, EU production is still faced with substantial location-related disadvantages. Among them is the higher price of natural gas which, as a major input factor in nitrogen fertiliser production, has strategic importance. Fertiliser producers are among the largest industrial consumers of natural gas in Europe (about 5% of the Union's total consumption), and therefore very much interested in the liberalisation of the common market for natural gas. The EU fertiliser industry also has an interest in the extension of EU environmental legislation to CEE countries (EC 1995).

Since the mid-1980s, Western Europe has moved from being a net exporter of fertilisers to being a major net importer today. Imports into the EU, amounting to

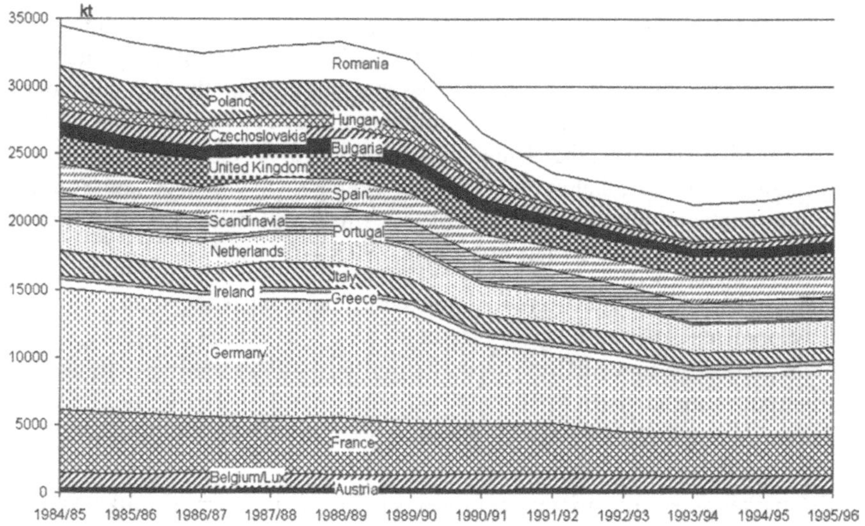

Source: FAO

Fig. 2. Fertiliser production in Europe 1984–96

13.4 million tonnes (1.8 billion ECU) in 1995/96, represent a high market share (nearly one quarter). Exports in 1995/96 amounted to only 4.4 million tonnes (0.6 billion ECU), whereas the total production of the EU fertiliser industry was some 50 million tonnes (gross product weights). The average unit values per tonne of imports and exports were nearly identical. Fertiliser exports consist mainly of multi-nutrient products, by-product sulphate of ammonia, and some nitrates and urea, the main markets being China – the world's largest fertiliser market, with a share of 27% of global consumption – Thailand and the Americas. On the import side, the most important suppliers of nitrogen fertilisers from abroad are CEE countries and the CIS, headed by Russia, Poland, Lithuania and Bulgaria. The market share of these imports has risen from 5% during the late 1980s to over 20% in 1995/96. The major suppliers of phosphate fertilisers are Russia, Morocco and Tunisia. In general, imports from the CEE countries and the CIS are expected at least to remain at these high levels, and are likely to increase. Surplus production capacity, the crisis in agriculture and a need for foreign exchange in these countries continue to result in large quantities being available for export, and pressure on the EU fertiliser industry will therefore probably persist (EFMA 1997a).

EU consumption of mineral fertilisers is currently made up of about one half nitrogenous fertilisers and about one quarter each phosphate and potash fertilisers. The latter are mainly offered as multi-nutrient products, whereas nitrogenous fertilisers are for the most part marketed as single nutrient fertilisers. Within the EU, France is the largest market for mineral fertilisers, followed by Germany, the United Kingdom, Spain and Italy, the highest rates of application being found in the Netherlands, Belgium, Finland and France. National differences in consumption and rates of application have much to do with the local soil and weather conditions and the kind of crops which these favour. Fig. 3 compares the average use of mineral fertilisers in the 15 countries of the EU. Generally speaking, the trend of falling rates of application per hectare has continued during the 1990s, although the opposite is the case for five of the fourteen markets shown. The largest falls can be found in Denmark, the Netherlands and Finland. As for Austria, the rate of application is one third below the European average. Only Greece and Portugal have lower rates of application. Main factors which will determine EU fertiliser consumption in the future are (EFMA 1997a, 1997b):

- restrictions on the production of oil seed crops, milk, sugar and wine
- developments in the world grain market
- grain production in the EU
- beef production in the EU
- EU and national environmental regulations, such as the Nitrates Directive
- developments in crop yields
- EU market prices for agricultural products falling towards world market levels
- agricultural compensation payments becoming de-coupled from output levels
- improved nutrient efficiency in organic manure and the implementation of better, more specific, fertiliser recommendations

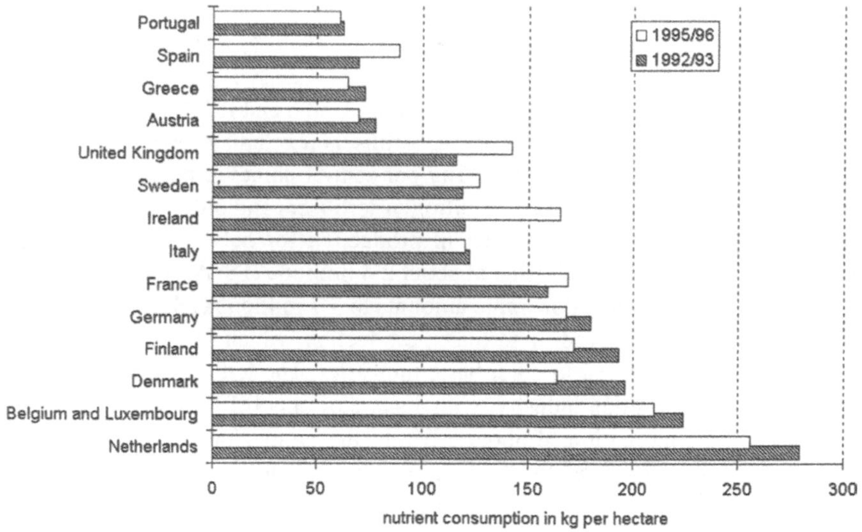

Source: EFMA 1997a

Fig. 3. Average consumption of mineral fertilisers in the EU 1992/93 to 1995/96

"The combined effect of these factors is currently expected to lead to a decline in the quantity of nutrients applied in the form of mineral fertilisers in the European Union" (EFMA 1997a, p. 20). In view of falling domestic demand and sustained competition from suppliers in regions with more favourable local conditions, further structural change seems probable, expressed in continued increases in concentration, closure of unprofitable plants and further reductions in the workforce.

2 Main Features of the Current Austrian Fertiliser Market

The mineral fertiliser market in Austria is characterised by a quasi duopoly on the supply side and strong export-orientation. Only two firms – Agrolinz Melamin GmbH, the largest producer, and Donau Chemie AG – account for virtually all domestic production volume. A very marginal market share is held by a few suppliers of organic fertilisers, occupying a market segment which has been a very small but solid niche since the mid-1990s (see Sect. 4.1.1 for details).

According to foreign trade statistics, about 80% of domestic production is exported, but it supplies about 60% of annual domestic demand. Including nutrient supply from organic fertiliser (mainly animal manure), domestic consumption of mineral fertiliser in 1996 amounted to 760 kt.[2] About half of domestic demand is

[2] The nutrient content (N, P_2O_5, K_2O) of organic manure in 1994 is estimated at 442 kt, while the nutrient content of mineral fertilisers was 231 kt (Dachler and Kernmayer 1997, p. 199).

currently geared to nitrogen fertiliser, a third to multi-nutrient fertilisers, the rest to phosphate fertiliser and potash fertiliser. About 20–30% of domestic demand is supplied by imports, mainly from Slovakia, the Czech Republic and Hungary (ÖSTAT) (information from Agrolinz Melamin Gmbh and RWA).

Total domestic production of mineral fertilisers (excluding limestone) amounts to about 1.2 million tonnes (1996). Fifty per cent of domestic production consists of multi-nutrient fertilisers, 47% of nitrogen fertilisers and only 3% of phosphate fertilisers. Potash fertiliser has been in production for only a short time. Total turnover in the branch is estimated at about 2.4 million ATS. According to official statistics, the gross production value amounts to 1.7 million ATS. About 600 people are currently employed in fertiliser production, which implies a rather low labour-intensity compared to other basic industries. The fertiliser industry plays a marginal role in Austrian industry, with a share of only 0.1% in terms of production value. Its contribution of about 1.6% to the production value of the whole chemical industry (2.6% relative to the production of chemical basic goods) is also fairly small. These indicators, however, underestimate its technological importance for the nitrogen industry as a whole. The fertiliser plants are an important link in the chain of industrial refinement from natural gas to a wide variety of nitrogen products for agricultural and industrial users.

Fertilisers are produced as single or multi-nutrient products, generally supplied in solid form as granules or prills. Liquid fertilisers are an exception and require special handling and storage. Although the components of finished fertiliser products are relatively simple chemicals, highly developed manufacturing technologies are employed in the production of fertilisers, and production plants are capital-intensive. The intermediate product in the case of nitrogen fertilisers is ammonia, which is produced by combining nitrogen extracted from the air with hydrogen from hydrocarbons such as natural gas, naphtha or other oil fractions, or from steam cracking. Approximately 85% of the anhydrous ammonia plants in the EU use natural gas. Further processing of ammonia produces simple nitrogen fertilisers such as urea and ammonium nitrate. Ammonia is also the main component of many multi-nutrient fertilisers and a broad product range of other nitrogen chemicals (EFMA 1997a). Agrolinz Melamin, the only domestic manufacturer of nitrogen fertiliser, is the biggest consumer of natural gas in Austria, with an annual amount of about 500 million m^3 (Agrolinz Melamin GmbH 1996b), 7.6% of total national consumption (BMWA and BLV 1995). OMV AG, Austria's major public petrol combine, which has interests in fertiliser production through Agrolinz Melamin, controls about two thirds of domestic natural gas extraction and a monopoly of natural gas imports. Information on the current cost price of natural gas to Agrolinz Melamin is confidential, but it can be assumed that it is significantly below the average cost price to other industrial firms. During the mid-1980s, the cost price was about one third lower than the industrial average (Aiginger and Bayer 1988).

Phosphate fertilisers are produced from rock phosphate, which has to be imported. To make phosphorus more readily available to plants and to concentrate the product, phosphate rock is processed with sulphuric, phosphoric or nitric acid. Acidulation with sulphuric acid produces either phosphoric acid, an intermediate product in the production of triple superphosphate, monoammonium phosphate, di-ammonium phosphate and multi-nutrient fertilisers, or single superphosphate. Acidulation with phosphoric acid produces triple superphosphate, while acidulation with nitric acid produces nitrogen-phospate slurries for use in the manufacture of multi-nutrient fertilisers (EFMA 1997a). Agrolinz Melamin uses an acidulation process to produce multi-nutrient fertilisers substituting sulphuric with nitric acid, which – together with urea – is an intermediate product of ammonia production.

On the whole, the fertiliser industry proves to be rather loosely integrated into the Austrian economy as such. With regard to the integration of both input and output, the industry is confined to a small number of subcontracting suppliers and clients. On the one hand, production is based on a few, relatively simple stages, with very low mineral and energy inputs, a large proportion of which must be imported. Changes in fertiliser production therefore have only a marginal effect on other domestic industries. On the output side, agriculture is the only sector which uses fertiliser.

With respect to the main environmental effects of fertiliser production, a distinction should be drawn between direct, more or less local, pressure on the environment, in the form of emissions and waste, and indirect pressure caused by the extraction and transport of raw materials (environmental *rucksacks*) as well as inappropriate fertiliser application. In general, excessive use of nitrogen fertiliser has a negative effect on groundwater (due to washing-out of nitrates), plant fitness, food quality, soil microorganisms and the ozone layer (due to nitrogen dioxide emissions from soil). Excessive use of phosphate fertilisers has a negative effect on surface water ecosystems by stimulating the growth of algae (eutrophication) (Gerhold 1990). Furthermore, almost all mineral fertilisers contain traces of heavy metals. Particularly problematic are the high cadmium content in phosphate fertilisers, the chloride content in potash fertilisers and the chromium content in phosphate fertilisers, which are produced from basic slag, a by-product of steel production (Dachler and Kernmayer 1997). The general decline in fertiliser use per hectare, combined with a fall in agricultural land use, have made an important contribution to reducing environmental pressure. In addition, efficiency has been significantly improved by technological progress and agricultural advisory services, primarily in response to stricter environmental regulations on protecting soil and groundwater.[3] A reduction in the environmental impact of the manufacturing process has been achieved both by improving its efficiency and by preventing emissions (see section 4.1.1).

[3] It should also be mentioned that the inappropriate use of animal manure puts similar pressure on the environment. Intensive livestock farming in particular gives rise to considerable quantities of animal manure, which bear no regional or local relation to the land area being farmed, and is far greater than that which can be utilised by the crops or absorbed by the soil.

3 Structural Change in Austrian Fertiliser Production Since the 1970s

The post-war industrialisation of agriculture in Austria – and elsewhere – has brought about a continuous increase in the consumption of mineral fertilisers, with an average long-term growth of about 12% per year (Fig. 4). This trend peaked at the beginning of the 1970s, followed by a dramatic decline after the first oil crisis, a considerable revival until the early 1980s and, from the mid-1980s to the present, a continuous fall, a consequence of heavy restructuring in agriculture. Today, domestic fertiliser consumption has fallen to the level of the early 1960s, more than three decades ago. It is obvious that structural change in agriculture must have had a powerful impact on domestic fertiliser production. In the long run, it has transformed the fertiliser industry from a former supplier of exclusively domestic demand to a largely export-orientated branch with a presence on highly competitive markets.

The long-term decline in fertiliser consumption can be accounted for by several more or less interconnected factors. It turns out that developments in Austria are somewhat similar to those in most European countries:

- Marginal returns from agricultural production fell.
- Financial pressure grew for public subsidy of high surplus agricultural production and price support for cereal production was gradually reduced.

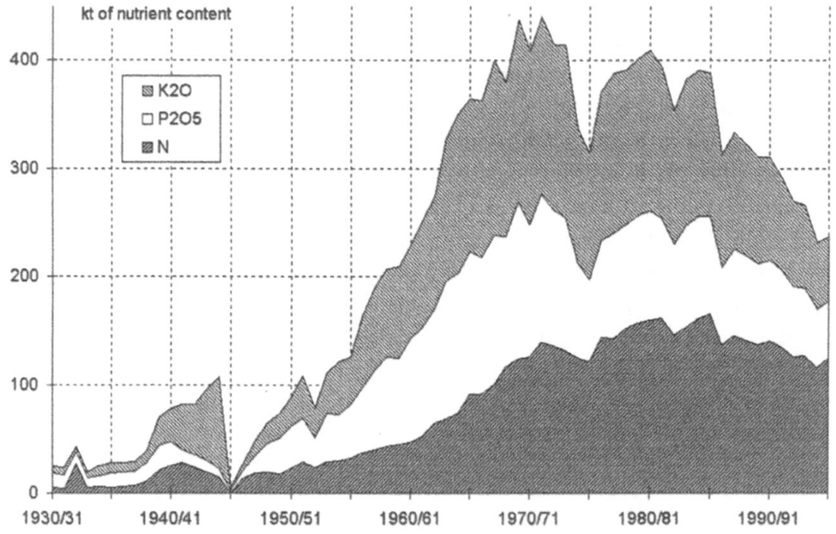

Source: Bundesamt und Forschungszentrum für Landwirtschaft

Fig. 4. Mineral fertiliser consumption in Austria 1930/31–1995/96

- US agricultural policy put competitive pressure on European agriculture (in the 1988 Uruguay Round, the US called for the complete elimination of all farm and export subsidies within ten years).
- Agricultural land use decreased, promoted by the introduction of set-aside schemes.
- Fertiliser consumption (measured by net nutrient content per hectare) became more efficient.
- Criticism of the environmental impact of intensive agriculture has increased since the early 1980s, incited mainly by the discovery of high nitrate contents in drinking-water, eutrophication of rivers and lakes through phosphates and the intensive use of pesticides.
- In 1990 an amendment to the Austrian Water Act (*Wasserrechtssgesetz*) came into force, prescribing a general licensing obligation for the use of fertilisers.
- After accession to the EU, a new incentive system for agricultural production was introduced in Austria, the "Austrian Programme for Environmentally Appropriate Agriculture" (*Österreichisches Programm für eine umweltgerechte Landwirtschaft, ÖPUL*) which, among other things, lays down limits on fertiliser use.
- The boom in organic farming since the beginning of the 1990s: by 1995, 7.3% of Austrian farms (6.5% of total agricultural land) was already adhering to the principles of organic farming, which inter alia almost completely renounce the use of synthetic fertilisers (BMLF 1997).
- In May 1986, following the examples of Finland, Norway and Sweden, a national fertiliser tax (*Förderungsbeitrag auf Düngemittel*)[4] was introduced in Austria, primarily in order to finance sustained surplus cereal production (via marketing and manufacturing subsidies), but also in order to give an incentive to the more effective use of artificial fertilisers (see Table 1). In fact, the tax had a greater than expected effect on fertiliser sales, as the tax offset the simultaneous long-term fall in international fertiliser prices, resulting in a continuous decline in consumption from 1986 (Gerhold and Steurer 1993). After Austria's accession to the EU, the levy was repealed because it did not conform to EU legislation. The abolition of the tax, however, failed to stimulate fertiliser use.

With respect to the parallel development of domestic production, three different periods can be observed (Fig. 5):

- a sharp decline during the mid 1970s as a consequence of the first oil crisis,
- an upturn until the end of the 1970s,
- a sustained decline from the beginning of the 1980s, interrupted only by an appreciable short-term recovery between 1986 and 1988.

In 1974, the physical output of fertiliser production reached its post-war high at a level of about 1.7 Mt. Since that time, output has fallen by 34% to about 1.2 Mt in

[4] also known as: *Bodenschutzbeitrag, Bodenschutzabgabe, Düngemittelabgabe*.

Table 1. Rates of and revenues from Austrian fertiliser tax 1986–94

	1986	1987	1988	1989	1990	1991	1992	1993	1994
	rates (ATS / kg net nutrient content)								
Nitrogenous fertiliser (N)	3.5	5.0	5.0	5.0	5.0	6.5	6.5	6.5	6.5
Phosphate fertiliser (P)	2.0	3.0	3.0	3.0	3.0	3.5	3.5	3.5	3.5
Potash fertiliser (K)	1.0	1.5	1.5	1.5	1.5	1.9	1.9	1.9	1.9
	Mio. ATS								
revenue	230	970	883	1.034	1.039	1.315	918	1.180	947

Sources: BMLF 1998

1996. However, the value of gross production in real terms did not follow suit for the whole period and may be summarised as follows: The value of gross production experienced a significant boost until the mid-1970s, declined during 1975 and 1976 under the medium-term effects of the first oil crisis, rose again considerably until the mid-1980s and has since then declined steadily to 1973 levels, as a result of several international falls in price. These were the consequences of the fall of the dollar and sustained reductions in energy prices, which had positive ramifications above all for the highly energy-intensive production of nitrogen fertilisers.

The sustained fall in production since the early 1980s was accompanied by thorough restructuring of Austria's entire state-owned industry until the mid 1990s and growing competition in Central European markets. The restructuring of Austria's formerly highly subsidised public basic industry combine resulted in a gradual separation and partial sell-off of petro-chemical and nitrogen production, which finally led to the consolidation of nitrogen production plants in the new company of Agrolinz Melamin in 1994. This new organisational solution has also brought about significant cuts in surplus fertiliser production capacity, which was created only a few years ago with the help of extensive, mainly environment-linked public subsidies and a guarantee of low purchase costs for natural gas input, but under massive protest from citizens' committees and the Green Party. Also in vain was a recommendation by the respected Austrian economic research institute, the Wirtschaftsforschungsinstitut, to avoid new investment because of the industry's lack of competitiveness (Aiginger and Bayer 1988). The fertiliser market in Central Europe has actually become more and more competitive. Since the 1980s the market has been determined by surplus capacity, significant falls in price and cheap supplies from CEE countries.

Besides the gradual rationalisation of production plants, Austria's fertiliser industry has also adapted to structural change with a significant shift into foreign markets (Fig. 5). However, a steady transformation has been in evidence since the beginning of the 1970s, from mainly domestic-orientated production to increased competition on international markets. Austria has shifted from being a former net importer to become a net exporter. Comparative advantages such as low energy

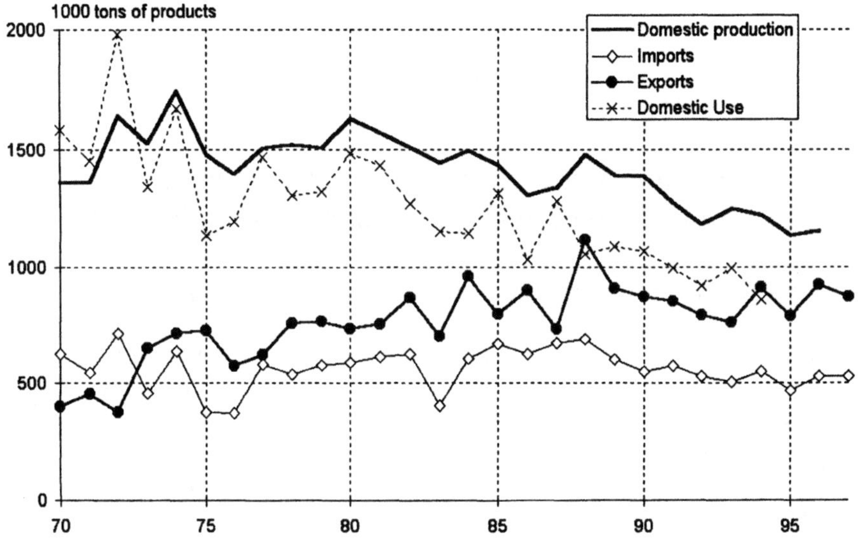

Source: ÖSTAT, own calculations

Fig. 5. Fertiliser production and foreign trade in Austria 1970–97

and transport costs have gained in importance. But more competitive markets are also less steady, which can be seen in the strong fluctuations in export sales during the 1980s. From the late 1980s to the early 1990s, a period characterised mainly by serious restructuring of the EU fertiliser industry and the agricultural crisis in CEE countries, Austrian exports have also fallen significantly. They started to rise again only after increasing stabilisation in the agricultural sector and the gradual adjustment of raw material prices to EU levels. Notwithstanding, the Austrian fertiliser industry has turned into a primarily export-orientated branch. Since the mid-1990s, export volumes have even exceeded total domestic consumption. Today, about three quarters of domestic production are exported to foreign markets in Southern Germany, Italy, Switzerland, the Czech Republic, the Slovakian Republic, Hungary and Slovenia, all of Austria's neighbours. The fact that both production plants lie directly on the Danube brings an important advantage in the competition for markets in CEE countries.

In view of the conclusions of the GATT negotiations, the expected reform of the CAP, the enlargement of the EU and the next WTO Round in 1999, it can be assumed that this trend of growing competition in a saturated market will continue. It seems obvious that, on the one hand, surplus agricultural production will be far more limited, the fall in the price of agricultural basic products will continue, agricultural earnings will be further diminished and the number of farms will fall. On the other hand, the efficiency of fertiliser use can still be improved significantly, subsidies for less intensive methods of farming will probably be increased and, last but not least, the emergence of new food technologies could reduce dependency on soil-based agricultural production. Moreover, the use of nutrient substitutes such as compost (organic waste), sewage, or the absorption of nitrogen by intercrops could gain importance in the future.

4 The Social Framework of Structural Change: The Stakeholders in the Decision-Making Process

4.1 The Relevant Actors, Their Interests and Strategies

According to Kenis' (1992) description of the sectoralisation of the economy, one might regard the fertiliser industry as a socially constructed sector or even a loose network, which includes not only the producers, but also a number of other core actors. The case study of the Austrian fertiliser industry – a small branch, which was incidentally relatively easy to survey, compared to other sectors – nevertheless shows a social structure involving a wide variety of actors and concerns: the industry has an oligopolistic structure of production companies and wholesalers in the centre, surrounded by intermediary agents, suppliers of input materials, public authorities, research organisations, other institutions and the aggregate demand of the farmers (Fig. 6).

4.1.1 The Production Companies

The mineral fertiliser market in Austria has always been characterised by very high concentration on the supply side. The fertiliser industry is a quasi duopoly, with two companies –Agrolinz Melamin GmbH and Donau Chemie AG – supplying nearly all of domestic production (Table 2). The former accounts for about three quarters of domestic production, whereas Donau Chemie AG makes up about one quarter[5].

After heavy restructuring within the persistently loss-making state-owned petrochemicals complex, *Agrolinz Melamin* was founded in 1994 as a new organisational structure for part of the nitrogen chemicals production complex. The company is 100% owned by Chemie Linz GmbH, which in turn belongs to OMV AG, Austria's major public petrol combine. Not only in Austria, but also abroad, vertical relationships with petrochemical companies are the norm in nitrogen fertiliser production. Agrolinz Melamin is the market leader in Austria and Southern Germany. Further important export markets are the neighbouring CEE countries and Northern Italy. During the 1990s, fertiliser trading companies were founded in the Czech Republic and Hungary. 54% of the company's turnover of 4.5 billion ATS

[5] The marginal quantities produced by a few other firms are not worth mentioning. One point of interest about these minor producers is their products, namely recycled waste from basic industrial production – which seems to point to a general trend towards turning recycled waste materials into goods for agricultural markets.

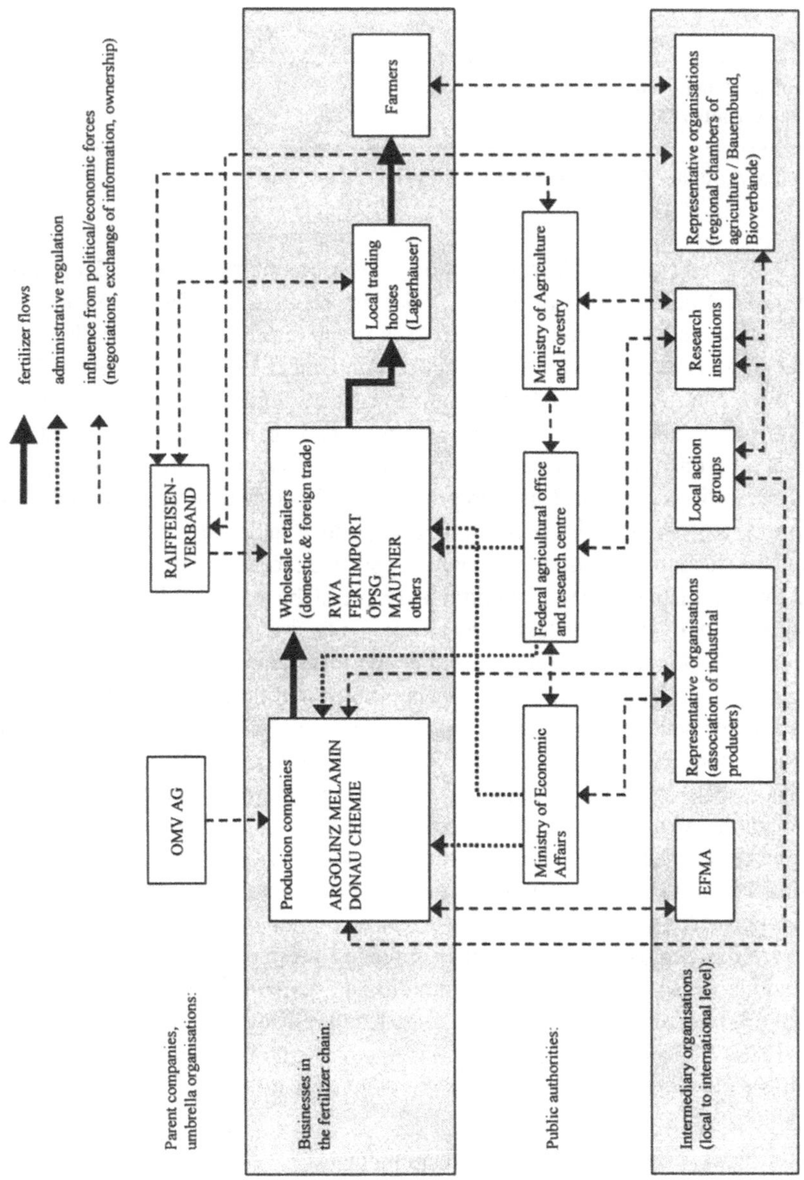

Fig. 6. The fertiliser network in Austria during the late 1990s

Table 2. Turnover, output and employees in Austria's fertiliser production 1996

	output (kt)	turnover (mio. ATS)	employees
Agrolinz Melamin GmbH	900	2 015	500
Donau Chemie AG	220	350	80

Sources: Agrolinz Melamin GmbH; Donau Chemie AG

in 1996 came from fertiliser production, 31% from melamin/urea, 7% from pesticides, 3% from feed additives and 14% from services. In 1996, annual fertiliser production, mainly nitrogen and multi-nutrient fertilisers with a high nitrogen content, was about 900 kt[6] (Agrolinz Melamin GmbH 1996a, 1996b; APA 1997).

At the end of the 1980s, technological restructuring in fertiliser production was by no means uncontroversial. Public opinion had been staunchly opposed to the plants for some time, because of their poor environmental performance, especially in the traditional processes used to produce sulphuric acid, which required serious improvement. A new clean air act in 1989, which included the strictest emission standards in Europe, was an important supportive factor. In principle, the industry had two alternatives: either to continue less profitable basic production using more environmentally friendly technologies, or a massive reduction in capacity. The Green Party, together with recognised economists and even the Minister of Environmental Affairs, favoured restructuring, but found little support. After negotiations lasting several years a new, greener plant was built, partially financed from by the national Environmental Fund (Aiginger and Bayer 1988, Nemeth 1992).

Today, the company presents itself as a champion of environmentally friendly production. Apart from the technological improvements, with noticeable benefits for the environment, the company implements a very open environmental policy, maintaining close contacts with both citizens and the public authorities. Environmental protection has become an important part of the corporate identity.

Agrolinz GmbH has succeeded in reducing the most significant atmospheric emissions (dust, nitric oxide, sulphur dioxide, carbon monoxide, ammonia) to about 6% of levels in the-mid 1980s. As a result of heavy capital investment in the late 1980s, Agrolinz Melamin today has one of the most environmentally friendly fertiliser plants in Europe. These measures have contributed to a remarkable im-

6 In this book, production data are usually given in product weights. It should be noted that official statistics and reporting systems on fertiliser demand, e.g. the FAO Fertiliser Yearbook, regularly show the net weights of fertilisers' nutrient content. The difference between product weight and nutrient weight results from various additives such as lime, dolomite, oxygen or hydrogen, mainly to thin the highly explosive nitrogenous compounds. These additives diffuse into the air, the soil or the plants. Nutrient content data, therefore, primarily indicate the agricultural effectiveness of fertilisers. While the nutrient content of nitrogen fertilisers amounted to about 20% in the 1950s, today it usually amounts to about 50% of the product weight, as a result of technical progress.

provement in air quality in Linz, the capital of Upper Austria, where the company has its production site. Furthermore, the company has undergone an eco-audit under "Responsible Care", an international chemicals industry initiative to promote higher environmental standards (Agrolinz Melamin GmbH 1996b).

Until recently, *Donau Chemie AG* belonged to the French Rhone-Poulenc chemistry group, Europe's biggest producer of phosphate fertiliser, which held 62% of the company, and to Creditanstalt-Bankverein AG – once Austria's flagship public bank, now merged into Bank Austria AG – which held 35%. In 1996, both majority shareholders got out of basic chemicals production and the company was taken over in a management buy-out (APA 1997). Besides fertilisers, the company also produces sulphuric acid, calcium sulphate, latex, sodium hydroxide, hydrochloric acid, chlorine, ferric chloride and carbide. Annual fertiliser production, mainly highly concentrated phosphate and multi-nutrient fertilisers without or with little nitrogen content, is about 200 kt (Donau Chemie AG 1996).

Until recently production of phosphate fertiliser was based on one of the company's core technologies, i.e. the production of phosphoric acid. The company was one of the first in Europe to successfully solve the massive waste problems in the technology. Phosphoric acid production creates a lot of gypsum, which must be disposed of with heavy costs. It was environmental legislation which, among other factors, provided the incentive to develop a new technology whereby gypsum is refined for use as a construction material, efficiently using heat from sulphuric acid. For some time, this technological innovation provided the company with a remarkable first mover advantage.

In 1993, however, production of phosphoric acid had to be halted, due to cost considerations and more restrictive emissions limits contained in new sewage legislation. Today, there are only two plants in the whole of Europe. It is more profitable to import phosphoric acid from countries with lower costs and less restrictive, if any, environmental regulation.

The environmental expertise induced by the measures was passed from the Austrian subsidiary company to France, which enjoys considerably less restrictive environmental legislation. However, the Rhone-Poulenc combine began to reduce basic production lines in order to concentrate on higher value products, such as pharmaceuticals and specialised applications. After the management buy-out, the company became a more flexible medium-sized business. In consequence, research and development investment has increased, e.g. in the hyperphosphates product line. Hyperphosphates are manufactured mechanically instead of chemically, which makes them suitable for organic farming.

By and large, fertiliser production has shifted from simple output orientation to smaller scale, more flexible production, responding to farmers' needs and adapting to changing environmental legislation. Today, environmental considerations mean that mineral fertilisers are used more efficiently. Mineral fertilisers are an essential cost factor in farmers' calculations and – especially after the restructuring which came with accession to the EU – farmers are very interested in further rationalising fertiliser use. With the general decline in agricultural production it is to be expected that domestic demand will, slowly but surely, fall in future. Thus domestic manufacturers will face even greater challenges in shifting to foreign markets, with competition from cheaper products.

The clean air policy in the 1980s brought a first wave of notable end-of-pipe improvements in existing facilities. During the 1990s, a second wave of environmental reinvestment followed, mainly due to new restrictions on emissions to water. This was accompanied by implementation of a fertiliser law which, *inter alia*, aimed at more restrictive controls of pollutants contained in the products themselves. Environmental modernisation of the industry has reached a level where clean technologies (closed substance cycles) and environmental management have been established (reporting systems, environmental declarations, ISO-certification, environmental audits, public participation). Still, this progress is broadly regarded as a competitive disadvantage vis-à-vis suppliers in countries with lower environmental standards. Competitive advantages are found in the relatively flexible production. Another hope for future business lies in the supply of environmental services, mainly in response to agricultural demand. Today, the fertiliser industry possesses notable environmental expertise, which might be commercialised more efficiently. Given the fact that fertiliser production is not the core business for any manufacturer in Austria, the search for new markets could become a high priority in future.

There are also apparently important intermediary agents in the fertiliser business. At international level, the *European Fertiliser Manufacturers Association (EFMA)* represents the major mineral fertiliser producers in Western Europe. Agrolinz Melamin GmbH, as the biggest domestic producer, is the only EFMA member in Austria. The association mainly supplies information and lobbying services to their members. Members' production capacity amounts to about 85% of total production capacity in mineral fertilisers in Western Europe. EFMA members promote working towards Best Available Technology (BAT), described in regularly updated EFMA booklets and in their internal Responsible Care programmes, and subscribe to the principles contained in the Business Charter for Sustainable Development. At national level, there is no specific association for the interests of fertiliser producers. However, domestic associations – the *Vereinigung Österreichischer Industrieller* and the *Bundeswirtschaftskammer* – also serve as representative organisations for the interests of fertiliser producers.

4.1.2 The Wholesalers

The most important clients for manufacturers are wholesalers. They serve not only as an interface with domestic agricultural demand, but play also an important role in foreign trade of fertilisers. In this function they are both partners and competitors for manufacturers[7].

---

[7] The interviews also included a managing director of Bodenkalk GenmbH, which is the clearing agency for the major part of Austria's limestone industry, and therefore responsible for nearly

The largest wholesaler, *Raiffeisen Ware Austria (RWA)*, accounts for at least two thirds of fertiliser trade in Austria (imports included). Thus, RWA and Agrolinz Melamin GmbH are the two most important fertiliser companies in Austria. Their performance and strategies are crucial to the development of the whole sector. RWA is a kind of clearing agency for local trading houses (*Lagerhäuser*). The distribution system is organised as a number of co-operatives. Each trading house is an autonomous co-operative society, many of whose members are farmers. The business activities of RWA include bulk marketing the whole range of agricultural inputs: fertilisers, pesticides, crops, fuels, agricultural machinery, farm equipment, and even construction materials. On the other side, RWA distributes basic agricultural products to the foodstuffs industry. Furthermore, the company supplies many services to farmers, such as consulting and public relations. After eastern Europe opened up, and in the course of integration into the EU, the company was reorganised in order to flatten hierarchies and to become more flexible. Today, the RWA is still a client-oriented service company for the local trading houses, rather than a central supply unit. The RWA is a member of the Austrian *Raiffeisenverband*, which represents one of the major conglomerates of Austria's national economy, with a corresponding influence on domestic political life.

Apart from RWA, there are other medium-sized private wholesalers (*Glatz, Mautner, Url, Beiselen*), who together account for about a quarter of total domestic turnover in fertilisers. Besides these, there is a handful of small importers, who mainly deal abroad.

In the 1980s, RWA and the two private wholesalers also founded a joint subsidiary in order to manage fertiliser imports. The organisation has become a kind of clearing house for the Central European market around Austria[8]. The building of a new transshipment terminal at Krems harbour on the Danube has been an important competitive advantage. The terminal, which is jointly owned by *Fertimport* and the local harbour company, provides a modern transshipment infrastructure including packaging plants and huge temporary storehouses. Adequate storage is necessary for the fertiliser trade, because of the relatively long storage periods. Mineral fertilisers are bulk goods which scarcely can be delivered on demand[9].

---

all limestone trade for agricultural applications. The production of limestone for soil improvement cannot be directly counted part of the fertiliser industry, because it does not supply nutrients for the plants as fertilisers do. However, its distribution to the farmers is often associated with distribution of fertilisers. Domestic annual turnover of limestone for soil improvement amounts to about 350 kt. About 30% of this market is supplied by Bodenkalk GenmbH. A further third of the market is supplied with lime residues from various basic industrial processes, like the cleaning of sugar mash, steel production, production of calcium cyanamide, production of cellulose or sewage treatment plants. Lime residues seem to be a fast-growing business field in all respects.

[8] Incidentally, the establishment of *Fertimport* also affected the domestic market, by increasing competition for domestic production companies.

[9] The EFMA (1997a) estimated that logistical costs, including handling, transport and storage, represent about 20% of the price of fertiliser to the farmer.

4.1.3 Public Authorities and Regulatory Instruments

Apart from plant licensing law, the most important regulatory instruments affecting fertiliser production are laws and decrees which determine limits on substances, both for the products and the pollution caused by their application:

Since 1985 there has been a fertiliser law in Austria which *inter alia* regulates controls on all fertiliser products. At the outset, a relatively unwieldy registration procedure with high application fees was implemented, mainly in order to combat growing pressure from imports. Then, in the run-up to EU membership, the system was adapted to the EU standard, involving type lists (*Düngemittelverordnung* 1994), which greatly simplified and harmonised registration procedures. According to the fertiliser law, the public controls on fertilisers and pesticides is a matter for the Federal Office and Research Centre of Agriculture (Bundesanstalt für Landwirtschaft), which is an agency of the Ministry of Agriculture and Forestry. In legislation and execution, the Ministry of Agriculture and Forestry is responsible for licensing agricultural inputs, in particular fertilisers, animal feed and pesticides. The Ministry of Agriculture and Forestry co-operates with the Ministry of Economic Affairs. In order to guarantee this high level of control, the public authorities conduct many field studies in close co-operation with universities and other research facilities.[10] Austria has retained the right to require registration for every new fertiliser product. Quality control is achieved in spot-tests on a wide range of parameters, such as brand name, nutrient content, pollutant content and tolerance of crops. The public authorities have a very good overview of the whole market, the products and their quality.

The Ministry of Agriculture and Forestry and the Ministry of Environment are also responsible for regulating polution from fertiliser application. The most important instruments are limits on substance levels in groundwater (*Verordnung betreffend Schwellenwerte für Grundwasserinhaltsstoffe* 1991), which is based on an amendment to the *Wasserrechtsgesetz* in 1990, and on nitrates in drinking water (*Trinkwasser-Nitratverordnung* 1989), which is based on Austrian food law.

4.1.4 Agriculture

Austria's agriculture is marked by small and medium-sized farms. According to the last survey of the agricultural holdings structure, there are about 263 000 agriculture and forestry businesses, which manage about 80% of Austrian territory. About two thirds of all farms are run as part-time farms providing subsidiary income. Whereas in 1960 almost a third of all employed persons were working in

---

[10] It seems that agricultural research has also made a considerable contribution to structural change in the fertiliser business, in particular, scientific efforts in analysing the metabolisis of nutrients in the soil and ground water cycles. For example, running improvements in the nutrient efficiency of cultivated plants is an important result of scientific experiment. It ought to be remembered that there is much organisational and personal intersection between the fertiliser industry, the agricultural universities and other agricultural research facilities.

agriculture, this amounted to only 4.4% in 1996. The annual migration rate from rural areas is currently about 5%. In particular, the accession of Austria to the European Union in 1995 has brought far-reaching change in the agricultural system. It has created a new competitive situation in the domestic market; introduced more market pricing instead of regulated prices, but also a protective network for the majority of agricultural products in the form of EU intervention, which ensures a minimum price; a comprehensive system of quotas, within which farmers are eligible for premiums; compensation payments for services rendered, especially a new agri-environmental aid system, by which environmental services, such as preserving the agricultural landscape, are rewarded. It can be assumed that the long-term easing of the Common Agricultural Policy will continue for the next few years. All in all, structural change during the last decades has brought a continuous decline in fertiliser consumption (BMLF 1997, 1998).

Today, organic farming's share in gross domestic product amounts to about 1.4%. The share of animal production in final agricultural production, amounting to 48 billion ATS, is about two thirds, plant production one third. In Austria, grassland areas (arable foraging, intensive and extensive grassland farming as well as alpine pastures and mountain meadows) cover about 58% of the total available agricultural land. Cultivation of these grassland areas is characterised by low livestock density and moderate use of mineral fertilisers and herbicides. The grassland areas are an essential factor in the rise of organic farming in Austria, which currently covers about 8% of the available agricultural land. Although support for organic farming might be regarded as one of the most promising opportunities for many farmers in Austria, its contribution to the general decline in fertiliser consumption ought not to be overestimated (BMLF 1997, 1998).

4.1.5 Non-Governmental Environmental Organisations

Undoubtedly, the green movement was an important factor in the rise of public opposition to the intensive use of fertilisers, pesticides and intensive agriculture as a whole. Green initiatives were also important for the successful development of organic farming in Austria. Today, environmental organisations are partly involved in agricultural administration, at least in matters like funding for landscape conservation or the programme and criteria for the ÖPUL fund.

However, there has been no more direct intervention by environmental actors in the structural change in Austrian fertiliser production. Production itself has never been the cause of serious environmental conflict. Only the debate on licensing Agrolinz Melamin's new fertiliser production plant brought direct criticism from environmental organisations and the green party, as mentioned above.

4.2 Patterns of Interaction

Altogether, it seems that there is little strategic co-operation between the various actors in the Austrian fertiliser business. The main medium for interaction is the market. There are no real networks. Most of the actors, however, seem to be well-informed about any activities due to the smallness of the domestic market. The two main players in the industry, the largest producer and the major wholesaler, do coordinate strategies, and there is coodination between the wholesalers themselves, which even produced a common subsidiary to manage foreign trade.

The failed agreements on pricing guidelines confirm the impression of little strategic co-operation. In fact, there is a common interest in fixing price margins, but individual companies generally break away from the pack. Stable co-operation is therefore more likely with foreign partners, mostly in order to open up new markets. This seems to be the result of polarisation in response to the two core actors, Agrolinz Melamin GmbH and RWA. Concerted action is common between these two companies, because of their strong economic interdependency. They regularly try to establish forms of "gentleman's agreements" in order to minimise the price exposure of trading houses and farmers.

This dependency between the two main actors also seems to be an important reason for the fact that political intervention or party rivalries seem to have little importance for development in the fertiliser industry. This might seem surprising, bearing in mind that the two enterprises represent in a sense the balance of economic power and influence between the two major political parties in Austria: on the one side, Agrolinz Melamin GmbH is at the heart of public industry, strongly dominated by the social democrats; on the other side, RWA belongs to the Austrian Raiffeisenverband, dominated exclusively by the conservatives. Whatever the reasons, coexistence between the two main players in the fertiliser industry is conducted in a market-based atmosphere, away from party political intervention.

Apart from the above, there are also institutional opportunities for coordination, organised mainly by intermediary agents and public authorities, such as standards committees, legislative committees or more or less static working groups on more technological issues, including the environment. These meetings seem to be a place for formal and informal exchanges of information and expertise, but not for common strategies or decision-making.

A further dimension of interaction within the fertiliser branch is the changing role of the public authorities. Their expertise and professionalism have grown remarkably, as bureaucracy and science have moved into agriculture, although the environmental movement has also played a part. As a result, market actors now try to keep stable and professional lines of communication open with the public authorities and vice versa, especially the most powerful actors.

4.3 The Stakeholders' View on Structural Change

If one asks experts about the effects of key external determinants of structural change in the fertiliser industry, price competition from abroad comes a clear first (Table 3). It is interesting that the importance of cheap imports seems to have been the dominant determinant over the whole period, even during the 1970s. This assessment points up two essential factors: the major influence of importers on the

Table 3. The influence of external determinants on structural change in the fertiliser market in Austria

Price competition from Eastern Europe	41	highly significant	37-45
Price competition from abroad in the 1970s	41		
Subsidies for set-aside land	32	rather significant	28-36
Increasing efficiency of fertiliser use	31		
Decreasing marginal returns in agriculture	30		
Implementation of consulting agencies for fertiliser application	30		
Fertiliser charge 1986 to 1994	28		
The crisis in public industries in the 1980s	28		
Introduction of ÖPUL after accession to the EU	27	significant	19-27
Implementation of a registration system	26		
Criticism of industrialised agriculture by the green movement	26		
The rise of organic farming	25		
Oil crises in the 1970s	25		
Results of the Uruguay Round	19		
Dollar price decline in the 1980s	17	less significant	10-18

aggregated ratings by experts (possible maximum = 45)

dynamics of the domestic fertiliser market and the necessity for domestic producers to shift to foreign fertiliser markets. According to the experts, then, the most important determinant for structural change is not agricultural, but from price developments in the European market.

As shown above, most of the other determinants, led by compensation for set-aside land, are considered factors of high to medium importance. In general, the market has greater influence than public controls such as emissions standards or licensing schemes. Even the most influential public measures – set-aside schemes, advice on fertiliser use and fertiliser taxes – have in effect been market measures.

As to the long-term impact of the fertiliser tax on restructuring in the Austrian fertiliser industry, the literature and the interviewed experts take a different view. There is evidence that farmers made large advanced purchases shortly before the tax came into force. There is no doubt that, in comparison to neighbouring countries, Austrian farmers had to pay higher prices for fertiliser. But the total additional burden was not high, because of the relatively low level of fertiliser use customary in Austria. It is hard to say that fertiliser use would have been higher without the tax. After all, the abolition of the tax after accession to the EU did not stimulate fertiliser consumption. However, the tax had at any rate important symbolic value.

Other important determinants were general criticism of the environmental pressure from agriculture and the accompanying rise of organic farming from the beginning of the 1990s. These green determinants do not however dominate in the assessment of the experts, their impact on the decline of fertiliser production is given no more weight than exorbitant oil prices in the 1970s.[11] Even less importance is attached to the results of the Uruguay Round and the decline of the dollar price during the 1980s.

5 The Influence of Environmental Policy

Environmental actors and policy can exert influence on the fertiliser industry at two different levels: on the production side, environmental interest focuses on the plant, whereas on the consumption side, the issue is reducing environmental damage caused by inappropriate use of mineral fertilisers. In this context, it should be mentioned that the damage from the intensive use of organic fertilisers can exceed the impact of mineral fertilisers many times over.

It has been shown that the structural change in the industry, namely the continuous decline in production, was primarily linked to the development of the European fertiliser market and the radical restructuring of Austrian and European agriculture as a whole. The environmental movement, and the growth in the importance of environmental policy supported the trend. But both the extent of this support and whether environmental policy could accelerate the decline remain unclear. In general, the environmental movement has also caused some rethinking in agriculture. The rise of organic farming and a profitable market for organic produce in Austria are notable indicators, but it should be remembered that many organic dairy farmers have never used mineral fertilisers.

As in many other countries, it was rising nitrate pollution of the groundwater which provoked the first widespread public criticism, followed by the heavy metals included in mineral fertilisers. Other problems, such as the accelerated growth of algae (eutrophication) caused by excessive use of phosphate fertilisers, or erosion, were more a matter of concern for experts. However, the gradual rationalisation of fertiliser use could not diminish the intensive use of animal manure as an organic fertiliser, mainly in regions with large livestock farms. Here the fertiliser industry is faced with a dilemma, because large-scale agricultural feed producers are important clients of industrial suppliers, but the main clients of feed producers, the large livestock farms, give a bad image to mineral fertilisers.

The experts see environmental research as the most significant mode of influence on structural change on the part of the various environmental actors (Table 4). The impact of activities by public environmental authorities, environmental politicians and non-governmental organisations is seen to be moderate. The influence of the media scores just over a half of the maximum 45 points, which means that assessments of media influence balance each other out among the interviewees. It should be noted that industrial representatives had a rather low opinion of

[11] Some experts also mentioned integration into the EU, the crisis in the Balkans (which caused a temporary halt of shipment at the Danube), increased use of animal manure, greater attention to crop rotation, the implementation of waste treatment legislation, and the rise of modern environmental technology.

the media, whereas their assessment of environmental research was the most positive of all.

6 Future Perspectives

It can be assumed that the set-aside policy will continue. The model will probably lead to still more liberalisation of the agricultural markets, with simultaneous government support for farmers' environmental services. This is also the clear target of Agenda 2000. This continuing shift may well result in greater social conflict over distribution concerning the future importance of rural cultivation, and it will also present a still greater challenge to farmers as actors in free markets searching for new business (e.g. food production, tourism, renewable resources). Whatever the case, it will not lead to rising fertiliser consumption in Austria.

Thus, the future of domestic production will be influenced more by the development of foreign markets, which means above all the development of agriculture in Eastern European countries. If the crisis in agriculture there can be overcome, demand for mineral fertilisers should increase to former levels. Today, average use is about one third below its level before the countries opened up to the West. The planned enlargement of the EU might also promote an upswing and even lead to a noticeable exports of agricultural goods to Central European countries. But in the long run, it may well be that fertiliser use in Eastern European agriculture will become as efficient as in the West.

Furthermore, it should be remembered that more restrictive legislation on waste treatment, together with progress in recycling technologies could lead to greater marketing of recycled bulk materials from basic production as secondary nutrients for agricultural use. If these products are offered at lower prices, they could represent serious competition to traditional mineral fertilisers. There are environmental arguments against this technology shift, however, and how realistic this scenario will prove remains to be seen.

All things considered, a serious assessment of the future changes in fertiliser production in Austria seems hardly possible. The European Fertiliser Manufacturers' Association even expects an end to the decline and a return to slow growth

Table 4. The influence of environmental actors on structural change (competitiveness, technical progress) in the fertiliser market in Austria

Research	36	rather significant	28-36
Public authorities	32		
Politicians	31		
NGOs, local action committees	30		
Media	26	significant	19-27

aggregated ratings by experts (possible maximum = 45)

within the next few years. Such optimistic forecasts should be handled with caution, however. With continuing over-capacity in Europe, further concentration is more likely, although the largest opportunities for rationalisation have already been exhausted. Future competition over sites in Europe will be determined very much by both transport facilities and energy prices. The best sites are those with the most comprehensive infrastructure coverage (seaport, railway and highways). The future development of transport prices will also play an important role. The current share of road transport in the modal split is definitely too high for bulk materials like mineral fertilisers. The coming liberalisation of European energy markets will provide a competitive advantage to companies in regions with lower energy costs, especially in production of nitrogenous fertilisers.

Considering the Austrian situation, closer co-operation between the two major producers should not be ruled out, as positive synergies are likely. But no realistic predictions can be made about the actual organisational arrangements of closer co-operation. Clearly, further concentration of the Austrian fertiliser industry could lead to a rather monopolistic situation, and therefore raises the issue of compliance with competition law. However, the complete disappearance of domestic fertiliser production seems be out of question, because of the high risks for domestic food supply where there is complete dependence on fertiliser from abroad.

References

Agrolinz Melamin GmbH (1996a): Bericht zum Geschäftsjahr 1996. Linz.

Agrolinz Melamin GmbH (1996b): Agrolinz Melamin GmbH – Geschichte des Unternehmens. Von der Stickstoffwerke Ostmark AG zur Agrolinz Melamin GmbH. Linz.

Agrolinz Melamin GmbH (1998): Annual Report 1997. Linz.

Aiginger, K., Bayer, K. (1988): Volkswirtschaftliche Bedeutung der Düngemittelproduktion in Österreich. Unpublished report by the Austrian Institute of Economic Research (WIFO) charged by the Umwelt- und Wasserwirtschaftsfonds. Vienna.

APA (Austria Press Agency) (1997): Unpublished special report charged by IFF Soziale Ökologie. Vienna.

BMLF (Bundesministerium für Land- und Forstwirtschaft) 1997): Lebensmittelbericht Österreich – Die Entwicklung des Lebensmittelsektors nach dem EU-Beitritt 1995. CULINAR Institut für Ernährungskultur im Auftrag des Bundesministeriums für Land- und Forstwirtschaft. Wien.

BMLF (1998): Bericht über die Lage der österreichischen Landwirtschaft (Grüner Bericht). Wien.

BMWA (Bundesministerium für wirtschaftliche Angelegenheiten) and BLV (Bundeslastverteiler) (1995): Energiestatistik 1994. Wien.

Dachler, M., Kernmayer, I. (1997): Düngemittelaufwand in Österreich. In: Köchl, A. (Ed.): Bodenschutz in Österreich. Bodenzustand, Entwicklungstendenzen, Schutzmaßnahmen. Bundesamt und Forschungszentrum für Landwirtschaft im Auftrag des Bundesministeriums für Land- und Forstwirtschaft. Wien, pp. 187–203.

Donau Chemie AG (1996): Geschäftsbericht 1996. Zwentendorf.

EC (European Commission) 1995): Panorama of EU Industry 95–96. Brussels.

EFMA (European Fertiliser Manufacturers' Association) (1997a): The Fertiliser Industry of the European Union: the Issues of Today, the Outlook for Tomorrow. Brussels.

EFMA (1997b): Forecast of Fertiliser Consumption in EFMA Countries to 2006/7. Brussels.

FAO (Food and Agricultural Organization): Fertiliser Yearbook. Rome. Various Issues.

Gerhold, S. (1990): Stoffstromrechnung: Düngemittel. In: Statistische Nachrichten 4/1990, 261–266.

Gerhold, S., Steurer, A. (1993): Ökologische Gesamtrechnungen: Stoffstromrechnung Düngemittel; Düngemittelabgabe. In: Statistische Nachrichten 4/1993, 285–289.

Jänicke, M., Mönch, H., Binder, M. u.a. (1992): Umweltentlastung durch industriellen Strukturwandel? Eine explorative Studie über 32 Industrieländer (1970 bis 1990). Berlin.

Kenis, P. (1992): The Social Construction of an Industry. A World of Chemical Fibres. Frankfurt a.M.: Campus.

Nemeth, D. (1992): The Austrian Case Study: The case of Chemie Linz, in: Le Blansch, K., Hildebrandt, E., Pearson, D. (1994): Industrial Relations and the Environment – Case Studies. European Foundation for the Improvement of Living and Working Conditions. Dublin, pp. 241–271.

Payer, H., Hüttler, W., Schandl, H. (1997): National Material Flow Analysis for Austria 1992 – Society's Metabolism and Sustainable Development. IFF Soziale Ökologie charged by the Federal Ministry for the Environment, Youth and Family Affairs. Vienna.

Interviews

A series of eight structured one-to-one interviews were conducted in June to July 1998 with most of the relevant actors in the political decision-making process affecting the Austrian fertiliser production sector. The – confidential – interviews were conducted with the following:

Ministry for Economic Affairs	*Dr. Reinhard Blauensteiner*
Federal Office and Research Centre of Agriculture	*Dipl.-Ing. Dr. Michael Dachler*
RWA Raiffeisen Ware Austria	*Ing. Erich Dorrer*
Donau Chemie AG	*Dr. Eduard Machat*
	Dr. Rainer Wiesböck
Friedrich Glatz GmbH	*Dipl.-Ing. Franz Schmid*
Agrolinz Melamin GmbH	*Dipl.-Vw. Günther Schwarz*
	Mag. Horst Eichinger
Bodenkalk Gen.m.b.H.	*Dipl.-Ing. Heinrich Telser*
Salinen Austria GmbH	*Dipl.-Ing. Klaus Tscherne*

Coal and Steel in Western Europe

Manfred Binder[1] and Simone Schucht[2]
[1]Free University of Berlin, FFU, Ihnestr. 22, 14195 Berlin, Germany
[2]Ecole des Mines de Paris, Cerna, 60 boulevard Saint-Michel, 75272 Paris cedex 06, France

In post-war Europe, technological change has traumatised the iron and steel industries and their most important input suppliers, the coal and iron ore mining industries. The close networks of producers and governments, with the European Coal and Steel Community (ECSC) as their centre, could not prevent, nor even predict, the disastrous developments. Even when they finally unfolded, reaction was slow and incoherent. But since the technological change at the heart of the crises has also led to increasingly fragmented networks, consisting of actors with increasingly different interests, coordination has become more and more difficult.

This chapter provides a brief overview of these developments. Most of the data used to illustrate this analysis refer to Germany and Luxembourg:

- German coal mining is strongly concentrated in the area around the river Ruhr, a tributary to the Rhine, in the West German state of North Rhine-Westphalia. Some smaller, but nonetheless important mining districts (Reviere) are Ibbenbüren (less than 100 km north of the Ruhrrevier), Aachen (less than 50 km south-west of the Ruhrrevier) and in the state of Saarland on the French border. The concentration of coal mining in north-western Germany has led to the concentration of steel and other industries here as well, and has made this region (the Ruhrgebiet) the most important industrial agglomeration in Germany. With a production of 41 Mt crude steel in 1994, Germany was the 5th largest steel producer in the world, after Japan, China, the USA and Russia. Europe is the largest world producer of high-grade steel, and Germany is the largest producer of crude and high-grade steel in Europe.
- Luxembourg's iron and steel production started in the mid-19th century, when it became technologically viable to process the country's vast deposits of minette, an iron ore with relatively low (less than 50%) iron content. All the other major inputs for steel-making, mainly coke, had to be imported from the beginning. With 400 000 inhabitants, the Grand Duchy of Luxembourg has the size and economic potential of a medium-sized city. Locating steel industries in such a small country resulted in national dependence on a single industry unparalleled in the rest of the world: in 1960, iron ore extraction and the production of iron and steel accounted for 34% of GDP. 40% of all workers or 18% of the entire working population were employed in this sector. The added value in iron and steel production was two and a half times as high as in all other manufacturing sectors combined (STATEC 1990). In the early 1960s, 97% of total industrial energy consumption was consumed by this sector, equivalent to 83% of all final energy consumption (or 73% of all primary energy requirements).

Section 1 briefly describes coal extraction, steel-making in an integrated steel mill, the major innovations since 1945 and the implications of the technologies for management and political regulation of these industries. The development of the industries is then described, based mainly on data from Germany and Luxembourg (Sect. 2). After some remarks on the environmental significance of coal and steel (Sect. 3), a short overview follows of the development of decision-making in European iron and steel industries in general, and in Germany and Luxembourg in particular (Sect. 4).

1 Innovation and Crisis

1.1 Traditional European Coal and Steel Industries

The technologies used in the coal and steel industries have important implications for managing these sectors. Broadly speaking, four stages in steel production can be identified (Stahlfibel 1989):

- *Extraction and pre-treatment of iron ore and coal (as coke).* Mining is an industry with considerable economies of scale and high labour intensity.
- *Production of pig-iron.* Iron ore, i.e. ferro-oxides, is mixed with an almost equal amount of coke and reduced in blast furnaces. The resulting "pig-iron" consists of about 95% pure iron, carbon being the main ingredient in the remaining 5%. This is by far the most energy intensive and polluting part of production. Pig-iron production is the only coal-consuming sector where coal can not be replaced by other fuels, beyond a few minor inputs of petrol, oil and gas. Steel producers have traditionally owned many coal pits, and in particular the majority of coking plants, in order to secure their supplies.
- *Production of crude steel.* Pig-iron is transformed to *crude steel*, which is at least 99% pure. Several different processes have been developed in the history of steel production in order to reach this level of purity (Bessemer, Thomas, open hearth (Siemens-Martin) processes, oxygen converters). Depending on the production technology, more or less scrap iron can be used as a substitute for pig-iron. Some processes, namely open-hearth and electric arc furnace (see below), can work only with scrap, and need no pig-iron at all.
- *Production of finished steel* such as steel plate, sheet steel, bars, rails, tubes, rods or profiles e.g. by rolling or galvanising crude steel: this is the most labour intensive part of steel production.

If the last three stages of production are concentrated on one location, the site is termed an "integrated steel mill".

Primary production of iron and steel is highly *material intensive*, and transport costs are a major part of the prices of the two main materials, iron ore and coal. It was natural, therefore, that the iron and steel industries grew up near natural deposits of these inputs, like the German Ruhrrevier in North Rhine-Westphalia, with its rich coal deposits, or the French region of Lorraine, the German Saarland and Luxembourg, with their iron ore (minette) deposits and coal mines.

The output of steel industries is still highly material intensive compared to the final products of steel consumer industries. Thus, many of these industries are also located near coal and steel sites, while most others still draw their steel from the nearest steel producer. Transport costs therefore limit the catchment area of each market, and in doing so limit competition between different steel regions.

Extreme *economies of scale* make only gigantic production facilities economically sensible, leading to a natural concentration of the respective markets with oligopolies or even monopolies at least regionally or even nationally. Like many other basic industries (e.g. electricity production or mineral oil refining), commissioning new plant or exploiting new finds of natural resources is not only expensive but *time-consuming*, needing some 5–10 years from inception to the start of production. On the other hand, most of the facilities can then be used for decades without major change. As a consequence, it might prove uneconomic to decommission a loss-making steel works, because most of the costs have already been paid (*sunk costs*). Thus companies will be keen to sell their output, even at way below their own total costs as long as they earn more than their marginal costs, in order to minimise their losses (Gieseck 1995). This might result in prices below production costs even for the most productive steel works, and to losses for all steel companies at times of over-capacity in competitive markets.

Since steel products are used mostly for construction and production of investment and durable consumer goods (mostly vehicles), steel demand is highly *dependent on the business cycle*, with a more or less severe steel crisis almost inevitably following in the wake of every recession, let alone depression.

Coal and steel companies therefore needed, and were very often able to restrict competition, either through private cartels or regulation by public authorities, who were more often than not eager to comply with the demands of the steel sector, given its great economic importance, high regional concentration and significant employment of well-organised workers (Conrad 1997, p. 19).

In pre-war Europe, the cartels were privately organised: the first European steel cartel (1926–31) broke down during the Great Depression, due to the inability of national business organisations to enforce decisions on individual companies. But the second steel cartel, started in 1933, was more successful (Jörnmark 1993): it was based on stronger national organisations with lean and informal international coordination by national product comptoirs using pre-determined quotas, uniform prices and sales conditions, sellers' licences and distribution cartels. It included not only continental Europe but also the United Kingdom (from 1935) and the USA (from 1938).

After World War II, the coal and steel markets were reorganised, now under public control. In 1951, the European Coal and Steel Community (ECSC) was founded. It prohibited private cartels and national policies which obstructed free markets, but supported co-operation between producers and made coordination of investment plans obligatory. In times of crisis, the High Authority (later the European Commission) was given the right – or rather the duty – to regulate the market strictly, e.g. by import restrictions, production quotas and minimum prices, in practice therefore installing a legalised Europe-wide cartel under public control (cf. Stehn 1993; Wieshoff 1986).

This solution was opposed by steel producers, because they preferred to regulate the market themselves. A private cartel would have given the steel industries the opportunity to earn monopoly rents by limiting steel supply. But because of the importance of steel for the production of military equipment, the political rationale was to prevent the control of coal and steel capacities by aggressive national governments, especially German ones (Conrad 1997 p. 47). The influence of the steel producers on ECSC decision-making was high and well institutionalised, while steel *consuming* industries or other stakeholders have never been much involved in the policy networks of European steel making (Conrad 1997, p. 50).

1.2 Technological Change and Economic Crisis

Four major innovations have shaped the development of the coal and steel industries in recent decades and have resulted in a severe crisis of over-capacity:

- the declining coal and steel intensities of industrial economies
- the decline in freight rates and the globalisation of raw materials markets
- the diffusion of oxygen converters in crude steel production
- the increasing competitiveness of mini-mills using electric arc furnaces

1.2.1 The Declining Coal and Steel Intensities of Industrialised Economies

There was a time when "*coal use*" was not a subgroup of, but rather a synonym for energy consumption: At the turn of the century, coal's share in the world's consumption of commercial primary energy amounted to 95% (Putnam 1953, p. 439, cited after Darmstadter 1971, p. 12). This certainly underestimates the role of – mostly non-traded – wood in the world's total energy consumption at the time, but for European industrial countries, traditional fuels were by then of virtually no significance. Today, less than 20% of the OECD's primary energy requirement is provided by coal. The decline of coal has progressed more or less steadily during the 20th century, but in Western European countries it began rather late (and in Eastern European countries even later, with the notable exception of the Soviet Union): In 1938, 91% of West European energy consumption was still based on coal (Darmstadter 1971, p. 89).

In West Germany, the decline was almost entirely a post-war phenomenon (Fig. 1): in 1950, coal and lignite still held an 89% share of the total energy requirement. Coal alone accounted for 72.3% (in 1952 even 73.2%), but since then its share has declined substantially, and by 1978 it had bottomed out at 17.8%. In the early 1980s, coal regained some of its lost market share as a result of the two oil

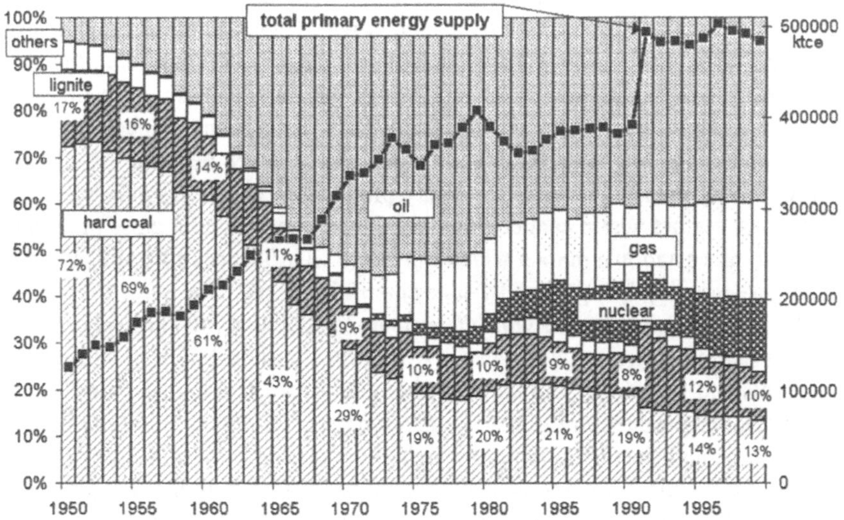

Source: Statistik der Kohlenwirtschaft. 1991–99: unified Germany

Fig. 1. Primary Energy Supply by Fuel Type in FR Germany (1950–99)

crises, but ceded it again in subsequent years. Energy consumption in the former German Democratic Republic was supported to a great extent by the huge deposits of lignite. Thus, with the reunification of Germany in 1990, the share of lignite in total primary energy consumption rose sharply to about 17%, but it has declined rapidly since then. Meanwhile, coal's share in energy consumption reached an all-time low of 13.3% in 1999 and is still falling.

The most important substitutes for coal have been petroleum products: German oil consumption grew by 16.3% annually (!) for about a quarter of a century and was therefore 32 times higher in 1973 than in 1950. The subsequent – somewhat limited – decline gave way to the rise of natural gas and nuclear energy, leaving an energy mix at the end of the century where petroleum is still the major energy source and coal is surviving in a slowly declining niche market.

For most uses, coal can be easily replaced. This has been especially important for users who consume only small amounts of energy and whose locations are widely dispersed, principally light industry, service providers and private households. In these sectors, the increasing competitive disadvantage of coal became most obvious in its high relative weight, the resulting transport costs, and – more importantly – the fact that petroleum products, gas and electricity can be handled and processed far more easily and cleanly than coal products.

Two other trends have also constrained growth in the coal industries. Firstly, with respect to road transport, there has been no serious alternative to the use of petroleum products. The transformation of crude oil in refineries yields not only gasoline but also a wide variety of other products (e.g. heavy fuel oil), and the respective fraction sizes cannot be altered at will. Owing to the enormous growth in road traffic since 1945, other petroleum products have come onto the energy markets in large quantities, as by-products of gasoline, and have substituted coal even in segments where they would not otherwise have been competitive.

Secondly, the only important use for the newly developed and heavily subsidised nuclear energy has been electricity production, which has also become one of the last sectors where coal could be used on a large scale, in close proximity to the coal mines and in a market which can easily be influenced by political pressure – because it is highly regulated by state authorities in any case.

The technological factors which contribute to structural falls in *steel production and consumption* are developments allowing for the substitution of steel by other materials and technological improvements within the steel industry itself, including increased production efficiency, improvements in steel quality and new construction methods using more light-weight products (Altgeld/Schneider 1992).

The introduction of continuous casting in the 1960s led, in the first place, to increased efficiency in production. Continuous casting is a technology which allows casting hot crude steel directly to billets, bypassing several of the labour and energy intensive production steps in traditional steel making, in which 15–20% of the initial crude steel was left as scrap to be recycled in the crude steel production process. This continuous production process has almost completely replaced the older discontinuous process of ingot casting. By reducing the amount of scrap produced, the same amount of rolled steel can be produced with 15% to 20% less crude steel than with the older process (McSweeney and Hirosako 1991). Input can therefore be reduced and the same quantity of final products produced with a lower quantity of crude steel (Hogan 1994; Gerstenberg et al. 1985). The reduction in output associated with this improved process is not therefore associated with a fall in demand. Table 1 shows the spread of continuous casting in Germany.

Improved construction methods in downstream industries have increased output relative to steel input and/or decreased scrap production and thereby the amount of steel input required. One example is the reduction in scrap arising from automobile manufacture (Hogan 1994). Since the mid-1970s, the importance of computer-assisted design has been growing. This has made it possible to optimise steel inputs and contributed to minimising the scrap produced. The improved strength

Table 1. Production of crude steel blocks for cast steel and contained continuous casting

	1964	1970	1975	1980	1985	1990	1995
Crude steel blocks [kt]	36 702	44 316	39 746	43 300	40 086	38 055	41 732
of which continuous cast [kt]	4 980	3 726	9 813	20 162	32 208	35 091	40 131
%age Total Production	1.4%	8.4%	24.7%	46.6%	80.3%	92.2%	96.2%

Source: Statistisches Jahrbuch der (Eisen- und) Stahlindustrie, various volumes

of steel has also contributed to a reduction in specific steel consumption by upstream industries (McSweeney and Hirosako 1991), for example the significant reduction in the gauge of sheet steel. Furthermore, the development of more light-weight steels has contributed to a reduction in crude steel production in terms of product volume (Hogan 1994; Bünnig et al. 1988).

Substitute materials competing with steel are non-ferrous metals – principally aluminium – and plastics. The input of steel to automobile production has decreased as the input of plastics and aluminium has increased. However, this decrease can be better explained by improvements in steel quality (light-weight steel) than by substitution of materials (Hogan 1994). Over the last 50 years, steel has also been replaced by concrete in the building sector. Moreover, instead of cast iron, ceramics are now used for high-temperature applications.

Typical changes in the structure of final demand also contribute to the decline of steel demand: During the early stages of industrialisation, steel demand rises at an above-average rate due to a high level of steel-intensive investment activity. As development progresses, the steel content of investment projects finally decreases, replacement and rationalisation investment out-weigh steel-intensive expansion investment, and a trend towards production which is more technology and knowledge intensive evolves (Gerstenberg et al. 1985; Gieseck 1995; Demgenski 1990).

1.2.2 The Decline in Freight Rates and the Globalisation of Raw Materials Markets

Until the mid 19th century, when industrialisation in Europe was in full swing, it was mainly spices, coffee, cocoa and other luxuries which were traded on international commodity markets. Since then, a continuous decline in freight prices has taken place, making more and more goods suitable for transoceanic trade. But it was not before the introduction of revolutionary new systems for ocean transport and handling bulk products at harbours (Radetzki 1990, pp. 17, 34) that iron ores with a high iron content from Brazil and cheaply mined coal from the United States became competitive in Europe.

This was not only an economic catastrophe for the European mining sectors themselves, e.g. for German coal mining and for Luxembourg's iron ore extraction, but it turned the logic of profitable locations for resource-reliant industries like the steel sector upside down: with domestic resources no longer attractive, it no longer made much sense to produce in Lorraine, for example, which has no access to major waterways to receive raw materials from overseas. The best places to produce steel under these circumstances are in the vicinity of a major coastal harbour. To be located in the German Ruhrgebiet is still not bad under the new circumstances, because of easy access to Dutch harbours on the river Rhine: works at Duisburg had transport costs comparable with Klöckner's Bremen site in the early 1970s (Jörnmark 1993 p. 202).

Thus, all aound Europe's coasts, new integrated steel plants were built, and started production during the 1960s and 1970s: Klöckner in Bremen (Germany),

Hoogoven in Ijmuiden (Netherlands), ARBED in Ghent (Belgium), USINOR in Dunkirk and DeWendel-Sidelor and USINOR in Fos outside Marseilles (France), Finsider in Cornigliano (near Genoa) and Tarento (Italy).

But most of the investors in these plants also owned traditional steel production capacity, and were not keen to throw them out of business. Therefore the new – and accordingly most advanced – integrated steel mills in Belgium, France and Italy specialised in the technologically more ambitious flat products (sheets, plates), trying to leave the market for long products open for their traditional producers in Lorraine and Vallonia, or the steel plants in Bagnoli and Piombino (near Naples), which still used ores from the Elba (Jörnmark 1993, pp. 174, 184).

1.2.3 Oxygen Steel

With the development of the Linde-Frankl process in 1928–29, cheap oxygen could be produced in large quantities. This paved the way to the development of the Basic Oxygen Process for converting pig-iron to crude steel, which was to revolutionise steel making in the coming decades (Jörnmark 1993, p. 168): While conversion needs 8 hours in a traditional open furnace, it takes only 45 minutes in a comparable oxygen converter, reducing costs and substantially increasing production capacity. Nevertheless, it took until the early 1970s for the last facilities using the older Thomas process, and until the early 1980s for the last open furnaces to be decommissioned in the European Community. The problem of economies of scale became even more severe with this new technology and its continuing improvement (Demgenski 1990, p. 100).

Of course, all new coastal steel mills were equipped with oxygen converters. The emergence of the Japanese steel industry is maybe the most striking illustration of this development, as it would have been impossible without these major innovations: Japan managed to become the world's leading exporter of steel and steel products, although they had no domestic supplies of raw materials. Japan's population and fast growing steel consuming industries are located almost exclusively along its coastline. In the 1960s, Japan started large-scale steel production at exactly the point where coastal areas had become the most advantageous locations and new oxygen plants were seen to be far more efficient than older ones.

But since producers at traditional steel locations wanted to stay in business, they had to change their crude steel production technologies as well. And all this increased capacity became necessary only because the new possibilities had to be used without losing sunk costs in traditional steel locations.

1.2.4 Electro Steel and the Rise of the Mini-mills

The innovations discussed up to now – the use of new raw materials and oxygen converters – did not challenge the paradigm of traditional steel making in integrated steel mills. Some of its key economic characteristics, like economies of scale and sunk costs were even exaggerated. But another wave of innovation established a competitive alternative to this kind of steel production: the improvement and diffusion of electric arc furnaces. Before 1950, electric furnaces had only been used in small segments of the market, but immediately after World War II, technological breakthroughs made them feasible for producing long products, and since the 1990s for some flat products also (Hogan 1994). With the use of electric furnaces, it is possible to produce crude steel from almost any combination of scrap or pig-iron (or direct reduced iron), while the scrap in the input to oxygen converters cannot exceed 20% (Stahlfibel 1989, p. 52). This makes it possible to produce steel without any blast-furnaces at all. Additionally, this technology is far more flexible in scale, allowing much smaller production units (mini-mills): "In the mid-1960s the minimill approach made it possible to enter the steel business for less than $5 million to produce 50–60 thousand long products annually." (Hogan 1994, p. 76) Since then, the optimal size of a mini-mill has increased substantially to about 1 Mt/a today, but is still substantially lower than that of integrated steel mills (some 3.5 Mt/a; cf. Conrad 1997, p. 20).

The aforementioned development of continuous casting has contributed to the success of mini-mills (Jörnmark 1993, p. 208): in the early 1960s, this major increase in efficiency could only be used in small production units, which gave them an important competitive edge. Today, almost all crude steel is continuously cast. Still, the improvement of casting technologies favours mini-mills the most: until recently, mini-mills could only produce an ever-increasing variety of long products, not so much because of deficiencies in the electric furnace process itself, but due to the economies of scale necessary for rolling mills to transform crude steel into flat products. This too is about to change, with the development of thin-slab casting (Hogan 1994 p. 79, 81; Conrad 1997: p. 44):

1.2.5 The Outcome: Over-capacity and the Increasing Heterogeneity of Interests

Given the technological achievements by the late 1960s, European steel industries would, if they had been able to start from scratch, have placed gigantic integrated steel works along the European coast to produce flat products, and built scrap-consuming mini-mills with electric arc furnaces in the vicinity of its consumers and/or scrap merchants to produce long products, and no mining sector would have been necessary whatsoever. Exactly this kind of structure was created in the 1960s and 1970s, *in addition to* the unexpectedly obsolescent one Europe already had, resulting in massive over-capacity.

Surprisingly, the crisis was itself a surprise for most of the actors involved (Jörnmark 1993, p. 200). Of course, the depression in the 1970s and the substantial de-coupling of steel use from GDP in industrialised countries could hardly be foreseen. But even in the 1960s, capacity was rising faster than demand – and planning for the 1970s was nothing if not ambitious! Investment and investment plans for steel would only have been sound if the additional steel could be exported, a risky assumption which proved to be illusory.

The innovations causing over-capacity also changed the institutions and networks, leaving different actors with increasingly divergent interests and options, making coordination increasingly difficult. Coordination was certainly never easy, even while coal and steel producers were a more homogeneous group, because they often lacked the means to enforce their decisions on individual producers (Jörnmark 1993, p. 73). But before the innovations described above, coordination was at least a kind of "common good" for producers using roughly similar production technologies and facing over-capacity (Conrad 1997, p. 32). But these innovations made it unclear what the "common good" actually was.

> The basic problem was that paradigmatic innovation(s) were fragmentizing old socio-economic coalitions, as the possibilities to reap the advantages of the new production functions were unevenly distributed. When the interests of coalition members grew increasingly divergent, this implied that the existing structural solutions were no longer functional. (Jörnmark 1993, p. 323)

Three main groups of actors and interests today can therefore be identified as (cf. Hudson and Sadler 1988; Jörnmark 1993; Eckart and Kortus 1995):

1. actors representing traditional, now less competitive, locations (coal and iron ore mining, inland steel production in Vallonia, Lorraine, Luxembourg, Saarland etc), whose aims are modernisation, market regulation and a gradual reduction in the capacities of *all* producers in the relevant market (a crisis cartel according to the existing institutional structure)
2. actors representing new, or at any rate competitive, integrated steel plants (coastal plant in the Netherlands, Belgium, France and Germany and to some extent the plant in the German Ruhrgebiet[1]), whose aims are more or less free markets (no crisis cartel but no challenge to the institutional structure)
3. actors representing the mini-mills, which are highly competitive in market niches (dispersed over Europe, concentrated in Northern Italy), aiming at free markets with neither a need for, nor any power to influence the existing institutional structure

2 The Development of the Coal and Steel Crises in Germany and Luxembourg

2.1 Coal Mining in the Federal Republic of Germany

2.1.1 Output and Employment

After World War II, the German coal mining industry needed some ten years to regain its pre-war level of production (Fig. 2). Despite the decline in its share of total energy requirements since 1953, German coal extraction expanded until

[1] The case of the Italian integrated steel works on the coast is an anomaly in this respect (see Jörnmark 1993).

1957. In 1958, excess supply in world energy markets led to increasing imports of oil and coal and to a slump in sales of domestic coal. But even this coal crisis of 1958/59 was not considered a sign of any long-term downward trend at the time but rather as a temporary disruption. As the annual report of the *Statistik der Kohlenwirtschaft* (1960, p. 35), an organisation of several employers' associations within the mining industry, put it: "It is in any case fairly safe to conclude that future coal mining will depend more on the size of the workforce and the management of working hours than on sales potential."

This optimism turned out to be misplaced. German coal mining has never become competitive anymore since then. In 1995, German coal cost 300 DM/t, while the international price for coal was about 70 DM/t (Simons 1997, p. 110). Coal is extracted from 900–1400 metres underground which makes mining extremely expensive (Vogl 1944, p. 241). Since the late 1950s, imported coal has been cheaper than domestic coal, due to the decreasing freight rates in overseas bulk transports described above (Sect. 1.2.2). In 1959 the German parliament limited coal imports to 6 Mt/a (*Zollkontingentgesetz,* in force until 1995). Other measures were taken in the following decades in order to sustain coal production (Neu 1995).

Step by step production declined by almost three quarters until 1996 equivalent to 3.2% annually (1956–99). Until the early 1980s, periods of rapid decline (1958/59, 65–67, 72–78) were always followed by years of more or less stable production at ever-declining levels (1959–64, 67–71, 79–82). Since the mid-1980s the decline has been continuous and heavy. Current plans are to reduce production

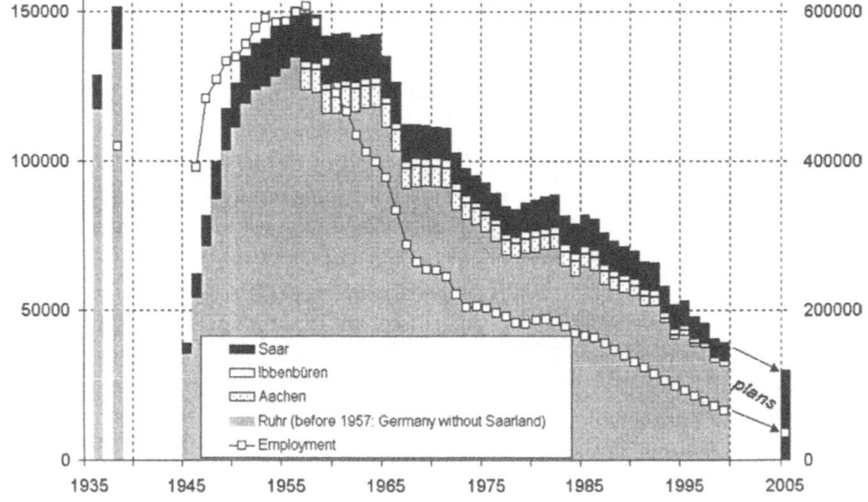

Source: Statistik der Kohlenwirtschaft

Fig. 2. Coal Extraction by Region and Employment in Germany (1936, 38, 1945–99)

to what is assumed to be a sustainable level of 30 million tonnes by the year 2005 (RAG 1997, p. 18), which means a further reduction of 24% or 4.4% per year (!).

The decline in jobs has been even more dramatic: Since reaching its peak in 1957, total employment in coal mining fell by 89% (5.1% annually) from 607 349 to 66 414 in 1999. A further reduction by 46% (9.7%/a) to 36 000 employees is expected. Employment has declined far more gradually than, and in advance of, production: about 66% of jobs in 1957 had been shed by 1973, i.e. during a period of full employment. After the onset of the economic crisis in 1973 employment declined by only 1.1%/a for almost a decade (1973–82). It subsequently has decreased by 64% (6.6%/a), however – now under conditions of high unemployment. Nonetheless, the reduction in employment has never been achieved by laying miners off, but generally through early retirement. This development has still contributed significantly to rising unemployment in mining districts, because the loss of available jobs has meant fewer opportunities for those seeking work and it has been estimated that, due to the economic difficulties faced by supply industries, at least one job is lost indirectly with every miner's job (Süddeutsche Zeitung 14.02.97).

While the decline in the *share* of coal in total energy requirement is a common phenomenon all over the world, the same is not true for the decline in the *absolute* quantities of coal produced or consumed. Figure 3 shows world production of coal, covering the period 1960–99 for OECD countries and 1971–99 for all others. Until 1989, world coal production grew fast but has remained fairly stable since

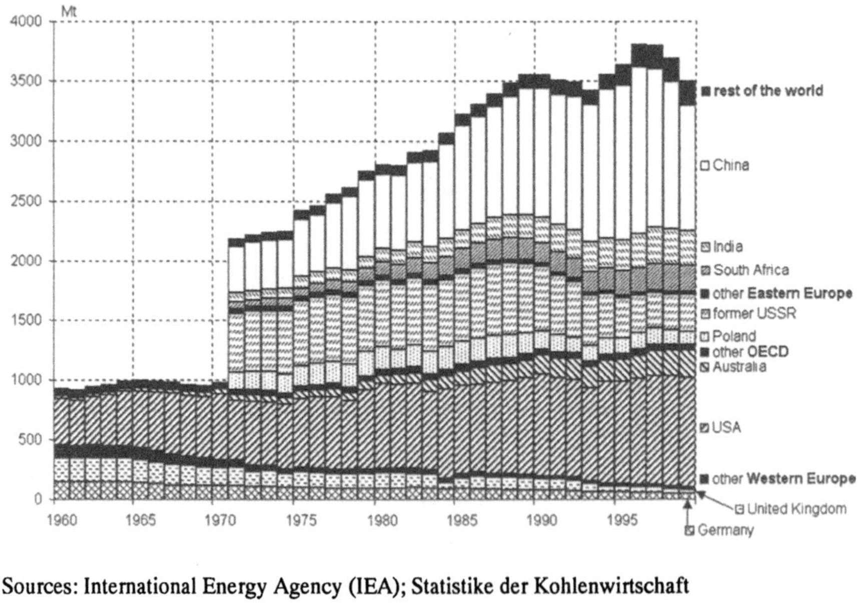

Sources: International Energy Agency (IEA); Statistike der Kohlenwirtschaft

Fig. 3. Coal extraction in OECD countries 1960–99 and in non-OECD countries 1971–99

then, due to the collapse of the Soviet mining sector and the decrease of the Chinese coal extraction by 25% 1996–99 after a period of extremely rapid growth of the Chinese coal industry. Even in the OECD countries, coal production is still growing: production in Australia and North America more than offsets the decline in European coal mining. Thus the long-term decline in German coal production is not typical for most important coal-mining countries.

However, relative to other European countries, Germany is not facing such an unusual decline at all – on the contrary (Fig. 4). With the exception of Spain, all major West European coal mining countries have cut back their coal sectors, most of them far more dramatically than Germany. Belgium recently closed down its coal sector, the Netherlands did so some twenty years ago. France has also showed a sharp drop in production, but without totally abolishing coal mining as yet. Starting from by far the highest per capita production in Western Europe, the coal sector in the United Kingdom has gradually declined to almost the per capita level of (unified) Germany. (All other coal mining Western European countries – Turkey, Ireland, Portugal, Sweden, Austria, and Italy – have never extracted more than 160 kg/cap annually during the last 35 years.)

2.1.2 Trends in German Coal Demand

While in the early 1950s, coal was the universal fuel type used for almost every purpose, its use has since become more and more restricted to two market niches

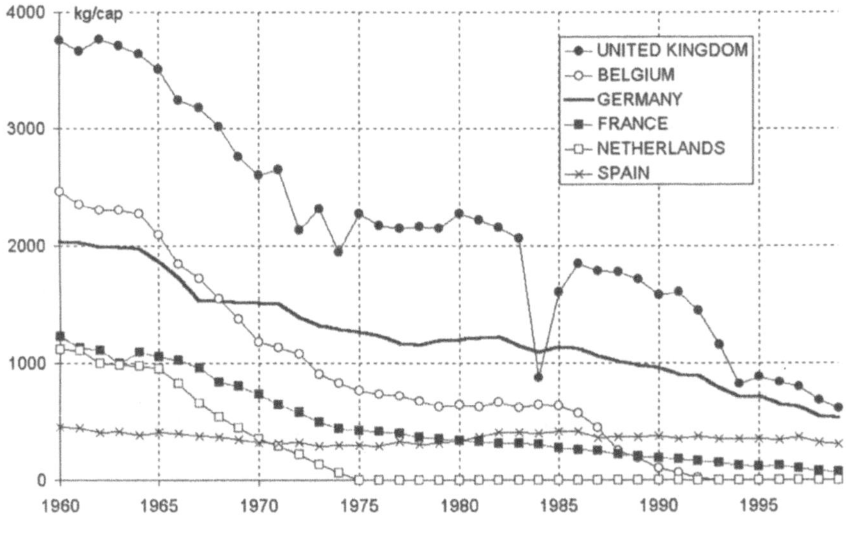

Source: IEA; German population includes both East and West Germany

Fig. 4. Development of coal mining in Western European countries 1960–99

(Fig. 5): large-scale electricity generation and pig-iron production, located in the vicinity of coal mines. All other markets have been ceded to the more convenient, cleaner and generally cheaper fuels: petroleum, natural gas and electricity:

1. The fastest-growing energy consumption sector – road traffic – could not increase sales in the coal sector because it has always been totally dependent on petroleum products. On the other hand, the former consumer, railroad traffic, has lost both its dominant position in the traffic sector and, by switching to diesel and electricity, its dependence on coal. Coal has not been used for transport since the 1970s.
2. The major uses of coal in households, services and agriculture ("non-industrial") have been heating, cooking and hot-water supply. Today, except for some – mostly poor – households, these needs are more conveniently met by oil, gas and electricity.
3. Industries other than iron and steel production ("total other industry") have not vanished as customers for coal, but even in these sectors more than 40% of the supply is used for electricity production.

Nevertheless, the development of domestic demand can only partially explain the fall in production, because it has remained fairly stable since the first oil crisis, due to the rise in demand for electricity generation and the virtually stable demand from iron and steel producers, which have in combination accounted for almost all domestic coal demand since then. But during the 1980s, German coal exports

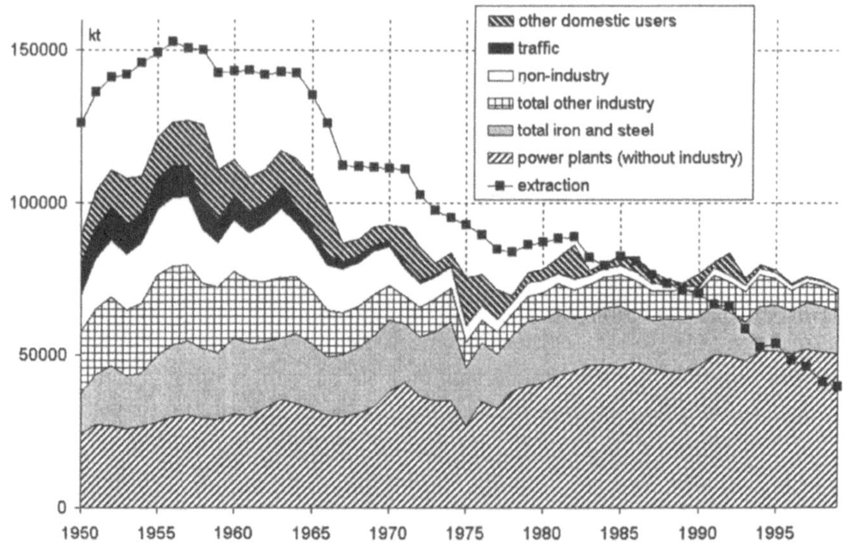

Source: Statistik der Kohlenwirtschaft; 1991–99 unified Germany

Fig. 5. Coal consumption by sector in FR Germany 1950–99

declined rapidly to insignificant levels and imports rose after thirty years of stagnation. This stagnation of coal imports was the result of trade restrictions which were slightly loosened in the wake of the second oil crisis and have been abandoned in recent years. German coal is still protected by subsidies, but coal *exports* are not subsidised any more, and the rising differentials between coal from Germany and abroad since the mid-1980s have made it increasingly expensive to finance German coal consumption without undermining the competitiveness of coal or electricity consuming industries.

In conclusion, let us take a closer look at those two domestic sectors which have been the only significant coal users in recent years:

- The only sector showing considerable growth in coal consumption has been power generation (Fig. 6). Even in this sector, coal has been taking a declining share in total production. This is mostly due to the introduction of nuclear energy, which has become the most important type of electricity production (30.8% in 1999, coal: 25.9%). Owing to efficiency gains, electricity production of coal-fired power stations has grown slightly faster than coal inputs.
- The production of pig-iron in blast-furnaces accounts for most energy consumed in the iron and steel industries (Fig. 7). Pig-iron production rose until the steel crisis in the mid-1970s and has remained at this level ever since. Energy consumption has tracked this development fairly closely, but with a continuous decline in energy intensity (energy input per unit pig-iron produced). Until the mid-1970s coal was substituted by oil and gas even in this sector, but this development has been partially reversed in subsequent years.

2.2 The World Steel Market

World *steel production* peaked in 1989. Since then the trend has been downward. Over the last two decades, two opposing trends have become apparent: increasing capacity and production in developing countries and declining or stagnating production in older industrialised countries (Cho 1992; Wienert 1996).

Up to the early 1970s, steel production in western industrialised countries grew continuously (Fig. 8). Following the first oil price shock (1973) production and demand dropped considerably. Since then the world steel market has been characterised by a restructuring-process (Cho 1992). The industrialised countries' share in the world steel production was about 80% in 1950 but only around 64% in 1974 and decreased further to about 50% in 1995 (Gieseck 1995; Statistisches Jahrbuch der Eisen- und Stahlindustrie 1976).

During the 1960s and 1970s, a number of developing countries invested a great deal of effort in creating their own steel industry. While, in the beginning, their steel production was based on relatively small plants and low capacity (Cho 1992), during the 1970s and 1980s some countries, such as Brazil and South Korea, built larger integrated steelworks and became serious competitors on the world market. Japan had already succeeded in building up a large steel industry

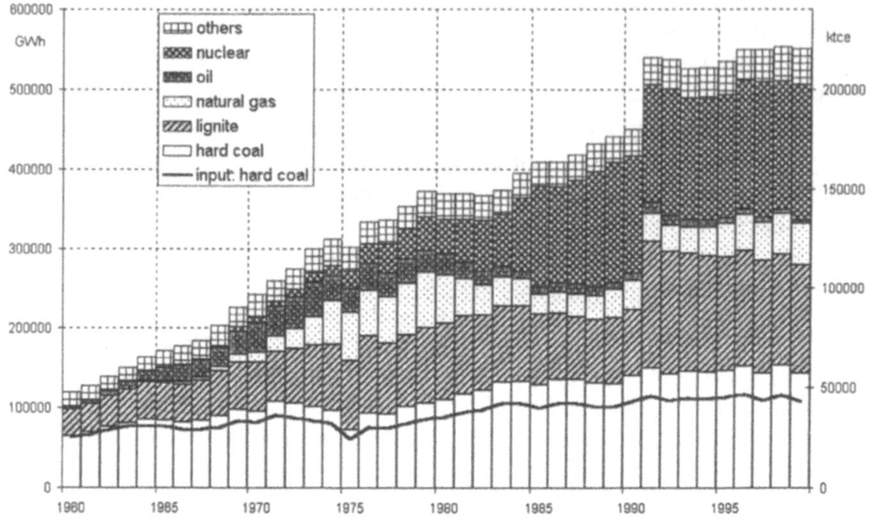

Source: Statistik der Kohlenwirtschaft; 1991–99 unified Germany

Fig. 6. Electricity production by fuel type and coal input in FR Germany 1960–99

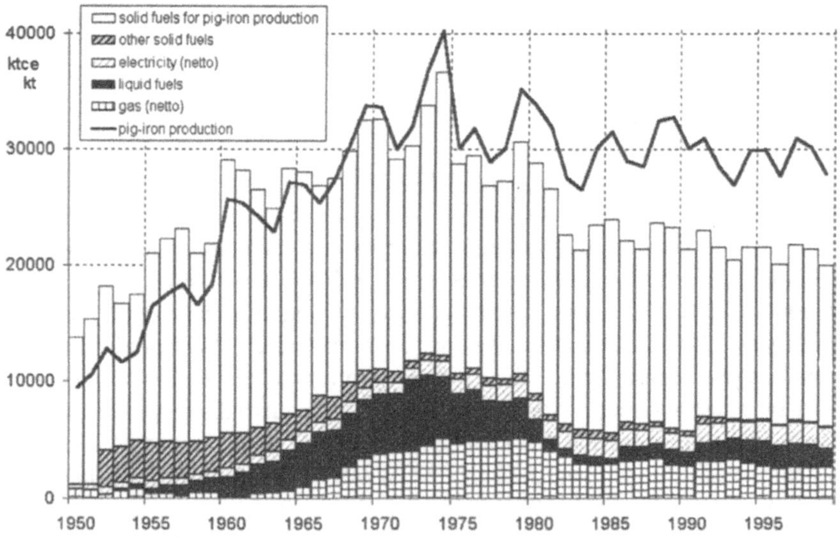

Source: Statistik der Kohlenwirtschaft; 1991–99 unified Germany

Fig. 7. Pig-iron production and energy consumption in iron and steel production by fuel type in FR Germany 1950–99

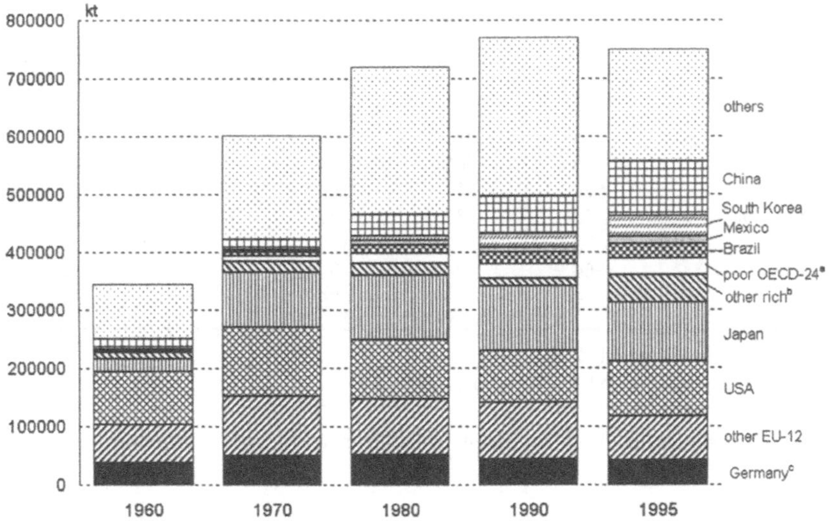

[a] OECD–24: OECD of 1993, i.e. OECD without South Korea, Mexico, Poland, Czech Republic
poor OECD–24: Greece, Ireland, Spain, Portugal and Turkey

[b] other rich: OECD–24 without poor OECD–24

[c] including Eastern Germany

Source: Statistisches Jahrbuch der (Eisen- und) Stahlindustrie, various volumes

Fig. 8. World crude steel production by countries 1960–95

and a considerable industrial agglomeration in the 1960s (Heinze et al. 1992; Hamm and Wienert 1990; Cho 1992). At the same time, steel production in the older industrialised countries was stagnating or declining. The number of countries with their own steel production capacity rose from 36 in 1950 to 54 in 1970 and has now risen to over 100.

These different trends also apply to *steel consumption*. Between 1977 and 1988, it decreased by 64 Mt in industrialised countries and grew by 96 Mt in developing countries (Kerz 1991). In the latter, steel capacity was built up in order to supply domestic demand and to exploit opportunities for export. This diminished the ability of older industrialised countries to export and, although western industrialised countries are still the leading exporters of steel, their market share has shrunk (Gieseck 1995; Wienert 1993a).

Within the EU – and Germany – apparent per capita steel consumption rose between 1960 and 1973 and has since fallen (Table 2). The same holds true for the USA. In Japan, steel consumption fell during the second half of the 1970s and recovered during the 1980s. In China, Mexico and Brazil, consumption rose once again during the 1990s. Total Western steel consumption grew between 1950 and 1973, and then became de-coupled from economic growth (Kerz 1991).

Table 2. Steel consumption per capita in selected countries 1970–94

in kg/cap	1970	1973	1981	1990	1994
Germany[a]	654	671	587	553	474
EC/EU[b]	489	504	399	376	362
Japan	675	805	561	802	636
USA	620	711	565	410	446
(PR) China	28	36	41	60	98
Mexico	82	95	148	96	128
Brazil	64	94	96	68	84
World	–	–	154	149	133

[a] Eastern Germany included from 1991 on.
[b] Up until 1980 EC 9, 1981 EC 10, afterwards EC/EU 12.
– no data available.
Source: Statistisches Jahrbuch der (Eisen- und) Stahlindustrie, various volumes

With the rising importance of the newcomers, export opportunities for the traditional steel producers were lost. The relative foreign trade position of industrial countries decreased relative to the 1960s, as did their share in world export trade. Owing to international differences in production costs, the industrialised countries lost market shares in mass production steel. With respect to high-quality and high-grade steel, however, where factors such as knowledge and human capital are more important, they remain the major producers.

Between 1978 and 1986, *crude steel capacity* shrank by almost 90 Mt in the EC–10, USA and Japan (USA: 43 Mt, EC–10 (1978–85)[2]: 34 Mt, Japan: 12 Mt). At the same time, capacity in Latin America (+21 Mt), Asia (+56 Mt[3]) and the (former) Soviet Union (+22 Mt) grew. It should be noted that capacity was still expanded in Western countries after demand fell (1974). The USA and Japan built up capacity until 1977 and 1978 respectively, the EC–10 until 1980. Owing to low demand, capacity utilisation fell. During the 1960s EC capacity utilisation reached 96%, but during the 1970s and the early 1980s it fell to below 70%.

The major factor behind the slump in *steel demand* in 1974 was the energy crisis (and the hike in oil prices) of 1973. In the wake of this crisis came a global interruption of 25 years of economic growth. The world depression had a direct influence on steel production and capacity utilisation. Between 1974 and 1975, steel consumption experienced a worldwide slump (Statistisches Jahrbuch der Stahlindustrie, various volumes): 59 kg/cap or 3.8 Mt in Germany, 76 kg/cap or 19.4 Mt within the EC, 131 kg/cap or 27.3 Mt in the USA and 108 kg/cap or 11 Mt in Japan.

---

[2] In 1986 the EC was expanded by two countries. Therefore, the decline for the EC between 1978 and 1986 was only 16 Mt.

[3] Middle East (+4 Mt), China (+24 Mt) and remaining Far East (+28 Mt), excluding Japan

A similar effect was apparent after the recession which followed the oil crisis of 1979. Steel demand reacts disproportionally to fluctuations in the economic cycle: periods of high growth are accompanied by steel-intensive investment in expansion. Periods of slower growth, however, bring with them reduced steel consumption and investment in rationalisation (Wienert 1985; Gieseck 1995).

For the long-term development of the steel industry, the general slow-down in GDP growth since the mid-1970s is particularly important. Moreover, apart from the general *reduction in GDP growth* since 1975, there has also been a discontinuity in the growth of significant steel users, such as the motor vehicle manufacturing and the industrial appliances sector (Demgenski 1991). But in general, declining per capita steel consumption is not explained by any declining importance of major steel users (Gieseck 1995) but rather by structural change within these industries, competition from substitute materials, more efficient production processes – such as efficiency improvements, where reducing waste permits a reduction in steel input in absolute terms – and longer product lives for steel-intensive products (see above, Sect. 1.2.1).

2.3 The German Iron and Steel Industry

Since the mid-1970s, the German steel industry has become a sector in crisis, experiencing a steady loss of economic importance. While the steel industry's share in total industry's gross added value amounted to 2.3% in 1960, it had decreased to about 1.6% by the mid-1990s (Statistisches Bundesamt). Furthermore, the steel industry is highly concentrated regionally and federal states such as the Saarland or North Rhine-Westphalia (NRW) – and here above all the Ruhrgebiet – have therefore been particularly affected by the developments in this industry.

With respect to *steel users* in Germany, steel is currently consumed in equal measure by the automobile industry, the mechanical engineering and the construction sectors. These figures represent an aggregate of direct purchases by the sectors from steel plants and their indirect purchases of semi-finished steel products from processing plants (Wienert 1996).

For its raw material *supplies,* the German steel industry – unlike that in most western industrialised countries – has maintained its coke coal mining. German iron ore reserves have become relatively uneconomic and can only be exploited with significant underground mining efforts. German ore exploitation has therefore been suspended and Brazil, Liberia, Australia, Sweden and Canada are now Germany's most important suppliers (Stahlfibel 1989).

2.3.1 The Decline in Production

Up until 1974 the German steel industry was characterised by enduring growth in crude steel production, allowing for variations due to the economic cycle (Fig. 9). After the first oil crisis, this trend turned around abruptly. Despite a slight recovery from the drop in 1975, crude steel production fell further and further during

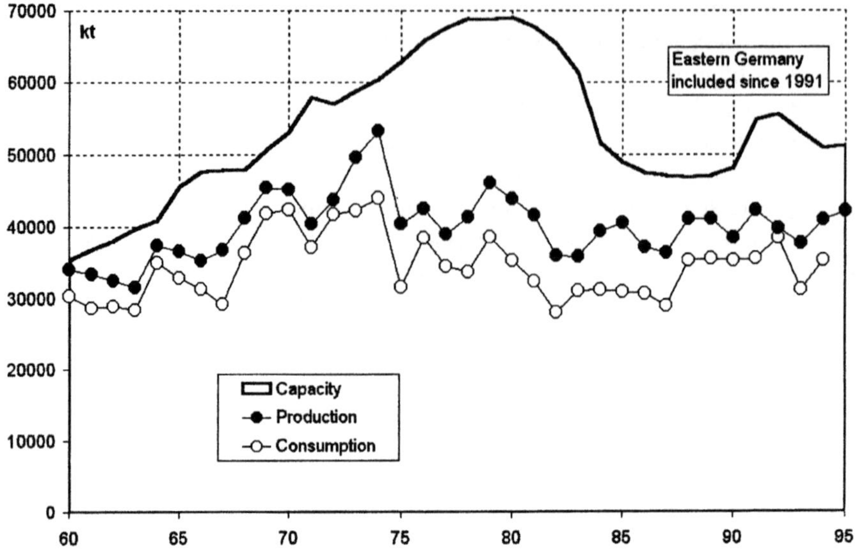

Source: Statistisches Jahrbuch der Stahlindustrie 1996

Fig. 9. German crude steel capacities, production and consumption 1960–95

subsequent boom phases for the economy as a whole. Thus, by the early 1980s, production had fallen to a level equal to that reached in the mid-1960s. Since the beginning of the 1980s, production has to a large extent remained stagnant. The development of German crude steel consumption runs, to a great extent, in parallel with production, with the exception of a sharper decline during the 1966/67 economic crisis and a rise at the end of the 1980s.

Instead, the development of crude steel production *capacities* has differed remarkably from that of production and consumption: In the years 1960–78 capacities doubled at a continuous pace, undisturbed by the slower growth of production and consumption in the 1960s and even the crisis in 1974/75. Finally in the early 1980s a substantial share of capacities were forced out of the market. The reunification of Germany led to a temporary increase in capacities with no visible impacts on production or consumption.

The modernisation of the steel industry described in Sect. 1.2 had made new capacities of oxygen converters and electric furnaces necessary and had thus contributed to the substantial over-capacities: While in 1960 almost all crude steel was produced in Thomas converters and open hearths, both processes were gradually abandoned during the next twenty years (Fig. 10).

The Saarland was affected earlier and more severely by the steel crisis than the Ruhrgebiet. At the end of the 1970s, the utilisation of its steel capacity was well below the German average (Bünnig 1983). When the Saarland became part of Germany in 1957, its steel industry was already characterised by poor competi-

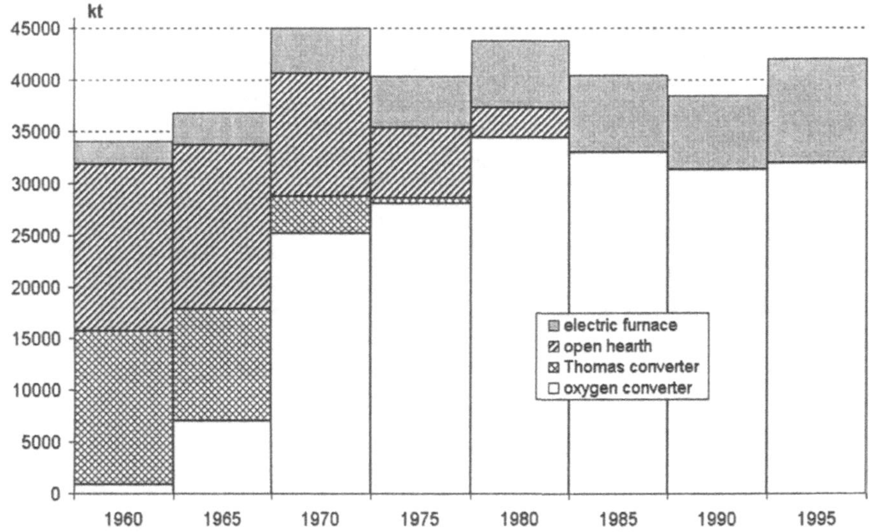

Source: Statistisches Jahrbuch der Eisen- und Stahlindustrie, various issues

Fig. 10. German crude steel production by process 1960–95

tiveness and low technological development. Thanks to state financial support and a strong economy, this underdevelopment and the need for change remained hidden until the end of the 1960s. Compared to the Ruhrgebiet, the Saarland was disadvantaged in the lack of waterways in its immediate vicinity, the lack of regional processing and manufacturing industries and the above-average share of mass production in its total steel production. Fragmented property rights resulted in small plants, while concentration, rationalisation and investment in modern technologies were not being pursued (Jürgenhake et al. 1988). It was not until 1977 that common restructuring strategies were developed. The Ruhrgebiet, on the other hand, owing to its local advantages and the steel industry's long-term orientation towards expansion has been seriously affected by the steel crisis only since the beginning of the 1980s (Bünnig 1983).

2.3.2 Economic and Social Impact of the Crisis

GDP and gross value added (GVA) by manufacturing industry developed more or less in parallel until the end of the 1970s (Fig. 11). Afterwards, growth in industrial GVA fell more sharply in times of economic depression. Gross value added by the iron and steel industry shows even greater variance around a stagnating trend: The peak values were reached in 1980 and 1992, but these were scarcely higher than those of the mid-1970s. Thus, its de-coupling from GDP is even more complete than that of manufacturing industry overall.

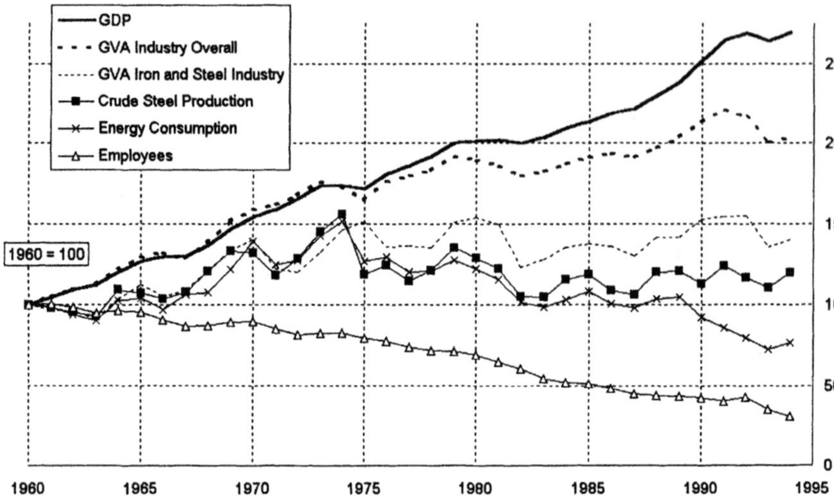

in 1991 prices.
Sources: Statistisches Bundesamt; Statistisches Jahrbuch der Eisen- und Stahlindustrie, various volumes; IEA

Fig. 11. Major trends in the iron and steel industry in West Germany 1960–94

The fall in employment within the German steel industry had already begun in the 1960s, i.e. at a time when the industry's economy was still healthy. At 421 000, employment was highest in 1961. By 1978 it had fallen to around 300 000 people and has now fallen by more than half again.

In the Saarland, employment in the steel industry fell from 53 000 in 1970 to scarcely 29 000 in 1986 (Jürgenhake et al. 1988). The rate of unemployment in the state of Saarland rose from 3.9% in 1974 to over 13% in 1986. In 1994, unemployment in the Saarland totalled 12% (Statistisches Jahrbuch 1995). Although the Ruhrgebiet was affected by the steel crisis later than the Saarland, its rate of unemployment has surpassed that of the averages for North Rhine-Westphalia and West Germany since the beginning of the steel crisis. In 1982, unemployment in the Ruhrgebiet was at 10.4%, compared to the West German average of 7.5%. This gap widened in subsequent years: In 1990 unemployment levels in the Ruhrgebiet and West Germany as a whole amounted to 11% and 6.6% respectively (Heinze et al. 1992).

2.4 Iron and Steel in Luxembourg

2.4.1 Major Materials Inputs and Outputs

Natural resources made Luxembourg the country with the by far highest steel production per capita. In 1907, domestic mining activity reached a peak, with 7.5

Mt of iron ore extracted, and production remained more or less stable until the late 1960s, interrupted only by the two World Wars and the Great Depression (Fig. 12). In the 1970s, iron ore mining declined rapidly and was definitively abandoned in 1981, due to the low iron content of the remaining resources, compared with iron ore from overseas. Since then, Luxembourg has had no natural resources for steel-making. But even in the 1950s and 1960s, increasing demand for iron ore could only be met through precipitously rising imports: since 1964, net imports (i.e. imports minus exports) have exceeded domestic production of iron ore.

Nevertheless, the most important inputs for the production of pig-iron – iron ore and coke – have always originated in the vicinity of Luxembourg. Until the early 1980s almost all imports of iron ore came from Lorraine, where France borders on Luxembourg (Fig. 13). However, production by the French mines, which extract the same sort of iron ore (minette) like the ones in Luxembourg, also declined, from 57 Mt in 1970 to 4 Mt in 1993, and the French iron ore mining sector will also be abandoned soon (Wirtschaftsvereinigung Bergbau 1994, p. 267). Therefore, Luxembourg has imported increasing shares of its domestic consumption from countries farther afield, such as Sweden or Brazil.

Coke imports have come almost entirely from neighbouring countries, mostly from Germany (Fig. 14). In the course of the gradual decline of German coal mining, increasing amounts of coal would have had to be transported greater distances (from the USA, South Africa, Australia). But this was not necessary, because coal demand in Luxembourg had fallen substantially since the early 1970s.

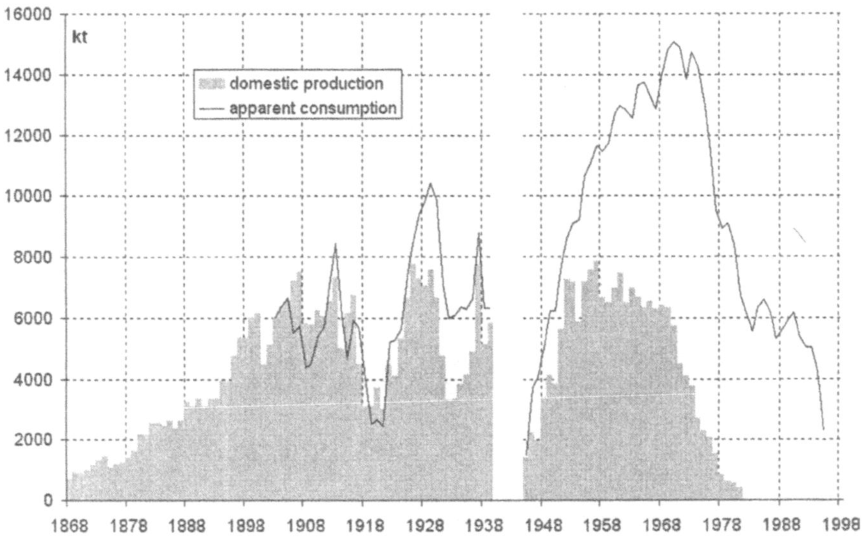

Source: STATEC 1990, 1997; no data available for 1940–44

Fig. 12. Production and consumption of iron ore in Luxembourg 1868–95

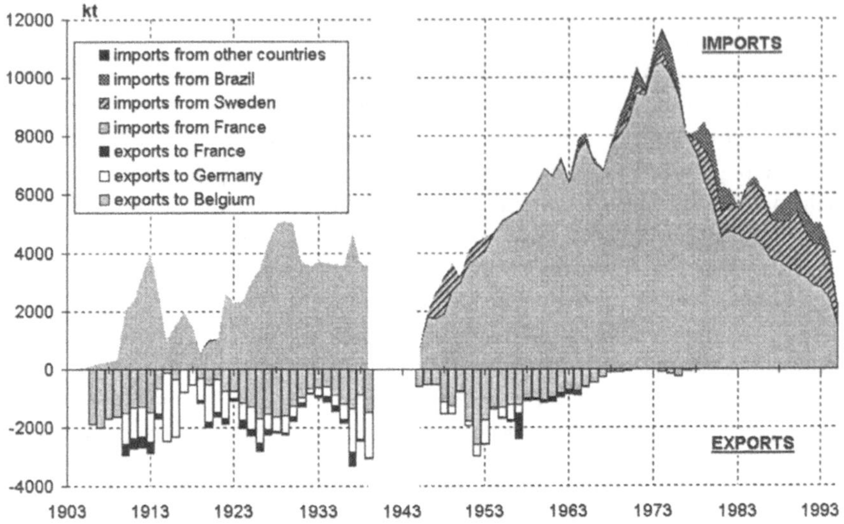

Source: STATEC 1990, 1997; no data available for 1940–44

Fig. 13. Luxembourg's foreign trade in iron ore by countries 1903–95

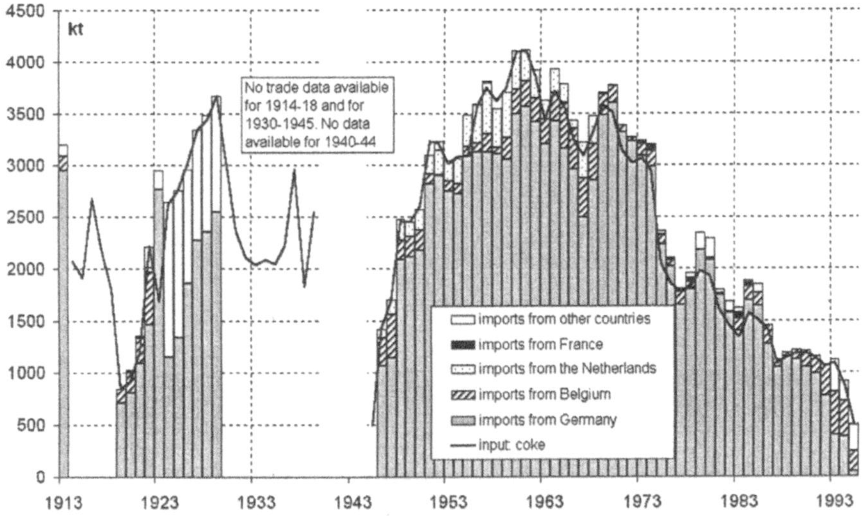

Source: STATEC 1990, 1997

Fig. 14. Luxembourg's coke imports by country 1913–95

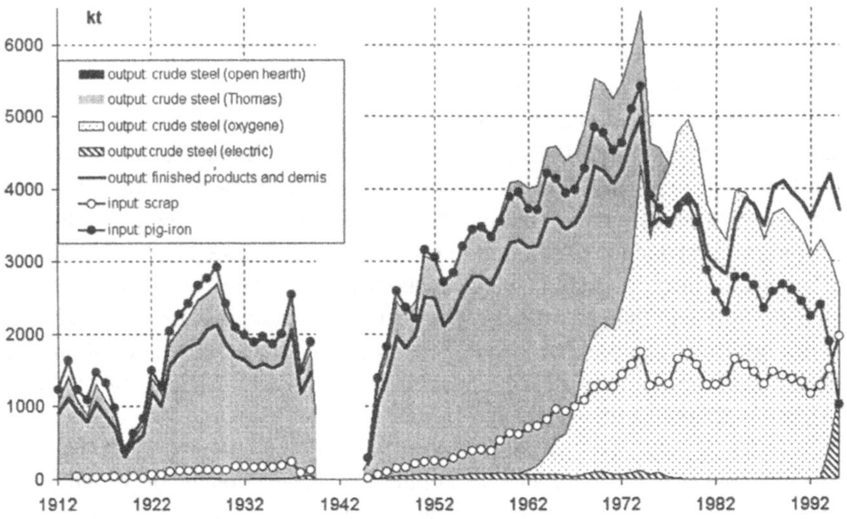

Sources: STATEC 1990, 1997; no data available for 1940–44

Fig. 15. Major inputs and outputs of crude steel production in Luxembourg 1912–95

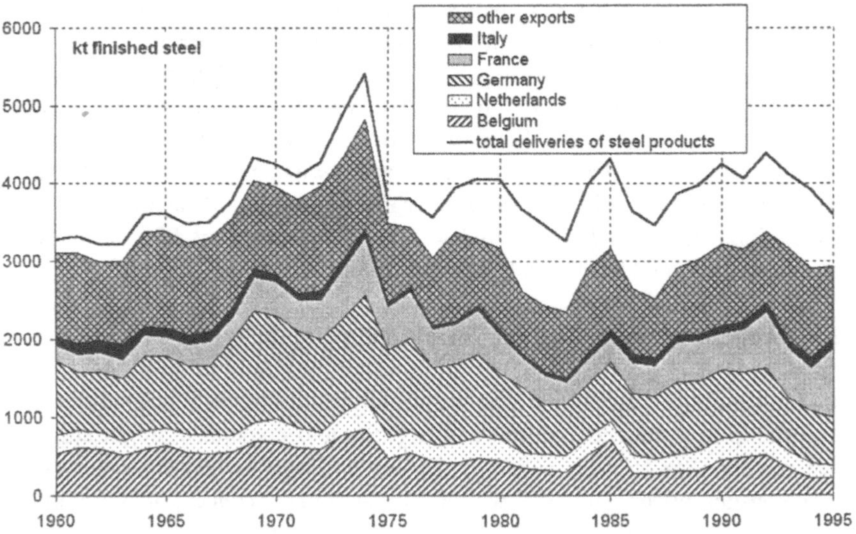

Sources: STATEC 1990, 1997

Fig. 16. Luxembourg's exports of finished steel by country 1960–95

The quantities of inputs and outputs in the production of *pig-iron* are very closely dependent: iron from iron ore becomes the major component of the output and cannot therefore be economised upon, while carbon is needed not only to heat the furnace but also to react with the oxygen in the ferric oxides which make up the ore. Basically, the trend in Luxembourg's pig-iron production parallels the trends in these inputs, with a steep rise until the late 1960s and a dramatic decline since then (Fig. 15). In 1970 there were 30 blast-furnaces in Luxembourg, of which 25 were actually in use, but in 1997 the last remaining furnace was decommissioned and production of pig-iron in Luxembourg did come to a final halt.

The scrap content in steel-making inputs has risen gradually since the 1950s, with a sharp rise in the early 1990s. The decline in *crude steel* production has at any rate still been sharp. After some 25 years of continuous growth of about 4% annually until 1973, crude steel production was halved to the level of the early 1950s in the following 25 years.

The trend in production of *finished steel* is a striking contrast: Until the mid-1980s, finished steel production tracked crude steel production almost exactly, at a slightly lower level. The difference between both time-series amounted almost precisely to the item "scrap arising in the steel industry" as documented by the ECE's Annual Bulletin of Steel Statistics, indicating that there was hardly any foreign trade in unprocessed crude steel. Since then, finished steel production figures have not declined any further, but instead risen slightly to a level even higher than that of crude steel production. In 1984, Belgium and Luxembourg agreed on co-operation between their steel industries, with ARBED specialising on rolling steel produced in Belgium by Cockerill-Sambre (Hirsch 1984).

Similarly to the situation with regard to input imports, neighbouring countries are also the main trading partners for finished steel, but now as importing countries (Fig. 16). Some 60% of all exports of finished steel have been delivered to Germany, Belgium, France and the Netherlands, with only minor variations in market shares. But from the mid-1970s the share of total exports in total deliveries declined from 93.5% in 1974 to 76.2% in 1983, leaving some 20% of all steel produced in the Grand Duchy for further processing.

2.4.2 Impacts on Value Added, Employment and Energy Use

The economic importance of the iron and steel industries has not declined as much as the slump in production of iron ore, pig-iron and crude steel might suggest. Unfortunately, data on value added by the iron and steel industries are not available in constant prices. Therefore we refer to the broader concept of "basic metal industries", which includes the production of non-ferrous metals (21% of this sector's value added in 1990; STATEC 1997). But the surprisingly close relationship between value added by basic metal industries and the production of finished steel (Fig. 17) indicates that the dynamics of this sector are dominated by steel production. Like the production of finished steel – and unlike all other products of iron and steel industries discussed – value added recovered after the mid-1980s

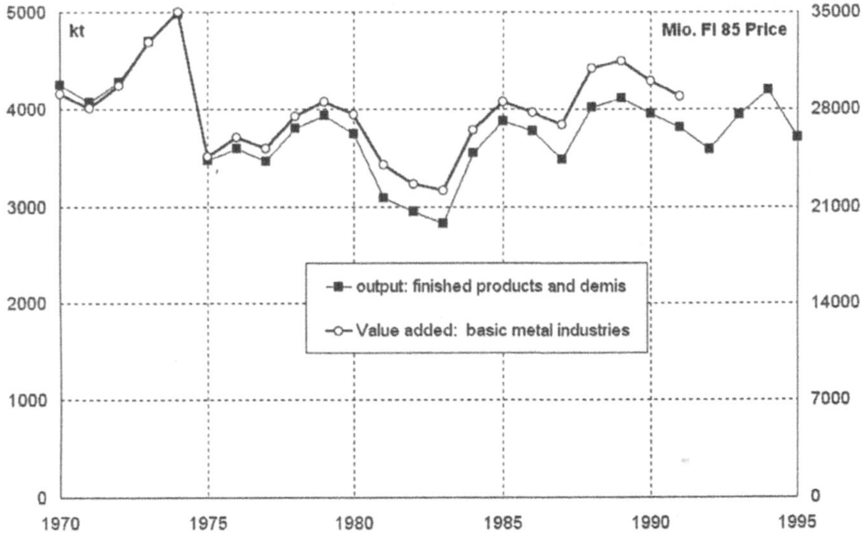

Sources: OECD: National Accounts; STATEC 1990, 1997

Fig. 17. Value added in basic metals and finished steel production in Luxembourg 1970–95

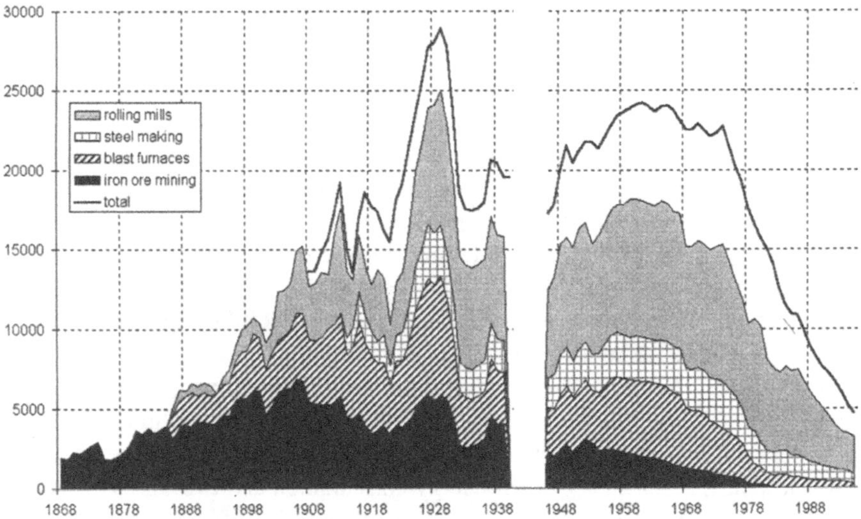

Sources: STATEC 1990, 1997

Fig. 18. Employment in the iron and steel industries in Luxembourg 1868–95

from the steel crisis which began in 1974, and has since maintained a more stable course. With hardly any growth in value added and a rapidly growing national economy, the sector's share in GDP continues a steady, but not dramatic decline.

This stands in contrast to employment (Fig. 18): between 1974 and 1995, the size of the workforce shrank by 79%, or 7.3% annually. Employment in the mining sector started to fall gradually after the first peak in mining production in 1906/07, one which has not been surpassed substantially since then. Employment in blast-furnaces has been declining since the mid-1960s, in crude and finished steel production since the mid-1970s.

3 Environmental Effects

For a long time, not only the economic importance and political power of the coal industry, but also the environmental problems of *coal production and consumption* have been most typical within industrial societies: this holds true especially for countries with their own coal production and coal-intensive economies. Compared with its substitutes, coal is a rather dirty fuel with considerably higher emissions of particulates, SO_2, NO_x and CO_2 per unit of usable energy. Not only coal consumption, but mining itself, causes a large number of local and regional environmental problems: extracting coal leads to the production of an amount of unusable overburden roughly as high as the coal extracted itself and to twice as great an amount of water (Wirtschaftsvereinigung Bergbau 1994, p. 171). Environmental modernisation has led to a substantial alleviation of most coal-related environmental problems. Nevertheless, reduction in coal consumption can still be regarded as a major improvement in the environmental situation even if this is compensated by a higher consumption of other fuels (cf. Franke et al. 1989).

The *steel industry's* significance for the natural environment lies in the amounts of raw materials, energy and water inputs, and the generation of solid and gaseous by-products. The steel industry is characterised by a continuing concentration of large plants in areas of high urban density. Steel production has never been the subject of an environmental debate, but has been subjected to environmental regulation, particularly as regards air pollution. It is also affected by standards requiring increased materials re-use and closed material cycles.

In 1993, the German iron and steel industry's final energy consumption amounted to 13.6 Mt OE or 5.6% of total final energy consumption (17.9% of industrial final energy consumption; IEA 1996). In Luxembourg, the steel industry is still responsible for 30% of final energy consumption and two thirds of industrial energy use (Fig. 19). The decline and final abolition of Luxembourg's pig-iron production has been a major relief to the environmental situation (Fig. 20). The steel industry's energy consumption has almost exactly followed the course of the declining production of pig-iron (cf. Fig. 7 for Germany). Coal has remained the most important energy source for steel production. With the end of blast furnaces and oxygen converters, electricity is about to become more important.

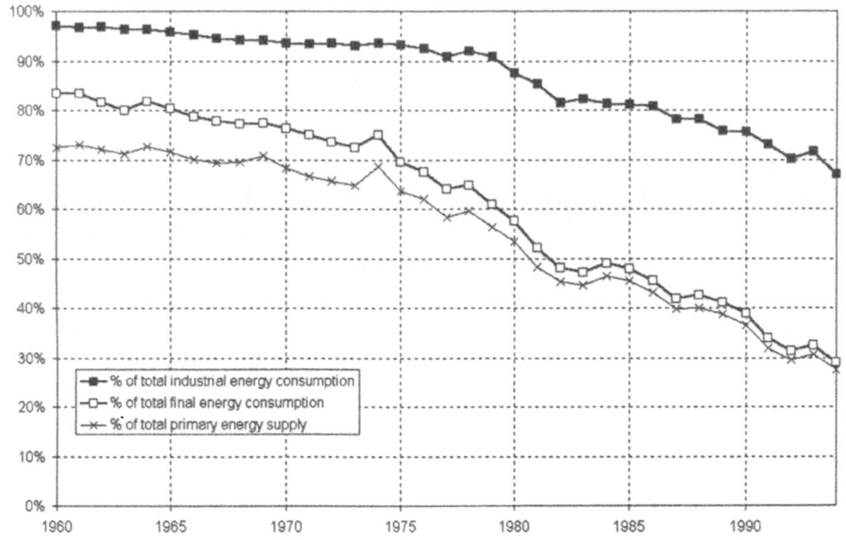

Source: IEA: Energy Balances

Fig. 19. Energy consumption of iron and steel production as a share of total industrial, final and primary energy consumption of Luxembourg 1960–94

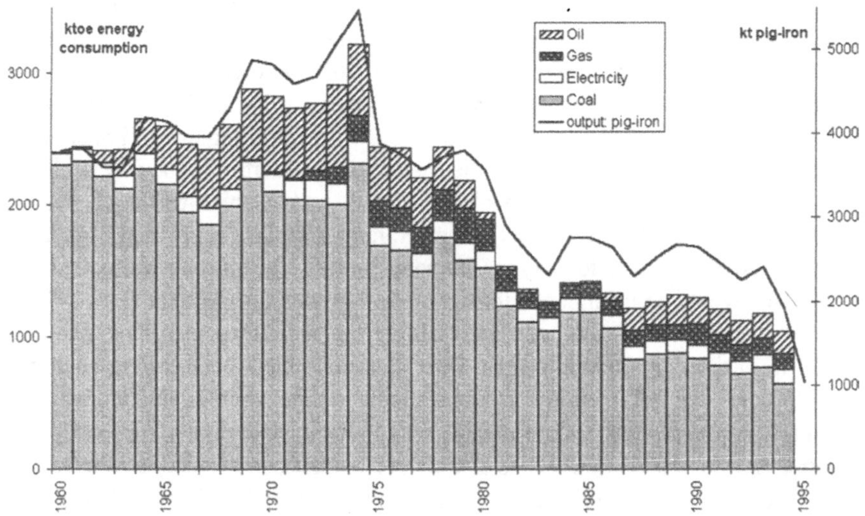

Sources: IEA: Energy Balances; STATEC 1990, 1997

Fig. 20. Energy consumption in the iron and steel industries and pig-iron production in Luxembourg 1960–94

In oxygen processes, the stages which consume most energy are the blast furnace, the coking plant and the hot-rolled strip plant (hot rolling mill), followed by the sintering plant and the cold-rolled strip plant. Electric arc furnaces consume considerably less primary energy than the blast furnace/oxygen converter route, owing to scrap being used as input. However, the most consumed form of final energy is electricity. As Germany uses a large proportion of fossil fuels in electricity generation, this process is significant in producing emissions of CO_2, SO_2 and NO_x, for example. According to the specific semi-finished steel product, steel's specific energy consumption varies between 20 and 26 MJ/kg.

Blast furnaces and steel works are responsible for huge amounts of solid residue such as slag – which is used to a large degree as input in the construction sector and cement production – and other wastes. Coking plants cause emissions of CO, HC, SO_2, NO_x, dust and H_2S, whereas sintering plants and blast furnaces cause emissions of CO, SO_2, NO_x, dust and dioxins. The problem of hazardous emissions, such as dioxins or heavy metals, is intensified when steel is recycled. This is due to the scrap used as input often containing impurities, for instance when alloys are recycled (Bandt 1991).

It can be assumed that the stabilisation and decline in German steel production has contributed to a decline in the environmental stress caused by the industry. However, improvements in its overall environmental impact are due in large measure to intrasectoral change, including end-of-pipe treatment, closing of material cycles (internal recycling of by-products), changes and reductions in inputs, as well as improved processes. The reduction in energy intensity can be inferred from Fig. 11 (see p. 264): It shows that energy consumption has declined considerably and continuousy, while steel production and value added in steel industries have remained fairly stable since the mid-1970s after increasing befor.

The replacement of older steel production processes by modern processes has had particular significance for air emissions. In West Germany, Thomas-steel production ceased in 1976, while the last Siemens-Martin furnace was closed down in 1982 (Statistisches Jahrbuch der (Eisen- und) Stahlindustrie; cf. Fig 10, p. 263). As the volume of exhaust gas per tonne of steel is four times higher for Thomas-converters than for oxygen-blowing processing, this transformation improved the overall environmental impact of steel production (Jeschar et al. 1996).

End-of-pipe technologies with increasingly better performance have been installed in various production stages. Dust emissions have been falling since the late 1960s, i.e. the fall in emissions had started at a time when production was still rising (Philipp et al. 1987). Considerable reductions in coke inputs, for example to sintering and in blast furnaces, has also reduced emissions of, among others, CO_2 and SO_x (Philipp et al. 1987; Philipp and Theobald 1993; Schade and Gliwa 1978; Schulz 1992). In Germany, specific dust emissions fell from 10 kg/t crude steel in 1960 to 2.8 kg/t in 1964, 1.4 kg/t in 1974 and 0.7 kg/t in 1992. CO_2 emissions were reduced from 2.6 t/t crude steel in 1960 to 1.4 t/t in 1993.

The closure of production cycles has led to a considerable reduction in water use. Not only water as a coolant but also water used in processing is re-used.

While water input amounted to 150 m^3/t crude steel in 1960, it has now been reduced to 31 m^3/t (Philipp and Theobald 1993). The rate of recycling exceeded 80% in 1987, and a rate of 98% is possible today (Philipp et al. 1992).

Recycling and re-use are also important with respect to solid residues. While German steel production at the beginning of the 1940s produced up to 1 200 kg/t crude steel of solid waste, integrated steel works currently produce an average of 575 kg/t (ibid.; Angrick 1993). Apart from increased recycling, the switch to inputs of richer imported ores has contributed to this improvement. The introduction of continuous casting was a further important development in this respect. Reduced production of slag can also be explained by changes in inputs, process optimisation and the introduction of secondary metallurgy (Philipp et al. 1992).

Add-on prevention technologies, however, have had a negative influence on waste arising from steel production. In Germany, the amount of dust and slag collected increased by 7.5 kg/t crude steel between 1960 and 1990. Currently, about 90% of the residues produced in the steel industry are re-used, of which 69% are sold to other industrial sectors and 21% are re-used directly within the steel industry. Ten per cent of the waste must be disposed of (ibid.).

4 The Management of the Coal and Steel Crises

4.1 Regulating Supply and Demand

4.1.1 German Coal Policy

In 1959, as German coal became uncompetitive, the German parliament limited annual coal imports to 6 million t (*Zollkontingentgesetz,* in force until 1995), but little was done to restrict competition from other energy sources, notably petroleum products: taxes on petroleum products were levied, but could not dampen rising demand, especially for gasoline, one of the major driving forces of growth in refinery output (see Sect. 1.2.1, p. 246). With subsidies for nuclear energy, the German government even brought a new energy source into competition with coal in one of its few remaining markets, electricity production.

Nevertheless, coal policy secured demand for German coal through agreements with electricity utilities (*Verstromungsgesetze* 1965, 1966 and 1974, *Jahrhundertvertrag* 1980) and steel producers (*Hüttenvertrag* 1969 and 1985). Since the latter had to compete with companies using cheaper coal, the price difference between German and overseas coal was compensated with subsidies. The higher prices for German coal used in electricity production were passed on to consumers via the so-called *Kohlepfennig* (since 1974), a "coal penny" levied on all electricity sales (including those not produced with coal). In 1994, the Federal Constitutional Court abolished this scheme, arguing that protection for German coal mining had to be financed by the state and not by electricity consumers (cf. Neu 1995).

Since coal mining has been dependent on state aid, the major decisions have been made by the German government in close co-operation with mining companies (owned in turn by the major customers for coal, i.e. steel and electricity producers), the governments of *North Rhine-Westphalia* and the *Saarland* and the miners' union (cf. Jörnmark 1993). With falling prices for competing energy sources during the 1980s, increased subsidies became necessary in order to finance the sale of coal mines, which had become less and less important to the economy and employment. The major stakeholders negotiated over the future of German coal mining at several "round tables" (*Kohlerunden*). Eventually, foreign trade restrictions were abolished, and a plan was agreed in 1997 to lower coal subsidies from 8.9 mio. DM to 5.5 mio. DM, reducing coal production from 50 Mt to 30 Mt by the year 2005 (Tagesspiegel 14.03.1997).

All in all, German coal mining serves as a prime example of environmentally beneficial structural change which has been resisted using almost every available means. Nonetheless, this industry's decline was not halted, but merely drawn out to an unbelievable extent. When, in the early 21st century at the earliest, German coal mining could be closed down for good, it will have taken more than half a century to halt a business which has been uneconomic the whole time and which has had significant impact on the entire national economy. The total disappearance of the sector is still a virtual taboo for all political parties, but with the planned falls in production, the influence of its political lobby may soon decline as well.

4.1.2 European Community Steel Policy

Even before the steel crisis of 1974/75, the steel industries maintained prices through agreements – supported by the European Commission as long they remained temporary (Wienert 1985). But with rising global demand and capacity utilisation of almost 90%, there was little need for regulation of the market, and European steel policy remained fairly liberal in the 1950s and 1960s (Stehn 1993).

After the beginning of the steel crisis, German producers first hoped to continue expanding capacity, owing to their high productivity compared to other European producers, and were opposed to French proposals for a common European steel policy. But with increasing competition from highly productive Japanese and subsidised Italian producers in 1975 and 1976, they finally agreed on intervention through the ECSC (Bünnig et al. 1983). With a 40% drop in steel prices and declining capacity utilisation from 87% in 1974 to 66% in 1976, European steel policy became more active, but still remained voluntary in most respects:

- The 1st steel plan (May 1975) asked companies for monthly reports on production figures, publication of quarterly (instead of annual) production projections by the Community, stricter controls on price reporting and a common calculation of transport costs for finished steel products. The 2nd steel plan (October 1975) asked for more detailed projections for different product groups. All these measures were voluntary (Kerz 1991, p. 52).

- The 1st steel crisis plan (December 1976) introduced voluntary production quotas by way of commitments by individual steel producers to limit their sales. At the same time, the EUROFER cartel of all European Community steel producers was founded, in order to make coordination of the market easier. This was also a response to the DENELUX cartel, founded in May of the same year by German, Dutch and Belgian-Luxembourgian steel producers, raising the spectre of a privately organised cartel (ibid., p. 78). The 2nd steel crisis plan (March 1977) added to the system of market regulation some voluntary price guidelines for all finished steel products and mandatory minimum prices for several long products (ibid., p. 54).
- The attempts to limit supply and keep prices at profitable levels above world market pricing led to rising imports. In order to halt this trend, the Commission set anti-dumping tariffs and started negotiations with foreign suppliers. In 1980, 13 countries, which were responsible for 85% of steel imports from outside the EU, had signed "voluntary restraint agreements" (Conrad 1997, p. 94).

But the results of these initiatives were disappointing: Italian and German minimills, in particular, refused to comply, and were even fined for disregarding minimum tarifs (Conrad 1997, p. 90). However, other companies undermined the agreements as well, new (and more productive) capacity continued to be installed, while obsolete capacity was often supported with subsidies aimed at modernisation, in the hope of increased demand sometime in the future (Gieseck 1995, p. 61), reflecting what had become a fundamental heterogeneity in steel producers' interests (see Sect. 1, p. 244). After a short-lived recovery in 1978/79, demand and prices fell again in the wake of the second oil crisis and a new voluntary agreement on production quotas failed. In 1980 the European Commission declared the situation to be a "manifest crisis" under Article 58 of the ECSC Treaty of Paris, which allowed the Commission to install a strict system of mandatory quarterly quotas for production and sales and compulsory minimum prices. Exports were also restricted, in order to prevent retaliatory measures by importer countries. The quotas were set in close co-operation with steel producers, in the person of the EUROFER-II cartel, which included the nineteen largest steel producing companies. EUROFER-II companies were even allowed to trade quotas (e.g. the agreement between ARBED and Cockerill-Sambre in 1984 mentioned in sect. 2.4.1). The system of production quotas was planned to last until 1985.

The Commission tried to restrict national subsidies which, according to the ECSC treaty, are only legal with the approval of the Community. In 1980, guidelines for approving national subsidies were adopted, which linked such approval to plans for reducing (and modernising) capacity. In addition, state aid had to be temporary and degressive. All plans were to aim at phasing out state aid by 1985 at the latest (Hudson and Sadler 1989, p. 33).

In 1985, deregulation of the European steel market began, although capacity was not reduced as much as the Commission had demanded and several illegal national subsidies persisted (Demgenski 1990; Conrad 1997). Production quotas were phased out gradually up to 1988. Since 1985, national subsidies can only be approved for research and development, environmental protection, assistance in decommissioning production sites or to promote regional development.

When, in 1990, demand for steel – and therefore prices – fell again, producers demanded that the Commission permit a temporary cartel, in order to be able to reduce capacity by 20% and to pay off dismissed workers with a thousand million ECU (Conrad 1994). But the Commission did not want another cartel or new production quotas, and tried to adopt indirect measures instead. In 1993, the Commission tried again to link approval of domestic state aid and ECSC grants with capacity reduction, but the steel companies were not prepared to commit themselves to sufficient decommissioning of production. National subsidies were paid out in any case (Conrad 1997).

During the steel crisis, the ECSC did not succeed in its goals. Most of the repeated attempts to link subsidies with capacity reduction failed, and made little difference where they did: for example, the United Kingdom paid out five times as many subsidies as Germany, but reduced only half as much capacity in the years 1975–91 (Fig. 21). Italy's enormous state aid even corresponded with a huge increase in capacity and production. From the point of view of a cartel pursuing the good of all its members by preventing price wars, capacity should have been reduced equally among companies or, if not, those who were forced to leave the market should have been compensated for their losses by the remaining suppliers,

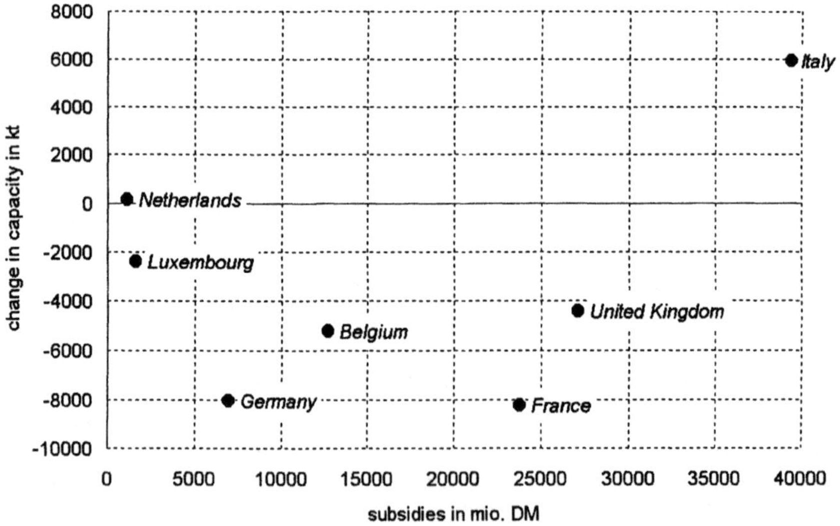

Sources: Berthold 1994; Statistisches Jahrbuch der Eisen- und Stahlindustrie; UN; own calculations

Fig. 21. Subsidies and capacity changes for major European steel producing countries 1975–91

who could now take advantage of the reduced competition. In fact, this was exactly the policy the European Community tried to follow.

But the European steel market no longer allowed for the implementation of an effective cartel: too many companies were profiting too much from breaking the rules, i.e. abandoning the common good. Mini-mills generally had nothing to gain from the cartel and were notorious for undermining agreements – agreements which they had not signed up to in the first place. But elsewhere also, differences in productivity – and in the ability to gain subsidies – had become too large to prevent companies and countries from taking the exit option, even if it was illegal to do so. Legally, the European Commission had every right and maybe also the means to implement an effective cartel (Conrad 1997, p. 142), but it had neither the will nor the political authority to do so.

4.2 Restructuring Companies

4.2.1 Coal in Germany

Up to the end of the 1960s, the German coal mining industry consisted of no less than 32 independent companies. Seven companies were owned by foreign companies (21.3% of coal extraction), five were state-owned (19.2%) and twenty privately owned (57.4%) (Statistik der Kohlenwirtschaft 1968), mostly by steel producers. In 1969, Ruhrkohle AG was founded, in order to merge the formerly independent coal mines of the *Ruhrgebiet* (76.3% of coal extraction in 1970), with the former owners becoming shareholders in the new company. This had two main effects: firstly, it became easier to concentrate coal mining activity at the most efficient coal pits, and secondly, the then profitable sectors of the coal mines' parent companies (namely steel production) were clearly separated from the losses of coal mining. This took place to such an extent that even most of the mining companies' land remained in the hands of its former owners – coal mines had traditionally belonged to major land owners in the respective regions – while only the obviously unprofitable coal mines were managed by Ruhrkohle AG afterwards – and subsidised by government.

During the 1980s, electricity generators became the dominant shareholders in Ruhrkohle AG (now RAG): in 1984, VEBA, the parent company of PreußenElektra, West Germany's second largest electric power company, and VEW (Vereinigte Electrizitätswerke) bought shares from the steel producers Mannesmann, Klöckner, Salzgitter and Krupp, in order to put together a majority shareholding of more than 60%, while RAG retained 10% of its own shares. RAG has not only been involved in German coal mining, which was organised into its subsidiary Ruhrkohle Bergbau AG until 1998, but also in mining engineering, trade, power generation, process technology, chemicals, plastics and the environment.

In 1998, Saarbergwerke AG (100% of production in Saarland) and Ruhrkohle Bergbau AG merged to form Deutsche Steinkohlen AG (DSK), which became a part of RAG. At the beginning of 1999, Preussag Anthrazit AG (100% of production in Ibbenbüren) also joined the DSK. Today, the whole German coal mining industry is organised in a single company.

4.2.2 Steel in Germany

German steel production is strongly concentrated in a few large companies. In 1995, the largest steel producers (measured in quantities of crude steel produced) were Krupp-Mannesmann, Krupp Hoesch, Preussag, Thyssen and ARBED S.A. The German steel industry has always been predominantly in private hands, unlike in other European countries.

The history of the German steel industry has been characterised by various forms of co-operation, dating back to the mid-19th century. Apart from horizontal and vertical integration, companies have been entering into co-operative projects since the early 1960s, all aimed at better exploiting their capacity through specialisation. One example is the setting up of the four *Walzstahlkontore* (Sheet Steel Boards) in 1967, which brought producers together to coordinate sales and investment. The European Commission supported the *Walzstahlkontore* until 1971. Afterwards, the companies were organised into four *Rationalisierungsgruppen* (Rationalisation Groups), two of which remained active until 1982, but none of them became strong enough to dominate the market (Gieseck 1995; Kerz 1991).

During the last decades, concentration in the German steel industries has gradually increased (Gieseck 1995):

- *Thyssen* took over *Hüttenwerke Oberhausen* (1966) and *Rheinstahl AG* (1973), which made Thyssen ECSC's biggest steel producer.
- *Hoesch* merged with *Dortmund Hörder Hüttenunion AG* (1966).
- The publicly owned *Salzgitter AG* took over *Ilseder Hütte* (1969).
- Luxembourg's *ARBED SA* took over *Stahlwerke Röchling Burbach* and *Neunkircher Eisenwerke*, both located in Saarland (1978). In 1986, they were nationalised under the name *Saarstahl Völklingen GmbH* and merged in 1989 with *Dillinger Hüttenwerke AG*, owned by the French *Usinor-Salicor*. After *Saarstahl AG* went bankrupt in 1993, *Saarstahl AG* and *Dillinger Hütte* were seperated again, but still hold shares in each other (cf. Saarstahl 1999).
- *Krupp AG* bought a majority share in *Hoesch AG* in 1991, and they merged to form *Friedr. Krupp AG Hoesch Krupp* in 1992.
- In 1994, the Belgian *Sidmar*, a part of ARBED, took on the majority of *Bremer Klöckner Stahl*. Since 1992, *Stahlwerke Thüringen* have also belonged to ARBED (cf. Arbed 1999).
- In 1994, the Belgian *Cockerill-Sambre* acquired a majority of the East German *EKO-Stahl* (Eisenhüttenstadt), and bought up the rest of the shares by the end of 1998. Since October 1998, *Cockerill Sambre* has been owned by the French *Usinor* (former *Usinor-Salicor*; cf. EKO 1999).
- *Thyssen* and *Krupp* merged to form *Thyssen Krupp AG* in March 1999 (backdated to October 1998; cf. Thyssen-Krupp 1999).

During the worst years of the steel crisis in 1975–88, German steel production was dominated by the privately owned integrated steel producers *Thyssen*, *Krupp*, *Hoesch*, *Klöckner*, *Salzgitter*, *Mannesmann*, *Dillingen* and *Saarstahl*. Most of them have diversified increasingly since 1970, usually into downstream activities like mechanical engineering and construction technologies. Today, steel production amounts to some 20–40% of these companies' total turnover. *Mannesmann* has made the most pronounced changes in this respect: until 1973, *Mannesmann* produced nothing but steel tubes, which account for only 12% of its worldwide sales today. Other production sectors are mechanical engineering (35%), automotive technologies (29%) and telecommunications (24%) (Mannesmann 1999; see also Kerz 1991; Hoffmann 1990).

4.2.3 Steel in Luxembourg

The steel industry in Luxembourg consists only of one company, ARBED (Aciéries Réunies de Burbach-Eich-Dudelange). ARBED is a privately owned company, but in 1985, the state of Luxembourg became the most important shareholder (today it owns 27.9% of the capital and 20.0% of the voting rights). The ARBED group is a multinational corporation which is seeking to integrate every stage of production, from extraction of iron ore and construction of steel plants to trade in finished steel products. Diversification into downstream activities is comparably unimportant.

In Luxembourg itself, ARBED stopped its integrated steel making and produces only long products. For this restructuring ARBED invested some 18 billion Luxembourg francs, mainly in order to build up three electric furnaces to process only scrap and to replace oxygen-blown converters which can only use limited amounts of scrap. Luxembourg is playing an increasingly smaller role in the ARBED group: today, most domestic production is organised in ARBED's long products sector with ProfilARBED as its leading company, which also comprises (inter alia) the German Stahlwerke Thüringen. In 1996, this sector had a share of 20.2% in ARBED's consolidated turnover. Much more important for ARBED are its flat products sector (43.3% of consolidated turnover), which includes LDD, a rolling mill at Dudelange (Luxembourg), but consists mainly of the Belgian SIDMAR, with its two integrated maritime steelworks at Ghent (Belgium) and Bremen (Germany). Other important fields of activity are stainless steel production in Belgium and an important stake in the Brazilian Belgo-Mineira group (32.0% of the capital and 35.3% of the voting rights) (ARBED 1997, 1997a).

4.3 Social Problems

The double crisis in coal and steel production had a major impact on the labour market. More than 700 000 jobs were lost in Germany between 1960 and 1997, most of them in the *Ruhrgebiet* and in *Saarland* (Fig. 22; for Luxembourg see Fig.

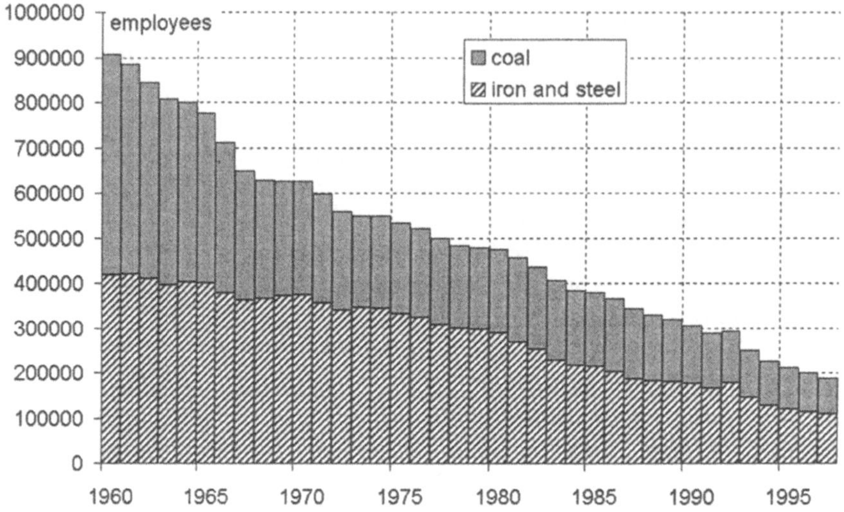

Sources: Statistisches Jahrbuch der (Eisen- und) Stahlindustrie; Statistik der Kohlenwirtschaft.

Fig. 22. Employment in the German coal and steel industries 1960–97

18, p. 269). Political action tried to help the miners and steel workers and to provide new jobs in the crisis regions.

In coal mining, most of the jobs were lost during periods of full employment, but since miners were the best-paid manual workers at the time, the shift to other sectors usually implied loss of income and thus the job losses provoked major protests aimed at preserving as many jobs as possible. With rising unemployment in the wake the first oil crisis in 1973 and the onset of the steel crisis in 1974/75, aid for dismissed coal miners and steel workers became more and more important. Redundancy payment schemes (*Sozialpläne*) had been common as early as the 1960s, for instance with respect to workers being dismissed due to technological change (Meyerwisch 1986), and became the most important measures assisting dismissed workers during the steel crisis (Arbeitgeberverband Stahl 1995/1996). Early retirement at 50, with 90% of annual salary has become the norm for German coal miners (Tagesspiegel 22.01.1995).

In Luxembourg, ARBED guaranteed job security and salary levels in 1974. Government, employers and trade unions set up the so-called "tripartite" conference in 1977, the results of which became law in December of the same year. This law established the Tripartite Coordination Committee, consisting of representatives from government, employers and trade unions, initially giving each of these groups veto rights on a number of economic policies in times of crisis (such as indexing wages and salaries, freezing profit margins and prices etc.). The Tripartite Coordination Committee agreed on public aid which helped to finance –

among other things – an early retirement programme and the "anti-crisis division" of ARBED, whereby some 3 500 steel workers were reassigned to public services, while retaining their former salaries. In 1981, the so-called solidarity tax was introduced to finance these measures (STATEC 1999; Hirsch 1986). But after 1981, the Tripartite Coordination Committee failed to reach any consensus, because the government followed an austerity policy with which the unions did not agree[4].

In fact, Luxembourg has not experienced many economic problems in recent decades, despite the steel crisis: since the mid-1980s, Luxembourg's economy has even been growing a little faster than in the 1960s, while almost all other OECD countries have had to get used to significantly lower growth rates (Fig. 23). This does not mean that the iron and steel industries have flourished: the economic success of Luxembourg since the 1980s is principally due to unparalleled growth in banking, insurance and other services, fostered by an increasingly diversified industrial structure.

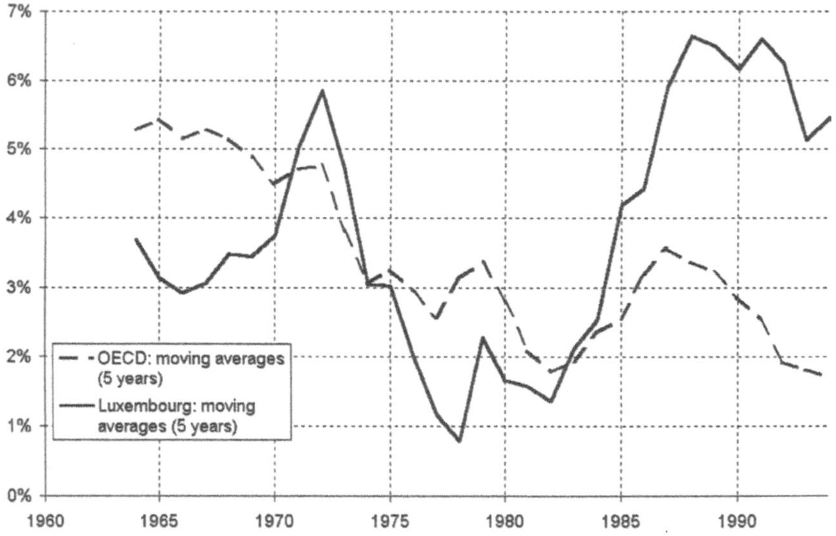

Source: OECD: National Accounts

Fig. 23. Annual GDP growth rates of Luxembourg and the OECD 1960–94

[4] "Tripartism has been most useful in obtaining the consent of unions to austerity measures. It never really succeeded in laying out an economic strategy, and its main achievement seems to consist in establishing a procedure by means of which the elements of an agreed strategy could emerge. The fact that no real strategy did emerge (except for the handling of social problems) might explain why the existence of Tripartism did not make any real difference to the way the government enacted policies or to the way industry sought to solve the problems. "(Hirsch 1986, p. 66)

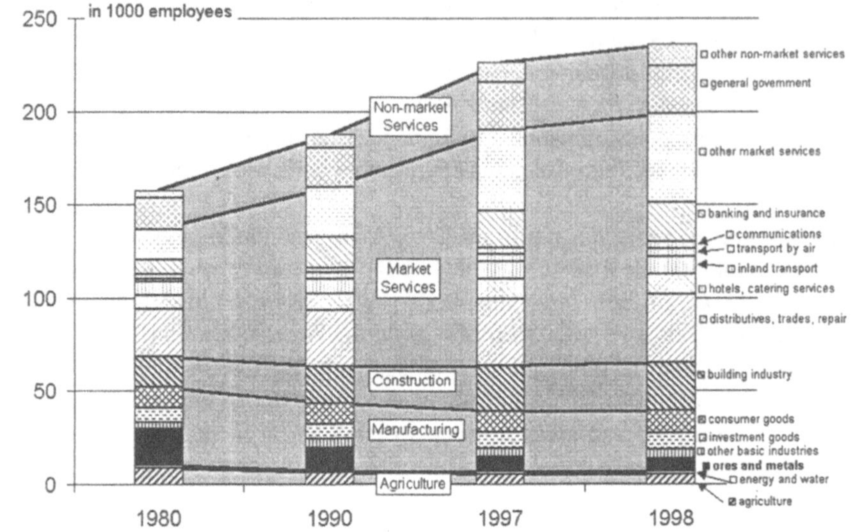

"Internal employment" includes commuters from adjacent countries, but excludes officials and agents of international institutions.
Source. STATEC 1999a

Fig. 24. Domestic employment by branch in Luxembourg 1980–98

Luxembourg's unemployment rate has never exceeded 3%, because employment has grown tremendously since 1980 (Fig. 24). Today, Luxembourg's labour force consists of over 50% foreign nationals, half of them living in neighbouring countries. Fast growth was not restricted to any single area, but occurred in a number of service sectors and in the building industry (cf. STATEC 1999; Afrough et al. 1997):

The rise of the *financial sector* began in the end of the 1960s. To some extent, the development has been a self-reinforcing process: according to interviews with bankers, the most important reason for establishing offices in Luxembourg has been the presence of other banks and trading partners (Nottrot 1985). This might also explain to some extent the growth in other service markets, like accountancy, auditing or office cleaning. The advantages of Luxembourg for banks include highly liberal legislation (including strict banking confidentiality, no minimum reserves requirement, no withholding tax), low costs (compared to London or Frankfurt), a central location in Europe, a multi-lingual population and a stable political and social environment. Lower taxes than in neighbouring countries have also been an important reason for the large proportion and high growth of *trade* services, especially since foreign trade within the EU was completely liberalised.

Not included in the employment data are the employees of many *international institutions* located in Luxembourg: the European Court of Justice, the European Investment Bank (EIB), the European Court of Auditors, the general Secretariat of the European Parliament and several important arms of the European Union Commission, including the Statistical Office (EUROSTAT) and the Publications Office, the Nuclear Safety Administration and the Directorate-General of "Credits and Investments". The growth of these institutions, of financial services and public investment have generated substantial demand for *construction* services.

The contributions of manufacturing sectors other than steel to employment and growth have been limited, despite major efforts since 1962, when the "framework law on economic expansion" was passed in order to curb the dominance of the steel industry. In 1977, the *Société Nationale de Crédit à l'Investissement* was founded, providing loans for modernising the production system and for research and development. The "new industries" promoted have created jobs, more in the former steel towns in the south of Luxembourg than in the service growth centre, the city of Luxembourg.

In Germany, federal and state governments work together on regional policies to support regions with structural economic problems (*Gemeinschaftsaufgabe zur Verbesserung der regionalen Wirtschaftsstruktur*). Initially, the Ruhrgebiet did not benefit from this programme, because of its high per capita income and its well-developed infrastructure (Hamm/Wienert 1990). But in 1979, industrial regions severely hit by structural change, like the Ruhrgebiet, were included in the programme, which has since supported medium-sized corporations as well as the service sector (NRW-Landesregierung 1979, Landesentwicklungsbericht 1988).

Long before this, the *Bundesland* North Rhine-Westphalia adopted several regional policy programmes for the *Ruhrgebiet*, for example the *Entwicklungsprogramm Ruhr* (1968–73), the *Nordrhein–Westfalen Programm 1975* (1970 – 75), the *Aktionsprogramm Ruhr* (1980–84), the *Zukunftsinitiative Montanregion* of 1987 and the *Zukunftsinitiative für die Regionen Nordrhein–Westfalen* of 1989. They sought to alleviate structural change and support diversification of the industrial structure by indirect measures such as infrastructure development and support for innovation and training (Kampherm 1993; Heinze et al. 1992; MWMT 1992). Similar programmes were implemented in *Saarland*, e.g. the *Zukunftsqualifizierungsprogramm Saar* of 1987 (Jürgenhake et al. 1988).

Bibliography

Afrough, A., Linkel, T., Plecher, K. (1997): Luxemburg. In: Haas, H.-D., Störmer, E., Wilking, J. (eds.): Saar-Lor-Lux. Wirtschaftsgeographische Aspekte einer Grenzregion im Wandel. München: Ludwig-Maximilians-Universität, Insitut für Wirtschaftsgeographie. (http://www.wigeo.bwl.uni-muenchen.de/~estoermer/exkurs.htm)

Altgeld, A. and Schneider, C. (1992): Stahl 2000 – Steel 2000. Technische Berichte, Heft 1/1992. Sonderdruck. Duisburg: Thyssen Stahl AG.

Angrick, M. (1993): Entwicklung des Standes der Emissionsminderungstechnik durch Investitionsvorhaben des Umweltbundesamtes. In: Moderne Techniken einer umweltfreundlichen Metallindustrie. Texte 39/93. Berlin: Umweltbundesamt.

ARBED (1997): Annual Report 1996. Luxembourg.

ARBED (1997a): ARBED. Luxembourg.

ARBED (1999): History. In: http://www.arbed.com.

Arbeitgeberverband Stahl (1995/1996): Jahresbericht 94/95–95/96. Düsseldorf.

Bandt, O. (1991): Ende der Fahnenstange. Mit dem Schrottrecycling sind erhebliche Umweltbelastungen verbunden. Müllmagazin, 2/1991, 35–39.

Berthold, N. (1994): Dauerkrise am europäischen Stahlmarkt – Markt- oder Politikversagen? Frankfurter Institut – Stiftung Marktwirtschaft und Politik. Bad Homburg.

Bünnig, J. (1983): Ursachen, Verlauf und regionale Auswirkungen der Stahlkrise. In: Bünnig et al. (ed.): Stahlkrise Regionalkrise. Duisburg.

Bünnig, J., Fobbe, G., Höfkes, U., Marx, W., Uske, H. (1988): Restrukturierungsprozesse in der Eisen- und Stahlindustrie. Ministerium für Arbeit, Gesundheit und Soziales des Landes Nordrhein-Westfalen. Düsseldorf.

Cho, T.-Y. (1992): Die Stahlindustrie Südkoreas im internationalen Vergleich. Ein Vergleich mit der westdeutschen Stahlindustrie. Göttingen.

Conrad, C.A. (1994): Die Industriepolitik der EU am Beispiel der aktuellen Krisenpolitik im Stahlbereich. List Forum für Wirtschafts- und Finanzpolitik 20, 1–4, p. 337.

Conrad, C.A. (1997): Europäische Stahlpolitik zwischen politischen Zielen und ökonomischen Zwängen. Baden-Baden: Nomos.

Darmstadter, J. (1971): Energy in the World Economy. A statistical Review of Trends in Output, Trade, and Consumption Since 1925. Baltimore, London: John Hopkins Press.

Demgenski, M. (1990): Die europäische Stahlindustrie im weltwirtschaftlichen Strukturwandel. St.Gallen: D-Druck-Spescha.

ECE (Economic Commission for Europe): Annual Bulletin of Steel Statistics for Europe, America and Asia. New York, Geneva (various issues).

Eckart, K., Kortus, B. (1995): Die Eisen- und Stahlindustrie in Europa im strukturellen und regionalen Wandel. Wiesbaden: Deutscher Universitätsverlag (DVU).

EKO (1999): Geschichte in Stahl. in: http://www.eko-stahl.de.

Franke, F.H., Guntermann, K.J., Paersch, M.J. (1989): Kohle und Umwelt. Essen: Glückauf.

Gerstenberg, W., Hummel, M., Knörndel, K.-D., Schedl, H. (1985): Subventionen in Europa – Konsequenzen einer Laissez-Faire Politik am Beispiel der deutschen Stahlindustrie. ifo studien zur industriewirtschaft Nr. 29. München.

Gieseck, A. (1995): Krisenmanagement in der Stahlindustrie. Eine theoretische und empirische Analyse der europäischen Stahlpolitik 1975 bis 1988. Berlin.

Hamm, R. and Wienert, H. (1990): Strukturelle Anpassung altindustrieller Regionen im internationalen Vergleich. Berlin.

Heinze, R.G., Voelzkow, H., Hilbert, J. (1992): Strukturwandel und Strukturpolitik in Nordrhein-Westfalen – Entwicklungstrends und Forschungsperspektiven. Opladen.

Hirsch, M. (1984): Ein Stahlproduzent ohne Heimat. Süddeutsche Zeitung, 8.3.84.

Hirsch, M. (1986): Tripartism in Luxembourg: The Limits of Social Concentration. West European Politics 9 (1), 54–66.

Hoffmann, K. (1990): Ein Mann will es wissen. manager magazin 2/1990, p. 46.

Hogan, W.T. (1994): Steel in the 21st Century. Competition Forges a New World Order. New York, Toronto: Lexington Books.

Hudson, R., and Sadler, D. (1989): The International Steel Industry. Restructuring, state policies and localities. London, New York: Routledge.

IEA (International Energy Agency): Energy Balances of OECD Countries. Paris, various issues.

IEA: Energy Statistics of non-OECD-Countries. Paris, various issues.
IEA: Energy Statistics of OECD-Countries. Paris, various issues.
Jeschar, R., Dombrowski, G., Hoffmann, G. (1996): Produktionsintegrierter Umweltschutz bei Industrieofenprozessen unter besonderer Berücksichtigung der Stahlindustrie. In: Brauer, H. (ed.): Handbuch des Umweltschutzes und der Umweltschutztechnik. Band 2. Produktions- und produktintegrierter Umweltschutz. Berlin, Heidelberg.
Jochem, E. and Bradke, H. (1996): Energieeffizienz, Strukturwandel und Produktionsentwicklung der deutschen Industrie. Jülich.
Jörnmark, J. (1993): Coal and Steel in Western Europe 1945–1993. Innovative Change and Institutional Adaptation. Göteborg.
Jürgenhake, W., Mengelkamp, W., Winter, B. (1988): Fallstudie "Saarstahl/Völklingen". Dortmund.
Kampherm, R. (1993): Strukturwandel im Ruhrgebiet. Kommunalverband Ruhrgebiet. Essen.
Kerz, S. (1991): Bewältigung der Stahlkrisen in den USA, Japan und der Europäischen Gemeinschaft, insbesondere in der Bundesrepublik Deutschland. Göttingen.
Landesentwicklungsbericht (1988): Landesentwicklungsbericht Nordrhein–Westfalen. Schriftenreihe des Ministerpräsidenten des Landes Nordrhein–Westfalen. Düsseldorf.
Mannesmann (1999): http://www.mannesmann.de.
McSweeney, C. and Hirosako, M. (1991): Understanding crude steel consumption. The perils of ignoring the role of technological change. Resources Policy 17, 258–270.
Meyerwisch, K. (1986): Sozialplanpolitik in der Eisen- und Stahlindustrie. Ein notwendiges Instrument zur sozialen Flankierung struktureller Veränderungen. In: Judith, R. (Ed.): 40 Jahre Mitbestimmung. Erfahrungen – Probleme – Perspektiven. Köln.
MWMT (Ministerium für Wirtschaft, Mittelstand und Technologie in Nordrhein-Westfalen) (ed.) (1992): Regionalisierung – Neue Wege in der Strukturpolitik Nordrhein-Westfalens. Düsseldorf.
Neu, A.D. (1995): Subventionen ohne Ende? Steinkohlenbergbau und Energieverbrauch in Deutschland. Kiel: Institut für Weltwirtschaft.
Nottrot, J. (1985): Luxemburg Beiträge zur Stadtgeographie einer europäischen Hauptstadt und eines internationalen Finanzplatzes. Innsbrucker Geographische Studien, Band 12.
NRW-Landesregierung (1979): Bericht der Landesregierung Nordrhein-Westfalen 1979. Schriftenreihe des Ministerpräsidenten des Landes Nordrhein-Westfalen. Düsseldorf.
OECD (Organisation for Economic Co-operation and Development): National Accounts. Paris.
Philipp, J.A. and Theobald, W. (1993): Stand und Aufgaben des Umweltschutzes in der Stahlindustrie. 5. Stahlkolloquium an der Technischen Universität Berlin. Berlin.
Philipp, J.A., Görgen, R., Henkel, S., Hoffmann, G.W., Johann, H.P., Pöttken, H.-G., Seeger, M., Theobald, W., Trappe, K., van Ackeren, P., Erve, S., Feierabend, K., Jansen, B., Maas, H., Nagels, G., Piotrowski, H. (1987): Umweltschutz in der Stahlindustrie – Anforderungen – Grenzen. Stahl und Eisen 107, p. 507.
Philipp, J.A., Johann, H.P., Seeger, M., Brodersen, H.A., Theobald, W. (1992): Recycling in der Stahlindustrie. Stahl und Eisen 112, p. 75.
Putnam, P.C. (1953): Energy in the Future. Van Nostrand.
Radetzki, M. (1990): A Guide to Primary Commodities in the World Economy. Oxford, Cambridge: Basil Blackwell.
RAG (1997): RAG Aktiengesellschaft: Geschäftsbericht 1996. Essen.
Saarstahl AG (1999): Geschichte Völklingen. In: http://www.saarstahl.de/

Schade, H. and Gliwa, H. (1978): Die Feststoffemissionen in der Bundesrepublik Deutschland und im Lande Nordrhein-Westfalen in den Jahren 1965, 1970, 1973 und 1974. Rat von Sachverständigen für Umweltfragen (ed.). Kohlhammer.

Schulz, E. (1992): Umweltschutz in der Stahlindustrie – Statement der Thyssen Stahl AG. Stahl und Eisen, Heft 5, p. 43.

Simons, J. (1997): Industriepolitik. Stuttgart: Schäffer-Poeschel (UTB).

StaBuA (Statistisches Bundesamt): Daten zur Bruttowertschöpfung und zum Bruttoinlandsprodukt. Wiesbaden.

Stahlfibel (1989): Stahlfibel. Verein Deutscher Eisenhüttenleute (ed.). Düsseldorf.

STATEC (Service Central de la Statistique et des Études Economiques) (1990): Statistiques Historiques 1839–1989. Luxembourg: éditpress Luxembourg.

STATEC (1997): Annuaire Statistique 1996. Luxembourg.

STATEC (1999): Luxembourg: An economic portrait (http://www.statec.lu/html_en/statistiques/portrait_economique).

STATEC (1999a): Statistical Information – Luxembourg (http://www.statec.lu/html_en/statistiques/luxembourg_en_chiffres).

Statistik der Kohlenwirtschaft e.V. (various volumes): Der Kohlenbergbau in der Energiewirtschaft der Bundesrepublik Deutschland. Essen, Köln.

Statistisches Jahrbuch der Eisen- und Stahlindustrie (various volumes): Wirtschaftsvereinigung Stahl. Düsseldorf.

Statistisches Jahrbuch (various volumes): Statistisches Jahrbuch der Bundesrepublik Deutschland. Wiesbaden.

Stehn, J. (1993): Stahlkrisenmanagement: Lehren der Vergangenheit für die Wirtschaftspolitik. Wirtschaftsdienst 1993/III.

Thyssen Krupp (1999): Zwischenbericht 1. Oktober 1998 – 31. März 1999. Düsseldorf (http://www.thyssenkrupp.com).

Vogl, R.J. (1994): Die deutsche Steinkohlevorrangpolitik vor dem Hintergrund eines europäischen Binnenmarktes für Elektrizität. in: Feldmeier, G.M. and Größl, M.M. (eds.): Auf der Suche nach einer Weltwirtschaftsordnung von morgen. Frankfurt/Main et al.: Peter Lang, pp. 231–256.

Wienert, H. (1985): Nachfrageschwäche und Staatsintervention – Zur Entwicklung der Stahlkrise seit 1975. In: Stahlkrise – Ist der Staat gefordert? RWI Essen. Berlin.

Wienert, H. (1987): Längerfristige Entwicklungstendenzen auf dem Weltstahlmarkt. Bestandsaufnahme, Perspektiven und einige stahlpolitische Folgerungen. RWI-Mitteilungen 37/38.

Wienert, H. (1993a): Stahlbericht 1993: Nachfrageeinbruch in Japan und Europa; zögerliche Erholung auf anderen Märkten. RWI-Mitteilungen 44, p. 143.

Wienert, H. (1993b): Schwerster Einbruch der Nachkriegszeit? Zyklische und strukturelle Elemente der gegenwärtigen Krise der westdeutschen Stahlindustrie. RWI-Mitteilungen 44, p. 243.

Wienert, H. (1996): Technischer und wirtschaftlicher Wandel in der Stahlindustrie seit den sechziger Jahren unter besonderer Berücksichtigung Nordrhein-Westfalens. Essen.

Wieshoff, D. (1986): Die Europäisierung der Stahlpolituk. Mitwirkungsmöglichkeiten der Arbeitnehmer und Auswirkungen auf die Unternehmensmitbestimmung. In: Judith, R. (ed.): 40 Jahre Mitbestimmung. Erfarungen – Probleme – Perspektiven. Köln.

Wirtschaftsvereinigung Bergbau e.V. (1994): Das Bergbau-Handbuch. Essen: Glückauf.

Aluminium Production in Germany

Simone Schucht
Ecole des Mines de Paris, Cerna, 60 boulevard Saint-Michel, 75272 Paris cedex 06, France

1 Structural Change in the German Aluminium Industry

1.1 Background Information on the German Aluminium Industry

In terms of quantity, aluminium is the most important world metal after iron and steel (Jochem and Bradke 1996). Over recent decades, however, the aluminium industry has undergone considerable structural change, resulting in a lasting decline in primary production, and secondary production has also recently sustained a fall in production.

In 1996, five *primary aluminium works* were still operating in Germany (Table 1), producing alloyed and non-alloyed primary aluminium (Wirtschaftsvereinigung Metalle 1993a). The number of operating plants has thus fallen by half since the end of the 1970s. In order to produce primary aluminium, German plants use the Bayer process to transform aluminium ore or bauxite into aluminium oxide (alumina), which then forms the input for electrolysis.

There are two methods of *aluminium recycling* (Eyerer et al. 1996): if collected separately, aluminium can be *remelted* into pure rollable alloys (wrought alloys). On the basis of pure, clean scrap, remelters produce very pure wrought alloys containing only small quantities of alloy elements (OEA 1993/94, 1995/96). Alloys of mixed quality can be *refined* with the help of salt. Refiners produce mostly casting alloys and can use all kinds of scrap. Large-scale recycling began in the early 80s, while the refining industry has been in existence since the 30s.

Table 1. German primary aluminium industry: companies and capacity in 1996

Company	Location	Capacity (kt)
VAW aluminium AG, Rheinwerk	Neuss	210
Hamburger Aluminium-Werk GmbH	Hamburg	120
Aluminium Essen GmbH	Essen	95
Hoogovens Aluminium GmbH	Voerde	78
VAW aluminium AG, Elbewerk	Stade	70
Total		573

Quelle: BT Drs. 13/6833

In 1997, *secondary aluminium* was being produced by 11 companies, most of them located in the state of North-Rhine Westphalia (VDS 1997; OEA 1997). The secondary companies have not published any data on their capacity, and production data are only available from refiners. The *production structure* of German refiners deviates significantly from the European norm. In Germany, most companies operate on a large scale with a capacity of over 20 kilotons per year (kt/a). Most companies in other European countries produce less than 20 kt/a, while more than 40% of refineries boast an annual production of less than 1 kt. According to the industry association, smaller melting plants are less often bound by environmental obligations, especially in foreign countries. At the end of the 1980s, 16% of European secondary production was reported to be saddled with few or no environmental costs, gaining thereby a competitive advantage (OEA 1989/90; Interview Kirchner[1]).

In 1996, the entire German aluminium industry – i.e. primary and secondary production and manufacturing – consisted of almost 600 companies employing 70 000 people (BT Drs. 13/6833). Data for the gross value added (GVA) are only available as a total for the whole of non-ferrous metal production and manufacturing, not for the aluminium industry alone. They are nonetheless used as a makeshift indicator here. In Fig. 1, the changes in GDP and in GVA for industry as a whole and the non-ferrous metals industry (in 1991 prices) are compared to the base year (1961). The figure shows faster growth of the GVA index for the non-ferrous metals industry than for GDP or the GVA for industry as a whole. The former is nevertheless notable for above-average fluctuation.

The German aluminium industry is privately owned. There has been no state regulation comparable to that of the steel industry at EU level, for example, nor national subsidies. Only in response to the high imports of primary aluminium from the Commonwealth of Independent States (CIS) at the beginning of the 1990s, which led to a considerable fall in prices, did the European Commission involve itself and introduce import restrictions in 1993.

1.2 Production and Consumption

Total aluminium production in Germany rose until 1989, apart for short-term falls in 1974 and 1979, due to the oil crises. Aluminium *consumption* has exceeded production by more and more, exhibiting higher growth rates (Fig. 2).

Primary aluminium production started to decline in the mid-1980s. This early decline was due rather to structural factors than the economic cycle. Maximum consumption was reached in 1995 (1.5 Mt), while production peaked in 1984 (0.8 Mt). With respect to primary aluminium, Germany has had a continuous surplus in consumption during the period under study.

1 Mr. G. Kirchner is managing director of the VDS (Association of German Remelters and Refiners; Vereinigung Deutscher Schmelzhütten) and the OEA (Organisation of European Aluminium Refiners and Remelters).

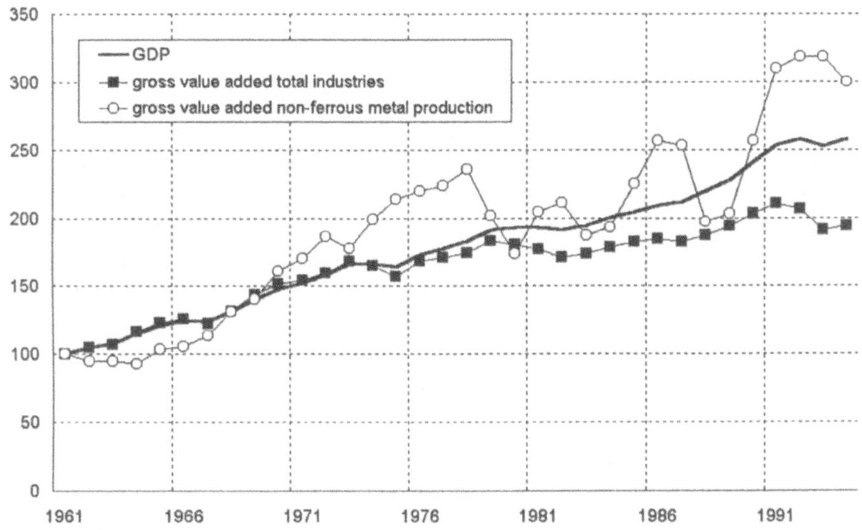

Source: Statistisches Bundesamt

Fig. 1. Gross domestic product, gross value added of industry overall and of the non-ferrous metal industry in West Germany 1961–94

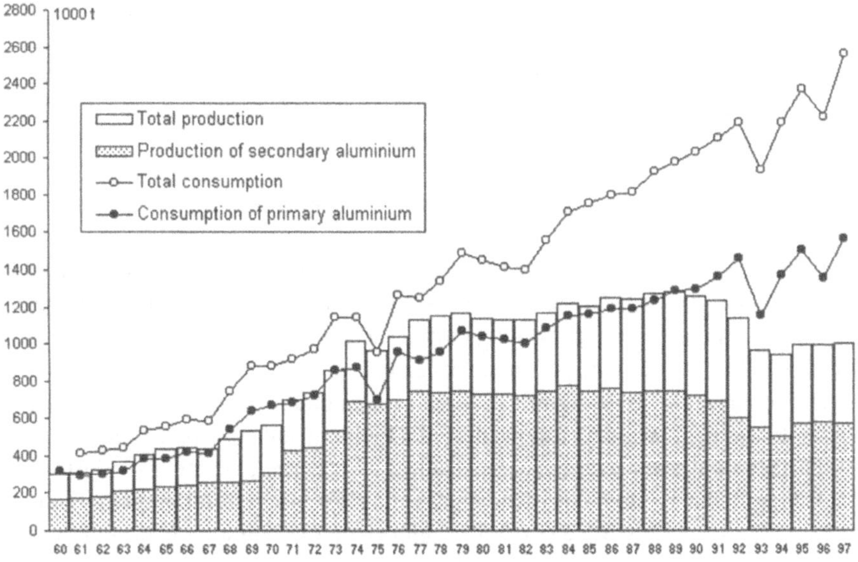

Source: Metallgesellschaft; since 1991 unified Germany

Fig. 2. Production and consumption of primary and secondary aluminium in Germany 1960–97

In contrast to primary aluminium production, that of *secondary aluminium*[2] exceeded consumption in Germany until 1980. Since 1983, rapid growth has led to increasing surplus consumption, which means that Germany has become a net importer. Apart from fluctuations in the economic cycle, consumption and production of secondary aluminium increased to 0.8 Mt and 0.5 Mt respectively by 1991 after which they fell. Between 1993 and 1995, however, consumption rose again, reaching hitherto unknown rates of growth.

The *shares* of primary and secondary aluminium in total metals production were subject to deep fluctuation. In contrast to other non-ferrous metals, such as zinc and lead (see e.g. Schucht 1996), recycled aluminium had already gained a high share (45%) of total aluminium production in the 1960s. After reaching its highest share (over 50%) in 1969, recycled aluminium rapidly lost ground, down to about 30% in 1973. This was probably due not to weakness in the secondary sector but rather to a renewed expansion of primary capacity up to the early 1970s. Secondary production share has since increased again. Since the early 1990s, the trend is remarkable for its strong annual fluctuations.

Due to limited employment data for the aluminium industry, it is impossible to establish time-series on *resultant social problems*. The few available data suggest that – during the 1990s alone – employment has fallen by about 10 000. These figures refer to the aluminium industry as a whole, i.e. to primary and secondary production and manufacturing.

From an environmental perspective, the structural change towards secondary aluminium is, in general, a positive development. However, there are some reservations with respect to the potential risk from certain residues: filter dust from secondary production, for instance, contains dioxins. There are nevertheless available technologies which produce almost no dioxin emissions. The relocation of primary production could still cause damage to the environment in newly exporting countries, which must also be evaluated.

1.3 Suppliers and Customers of the Aluminium Industry

Aluminium production is closely linked to other branches of industrial production, e.g. car manufacturing, the building and construction sector, mechanical engineering, the metals sector, the packaging industry and electrical engineering. Germany has the highest world-wide per capita consumption of aluminium (Joliet 1991).

In Germany, the most significant *uses* of aluminium – besides export – are in the motor vehicle industry, followed by the building and construction sector (Fig. 3). Further significant sectors are mechanical engineering, the metals industry, the packaging industry and electrical engineering (Wirtschaftsvereinigung Metalle 1993b; BT Drs. 13/6833). The relative importance for aluminium consumption of most industries has remained largely stable over the period shown.

---

[2] Data on secondary aluminium production cover only production by the refiners. Production data for remelters are not available from official statistics (Interview Kirchner).

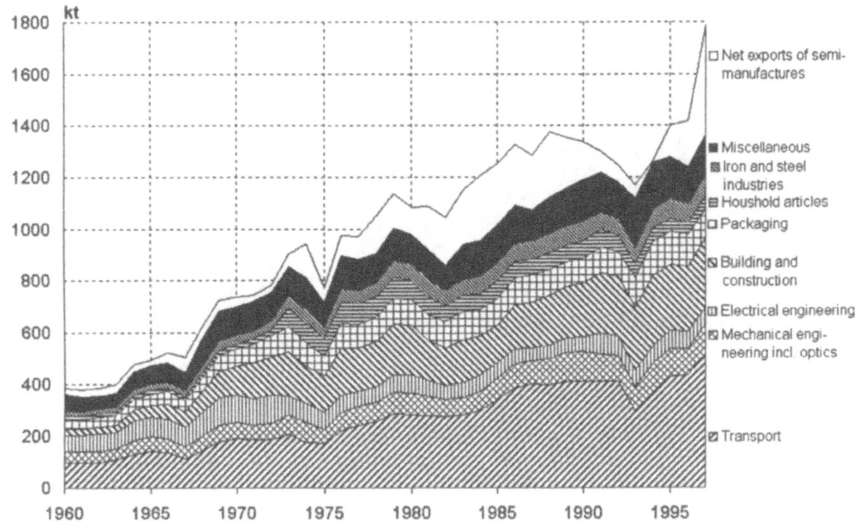

Source: Metallgesellschaft; since 1991 unified Germany

Fig. 3. German aluminium deliveries by uses 1960–97

As for suppliers, aluminium oxide and aluminium hydroxide (from bauxite or other aluminium ores) were produced at three plants in 1993[3] (Wirtschaftsvereinigung Metalle 1993a). At the end of the 1980s, however, only one German company was producing aluminium oxide with sufficient quality for electrolysis (Seebauer 1989).

The German aluminium industry processes aluminium ore or bauxite from Australia, Guinea, Sierra Leone (until 1994) and Guyana (Fig. 4). Re-exports are negligible. The fall in bauxite imports began earlier and was stronger than the fall in primary aluminium production. Until the end of the 1980s, increasing quantities of alumina were imported from overseas (Australia, Jamaica, Guinea, Surinam and Mexico). Thus, the first stage of production was partially relocated to bauxite mining countries (Fig. 5). Finally, alumina imports decreased as well. In the early 1990s, Germany became a net exporter of aluminium scrap, despite increasing imports from Eastern Europe (Fig. 6). There is hardly any trade in aluminium scrap with non-European countries.

From the indicators of environmental damage shown in Table 2, it can be concluded that the overall environmental impact of secondary aluminium is lower than that of primary aluminium[4]. However, a complete substitution of primary

[3] Aluminium Oxid Stade GmbH, Martinswerk GmbH (Bergheim) and VAW aluminium AG (Bonn).

[4] An environmental assessment has additionally to take into account the consequences of bauxite extraction. These are the destruction of the landscape and lowering of the water table.

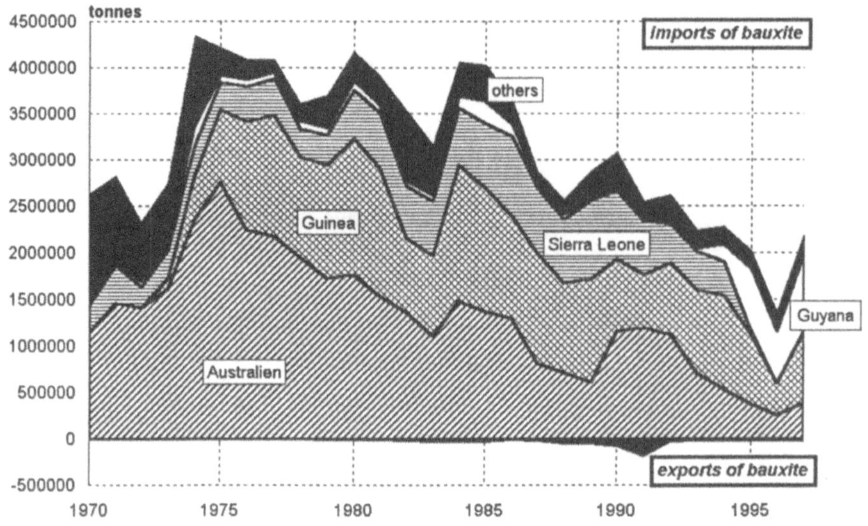

Quelle: Metallgesellschaft; since 1991 unified Germany

Fig. 4. German foreign trade with bauxite 1970–97

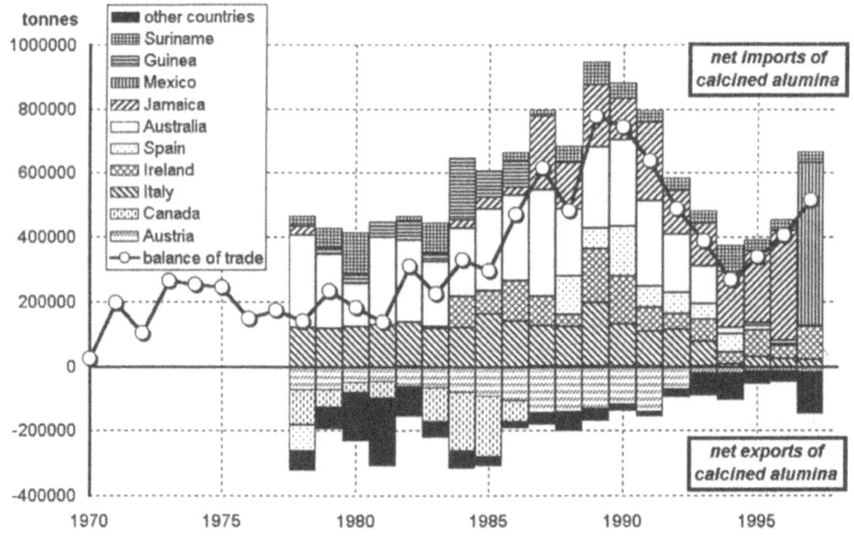

Source: Metallgesellschaft; since 1991 unified Germany

Fig. 5. German foreign trade in calcined alumina 1970–97

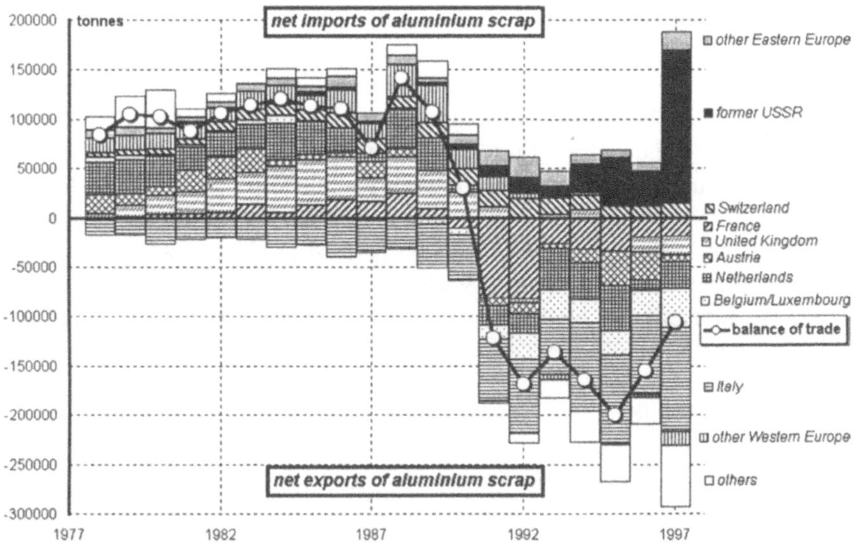

Source: Metallgesellschaft; since 1991 unified Germany

Fig. 6. German foreign trade in aluminium scrap 1970–97

aluminium is not possible, as secondary aluminium cannot be used in every application (Sames 1993): alloy elements either accumulate in the secondary cycle and can exceed acceptable limits or they can render the metal unsuitable for certain applications (Neumann 1992).

Table 2. Comparison of environmental impact per tonne of aluminium produced

	Primary aluminium	Secondary aluminium
cumulative[a] energy use	163 730 MJ[b], about 80% electricity for electrolysis	19 505 MJ, about half of it for melting
solid residues	3 650 kg, most of it red mud which has to be disposed of	386 kg, partly reusable
atmospheric emissions[c]	204 kg, 85% from fossil fuels	12 kg

[a] The sum of the energy consumption of all production steps.
[b] The specific energy consumption of modern electrolysis furnaces as used in Germany amounts to about 56 000 MJ/t aluminium (Krone et al. 1990; OEA 1989/90).
[c] As dioxins and other toxic chemicals belong to atmospheric emissions, a merely quantitative measure is not sufficient.

Sources: Krone et al. 1990; Aluminium-Zentrale; Eyerer et al. 1996

2 Explanation of Structural Change

The decline in production of primary aluminium is typical within Western Europe. In some countries the drop in production started later than in Germany but it applies to most of the countries. European production has declined after 1990. Norway presents an exception due to the availability of cheap power from hydroelectric generation: Norwegian production increased until 1991 and has shown slight fluctuations on a high level of production since then (Metallstatistik 1996). In some countries outside of Europe, however, the production of aluminium has been increasing, especially in Australia, Brazil and Canada, Venezuela, Bahrain and India (Jochem and Bradke 1996; Boesken 1992b).

2.1 Location Factors

The following location factors are relevant for *primary metal works* (Wettig 1980): good traffic connections for the transport of raw materials, fuels and finished products; the supply with energy and water; the location in close vicinity of customers and environmental requirements.

The production of primary aluminium is estimated to have directly spent 22% of the production value on electricity in 1985. Compared to that the overall average for electricity costs in the mining and manufacturing industry was at only 2% (Haas 1990). The supply with cheap energy, and particularly the close vicinity to power stations, is the most emphasised location factor.

Although the German aluminium industry itself often points out higher environmental protection costs than in other countries, these costs or related requirements are not specified. Indications for stricter requirements compared to other countries exist only with respect to air pollutants (BT Drs. 13/6833).

Relevant factors for the *secondary production* are: the value of the material; the availability of scraps and residues; the energy costs; the costs for waste disposal and the proximity to centres of the metal manufacturing industries (Neumann 1992). For the secondary aluminium industry the energy prices are only of indirect significance because high costs make the secondary production more attractive relative to the primary production (ibid.; Kirchner 1989).

Environmental protection requirements and the costs resulting from those are relevant for the industry (OEA 1989). The requirements that deviate from those of other countries are said to be of particular significance for the competitiveness of the German companies. The share of environmental protection costs of the entire processing costs was estimated at 15% at the beginning of the 90s (Konzelmann 1992). An evaluation of environmental investment of industrial sectors (StaBu FS 19,3) shows that the secondary aluminium industry's costs for environmental investment out-weighed that of other industrial sectors. Within secondary aluminium production the refiners bear higher environmental protection costs than the remelters. This is partly due to different production processes used. The highest costs appear in three areas (Kirchner 1989; Interview Kirchner):

- with respect to salt slag (Salzschlacke) that is not allowed to be deposited in Germany and thus has to be processed[5]
- emissions, where German abatement standards are higher than those of other countries and
- filter dusts that have to be treated as hazardous waste and have to be deposited in landfills at high costs as they contain dioxin; compared to other countries costs are higher because the abatement standards and, thus, the amounts of dust collected are higher

2.2 The Decline in German Primary Aluminium Production

Since the end of the 1970s, five German primary aluminium works have been closed down. The total capacity of these plants was 277 kt (Table 3). They were rather small, but still represented 1/3 of total capacity.

The decline in *primary aluminium production* in Germany can be explained by the following reasons:

- Technological developments have made it possible to produce secondary alloys of a quality equal to that of primary alloys, and secondary aluminium is therefore substituting in applications which had formerly been the preserve of primary aluminium (Jochem/Bradke 1996).

Table 3. Closure of primary aluminium plants

Enterprise	Location	Reason for Closure	Capacity (kt)	Year
VAW aluminium AG Erftwerk	Grevenbroich	capacity too low, electricity costs too high	32	1978
Alcan Aluminiumwerke GmbH	Ludwigshafen	increase in costs for electricity	44	1987
VAW aluminium AG Lippewerk	Lünen	antiquated plant, energy too expensive, capacity too low[a]	47	1989
Aluminium Hütte Rheinfelden GmbH (Alusuisse)	Rheinfelden	decrease in aluminium prices, increase in production costs	64	1991
VAW aluminium AG Innwerk	Töging	plant too small[b], electricity costs too high	90	1996
total:			277	

[a] At the time the plant was closed, capacity only amounted to 20 kt.
[b] In the end capacity amounted only to 30 kt. One of the two furnaces in Töging had already been closed down in October, 1993 (Kühnle 1994).
Source: BT Drs. 13/6833.

[5] There have been efforts to develop processes for melting without salt. However, these technologies can only treat specific forms of scrap (Kirchner 1989; Interview Kirchner).

- High environmental costs have repeatedly been pointed out, but they are neither specific nor brought into direct connection with plant closures (see e.g. Boesken 1992a; Steil 1993).
- Instead, high electricity costs are most often cited as the main reason for plant closures. Notwithstanding, profitable production was possible until the mid-1980s, as metal prices were still high, exchange rates profitable and contracts with electricity suppliers more favourable (Seebauer 1989).
- At the beginning of the 90s, heavy imports of cheap aluminium from the CIS countries led to over-supply and a drastic drop in prices. In 1988 the average price of aluminium was 4.20 DM/kg, but it had fallen to 1.88 DM/kg by 1993. The price then rose again to 2.58 DM/kg in 1995 (Metalgesellschaft 1996). The price drop affected other countries too: in 1994, 69% of capacity in Western countries was running at a loss (Kühnle 1994).
- Germany was strongly affected by the recession in 1992, which was aggravated by the burden of reunification (Schirner 1993). Falling prices and the devaluation of the dollar added to the industry's problems.
- All smelting works had had to sustain heavy losses and restrictions on production became inevitable. Capacity utilisation was only 85%. As domestic demand was still rising in 1992, imports of primary aluminium reached 1 million tonnes for the first time (Schirner 1993; Metallgesellschaft 1996). After export demand had already begun to fall in 1991, domestic demand for aluminium finally declined as well in 1993. This was due to the fact that economic growth was slowing (Dermer 1993) and, furthermore, export opportunities had diminished and pressure from competitors was growing.

2.3 The Decline in German Secondary Aluminium Production

The lack of competitiveness in the German secondary aluminium industry is shown by rising imports. While demand rose, secondary production fell, compensated for by imports. In 1995, almost 100 kt of casting alloys were imported, compared to only 3 kt in 1992. Net exports of scrap, however, reached a record high and, as a consequence, supplies of this raw material ran short (OEA 1995/96). Imports have been increasing since the 1980s. Imports of secondary aluminium grew from 57 kt in 1982 to 94 kt in 1985 and about 160 kt in 1992. In 1995, for the first time, more than 200 kt of secondary aluminium were imported. Owing to this development, the industry was unable to sell its products at sufficiently high prices to cover its costs (Konzelmann 1992; Metallgesellschaft). The fact that the German refining industry did not manage to increase its market share is attributed by the Secondary Aluminium Association to high wage costs and taxes, high environmental costs and exchange rate fluctuations (OEA 1995/96).

Competition for raw materials and impediments to international trade of secondary material, for instance by the adoption of EU regulations into the "Basel Convention", are cited as reasons for both the aggravated situation on the scrap market, increased costs and falling margins[6]. In general, problems of availability, as well as rising prices for high quality scrap can also be explained by the fact that the manufacturing industry itself (such as semi-finished manufacturing plants) uses secondary material in some of its production processes. These industries are thereby competitors for raw materials (OEA 1992/93; Kirchner 1989), and the same holds true in the primary aluminium industry (OEA 1989/90).

Despite increasing production up to the beginning of the 1990s, secondary aluminium had already come under considerable pressure in the 1980s. After 1983, the industry was already facing a continuously worsening situation: margins were diminishing further and further. In part, this could be traced back to an excessive supply and to continuously rising costs, which could not then be passed on via pricing (OEA 1989, 1992/93).

Problems related to shrinking margins were also important at the end of the 1980s (OEA 1989/90; Boesken 1990). Whereas scrap prices were pretty much the same all over Europe, differences in metal prices were regarded as an indicator of highly divergent production costs. In this context, environmental costs – along with wage costs – were assumed to hold increasing importance. The European association of secondary aluminium producers claims that in Germany, environmental costs are comparatively high – 15% of total costs to companies in 1990, at times even higher (OEA 1989/90). Continuously rising costs are also explained by German and European regulations on waste (OEA 1989, 1993/94).

The trend in production had slowed down considerably after 1988. The weakness in demand from individual sectors took its toll on at least sections of secondary production (OEA 1988/89; Kirchner 1989). Of considerable importance for good performance cycle by secondary aluminium within the economic is the car industry. Alongside the large number of cars sold, the continuously increasing use of cast aluminium in automobiles is a significant factor. The car industry consumes more than 50% of secondary aluminium production in Germany (OEA 1989). Further industrial sectors important for casting alloys are mechanical engineering, the electrical industry and the building and construction industry. Thus the economic crisis in 1992 and the consequent slump in car production and other relevant sectors led to declining consumption of secondary aluminium. Only the construction sector was spared from this decline (OEA 1992/93).

The heavy slump in demand up to the end of 1992 led to bankruptcies. The losses, however, were not reflected to a significant extent by production until 1993[7]. In subsequent years, more companies went into liquidation. In 1992/93 and at the end of 1996, even some large business groups were affected by bankruptcies (OEA 1992/93; Interview Kirchner).

Secondary aluminium works were eventually also affected by the drop in prices in 1990 sparked off by high CIS exports, albeit only indirectly. As there is usually

---

6 Margins are defined as the difference between the price for raw materials (e.g. scrap) and the metal price.

7 The loss in production was partly compensated for – e.g. through the take-over of one plant by other companies, or through greater use of other companies' capacity (Schirner 1993).

an interdependence between the prices for primary and secondary aluminium, the price collapse for primary aluminium caused prices for secondary aluminium to fall to a historical low. Until May 1993, the price fell to its lowest value since the beginning of the 1970s. On the other side, costs (scrap prices) increased, reducing margins and leading to losses (OEA 1993/94; Schirner 1993).

3 Outlook

From a technological point of view, the industry sees no reason for further restructuring. Besides the local disadvantage of high domestic costs and the problems resulting from heavy CIS exports, the industry sees a problem in the fact that some competitors, even within Europe, are protected by their governments and do not have to operate under market conditions. According to the industry, all these problems together have weakened the competitiveness of the German primary industry (Schirner 1993). Other writers emphasise that global prices for energy and raw materials are too low, as they can only be achieved by overexploiting natural resources and exempting consumers from social costs (Kühnle 1994).

Nevertheless, there seems to be no question that further restructuring is to be expected, to which the European industry will have to adjust. It is also expected that financial pressures will lead in the long run to a concentration of primary production in regions with favourable local factors, such as raw material deposits reducing transportation costs and a cheap energy supply (Neumann 1992). In Germany, restructuring in this sense is currently underway: the domestic capacity of primary aluminium plants is being reduced and processing and manufacturing capacity is being expanded. Whereas some writers argue that it will be mostly plants in Venezuela and Canada which will close the gap in the metal supply (ibid.), others consider East European countries to offer better prospects for German investment than Southern Europe, Mexico or Brazil (Pawlek 1994).

Several writers see an opportunity for the German aluminium industry in exploiting the competitive advantages of German production processes by conquering markets for high quality products. With regard to standard products, however, a transfer of production to Eastern Europe should be considered. As the domestic aluminium industry operates to high environmental standards, the German industry considers it sensible to invest in modern plant in Eastern Europe. As a strategy for Western Europe, they propose to further close material cycles for aluminium (Schirner[8] 1993). For environmental reasons they also emphasise the increasing necessity of exporting modern production technology to developing and newly-industrialised countries (Neumann 1992).

The almost continuous – apart from fluctuations in the economic cycle – strong growth in (primary and secondary) aluminium consumption in Germany indicates that there is still a good market for aluminium. It also indicates that, for the aluminium industry – in contrast to the steel industry – changes in final demand were

[8] Member of the board of directors at VAW and VIAG.

of minor significance to production development. Instead, national and international restructuring within the aluminium sector are the variables explaining falls in production.

References

Aluminium-Zentrale e.V. im Gesamtverband der Deutschen Aluminiumindustrie (GDA): Aluminium und Rotschlamm. Düsseldorf.

Boesken, D.H. (1990): Federal Republic of Germany. Aluminium 66 (7/8), 631–635.

Boesken, D.H. (1992a): Strukturwandel in der Aluminiumindustrie. Aluminium 68 (5), 357.

Boesken, D.H. (1992b): Federal Republic of Germany. Aluminium 68 (8), 626–628.

BT Drs. 13/6833: Deutscher Bundestag. 13. Wahlperiode. Drucksache 13/6833. 28.1.1997.

Dermer, D. (1993): Review 1992 – Outlook 1993. The basic facts about the European aluminium market. Aluminium 69 (11), 956–958.

Eyerer, P., Schuckert, M., Pfleiderer, I., Bohnacker, A., Kreißig, J., Harsch, M., Saur, K. (1996): Stoffbilanzen als Grundlage für die technische, ökonomische und ökologische Beurteilung von Produktionsprozessen und Produkten. In: Brauer, H. (ed.): Handbuch des Umweltschutzes und der Umweltschutztechnik. Band 2: Produktions- und produktintegrierter Umweltschutz. Berlin et al.

Haas, U. (1990): Standortperspektiven stromintensiver Produktionen in der Bundesrepublik Deutschland. Mögliche Konsequenzen umweltschutzinduzierter Strompreise bei Aluminium und Chlor. Weltwirtschaftliche Studien 22. Göttingen: Vandenhoeck & Ruprecht.

Jochem, E. and Bradke, H. (1996): Energieeffizienz, Strukturwandel und Produktionsentwicklung der deutschen Industrie. Monographien des Forschungszentrums Jülich. Band 19. Jülich.

Joliet, H. (1991): Aluminium heute. In: Schäfke, W., Schleper, T., Tauch, M. (ed.): Aluminium. Das Metall der Moderne. Gestalt – Gebrauch – Geschichte. Köln.

Kirchner, G. (1989): 1988 war ein gutes Jahr für die Sekundäraluminiumindustrie. Aluminium 65 (7/8), 783–785.

Konzelmann, G. (1992): Die deutsche Aluminium-Sekundärindustrie gestaltet Recycling umweltverträglich. Aluminium 68 (12), 1058–1061.

Krone, K., Krüger, J., Orbon, H., Sommer, H.W., Vest, H. (1990): Ökologische Aspekte der Primär- und Sekundäraluminiumerzeugung in der Bundesrepublik Deutschland. Metall, 44 (6), 559–569.

Kühnle, H. (1994): Aluminium global. Der Weltmarkt und die sozial-ökologischen Folgen. Wechselwirkung. No. 66, 33–36.

Metallgesellschaft AG and World Bureau of Metal Statistics (various volumes): Metallstatistik.

Neumann, J.C. (1992): Recycling und Energiesparen. BWK 44 (9), 392–398.

OEA (Organisation of European Aluminium Refiners and Remelters) (1997): Membership List January 1997. Düsseldorf.

OEA (various volumes): Aluminium Smelters Europe Japan USA. Organisation of European Aluminium Refiners and Remelters. Düsseldorf.

Pawlek, R. (1994): The Primary Aluminium Industry at the Turn of the Year. Aluminium 70 (1/2), 10–20.

Sames, C.-W. (1993): Die Kreislaufwirtschaft und ihre Folgen – einige quantitative Abschätzungen. Metall 47 (2), 171–174.

Schirner, J. (1991): Federal Republic of Germany. Aluminium 67 (7/8), 624–629.

Schirner, J. (1993): Federal Republic of Germany. Aluminium 69 (8), 696–701.

Schucht, S. (1996): Emission Reduction Through Restructuring of the Non-Ferrous Metal Industry in the Ruhr Area. A Historical Record. Working Paper. WP–96–36. International Institute for Applied Systems Analysis (IIASA). Laxenburg/Austria.

Seebauer, H. (1989): Federal Republic of Germany. Aluminium 65 (7/8), 669–674.

Seebauer, H. (1992): Ein Silberstreif am Horizont. Aluminium 68 (10), 813.

StaBu (Statistisches Bundesamt) (1996): Monatsbericht im Verarbeitenden Gewerbe sowie im Bergbau und in der Gewinnung von Steinen und Erden. Statistisches Bundesamt. Wiesbaden.

StaBu FS 7, 2: Statistisches Bundesamt. Fachserie 7, Reihe 2. Außenhandel nach Waren und Ländern (Spezialhandel). Various volumes. Wiesbaden.

StaBu FS 19, 3: Statistisches Bundesamt. Fachserie 19, Reihe 3. Investitionen für Umweltschutz im Produzierenden Gewerbe. Various volumes. Wiesbaden.

StaBu: Bruttoinlandsprodukt, Bruttowertschöpfung und Produktionsindex. Statistisches Bundesamt Wiesbaden.

Steil, H.-U. (1993): Der Einfluß gesetzlicher Umweltschutzvorgaben auf das Recycling metallhaltiger Rückstände. Metall 47 (5), 480–485.

VDS (1997): Mitgliederverzeichnis Februar 1997. Vereinigung Deutscher Schmelzhütten. Bundesverband Deutscher Aluminium-Schmelzhütten. Düsseldorf.

Wettig, E. (1980): Erzeugung und Verarbeitung von NE-Metallen in der Bundesrepublik Deutschland. Standorte und Entwicklungstendenzen. Stuttgart.

Wirtschaftsvereinigung Metalle (1993a): Mitglieder- und Produktionsverzeichnis. Directory of German Non-Ferrous Primary and Secondary Smelters. Düsseldorf.

Wirtschaftsvereinigung Metalle (1993b): NE-Metallstatistiken 1993. Bundesrepublik Deutschland. Haus der Metalle. Düsseldorf.

Cement Production in Austria

Walter Hüttler[1]
IFF, Schottenfeldgasse 29, 1070 Vienna, Austria

One of the few European industries to experience a decline in material production has been the Austrian cement industry, with a fall of 40% between 1974 and 1998, accompanied by a decline in domestic cement consumption of about 20%. Austria's cement industry therefore presents itself as an interesting field for analysing structural change and green restructuring.[2]

During recent decades, the Austrian cement industry has been influenced by several fundamental changes:

- shifts in energy costs in the 1970s
- stagnation in the construction sector in the 1980s
- expansion of foreign trade with Eastern Europe
- the break-up of the Austrian cement cartel
- concentration of cement production within a few international enterprises
- more stringent general environmental regulations during the 1990s

1 Background Information on the Austrian Cement Industry

1.1 General Characteristics of the Cement Market

Compared to other basic industries, the cement production and the market have a number of interesting characteristics, relating to possible driving forces for structural change as well as to industrial restructuring strategies:

- The demand side is determined only by the construction sector, which depends to a great extent on public investment.
- Centralisation and relocation of cement production are restricted by economic limits on transport distances (150 km), resulting from the relatively high share

---

[1] I should like to thank Karin Baumgardinger, who collected data and background information.

[2] The data for the case study refer to the Austrian cement industry as defined by the Austrian Central Statistical Office and the Austrian Association for Building Materials and Ceramic Industries, both the main sources of data in this area. As a consequence of Austria's integration into the EU, Austrian industrial statistics are in the process of being converted to the EU system for recording economic statistics. Some of the Austrian Central Statistical Office's primary statistics (e.g. foreign trade) are therefore only available for years up to 1994 or 1995.

of transport costs in the total production costs for cement.[3] The Austrian cement industry is therefore more or less evenly spread across the regions. Every province except Burgenland has at least one production site, which means that regional consumption can be served by cement companies within the corresponding region. For instance, the Vorarlberger cement company Lorüns supplies 80–85% of total cement consumed in Vorarlberg. In general, foreign trade in cement plays a minor role. However, since the opening up of Eastern Europe, the Austrian cement market has become quite attractive for Central and Eastern European exporters as long as costs for energy, labour and transport in these countries remain significantly under the Austrian level.

- Substitution of cement itself is not possible: concrete production needs a certain proportion of cement; substitution and material efficiency strategies as well as product innovations focus on concrete.

In 1993, the Austrian cement industry contributed 0.13% to Austrian GDP or 0.6% to total industrial added value. These figures correspond to the cement industry's employment share (0.6% of total industrial employment and 0.1% of total employment). Thus the Austrian cement industry appears to have relatively little impact on the national economy. At regional level, however, cement companies have important functions in employment and economic stability.

Austrian per capita consumption of cement in 1998 was 600 kg. In comparison with other European countries this is in the high mid-range. According to CEMBUREAU (the umbrella organisation of the national cement organisations), average European per capita cement use in 1991 was 479 kg (Austria 663 kg), while the global average was about 230 kg/cap (VÖZ 1992).

Consumption of cement (Table 1) is determined largely by demand from the construction materials industry and the construction sector as a whole. The most important area within the construction sector is structural engineering (house building), followed by civil engineering projects like hydroelectric power plants and sewage treatment facilities (APA, 29-1-97). The construction sector is highly

Table 1. Demand-side structure of the Austrian cement market

share	customers
45%	ready mix concrete factories
20%	prefabricated concrete products industry
20%	small customers (bags)
15%	other (deliveries to construction sites, plaster and mortar)

Source: Fritzl 1997

[3] According to Roubin (1997) transport accounts for 16% of total costs, wages and energy constitute about one third each and investment contributes to one sixth of the costs structure for Austrian cement producers.

dependent on changes in economic activity and cement consumption varies accordingly. Due to the relatively high costs of storing excess stock, cement production varies seasonally, as construction activity falls during the winter (Hauser 1994). Most cement production companies are located in the immediate vicinity of their own quarries.

1.2 Technological and Environmental Issues

A form of cement was known and used for construction 2000 years ago by the Romans. Mixed with water and then combined with sand and gravel, it hardens to concrete or mortar. Industrial production of cement in the Austro-Hungarian Empire started in the middle of the 19th century and reached a volume of 600 kt by 1900, rising to 1 430 kt in 1912.

Limestone, clay and plaster are used as raw materials. Special types of cement (e.g. portland cement) require the addition of blast-furnace slag. The raw materials are ground up and burned at temperatures of 1 400 – 1 450°C. The technical infrastructure needed for cement production consists of conveyor equipment, grinding mills and burning kilns, as well as storage facilities and distribution vehicles.

Cement is one of the strategic materials for industrial economies (Hüttler et al. 1996). About 40% of material inputs to the construction sector is concrete. The production of cement is a comparatively energy-intensive process. Relative to value added, emissions of CO_2 and SO_2 are four to six times above the industry average.[4] About 1 700 kg of raw materials (excluding fuels), water and air are needed to produce one tonne of cement. In addition, more than 3 000 MJ of energy are needed for combustion processes and to run equipment such as conveyor belts (Table 2).

The main fuels for the cement industry, coal and oil, make up about 80% of energy input (excluding electricity; Table 3). Due to rising energy costs in the 1980s,

Table 2. Material and energy intensity of cement production

material intensities	(kg/t)	energy intensities	(MJ/t)
primary raw materials	1 275	combustion energy	
secondary raw materials	161	fossil fuels	2 253
water	190	waste	392
air	47	mechanical energy (technical facilities)	406
total material input	1 673	total energy input	3 051

Source: Knoflacher et al. (1995)

[4] The emission intensity (CO_2 per value added) refers to CO_2 emissions from energy requriements and excludes process emissions. Data for 1990 (BMU 1995; ÖSTAT; Hackl und Mauschitz 1995; own calculations).

Table 3. Energy input to the Austrian cement industry 1993 (excluding electricity)

	[GJ/a]	[%]
Coal	4 808 138	38%
Coke	547 003	4%
Oil	4 955 654	39%
Natural Gas	577 829	5%
Waste	1 819 254	14%
Total	12 707 878	100%

Source: Hackl and Mauschitz (1995)

the cement industry began burning waste tyres to generate heat. In 1993, the following waste materials were incinerated by the cement industry (VÖZ 1993):

- waste tyres (82% of domestic yield)
- waste oil/solvents (55% of domestic yield)
- paper sludge (11% of domestic yield)

This gives rise to a fossil fuels substitution rate of 15% (Knoflacher et al. 1995). In 1996, four production sites started trial projects for the use of waste plastics as fuels within the cement production process (VÖZ 1996).

The increased burning of waste materials in cement production focused attention on the issue of atmospheric emissions. After years of discussion, environmental regulations were passed in 1993, setting obligatory emissions standards for the cement industry (*Verordnung über die Begrenzung der Emission von luftverunreinigenden Stoffe aus Anlagen der Zementerzeugung, BGBl. Nr. 63/1993).* The new ordinance caused the cement industry to make significant investment in environmental protection equipment (mainly filter and measurement technology), amounting to 120–150 million ATS annually (Raffel 1995). The ordinance sets partial time limits for upgrading production facilities by 1998 (Sedlak 1995).

Environmental benefits have resulted from the decline in specific energy input into the cement production process, as well as from the application of end-of-pipe technologies such as filters.

The decline in CO_2 emissions indicates a reduction of the specific energy input into cement production. Whereas in 1950, the production of one tonne cement caused emissions of 1 100 kg CO_2, in 1990 it gave rise to 620 kg CO_2, a reduction of over 40%. Notwithstanding, the cement industry remains Austria's biggest emitter of process-related CO_2 emissions (APA, 17-1-92, 19-5-92).

Due to changes in manufacturing technology and logistics, most emission factors (emissions per tonne cement) exhibit downwards trends in the last few years (Table 4). At the same time, the cement industry's share in Austria's total atmospheric emissions is increasing. Other economic sectors and domestic households have been more successful than the cement industry in reducing emissions to air.

Table 4. Change in emission factors 1988–92

	change 1988–92	share[a] 1988	share[a] 1992
dust	-37.75%		
NO_2	-5.78%	2.76%	3.23%
SO_2	-26.19%	0.96%	1.36%
HCl	+4.64%		
HF	±0.00%		
TOC	-2.36%	0.05%	0.06%
CO	-13.96%	0.74%	0.81%
CO_2	-6.42%	5.34%	5.41%
metals	+3.60%		
zinc	+11.87%		

[a] share of the cement industry in total emissions
Source: Hackl und Mauschitz (1995)

2 Structural Change in Austria's Cement Industry

2.1 International Trends in Cement Production

In order to analyse cement production at international level, cement production data for 26 countries were grouped according to regional categories, thereby indicating the different patterns of cement production development for the period of 1960 to 1993 (Fig. 1). The results show that the development of the Austrian cement industry is typical for highly industrialised countries.

While cement production expanded in general during the 1960s, energy cost shifts in the 1970s caused an economic decline overall, along with considerable turbulence within the cement production industries themselves. Only within Eastern European planned economies and Southern European countries did cement production continue to grow during the 1970s. With regard to Eastern European countries, it can be assumed that energy costs which did not reflect current market levels were an important factor in this growth. For both Eastern and Southern European economies, a kind of backlog in demand for infrastructure expansion could be another major factor in the continuing growth.

While the opening up of the East promoted construction activity and therefore demand for cement in Western Europe, cement production in Eastern European countries went through a marked slump at the end of the 1980s.

Central and Western European economies suffered a continuous decline in cement production in the 1980s, followed by a short construction boom, beginning in the late 1980s and ending in the early 1990s. In contrast, the decline in Northern European countries has followed a more consistent trend over the past two decades, leading to current production volumes below those in the 1960s.

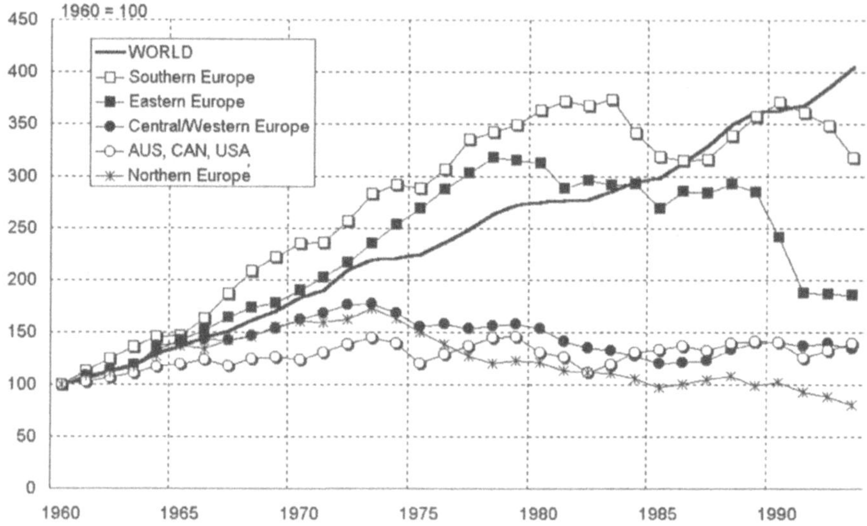

Source: UN; own calculations. Complete figures for cement consumption for the period of 1960–1993 are not available. Central and Western Europe: Austria, Belgium, Denmark, France, Germany, Netherlands, Great Britain; Southern Europe: Spain, Greece, Italy, Portugal; Northern Europe: Finland, Iceland, Norway, Sweden; Eastern Europe: Bulgaria, Czechoslovakia (resp. Czech Republic and Slovakia), Hungary, Poland, Romania, Yugoslavia; Non European Countries: Australia, Canada, United States.

Fig. 1. Cement production world-wide and in European regions 1960–93

In economies with relatively high per capita income like those in Central, Western and Northern Europe, as well as Canada and the United States, cement production has been declining since the 1970s; the same is true for Southern European economies since the late 1980s, whereas total world production of cement exhibits a continuous rise.

The European cement industry is comprised of 123 enterprises with about 300 production sites. The leading European companies like Blue Circle (United Kingdom), Dyckerhoff, Heidelberger (both Germany), Holderbank (Switzerland), Italcementi (Italy), Lafarge (France) and Scancem (Sweden) are also main players on the global market. During the recent years they have intensified their expansion strategies, which results in a continuing concentration process concerning both enterprises and production sites (Europäische Kommission 1997).

2.2 Trends in Cement Demand, Production and Foreign Trade

Production and consumption of cement in Austria exhibited virtually identical trends between the 1950s and the start of the 1990s (Fig. 2). Due to the cement cartel – established in 1951 – production volumes were strictly regulated accord-

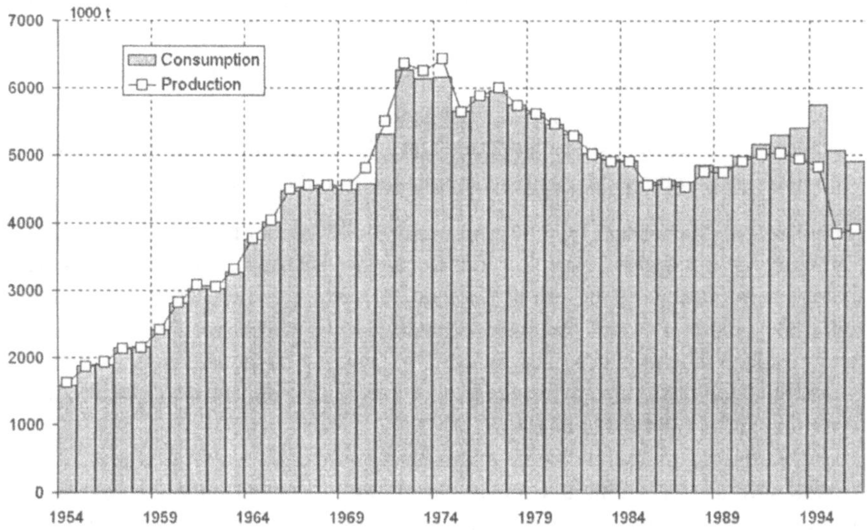

Sources: ÖSTAT; VÖZ

Fig. 2. Austrian cement industry – production and consumption 1954–98

ing to actual demand from the construction sector. In 1974, production of cement reached its all-time high.

Between 1974 and 1998, cement production declined by 40%, whereas consumption decreased by 21%. Due to population growth, the relative change in the per capita values is greater than the absolute change (Table 5).

Structural change and ecological restructuring in Austria's cement industry since the 1970s were caused by quite different driving forces and can be divided into three stages (Table 6).

1970s. During the 70s the most important influences were significant energy cost shifts. In order to meet the rising energy costs the increase of energy efficiency as well as the substitution of fossil fuels by waste materials were main strategies of the cement industry.

Table 5. Per capita consumption and production of cement 1974 and 1998

(kg/cap)	1974	1998	*Change*
Consumption	810	599	*-26%*
Production	847	480	*-43%*

Source: ÖSTAT, VÖZ

1980s. The 1980s were characterised by a continuous stagnation of the construction sector and consequently, a decline of cement consumption. The main reasons were (Fachverband der Stein- und keramischen Industrie, various issues):

- a decrease in public investments (particularly in the construction of dwellings) due to the restrictive financial policy since 1980 ("Budgetsanierung") and
- a decrease in private investments in construction due to high interest rates.

There was no Keynesian type policy (e.g. with additional public investments into the construction sector). After the review of the Austrian traffic concept in 1984 plans for the further expansion of the road network were reduced and investments into the road network declined constantly (Fachverband der Stein- und keramischen Industrie 1985, BMwA 1997). Due to budget restrictions, the construction sector has once again suffered from annual reductions in public sector spending since 1994 (Czerny 1996, 1997).

Furthermore, structural change within the construction sector itself has caused – ceteris paribus – a decline in cement consumption. Cement-intensive construction lines such as carcass (shell) and civil engineering projects (power plants, road-building etc) are declining (1996: -9%), while demand for renovation work is constantly increasing (1996: +2%). Due to the low specific cement input in renovation of old buildings, the cement industry cannot profit fully from growth in the construction sector (VÖZ 1992; Czerny 1997).

1990s. Main factors which had severe consequences for the Austrian cement industry during the 1990s were the opening of the East and the transformation of the Central and Eastern European economies, Austria's integration into the EU, the end of the cement cartel and a number of new environmental regulations.

Table 6. Driving forces for structural change and ecological restructuring

	Driving forces	Consequences and strategies
70s	energy cost shifts	increase of energy efficiency substitution of fossil fuels by waste materials
80s	stagnation of the construction sector	continuous decline of demand and production
90s	opening of the East	increase of cement imports due to price differences, partly limitatio of cement imports
	Austria's integration into EU	end of the Austrian cement cartel, restructuring of the cement industry, concentration process and reduction of production capacities, decline of cement prices stagnation of the construction sector due to budget restrictions since 1994 ("Maastricht criteria")
	intensified environmental regulations	investments in environmental protection facilities monthly information about emissions

The opening up of Eastern Europe has boosted foreign trade significantly, to the extent that in 1995, cement imports made up 25% of domestic consumption. Eastern European companies account for 80% of cement imports to Austria. The high level of imports from Italy seems to be a temporary phenomenon, due mainly to the devaluation of the Italian Lira during this period. Italian cement was about 30% cheaper than domestic cement (Doppler 1996). Cement exports by Austria play a minor role, since they involve only 3% of domestic production.

The increasing imports of cement from Eastern European countries were based on agreements between the "reform countries" and the EU which aimed at establishing a free trade area, step by step within 10 years as of 1992 (Doppler 1996; Steiner 1998). In the early 1990s, cement producers in Central and Eastern European countries were favoured by lower wages, lower environmental standards, lower energy costs and lower transport costs.

Consequently, the price of cement imports from these countries was considerably lower (about 40%) than cement produced in Austria (Kramer 1995). The most important factor seems to be lower wages: in 1992, wages in the CSFR were only 7% of those in Austria. In contrast, energy soon rose to western price levels: in 1991, heavy fuel in the CSFR cost 90%, diesel 83% of Austrian prices. Railway transport in the former CSFR was about 50% cheaper than in Austria (Doppler 1996). Imports from CEE-countries mainly concern cement, not ready-mixed concrete. However, prefabricated concrete products are imported from Austrian subsidiaries in Slovakia (Artner 1998).

The government passed regulations on the limitation of cement imports, in which permitted import volumes were subject to tough negotiations between the cement industry and the construction sector. Furthermore – in 1993 – the cement industry made great efforts to reach self-restricting agreements with Eastern European cement companies concerning cement exports to Austria.

With Austria's integration into the EU, it became obvious that the cement cartel which regulated prices and production quantities could not continue to exist in its current form. The cartel was transformed into a "structural crisis cartel", which entailed balancing out regional production volumes. This structural crisis cartel was approved – as a temporary solution – by the EU authorities, and remained in existence until January 1995. The main objective was to restructure the cement industry and reduce production capacity by 15% (Doppler 1996; Hauser 1994).

Under the original cartel, the cement industry was very inflexible. In comparison to Germany, Austrian cement production is still considerably less productive: production per employee is about 3 000 t/a in Germany, whereas in Austria it has increased from 1 600 t/a in 1995 to about 2 400 t/a in 1998 (Bundesverband der Deutschen Zementindustrie 1998, VÖZ 1998).

The structural change in the Austrian cement industry has been accompanied by a sharp decline in cement prices by a third within recent years, from about 1 000 ATS to about 600 ATS per tonne for imported cement and about 700 ATS per tonne (at least) for cement produced in Austria (Steiner 1998; Ambrositsch 1998; Doppler 1996).

2.3 Economic and Social Problems

At the end of the 1980s, 11 cement companies ran 16 production sites, whereas by 1998 the number of companies had decreased to 9, with 12 production sites still in use. 3 of these remaining production sites are currently operating only as grinding mills, which means that nowadays cement burning is restricted to 9 sites. Because some production sites were located relatively close to one another (e.g. in Upper Austria), the concentration process has not increased transport distances beyond the levels mentioned above (100–150 km). This process has been accompanied by a constant decline in employment in the Austrian cement industry – particularly in the 1990s – with a 50% fall since 1970 (Fig. 3). Gross production value in real terms shows a trend rather similar to that of physical production. Price reductions during the last few years have produced a particularly sharp decline in gross production value (about 50% since 1974). These price reductions were part of a strategy to recover market share and a consequence of the end of the price cartel. The unit value of cement has correspondingly declined by about the same amount.

With respect to the value produced by the Austrian stone and ceramics industry, the share from cement production has declined from 32% in 1950 to 12% in 1995 (Fachverband der Stein- und keramischen Industrie 1997). This indicates a significant change in the internal structure of the building materials industry. One reason for this is an increased variety in the construction materials used.

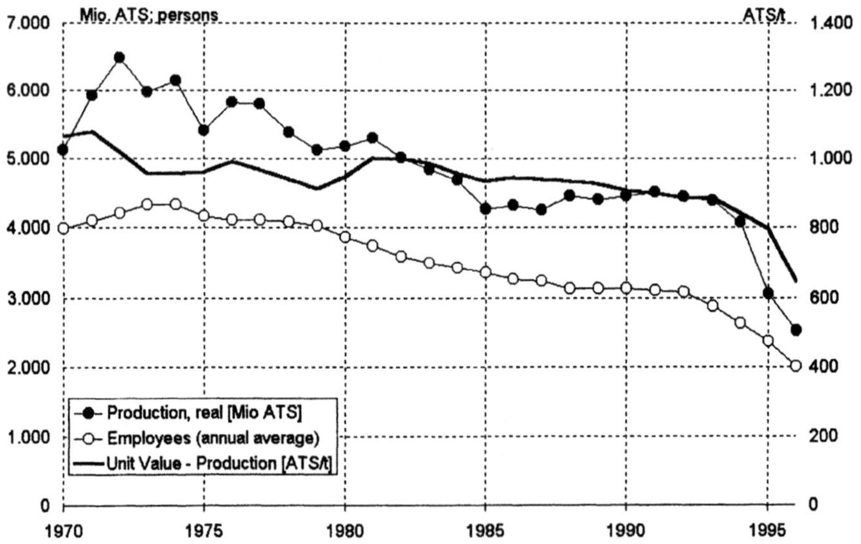

Sources: ÖSTAT; Fachverband der Stein- und keramischen Industrie

Fig. 3. Austrian cement industry – major trends 1970–96

3 The Social Framework of Structural Change

3.1 Actors

The cement market in Austria is characterised by concentration on the supply side (oligopoly) and a wide range of construction companies on the demand side. Among the nine cement producers, *Lafarge Perlmooser* occupies the dominant position, while other producers hold market shares of 4–12%. (Table 7).

The structure of ownership within the Austrian cement industry has changed significantly in recent years. Only a few years ago, when the Austrian cement cartel was still operating, the companies were by and large in (domestic) private ownership, each of them with their own production site. Today, the big multinational cement companies – as in other countries – have brought their influence to bear in Austria as well:

- Perlmooser – the market leader with a share of about 40% of domestic production – was taken over by Lafarge (France) in 1993. Meanwhile, 100% of Perlmooser's capital stock belongs to Lafarge Copée, the world-wide number two cement producer. Lafarge also has a share in the Kirchdorfer Zementwerk.
- Heidelberger Trust (Germany) has taken over SPZ Zementwerk Eiberg and has a share in the Leube Zementwerk and the Gmundner Zementwerke.
- Finally, world market leader Holderbank (Switzerland) has taken over the Vorarlberger Zementwerke Lorüns and is currently leasing a production site from Perlmooser.

Since the end of the cement cartel in 1993 and Austria's integration into the EU in 1995, these three actors have played a key role in the Austrian cement industry. Lafarge concentrates its activities on the east of Austria, whereas Heidelberger and

Table 7. Companies in the Austrian cement industry (1998)

market share[a]	
38%	Lafarge Perlmooser
12%	Wietersdorfer & Peggauer Zementwerke
10%	Vorarlberger Cement Lorüns
9%	Gmundner Zementwerke
9%	Zementwerk Leube
8%	SPZ Zementwerk Eiberg
6%	Kirchdorfer Zementwerk Hofmann
4%	Schretter & Cie
4%	Wopfinger Baustoff Industrie
100%	Total

[a] by production volume
Sources: VÖZ 1998, Roubin 1998

Holderbank focus on Austria's western regions (Fritzl 1997, Ambrositsch 1998). Wietersdorfer & Peggauer and Wopfinger are still independent private companies.

In 1998, nine companies were producing 3.9 Mt cement at twelve sites. At that time, about 1 600 people were employed in the cement industry, which works out at 2 425 t cement per employee. After average capacity use had fallen to 60% in the middle of the 1990s, the figure is rising again (ÖSTAT; VÖZ; APA).

Together, the 9 companies form the *"Vereinigung der österreichischen Zementindustrie – VÖZ"* (Association of the Austrian Cement Industry), which provides services to members and customers of the Austrian cement industry as well as representation in industrial policy and dealings with the public authorities, the media and environmental actors. The VÖZ co-operates in particular with the most important demand side actor, namely the concrete industry, by mounting joint public relations campaigns, organising events and providing technical information on cement and concrete. Moreover, the VÖZ is the link to CEMBUREAU, the cement industry's representative organisation on an international – especially European – level.[5]

The Austrian cement industry is also well represented in the chamber of commerce, which supports the interests of the cement industry in semi-institutional bodies within the *"Sozialpartnerschaft"* (partnership between trade unions and the chamber of commerce) dealing with delicate issues like restrictions on cement imports from Eastern Europe. Both VÖZ and the chamber of commerce are involved in the legislative process, including implemention of EU directives into Austrian law. In general, the representative organisations are already integrated into the preparation of environmental regulations, and can at least comment on proposals by the administration. In recent years, the trade union for the building sector has become increasingly important, as employment in the sector has fallen significantly.

The main demand side actor within the concrete sector is the ready-mix sector[6], which now accounts for nearly half of all cement consumption. The rest of consumption is made up by the pre-cast concrete industry[7], private customers and direct deliveries to construction sites.

In contrast to other countries like Germany, the Austrian cement industry did not move into the strategically important concrete production for decades, keeping a long-tim gentlemen's agreement. Only with the turbulence of the free market, did some companies become active in concrete production.

---

[5] CEMBUREAU is based in Brussels and has 24 members in Europe (19 members: 12 national cement industry associations and 7 cement companies of the European Union and the European Economic Area countries plus Switzerland and Turkey; 5 associate members: national cement associations of the Czech and Slovak Republics, Hungary, Poland and Estonia). The members account for over 15% of total world cement output.

[6] The ready mixed concrete sector is represented by the *Güteverband Transportbeton* (Quality Union Ready Mixed Concrete).

[7] The pre-cast concrete sector is represented by the *Verband der Österreichischen Beton- und Fertigteilwerke – VÖB* (Association of the Austrian Concrete and Finished Products Industry).

Within recent years, importers have gained an important position in the Austrian cement market. Their profit margins have had a powerful influence on the attractiveness of imported cement.

Laws and regulation in the cement industry are generally drafted and coordinated by administrative bodies within the Ministry for Economic Affairs and the Ministry for Environmental Affairs. Regulations on extraction of raw materials, licensing of production facilities and environmental issues (e.g. emissions, waste incineration) are important to the cement industry. Execution of these regulations are generally a matter for regional and local authorities.

In the course of public debate on emissions from waste incineration related to cement production, media as well as research and scientific institutions played important roles. Research institutions can be divided into three groups: the Cement Research Institute, a part of the Association of the Austrian Cement Industry; scientific groups, e.g. connected with the Technical University in Vienna, which largely carries out studies commissioned by the cement industry; thirdly, the "environmentalists", consisting of the Austrian Institute for Applied Ecology (NGO) and the Federal Environment Agency (a government organisation until the end of 1998).

To complete the list of important actors, one should mention regional and local action groups, which generally grew up during licensing procedures for new facilities or in relation with the incineration of (hazardous) waste in cement burning kilns. In recent years, several delicate licensing processes have been managed with innovative mediation procedures, integrating public authorities, local action groups and companies (see Sect. 4).

Fig. 4 shows the most important actors in the Austrian cement industry and its surroundings. The actors belong to three categories:

- companies, owners and customers
- public authorities
- intermediary organisations

While the first group operate mainly through markets, public authorities exert power through legislation and its execution. The third group, the intermediaries, frequently operate between companies and public authorities. Their usual means of communication are lobbying, information, and discussion.

Within the various administrative hierarchies, one can group together actors and their specific networks:

- international / EU level: multinationals, EU administration, CEMBUREAU
- national level: cement producers, their owners and customers – the construction (materials) industry, Ministries for Economic Affairs and Environmental Affairs, association of the Austrian cement industry, chamber of commerce, environmental research institutions, the media
- regional and local level: individual cement production sites, licensing and mining administration, environmental research, media, local action groups

3.2 Interests and Strategies

The energy price hikes in the 1970s were confronted with increased energy efficiency and fuel substitution. Some companies shifted from natural gas to coal, others focused on substituting primary fuels with waste materials like scrap tyres. Technological change during the 1970s was mainly driven by rising energy costs.

The decline in cement consumption as a consequence of stagnation in the construction sector during the 1980s caused a corresponding decline in production. Due to the cement cartel, price and production volumes were strictly regulated, and therefore the decline in production was distributed relatively evenly throughout all companies, and had no broader effects like company closures.

When Eastern Europe opened up to the West and Austria acceded to the EU, the structures of the Austrian cement industry were rigid, due to the cement cartel. The companies were used to a closed market, had no external communications and were not flexible enough to meet the challenges of a free market (Fritzl 1997). Consequently, they were ripe for take–over by multinational enterprises. Currently, only three out of nine companies are still fully independent – i.e. privately owned in Austria (Schretter, Wietersdorfer, Wopfinger). The six others are partially or fully owned by a multinational.

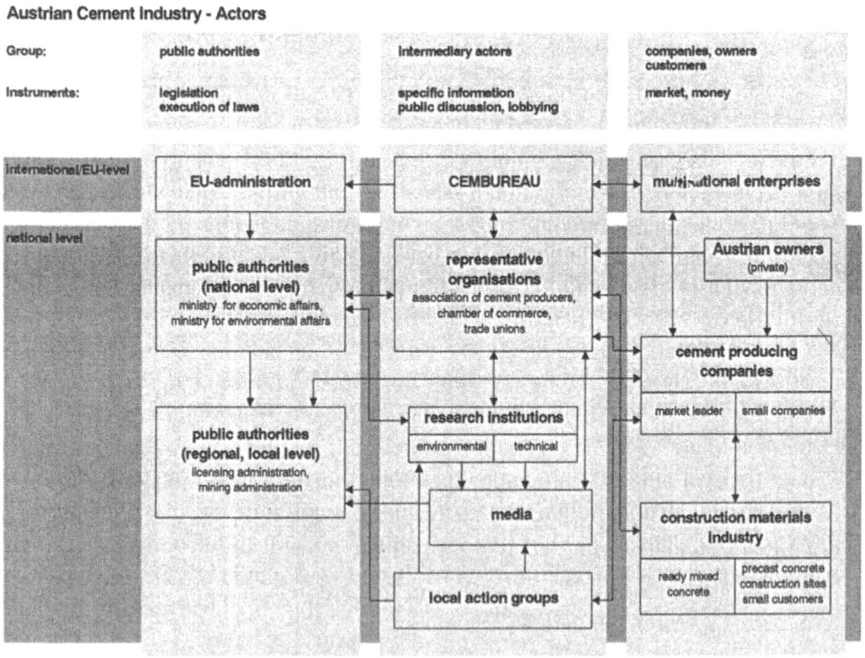

Fig. 4. Austrian cement industry: actor groups at national and EU-level

Faced with the current structural crisis in the Austrian cement industry, companies may pursue several strategies in order to survive. One of them is diversifying product lines (construction materials). Some companies are now engaged in sand and gravel quarrying as well as concrete production, particularly companies involved in ready-mix concrete, prefabricated concrete products and plaster.

A few companies have developed strategies to manufacture hi-tech niche products. "High-performance concrete" was designed for use under extreme chemical conditions, is especially dense and is believed to last twice as long as standard concrete (Nischer 1997). Special qualities are also required in shot concrete, used for lining tunnels (VÖZ-JB 1995, Artner 1997). The companies' R&D departments are expanding, and in 1995 the *Lafarge Group* located its new "Central European Technical Center" in Vienna (Perlmooser Report 1995). As a further example, the small Tyrolean company *Schretter & Cie* produces specialist materials for tunnel construction – some of them patented – and has a market share of 40% for this product line in Austria. With an export share of about 20%, the company is a long way ahead of the current sectoral average of 3% (Fellier 1997).

In order to recover market share, the Austrian cement industry has intensified its efforts in quality management, service and consulting. Furthermore, the cement industry offers customers training courses in the use of cement and concrete (VÖZ-JB 1995).

Depending on the form of ownership, large and small cement companies are resorting to very different strategies to meet the challenges of structural change. Large multinational enterprises like Lafarge show less interest in diversification, and concentrate on the core activities of their new subsidiaries. For example, Lafarge cancelled a number of activities, e.g. production of compound materials from plastics and steel, after taking over Perlmooser.

The current key strategies at Lafarge Perlmooser can be described as follows (Kriegner 1998; Lafarge Perlmooser 1997; Collomb 1997; Anonym 1997a):

- marketing, product innovation, advisory services and consulting (e.g. training colleges for customers)
- integration of the Austrian subsidiary into the multinational structure, transfer of expertise in both directions
- expansion and vertical integration (in the sand and gravel and ready mixed concrete sectors) within Austria and the surrounding Central and Eastern European countries

Expansion within Eastern Europe is one key strategy of the big multinationals. The Lafarge technical centre for Central Europe in Vienna is now responsible not only for production sites in Germany, Austria, the Czech Republic and Poland but also in Russia, Belarus and the Ukraine (Lafarge Perlmooser 1998).

On the other hand, small cement producers in private ownership, like Wietersdorfer & Peggauer and Wopfinger are developing their diversification strategy. Producing cement is only one of their business areas. They try to have at least one other string to their bow by setting up joint product lines in other construction materials. Vertical integration through participation in the strategically important production of ready mixed concrete less important to them than to market leader Lafarge Perlmooser.

Small companies concentrate on developing their regional markets (within a radius of about 150 km). They emphasise flexibility as their strategic advantage over larger enterprises and focus on their local customers, for whom they provide special services and guaranteed quality (cf. Steiner 1998). They are not therefore interested in taking over cement companies in Central and Eastern Europe, with the exception of production sites within their potential local markets.[8]

Large multinationals sometimes experience conflicts of interest: market leader Lafarge Perlmooser traditionally heads the board of directors of the association of Austrian cement producers. As such, Lafarge Perlmooser has to argue against cement imports from Central and Eastern Europe, whereas the company itself has already expanded its activities throughout CEE countries, and the Lafarge management does not want to exclude future options for importing cement from Central and Eastern Europe to Austria.

The trade unions (*Gewerkschaft Bau-Holz*) played an important role within the social management of the cement industry's restructuring during the last years. In order to reduce the social problems caused by factory closures and reduced staffing levels, the unions were involved in drawing up compensation plans, which were negotiated locally for single companies and included voluntary part payments, financial assistance (70–100 000 ATS per employee), help in finding new jobs in co-operation with the employment exchange, and retraining initiatives.

The typical procedure was that after the trade union was informed of planned staffing reductions by the works council (*"Betriebsrat"*), the president of the trade union started negotiations with the company management.

4 The Influence of Environmental Policy

Environmental issues have been crucial to structural change in the Austrian cement industry during the last decade. Public and scientific debate focussed on emissions from cement production, particularly on NO_X and dust, later also on the issue of additional emissions to air from incineration of (hazardous) waste in cement burning kilns.

The increasing substitution of primary fuels (fossil fuels) by waste materials was one strategy for meeting rising energy costs.[9] However, increased waste incineration within cement production focused public attention on the subject of air emissions in the late 1980s. In 1989 the Ministry for Economic Affairs proposed regulation on emissions from cement production and requested expert opinion

---

8 Wietersdorfer & Peggauer in the southern Austrian province of Carinthia, for example, are trying to take over a production site in neighbouring Slovenia (Anonym 1997b; Artner 1998).

9 22% of fossil fuels are substituted by secondary energy sources which, in 1996, consisted mainly of scrap tyres, waste oil, paper sludge and plastics waste (Hackl and Mauschitz 1997).

from 30 institutions. In the course of delicate negotiations, especially between public authorities (namely the Ministries for Economic Affairs, for Environmental Affairs and for Agricultural Affairs) the process nearly broke down. Eventually, the cement industry – which needed predictable and generally applicable environmental regulation – took the initiative and suggested a compromise, which was finally approved in 1993. The regulation set compulsory emissions standards and included grace periods for upgrading production facilities (by 1998).[10]

The new regulations caused the cement industry to make significant investment in environmental facilities (mainly filter and monitoring technology). The interviews with cement industry representatives clearly revealed that the ambitious environmental standards in Austria foster technological development and have consequently had positive effects on the competitiveness of the Austrian cement industry.

Waste incineration within the cement industry is still a controversial issue. The cement industry enjoys more favourable legal limits on waste incineration than those in municipal solid waste incineration plants (Reiter and Stroh 1995).[11] Environmental groups, like the Austrian Institute for Applied Ecology and the waste incineration lobby have both called for an end to this differential (Konrad and Jorde 1994; Anonym 1998b). The cement industry argues that waste incineration within the cement production process has no net negative impact on air emissions, and saves non-renewable resources and landfill (Hackl and Mauschitz 1995, 1997; Knoflacher et al. 1995).

The use of end-of-the-pipe technologies within the cement production process was not a specifically Austrian development, but has affected the increased substitution of fossil fuels with waste materials. In 1995 specific regulations on air emissions from the cement industry have existed not only in Austria but also in Germany, and have been discussed in Switzerland (Kuhlmann 1995; Tellenbach 1995).

Greater environmental regulation, however, also accelerates the concentration of the cement industry. It is argued that investment in environmental facilities and increasing energy efficiency are only profitable in larger production units. However, the small company Kirchdorfer has launched a pilot plant, including a special process for reducing nitrogen oxides. The pilot project, which is the only one of its kind in the world, is also a consequence of the production site's location: it is only 500 meters from the centre of the village (APA 1998; Anonym 1998a).

One of the most important issues recently, the proposed EU-directive on incineration of hazardous waste, has considerable consequences for the cement industry.[12] The implementation in national law was passed in 1998 and came into force

[10] Verordnung über die Begrenzung der Emissionen von luftverunreinigenden Stoffen aus Anlagen der Zementerzeugung BGBl. Nr. 63/1993; the process is documented in Sedlak 1995.

[11] The cement production process itself generates NO_X emissions above the permitted limits for incineration plants.

[12] Proposal for a Council Directive amending Directive 94/67/EC on incineration of hazardous waste COM(97) 604 final – 97/0314(SYN).

in February 1999. According to this regulation, cement producers are allowed to substitute up to 40% of fossil fuels by hazardous waste. In general, the cement industry is subject to the same air emissions limits as incineration plants, with the few exceptions being due to specific factors in cement production. NO_X-emissions must be reduced by 50% until 2007, but the cement industry welcomes the fact that the regulatory framework is now more predictable (Anonym 1999).

There are highly specific modes of interaction between the cement industry and environmental actors: While representative organisations (association of the Austrian cement industry, chamber of commerce) communicate with national public authorities (ministries), each single company must go its own way at regional and local level. There are frequent reports about close interaction between licensing administrations, local action groups, and cement companies.

Also worth mentioning is the Leube mediation process on plastic waste incineration as a fuel substitute, which involved local residents, civic action groups, local government and experts, as well as the cement company itself, and produced a formal contract between the civic advisory board and the cement company. This win-win solution involved early integration of residents and civic action groups, with better access to information than in a traditional administration process on the one hand, and led to greater acceptance and a better image for the cement company on the other hand (Hittinger 1997; Anonym 1997c,d).

5 Conclusions and Perspectives

Structural change with environmental benefits has taken place in the Austrian cement industry itself, as well as in its relationships to the construction materials industry as a whole (Table 8). Moreover, structural change within the construction sector has directly affected demand for cement. The following phenomena in inter-sectoral as well as intra-sectoral structural change can be identified:

- entry of multinational cement enterprises into the Austrian market after the end of the Austrian cement cartel
- declining physical production due to increased imports and changes in demand from the construction sector
- concentration and reductions in production capacity

Table 8. Environmental benefits of structural change in the Austrian cement industry

Environmental benefits due to	Driving Forces	Policy
decline of production	decline in consumption (80s) imports from CEE (90s)	no Keynesian intervention temporarily import restrictions
technological change	energy costs (70s) environmental regulations (90s)	no policy environmental policy

- declining value produced, falling employment
- technological changes to reduce costs and meet environmental regulations
- strategic co-operation in Eastern Europe
- diversification of product lines, concentration on niche products and exports
- intensified R&D and human resource management

The state of the Austrian cement industry since 1997 is characterised by a slight downward trend in consumption and production, and a slight rise in imports, mainly from Central and Eastern European countries. The share of cement imports from these countries amounts to 84% of total imports.[13]

All representatives of the cement industry agree that concentration will continue in the coming years. In the medium to long term, cement consumption in Austria will fall. Imports from Central and Eastern Europe will also fall significantly as soon as the costs and environmental regulations in those countries reach EU levels. An important factor is the liberalisation of the electricity markets within the EU, which will result in lower prices, particularly for bulk purchasers.

The analysis of structural change in the Austrian cement industry shows examples of innovative conflict management at various levels. These examples can be considered successful where it was possible to minimise the effects of drastic job-cutting through compensation plans. Negotiations between management and trade unions at an early stage were helpful.

Innovative procedures for conflict solution during enlargement and or upgrading of production sites (such as waste burning) can also be regarded as models. Mediation which compulsorily involved residents, local administration and management proved effective.

Moreover, the involvement of different public authorities and representative organisations at an early stage of the legislative procedure must be viewed as basically positive. In this case, a close network eveolves between administration, industry, researchers and the interested general public. Conflicts of interest between administrations have, however, proved to be problematic and delayed the process in the past.

References

Ambrositsch, H. (1998): In der Mühle der Zementriesen. Oberösterreichische Nachrichten, 18.8.98, 10.

Anonym (1997a): Ganzheitliche Strategien – Die Zementproduzenten steigen verstärkt in die Betonerzeugung ein – und umgekehrt. Bau- und Immobilien Report No. 7/97, 30–33.

Anonym (1997b): Wietersdorfer: Positiver Schiedsspruch. Bau- und Immobilien Report No. 7/97, 25.

[13] The major exporters are Slovakia (40% of total imports) and Romania (29%), followed by the Czech Republic (11%), Slovenia (5%), Hungary and Bulgaria (1% each) (VÖZ 1998, 1999).

Anonym (1997c): Ein Sieg der Dialogbereitschaft – Freiwilliges Bürgerbeteiligungsverfahren bei Müllverbrennung im Zementwerk Leube. Umweltschutz No. 9/1997, 18–19.

Anonym (1997d): Rudi Rastlos – Leube-Chef Zrost. Bau- und Immobilien Report No. 11/97, 28.

Anonym (1998a): Öko-Pionier. Bau- und Immobilien Report No. 10/98, 5.

Anonym (1998b): Der Kampf um den Müll – Zementindustrie kontra Entsorger. Bau- und Immobilien Report 9/98, 40.

Anonym (1999): Endlich Rechtssicherheit im Verbrennungsdschungel. Das Zement-Telegramm No. 1, Jänner 1999. Wien.

Artner, F. (1998): Staubig und ätzend. Bau- und Immobilien Report No. 10/98, 8–12.

APA (Austria Presse Agentur) (1997): Recherche "Zementindustrie", Juli 1997. Wien.

APA (1998): Umweltfreundlichere Kirchdorfer Zementproduktion. Pressemeldung vom 3.7.98. Wien.

Artner, F. (1997): Forscher vielbeschäftigt. Bau- und Immobilien Report No. 7/97, 36.

Braun, H. (1994): Produktion, Energieeinsatz und Emissionen im Bereich der Zementindustrie. Zement und Beton 1/94.

BMU (Bundesministerium für Umwelt) (1995): Nationaler Umwelt Plan Österreich. Wien.

BMwA (Bundesministerium für wirtschaftliche Angelegenheiten) (1997): Statistik Straße und Verkehr. Wien.

Bundesverband der Deutschen Zementindustrie (1998): Zahlen und Daten 97/98. Köln.

Collomb, B. (1997): Die Zukunft der Perlmooser. Interview in A3Bau, No. 10/1997, 78.

Czerny, M. (1996): Weitere Abschwächung der Baukonjunktur in Westeuropa. WIFO Monatsberichte No. 2/1996, 117–121.

Czerny, M. (1997): Bauwirtschaft erholt sich in Westeuropa 1997 langsam. WIFO Monatsberichte No. 2/1997, 88–93.

Doppler, K. (1996): Die Zementindustrie in Österreich angesichts des EU-Beitritts und der Ostöffnung. Diplomarbeit an der Wirtschaftsuniversität Wien.

Europäische Kommission (1997): Panorama der EU-Industrie 1997, Band 2. Luxemburg

Fachverband der Stein- und keramischen Industrie (1984–98): Jahresbericht. Wien.

Fachverband Stein- und keramische Industrie (1984–97): 50 Jahre Stein & Keramik 1947–1997, Wien.

Fellier, W. (1997): Schretter & Cie – Nur ein bißchen anders. Bau & Immobilien Report, No. 7/97, 44.

Fritzl, M. (1997): Der Erneuerer – Die Zementindustrie auf dem Weg vom Kartell zum liberalen Markt. Bau- und Immobilien Report No. 7/97, 27–29.

Hackl, A. and Mauschitz, G. (1995): Emissionen aus Anlagen der österreichischen Zementindustrie. Wien.

Hackl, A. and Mauschitz, G. (1997): Emissionen aus Anlagen der österreichischen Zementindustrie II – Jahresreihe 1994–1996. Wien.

Hauser, U. (1994): Verteilungswirkungen von Handelshemmnissen als Bestimmungsfaktoren der Handelspolitik dargestellt am Beispiel der Kontingentierung von Zement. Diplomarbeit an der Wirtschaftsuniversität Wien.

Hittinger, H. (1997): Differenzen beigelegt – Im Umweltmediationsverfahren "BürgerBeirat Gartenau" fanden alle Beteiligten eine zufriedenstellende Einigung. Müllmagazin No. 3/1997, 6–8.

Hüttler, W., Payer, H., Schandl, H. (1996): National Material Flow Analysis for Austria. Society's Metabolism and Sustainable Development. Federal Ministry for the Environment, Youth and Family Affairs. Vienna.

Knoflacher, H.M., Tuschl, P., Medwedeff, A. (1995): Die ökologische Standortbestimmung des in österreichischen Werken hergestellten Bindemittels Zement. Wien.
Konrad, W. and T. Jorde (1994): Abfalleinsatz in der Zementindustrie. Österreichisches Ökologie-Institut, Wien.
Kramer, H. (1995): Perspektiven der Stein- und keramischen Industrie Österreichs. Branchenanalyse des Österreichischen Instituts für Wirtschaftsforschung (WIFO) im Auftrag des Fachverbands Stein- und keramische Industrie. Wien.
Kriegner, M. (1998): Ein weiter Weg. Interview. Bau- und Immobilien Report No. 9/98, 25–26.
Kuhlmann, K. (1995): Sekundärstoffeinsatz in der deutschen Zementindustrie. In: Zement und Umwelt – Beiträge zur Enquete der österreichischen Zementindustrie. Wien, 88–97.
Lafarge Perlmooser (1997): Zeitung für Kunden und Mitarbeiter No. 155/1997. Wien.
Lafarge Perlmooser (1998): Zeitung für Kunden und Mitarbeiter No. 159/1998. Wien.
Nischer, P. (1997): Beständig und dauerhaft. Bau- und Immobilien Report No. 7/97, 34.
ÖSTAT (Österreichisches Statistisches Zentralamt) (1970–1994): Außenhandelsstatistik. Wien.
ÖSTAT (1970–1994): Industriestatistik 1. und 2. Teil. Wien.
Perlmooser Zementwerke (1970–1996): Geschäftsberichte, Wien.
Raffel, G. (1995): Situation der österreichischen Zementindustrie. In: Zement und Umwelt – Beiträge zur Enquete der österreichischen Zementindustrie, 5-11. Wien.
Reiter, B., Stroh, R. (1995): Behandlung von Abfällen in der Zementindustrie. Umweltbundesamt Monographien Band 72. Wien.
Roubin, E. (1997): Zur Lage der österreichischen Zementindustrie. Das Zement-Telegramm No. 5, November 1997.
Roubin, E. (1998): Personal communication (July).
Sedlak, R. (1995): Die Emissionsverordnung für Anlagen der Zementindustrie. In: Zement und Umwelt – Beiträge zur Enquete der österreichischen Zementindustrie. Wien, 12–15.
Steiner, S. (1998): Internationalisierung der österreichischen Industrieunternehmungen unter besonderer Berücksichtigung der Baustoffindustrie vor der Hintergrund der Transformation in Zentral- und Osteuropa. Diplomarbeit an der Wirtschaftsuniversität Wien.
Tellenbach, M. (1995): Vorgesehene Regelungen für die schweizerische Zementindustrie. In: Zement und Umwelt – Beiträge zur Enquete der österreichischen Zementindustrie. Wien, 34–38.
UN (United Nations): Industrial Statistics Yearbook. New York.
VÖZ (Vereinigung der österreichischen Zementindustrie) (1990–1998): Jahresberichte. Wien.
VÖZ (1994): 100 Jahre Vereinigung der österreichischen Zementindustrie. Wien.

List of Interview Partners and Information Persons

Gewerkschaft Bau-Holz	*Mag. Herbert Aufner*
CTEC – Centre Technique Europe Centrale Lafarge	*Dr. Werner Bittner, managing director*
Lafarge Perlmooser	*Dipl.-Ing. Frederic Fleuret, industrial manager*
Kirchdorfer Zementwerke Hofmann	*Dipl.-Ing. Dr. Hans-Jürgen Gulas, plant director*
Österreichisches Ökologie Institut	*Dipl.-Ing. Tristan Jorde*
Fachverband der Stein und Keramischen Industrie, Wirtschaftskammer	*Dr. Kollinger*
VÖZ – Vereinigung der Österreichischen Zementindustrie	*Dr. Ernst Roubin, managing director*
Wietersdorfer & Peggauer Zementwerke	*Dipl.-Ing. Jochen Ziegenfuß, managing director*
Verband Österr. Beton- und Fertigteilwerke	*Dr. Günther Zieschank, managing director*

Cement Production in Denmark

Børge Klemmensen[1]
Department of Environment, Technology and Social Studies, 4000 Roskilde, Denmark

In 1990, the Danish cement industry was at almost the same level of production as it had been 30 years earlier. In between, production had doubled to an all-time high in 1973, only to decline again in the wake of the two oil crises. But since 1990, and despite ever-constant domestic demand, production has risen again, in 1996 reaching almost the peak level of 1973. In between these two peaks, radical changes and restructuring have taken place in the industry.

1 The Danish Cement Industry Network

1.1 The Cement Production Monopoly

Ålborg Portland Ltd. is the only producer, the entire production taking place at one site located in Rørdal, a suburb of Ålborg in North Jylland. Ålborg Portland has dominated the industry since the end of the Second World War, at which point they owned 4 out of 5 factories with more than 85% of production capacity. Even at this early stage, two thirds of this capacity was already located at the Rørdal factory. This dominance meant that the cement industry – thereby first and foremost the Ålborg Portland company – was subject to regulation under anti-monopoly legislation. The monopolies board was concerned about the supply situation, wishing to ensure that cement was easily available across the country, and about pricing, which it kept under its control until well into the 1990s.

Prices were negotiated annually between Ålborg Portland and the monopoly board until the end of the 1960s, when the pricing system was changed to a 3-year model. After severe problems associated with the dramatic increase in energy prices after 1973, as well as the slump in building activity, which brought the first warnings of reductions in cement consumption, the 3-year system was finally able to establish itself and operated smoothly from 1977 onwards.

As regards the level of the prices, it was agreed that they should cover the actual costs, unit costs and capacity costs, and that changes herein would be the main component affecting the price of cement. Problems arose over the depreciation allowance and the level of profits, whereby the company exerted pressure in terms of its "ability to invest in new technology". This was again important, in that only

[1] in collaboration with and with strong support in the research from Malene Linderoth and Maria C. Nielsen, M.Sc.-students at TEK-SAM

low costs could secure prices which were comparable to – or even lower than – those in neighbouring countries, from where imports could otherwise represent a threat to the monopoly. And imports had always been at very low level indeed – about 1–2%, normally controlled by Ålborg Portland itself, as they were made to cover production fall-out, strikes etc. The company had its way to a great extent, but only in 1977 was the system so structured that productivity gains were reaped fully by the company. This gave the company the full benefits of the fight against the energy price explosion which, between the first energy crisis in 1973 and the second in 1979, doubled the energy component in production costs from 16% to some 33%.

At the same time, the company had already managed to solve the industry's capacity and restructuring problems by about the mid-1970s, closing all three of its other factories and gaining control over the factory of its only competitor – which it subsequently closed down in 1978. Thus the benefits of restructuring the industry also flowed to the company, helping it survive a double attack in 1979/80: soaring energy prices and a dramatic drop in domestic consumption from summer 1980, almost halving demand to a level of some 1.2 million tonnes – a level at which, with minor variations over the years, it has more or less remained.

In 1990 Blue Circle Ltd., Britain's largest cement producer, bought a 50% stake in Ålborg Portland. This was seen as a strategic alliance, ensuring co-operation on technology, cost control and quality, as well as helping market orientation, with special emphasis on further expansion of exports of high-quality white cement in international markets and finding customer for surplus standard cement production not required by the domestic Danish market.

1.2 Industrial Partnerships and Networks

The 50% stake was sold to Blue Circle by Ålborg Portland's parent company, Ålborg Portland Holdings Ltd. This holding company is owned in turn by Potagua Ltd, whose portfolio includes F.L.S. Industries, a global supplier of *cement production equipment*, developed in close co-operation and consultation with cement producers over the last 100 years. Due to the nature of the raw materials available in Denmark, this technology was developed predominantly for "wet" production. The information exchange has clearly been in both directions, in that discoveries and developments, achieved during the design and adaptation of cement production facilities for customers elsewhere in the world, have been made available to the group's factory (or factories) in Denmark and used to optimise production or upgrade technology, keeping the company aware of opportunities for technical improvement. On the other hand, it has linked technological progress in the cement company to that of the group itself, possibly restricting access to alternatives and thereby making production more costly than necessary.

The company has, since the start of production at the Rørdal site in 1889, *controlled the principal raw material components* of cement – chalk, clay and sand. Chalk constitutes over 80% of the raw material, and the Rørdal factory is more or less sitting on huge deposits of very high-grade chalk, plentiful enough to guarantee production at current levels well into the 21st century. The factory owns the land over the chalk deposits and handles the excavation itself.

Sand is pumped up by ships from a nearby shoal. Sand, along with fly-ash, gradually substituted for the use of clay until the latter was finally abandoned in 1988. (This was what helped establish the "semi-dry" process, which was implemented in the same year, see Sect. 2.2)

The fly-ash is supplied primarily by Danish power stations, which have been forced to remove particulates from exhaust smoke since 1980/81. That has resulted in a "production" of 800 – 1 500 kt/a, depending on the level of electricity production. In 1978, Ålborg Portland Ltd. formed a company, "Danaske I/S", in 50–50 partnership with the Danish power stations. It has managed to sell and thereby recycle almost all the fly-ash produced and, since 1981, Ålborg Portland itself has consumed more than 200 kt/a, with increasing consumption in recent years due to the production of new types of cement containing 25% fly-ash compared to 12–15% previously. Another 250 kt/a is sold to Swedish and Norwegian cement factories and to the Danish pre-cast concrete industry.

In recent years, another symbiotic relationship has developed with the power stations with respect to gypsum, which is used to control the hardening process in concrete production. Gypsum arises from waste gas de-sulphurisation in the coal-fired plants which have supplied the bulk of electricity and district heating in Denmark since the mid-1970s. Chalk is sprayed onto the smoke in the smoke-stack, binding the sulphur to gypsum. Ålborg Portland now supplies chalk to a large power station nearby in return for the gypsum produced there. A waste product of power stations has become a raw material for cement production – and Ålborg Portland has gained control of this link in the raw material supply chain as well.

Cement for the *domestic market*, of which some 82% is sold in bulk, is delivered directly to the customer by Ålborg Portland itself, either by lorry from a nearby silo (terminal) or directly from the factory itself to the customer's own silo. The company has set up its own storage facilities in 7 different cities and towns, supplied by one of the company's own ships. From the terminal, lorries (some of which are from Ålborg Portland itself) then deliver the cement to the customers. The silo terminals are spread across the country so that all regions are covered, securing continuous and easy supply of cement. The remaining 18–20% of the cement is delivered in sacks, and that segment is normally collected by the customers themselves.

A 100% subsidiary of Ålborg Portland is the Danish Eternit Factory Ltd., which produces corrugated roofing plates, originally made of asbestos cement but, since the mid-1980s, now produced without the asbestos. This has reduced the cement content, and the factory now accounts for some 5–6% of the domestic cement sales, some 75 kt in 1996.

Another subsidiary is "Spæncom" Ltd., where Ålborg Portland holds 28% of the shares. It produces pre-cast concrete, manufacturing parts for the construction industry, and has set up a subsidiary in the former East Germany to exploit the construction boom in this area following German reunification. In 1995, this sector accounted for some 20% of domestic sales or 265 kt cement. Exactly how much of this is due to "Spæncom" Ltd. is unknown, but it is significant, and may be up to 20–25%.

Furthermore, Ålborg Portland has subsidiaries within the readymix sector, through "Færdigbeton Ålborg" Ltd., which is a wholly owned subsidiary and which has decisive influence on another major player in the market, Unicon Ltd. The sector consumes about a third of total domestic sales, or some 400 kt of cement in 1995. Control of this sector is therefore extremely important, in a market where cheap cement is increasingly being offered by nearby CEE-countries such as Poland and Latvia. Through this company, Ålborg Portland also has a 50% involvement in cement and concrete delivery to the bridge over and the tunnel under the Great Belt, which have been under construction since 1991, but which stopped requiring cement in 1994/95. This model of co-operation has been continued in the Sweden bridge project, now under construction. The partner in this case is the main Swedish producer of cement.

Finally, the company also has a stake in the *export market,* with a 100% subsidiary in the USA and a 20% share in the largest US producer and distributor of "white cement", Lehigh White Cement Company Ltd. White cement is the most important export product for the company, which covers about a third of the world market for this product, the USA being its single most important market.

2 Innovation in the Cement Industry

2.1 Product Innovation

Cement is a binder with a very precise quality definition – its strength after 28 days is the decisive property. Additionally, it should be able to resist degradation under specific conditions, first and foremost under continuous contact with (sea) water, which is the case with all bridges, harbours, tunnels etc. These standards must be strictly observed for any random tonne of cement in each quality category. And it is uniformity, not improvements, which are important to quality (Justesen, CtO, interview January 1998). This is the framework within which innovation in cement production takes place.

Innovation can therefore run along two lines – both of which concern increased profitability and survival in the market, of course. One is to develop a product which supplies a niche market which can be held for at least some time. The other is to develop – or modify – standard products to reduce production costs without affecting quality, and thereby gain a competitive edge over competitors. Ålborg Portland has done both over the years. There are, however, no radical innovations on the product side in the cement industry. Product modernisation – or innovation – is determined more by the need to reduce costs by reducing the energy content per kg of cement than the need to introduce products with major new features or applications. (Important research is being conducted in the concrete production sector which, together with research into cement properties, has led to a reduction of 20%+ in the cement content of concrete, see Table 2, p. 333).

The most important niche product, in production since the 1960s, is white cement, used for decorative and other purposes where the appearance of standard grey cement is to be avoided. The "secret" of the white cement lies in its raw materials: only the whitest chalk fractions are used, together with light or uncoloured sand and kaolin. In addition, white cement is sulphate resistant and therefore suitable where buildings are in contact with sea water or soil, for example. Most of the white cement produced has always been exported and, even in the late 1970s it totalled some 80 kt/a. In the difficult years of the early 1980s it became the company's "cash cow", and has remained so ever since. In 1998 it received a further boost, taking production up to some 600 kt/a, over 50% of total export volume but accounting for around two thirds of export income. The main modernisation efforts with respect to white cement are directed at reducing energy consumption, as this is considerably (about 40%) higher than that for grey cement.

Back in the early 1970s, the company started developing a sulphate resistant cement, based on cheaper raw materials than white cement and targeted at the construction of bridges, harbours etc. The market for this product did not exist until the late 1970s, but further development has taken place and it has also been specially adapted to meet the needs of major construction projects, e.g. for the building of the Great Belt bridge in the first half of the 1990s.

Back in the 1960s Ålborg Portland had already conducted the first feasibility tests for adding fly-ash (the particulate residue of purifying exhaust smoke, predominantly from coal-fired power plants) to the production process for cement clinker. The stimulus was reduction of energy costs, and this idea was therefore taken up more seriously again after 1973, in the wake of large rises in energy prices. The idea was to use the fly-ash to replace some of the clay input, which made up some 10–15% of raw materials at the time and, simultaneously, to use its unburned coal content as a fuel for heating the ready-mixed raw materials in the oven. The reduced clay content meant a reduced energy requirement for the process, as clay has a high water content, while fly ash is a dry material. In 1977 the tests were concluded and the use of fly-ash introduced in a single oven – from 1980 it was used in all ovens. Afterwards, fly-ash was introduced as a further component of cement, this time as an additive after incineration – i.e. during the clinker grinding process. From 1984, all Standard Portland Cement leaving the factory contains about 25% fly-ash.

In 1992, a new base cement was launched as the latest result of the continued search for a reduction in energy consumption and thereby in overall production costs. This cement has a more narrow field of application, but it is easier to use in these cases. The reduced energy input is achieved through lower oven temperatures when burning the clinker and by adding further fillers during the grinding process, reducing the clinker content per kg of the new cement.

2.2 Process Innovation

It should be remembered that, since its beginnings in 1889, Ålborg Portland has been member of a group which also includes an engineering and manufacturing company producing equipment for cement production. This means that technological developments are made in parallel with the engineering company, and that the results would normally be used for the next delivery of equipment to a foreign customer of the engineering company. But it also means that there is permanently a simple form of contact with regard to technological change in the industry, and that the factory will be involved in a number of tests on new ideas and modifications to existing practices that a normal cement production company would not.

A number of technical modifications have been associated with the aforementioned product modernizations and, as consistent quality is crucial, these modifications have to be thoroughly tested before being allowed to become part of the standard production process.

However, the real changes have been associated with reducing energy consumption which, at some stages in the mid-1970s and at the beginning of the 1980s (after the two major energy prices rises) made up 33% of total cement production costs, thereby becoming the single most important cost factor.

The efforts at Ålborg Portland can be divided into three main categories: changing the fuel, changing the raw materials, heat recovery.

Changing the fuel. The main shift here was back in the 1970s, when the factory substituted oil with coal, which remained at a considerably lower price than oil. Furthermore, as mentioned above, the fly-ash represented both a raw material modification and the introduction of a new fuel – its residual coal content.

An attempt to include scrap tyres in the process has recently been aborted, due to waste taxation in Denmark, which makes it more profitable for the tyre and scrap industry to export the tyres than to burn them in Denmark. They are now burned in the Swedish cement industry. From an environmental perspective, the additional transportation caused by the tax system appears indefensible.

The latest development is an agreement with the regional authorities to burn household and other non-hazardous waste in the ovens, exploiting its greater energy content, due to the increased incinerable element in this waste. The agreement provides for burning 100 kt/a, with an initial pilot scheme covering 25 kt/a. The actual maximum capacity would be double this figure, 200 kt/a. Every kilogram of waste replaces 0.5 kg coal, as the first agreement covers the fractions of the household and industrial waste, including sludge from waste water treatment plants, with the highest energy contents.

Changing the raw materials. With regard to raw materials, the single most important shift was the establishment of the semi-dry process in 1988. This is a well-known process, and may even be the most common, where limestone is the raw material for chalk. But Ålborg Portland's chalk comes from deep layers of natural chalk with a water content of 28%. After the introduction of fly-ash, and having replaced clay with sand – now including waste sand from the engineering industry – it became possible to introduce a new oven, based on the semi-dry principle, which used hot air from the burning process to "pre-dry" the chalk component, which remains the same chalk extracted on the factory's premises. For some formulae, especially base cement, the composition of raw materials has made it possible to reduce the oven temperature when burning the clinkers.

From 1993 waste paper from a paper recycling plant has been burned in the ovens at a rate of 55 kt/a. Besides its energy content, the waste paper also contains chalk, and as a waste product, it would cost the paper company disposal charges at the landfill. The exact figures are not available but, for the cement company, the deal means free fuel and raw materials, as well as a possible additional payment for handling the paper plant's waste problem.

Recent years have also seen aluminium silicate used in the clinker production process. It is a catalyst in oil refining and thus a waste product from the oil industry, where it replaces china clay and sand in final cleaning and filtering of both the mineral and vegetable oils. This, apart from alleviating a waste disposal problem, has also meant a cheap supply of raw materials for Ålborg Portland, as well as some savings on its energy bill.

Other shifts in raw materials are related to the stage of the production process which follows the burning of the clinker. In the grinding process, additional fly-ash is added to the grey cement, as well as gypsum, which gives better control of the hardening process. For some of the grey cements other types of filler are also used. This means that the clinker content, and thereby the energy content per tonne cement, is decreased.

Heat recovery. Heat recovery principally means recovering heat from flue gas – dust and SO_2 can be removed at the same time. A major step was taken in 1988 with regard to the semi-dry process, where recovered heat is now used to dry the "raw meal" – the ready mixed raw materials – before it enters the calcination stage. The temperature required to dry the raw materials is 750°C.

The most significant step, however, was taken in 1991, when a heat recovery system was fitted to the four ovens producing white cement. The heat recovery system was developed in close co-operation between Ålborg Portland and FLS-Industries, and is held world-wide to be a major advance and a unique technical solution. The heat recovered here is supplied to Ålborg's municipal district heating system, and now supplies 40 MW, equivalent to 20% of the fuel used to produce the white cement and enough to supply heat for 20 000 – 22 000 households, or some 15% of total heat consumption by the Ålborg district heating system. For the factory it means that some 20% of its energy costs are recovered, as the factory is naturally paid for the energy it supplies to the district heating system.

Heat recovery is now being extended, almost doubling the heat supplied to 70 MW. This development is due partly to the increase in white cement production (to 600 kt/a or by 11%), and partly to improved boiler efficiency. The introduction of heat recovery from the new scrubbers which purify the flue gases has also contributed towards the increase.

3 Structural Change, Environmental Benefits and Social Impact

3.1 The Duration and Nature of the Decline in Production

Cement production in Denmark reached an all-time high in 1973. This was the year of the first oil crisis, which resulted in dramatic rises in the price of oil and highlighted the dependence on oil as a source of energy which gave the oil-producing countries such a strong influence on western capitalist economies. During 1974, cement production fell by 15%, falling by a further 10% in 1975. This 10% was recovered in the following year and production stabilised, even recovering further in 1978 before the onset of the second oil crisis in 1979. Between 1979 and 1980, production was in virtual free-fall, and this continued into 1981. Within two years, production plummeted by about 35%. It remained at this level all through the 1980s. Only in 1993 could real signs of change be observed (Fig. 1).

Production followed domestic market developments very closely, as there were scarcely any significant imports until the late 1980s, accompanying the changes in

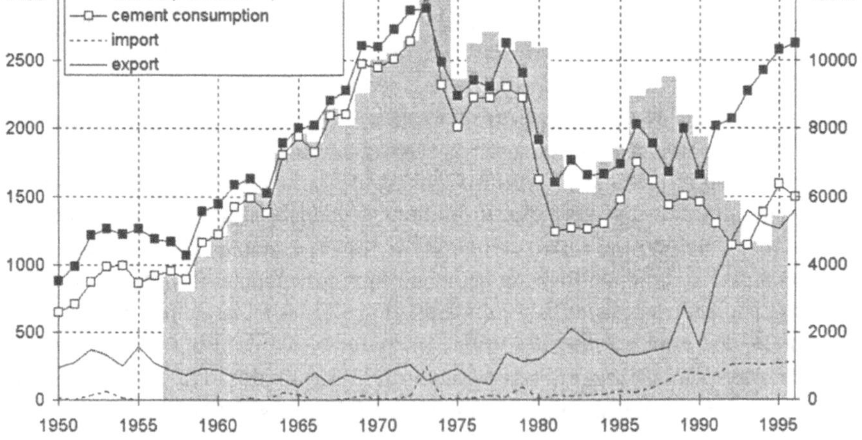

Source: Trade and Industrial Statistics

Fig.1. Cement production, consumption, foreign trade and area covered by new buildings in Denmark 1950–96

Central and Eastern Europe. The decline in production was therefore merely adapting to the decline in the domestic cement market. This market remained at 1.2–1.3 Mt, the level reached in 1980/81 – some 45% below that of 1978 – recovered after the shock of 1980/81 to reach 1.5 Mt (1984/85), fell again to below 1.2 Mt (1991/92), and has now recovered to 1.5–1.6 Mt (1996/97).

Total production, however, began to increase in 1991 and, in 1997, reached a level not far below the all time high of 1973. This discrepancy is explained by exports, which have grown strongly during the 1990s, to the extent that exports now account for 50–60% of production, as opposed to 1973, where exports made up only 5%.

The reason for the growth in exports lies first and foremost in white cement. Its exports rose from 117 kt in 1981 to 536 kt in 1996, and will rise still further in 1998. But increased exports of standard grey cement have also greatly influenced developments, above all from 1992/93 onwards.

The decline in production was therefore particularly sharp in 1980/81 – the decisive shift – corresponding to a change in demand on the domestic market, upon which the production companies were wholly dependent. Domestic demand has remained at this low level, and the company has gone to the world market to find customers for the production lost in 1981. And management has been successful, taking production back to pre-crisis levels.

3.2 Demand Side Reasons for the Decline in Cement Consumption

The fall in the number of new buildings (in terms of area covered) completed per year is very significant, especially in 1979–81, as can be seen from Fig. 1. The curve showing cement supply to the domestic market demonstrates the parallel development of supply and demand. The supply side adjusts rapidly to changes in the market, which corresponds well with the fact that the product cycle of cement is very short indeed and that the supply side is dominated by a single producer.

In addition, business cycle fluctuations have become far more marked in the wake of the first oil crisis. Although the cycle runs more or less parallel, some dislocations can be seen. One starts in 1981, where demand recovers slightly, peaks in 1986 and falls to a new low (at the level of 1959) in 1993. The other is since 1993, where demand begins to recover again.

The first of these periods, 1981–93, shows cement supply and construction running parallel to one another, but where the line for cement supply was previously above or at the level of the one for m^2 completed, it then falls below. The reason for this can be found in distributional changes among the four construction groups shown – cottages, non-business, business and residential. Business construction remains at the previous level or grows in relative terms (in 1984/85 even in absolute terms), non-business construction also remains more or less stable, but residential construction is down to half or even less at some stages. Business and non-business construction appears to require less concrete – and thereby less cement – than residential building.

In the second period the exact opposite is the case – the supply curve rises considerably more sharply than that for construction. The reason can be found in major civil engineering projects – the Great Belt and Oeresund bridges. The Great Belt bridge and its associated railway tunnel alone used some 350 kt of cement in 1993 and 1994.

The effects of this unique situation can also be seen in Table 1, which shows domestic cement consumption from 1977, prior to the profound changes in 1980/81, up to the present day. The on-site consumption of bulk cement rises very sharply in 1991 and the next couple of years, precisely due to the Great Belt project. Apart from this, ready-mix concrete has supplied about a third of bulk cement consumption over the whole period, with a slight increase during the turbulence of 1980/81, but remaining principally stable during the entire 25–30 years.

Pre-cast concrete has been an area of steady growth, from 10% in 1977, 12–13% in 1981, to 21% in 1995, at which level it has more or less remained. Flagstones has been a fairly stable area as well, edging downwards a little, ending at some 6% lower on 19% in 1995. The decline in demand for roofing plates is due to a shift to asbestos-free plates in the mid-1980s, through which the plates became lighter and contained considerably less cement. Finally, bagged cement experienced more or less the same decline as flagstones: consumption started at 23% and now stands at 18%.

Part of the explanation for the decline in flagstones' share of cement consumption is that there is 20% less cement in any given type of concrete today (Table 2). This average estimate is even exceeded for what is a large share of concrete. A new type of cement developed in 1992 (base cement, see section 2.1, p. 326), which meets the basic requirements for concrete, including most of those for flagstones, contains more microfiller, which reduces the clinker content per kg ready cement (and clinker can be burnt at a lower temperature, therefore at a lower energy cost than standard grey cement previously).

Table 1. Cement consumption in Denmark by purpose 1977–95

	purpose	percent 77	80	81	86	91	95
bulk cement	at site	1	1	1	<1	8	1
	ready mixed	33	36	34	37	32	33
	flagstones	25	22	22	21	17	19
	pre-cast concrete	10	13	12	17	18	21
	roof plates	8	6	8	6	4	5
	miscellaneous	<1	1	3	1	3	3
bulk cement total		77	79	80	82	82	82
bags total		23	21	20	18	18	18
cement total in Mt		2.1	1.6	1.2	1.7	1.3	1.2

Source: Cement Producers' Technical Information Office

Table 2. Reduction of cement per m^3 concrete in Denmark 1977–95

greater and more uniform strength	2–4
fly-ash substituting cement	8–10
micro-silica substituting cement	3–5
addition of dehydration agents	1–3
improved control of the concrete mixing process	1–3
total reduction of cement in 1 m^3 concrete	ca. 20

Source: Ålborg Portland (1992)

It should be noted that other cement-based products apart from roofing plates have experienced a reduction in quantities of building materials used, e.g. the precast concrete sector. The construction elements are built lighter to save materials and money, and to make them easier to handle. On the other hand, a major push for substitution of concrete by totally different materials has not appeared from any quarter.

3.3 Environmental Benefits of Reduced Production, Innovation and Ecological Modernisation

As the main environmental impact from the production of cement is caused by the actual amount of energy required, principally for burning clinkers, which takes place at 1 450 – 1 700 °C, a drop in production means a parallel reduction of environmental impact. And the environmental impact consists mainly of the emissions to air caused by burning fossil fuels – 500 kt/a, of which coal and pet-coke are the main components, contributing some 260 kt/a and 160 kt/a respectively.

Over the entire period since 1979, the amount of white cement produced has been growing, as has its share of total production. As the energy consumption in J/g for white cement is more than twice that of grey or standard cement, this factor is of great importance to the overall environmental impact of cement production. In 1979, white cement accounted for about 3.4% of total production, while in 1996 its share had risen to 20.8% (Table 3).

The environmental benefits of the fall in domestic consumption in 1980/81, from 2.1 Mt to 1.2 Mt, was mainly due to the equivalent reduction in energy input, 44% or around 6 million GJ. As CO_2 emissions at that time, for example, were about 1.1 tonnes per tonne standard cement (Table 3), the reduction in CO2 emissions was over 1 Mt.

A number of innovative investments have partly kept these benefits – and partly even enlarged them further – as the production level went up again in the following years. The company itself points to substantial investment in production technology – about 1 billion DKK or $140 million over the last 20 years – which has benefited the environment. As the major environmental impact is associated with energy consumption, which is again a major cost component in cement production overall, any investment made to reduce costs through savings on energy will automatically constitute environmental investment. Thus all environmental investment by the company has been a win-win situation, which of course does not reduce the significance of these investments for the environment. But it should be remembered when the company tells the story about this huge 1 billion DKK environmental investment. These investments have made the energy efficiency continue to improve from the 1981 bottom level of production.

The reduction in energy use between 1981 and 1996 comes to 51% or 3.85 PJ. This is a larger relative reduction in energy consumption than the fall in 1980/81. In absolute terms, the improvement is also significant, although less so, as the base position – supply volume in 1981 – was already 44% lower. For the whole period 1979–96, Table 3 shows that energy consumption in J/g has been reduced by 13.1% for white cement and 34.5% for standard cement.

This development has positive implications both for domestic cement consumption, and for export production, which comprises almost all white cement and a substantial proportion of standard cement as well, taking the overall export share of production to over 50% (56% in 1992 and 54% in 1996). With the inclusion of exports, production was higher in 1996 than in 1979 (2.6 Mt as opposed to 2.4 Mt), but the energy consumed for the 2.6 Mt in 1996, where white cement made up 26% of production, was 2.1 million GJ, or 13% less than for the 2.4 Mt in 1979 (when white cement accounted for 3.5% of production).

Table 4 shows the considerable effect of the heat recovery technology, which has been introduced for the white cement ovens. The heat is delivered to the dis-

Table 3. Energy consumption and flue gases in Danish cement production 1979–98

	1979	1981	1983	1989	1992	1996	Reduction absolute	in %
white cement								
production (kt)	83	117	294	405	426	547		
energy used								
in the ovens (J/g)	10 251	n.a.	n.a.	n.a.	n.a.	8 909	1 342	13.1%
in clinker grinding (J/g)	n.a.	n.a.	n.a.	176	115	99	77 (89–96)	43.8%
grey cement								
production (kt)	2 326	1 486	1 363	1 594	1 646	2 082		
energy used								
in the ovens (J/g)	6 340	n.a.	n.a.	n.a.	4 155	n.a.	2 185	34.5%
flue gases								
CO_2 (kg/t cement)	1 145	n.a.	n.a.	n.a.	1 014	1 001	144	12.6%
SO_2 (g/t cement)	n.a.	n.a.	n.a.	n.a.	1 244	935	309	24.8%

Sources: Own calculations based on company annual reports, green annual reports 96 and 97 and interviews at the company January 1998

trict heating system for the city of Ålborg, on the outskirts of which Rørdal, and the cement factory, are located. In 1996, 20% of exhaust heat was recovered and, after the major extension added in 1997, recovery rates will rise to about 25% from 1998. Although this does not reduce the energy used for production, it reduces the energy required by the district heating system, where it will be meeting between 20% and 25% of the total energy requirement. In environmental terms, benefits come from the expanding export component of production. (Whether all recovered heat actually yields environmental benefits depends upon the other sources of energy for the district heating system, i.e. the power stations, and the opportunities for finding alternative uses for their waste energy. During the summer period in particular, with low demand for heating, there may be problems.)

As regards flue-gases, where CO_2 is the most significant, since other gases can be extracted before emission, the development of environmental benefits is coupled almost directly to the fall in domestic demand, as the relative improvement here – an overall reduction of some 12% – is more modest, although still significant. Part of this fall in demand is due to cement being substituted by filler materials in the concrete production sector. The sizeable fluctuations in emissions of other flue gases are due to intermediate measuring problems and the installation of new scrubbers, which have been in place from the beginning of 1998.

Among these filler materials is sand taken from the bottom of the Limfjord – at Hals Barre, where the fjord connects with Kattegat. Because the fjord sands up at this point, the removal of a substantial part of the sand from here maintains (and improves) access to the fjord – and thereby to the city of Ålborg, including the factory itself – for the ships which bring in coal and take away cement for export. The environmental benefits here are limited to savings in the energy which would otherwise be required to keep the access deep enough, and to the reduction in the need for fresh sand resources elsewhere in the region.

The company was awarded the Northern Jütland County 1994 Environmental Award for its overall environmental performance, whereby the energy savings, including the heat recovery system and the reuse of waste materials were the main factors behind the award. The award also tells a tale of fairly amicable relations with the environmental authorities.

Table 4. Heat recovery from white cement production

	1992	1996	1998[a]
Delivered to Ålborg city district heating (GJ)	577 191	968 458	1 350 000
Total used (GJ)	3 959 717	4 872 540	5 359 794
Heat recovery in %	14.6%	19.9%	25.2%

[a] estimate after further technical changes

Sources: Own calculations based on company annual reports, green annual reports 96 and 97 and interviews at the company January 1998

3.4 Social and Economic Impact

The most important consequence of the decline in production has been the reduction of the workforce, which stood at 2 473 in 1973, before the first oil crisis, and had fallen to 1 433 in 1981, after the full impact of the second oil crisis – a reduction of more than 1 000 employees or 40%. Before its final cuts in the workforce in 1981, the company operated a system of periodically shutting down completely for some weeks while keeping the employees on standby, for if and when the market grew and demand rose again. This was made possible by Danish unemployment benefit regulations, which allow companies to "park" the workforce on benefit for short periods. But that was abandoned after a year, as market prognoses remained pessimistic.

The 1981 level of employment was then maintained during the 1980s, until a new cut in the workforce took effect in 1991, a trend which continued throughout the 1990s, with employment in cement production falling to 834 in 1996. This was a drop of 477 relative to 1990, another 35%. Moreover, the number of employees in 1996 had fallen by two thirds compared to 1973, while production was at 90% of 1973 levels.

These figures indicate the two key factors in the social consequences of changes in Danish cement production. Firstly, it is clear that a dramatic increase in productivity has taken place. A third as many employees in 1996 produce about 90% as much as in 1973, which in turn points to the second key factor, a profound restructuring of the Danish cement production industry. This had already been set in motion in 1972/73 and carried out for the main part in 1975, when two of what were then five production sites were closed, while a third site switched to an alternative production. That was therefore the year when the largest single reduction in the number of employees took place. The workforce was reduced by 400, some 15%, during 1975.

This profound restructuring was able to take place because, from 1974, all major production sites were controlled by one company, Ålborg Portland A/S, although one site was still owned by another company. By the end of 1978, the fourth site, belonging to the other company, was shut down and, apart from very small quantities produced from time to time at the third site, all production was concentrated at the Rørdal Site from 1979 onwards. The company which had owned the fourth site got out of cement production altogether.

Only in one of the cases could the closure be said to have had a noticeable regional impact: the Dania site, close to the small town of Mariager, as there is no obvious industrial base here apart from the cement factory. An immediate closure in 1975 would have increased the rate of unemployment in that area to over 30%. After ceasing cement production, it was kept running for bauxite production and some test production runs for the semi-dry cement process. Employment fell gradually, and the site was shut down for good in 1984, by which time there was only a small number of employees left. The gradual closure together with fairly rapid and substantial industrial development in the nearby town of Hadsund made the regional impact of the shut-down far smaller than had been anticipated in 1975.

In general, the economic and social consequences of the decline in cement production were not substantial at national or regional level. The industry is not large enough to cause serious problems at these levels, and the actual existence of cement production in Denmark has never been in question. The consequences primarily affected the employees and the two companies which were active in the industry at the start of the restructuring.

3.5 Summing up environmental benefits and social impact

The decisive fall in Danish cement consumption came in 1980/81. It was influenced by the recession, following the second oil crisis in 1979. In 1980/81, investment fell by 35%, Denmark changed from a net importer into a net exporter and overall growth fell to 0. Residential construction fell by two thirds from 1979 to 1982 and has never regained 1979 levels, whereas construction for business and for the public sector fell by 40%, but had recovered to former levels by 1984. Large civil engineering projects influenced cement consumption in 1993–95, and still do, to some extent. But cement consumption has remained stable at 1981 levels, despite the fact that the general economic situation has changed completely, with all economic indicators pointing to a strong economy, because housing construction still has not recovered, and only in 1995/96 did a slight upturn become visible, one which has persisted in 1997.

Another reason for continuing low cement consumption lies in research into concrete technology, which has brought, for example, a cumulative reduction of 20%+ in cement used per unit concrete, relative to the situation before the decline. Material substitution has also had some effect, albeit with less significance for the level of cement consumption than might be expected. It is very hard to quantify the influence of this factor.

The dramatic rise in oil (and subsequently all energy) prices has also meant a substantial increase in the costs of the very energy intensive cement production. A long-term, still ongoing, cost-reduction exercise was initiated within the industry. It has targeted energy-saving measures, which has in turn resulted in substantially reduced impact on the environment. For grey (standard) cement, energy-efficiency has been raised by around 35% per tonne cement, by about 15% for white cement. Taking into account the large heat recovery system associated with white cement production, the energy efficiency of white cement is now even slightly higher than that of grey cement.

Improved energy efficiency represents a competitive advantage in the international cement market and has led to steadily growing exports of both white and grey cement, especially during the 1990s, taking factory production back to pre-1979 levels. Thus the environmental benefits of energy-related cost reductions are even more important from a local, national and global environmental perspective.

4 Social Management of Structural Change

4.1 Three Phases of Restructuring

The restructuring of the Danish cement industry had three main phases. Phase I was in the 1970s, where production sites were cut to one factory and all production was under the control of one company – Ålborg Portland Ltd. – by the end of the decade (in reality from 1974). This restructuring was influenced by the first energy crisis in 1973/74, but was primarily the result of the sector adjusting capacity and cutting costs.

Phase II took off in the wake of the second energy crisis in 1979, where energy costs rose to account for ca. 33% of cement production costs. The energy crisis was also a key factor in a virtual collapse of the construction sector and the resulting 50% reduction in demand for cement within a year. An immediate consequence was lay-offs and a complete halt to production at three of the ovens. But it also signalled the start of a decade-long battle during the 1980s to reduce energy costs per unit of cement, in order to remain (or become) competitive and stay in business. That meant huge investment and significant technological developments.

Phase III began during Phase II, but picked up at the start of the 1990s, as Blue Circle Ltd. of the UK bought up over of 50% of the cement company's shares. Together with a new management in 1993, this meant that capacity use at the plant was optimised thoroughly, and as domestic consumption remained low, this was again based on expanding exports, mainly of white cement, but also of grey cement. This phase has been very much about raising the overall efficiency of the plant, focusing on organising labour and out-sourcing, while maintaining and also extending the savings in energy costs from Phase II. Growing green taxation in the 1990s affected the profitability targets of Phase III, and the company put considerable efforts into preventing the CO_2-tax, which took effect for industry in 1996.

We will focus on Phases II and III as Phase I involved little which was central to the issues studied in this book.

4.2 The Actors and Their Interests

4.2.1 The Company

The company Ålborg Portland is the main actor throughout these phases. Backed by its parent company, Portland Holding Ltd., it had links to the Engineering Company, producing and exporting cement production equipment world–wide. But from 1990, i.e. in Phase III, with the take-over by Blue Circle, ownership becomes more significant to the company's strategies and actions for expanding its business opportunities. The company's interest is to remain in business and make that business as profitable as possible. This constant global priority is expressed differently in Phases II and III.

Phase II: Energy cost reductions and production capacity adjustment. In all of Phase II, the key issue was reducing energy costs and adapting to the continual sharp fall in domestic demand. Capacity was adjusted in two steps. At first, the company adopted a "wait and see" strategy, in the hopes that the slump would prove short-lived. When, after the first couple of years, that hope evaporated, three ovens were permanently closed, thereby reducing capacity to match market demand, including the – expanding – export market for white cement.

The savings in energy costs had three main components. The first one was *raising the energy efficiency of existing processes*. Raw material substitution became important in this respect. Initial tests, made in 1973, on replacing the 12% clay input with fly-ash from flue-gas cleaning in power generation, were taken up again, and by 1980 this substitution was complete for the grey cement lines.

The second component was building a new, high capacity oven for grey cement, based on the *"semi-dry" process* as opposed to the "wet" process used till then. Substituting clay with a dry material is crucial for implementing the "semi-dry" process, which requires removing less water from the raw materials in the pre-heating process, from which the main energy savings are gained. Since 1988, all grey cement has been produced on this new, single oven.

The ovens used for pre-heating were not decommissioned, but adapted for producing white cement, which still requires a wet process, to exploit the unique "whiteness" that the chalk in the company's chalk mine offers. This brings us to the third component, as *heat recovery* became the major energy efficiency initiative for white cement production. With constant expansion of white cement production, heat-recovery has also grown from more than 20% to over 25% of the energy fed into the process as primary energy input – 42 MW to 70 MW in absolute terms – most recently through improvements in the system in 1997. This heat covers more than 30% of heat required by Ålborg's District Heating system, to which the recovered heat is delivered.

The energy savings continue to be an important issue for the company, which is looking for more possibilities in this area. However important these initiatives have been, it should be noted that the most important contribution towards energy cost savings in the 1990s was the considerable drop in energy prices over the whole decade. But this may change again at any time, and energy which is no longer consumed provides the most lasting cost reduction.

Phase III: Management revamp and focus on customers. While continuing to benefit from cost savings on energy, phase III saw the previous technologically oriented approach being replaced by a strongly management and marketing oriented approach, which represented important shifts in the company strategy. In a January 1995 "Newsletter", the newly appointed managing director announced "new approaches" with "focus on our customer" as the key to the company's strategic reorientation. The whole organisation was directed at this strategy, and at expansion in the export markets as well as keeping control of and capturing any possible growth in the home market.

We shall concentrate on the influences on the production side, leaving aside the actual marketing initiatives and their content for now, apart from noting that the expansion of production by 30% or some 700 kt from 1991 to 1996 was sustainable in 1997, after the bridge and tunnel construction on the Great Belt was finished. The marketing initiatives had therefore succeeded in this sense.

Production has been maintained, using capacity for producing grey cement in the semi-dry process to the full, and somewhat expanding production of white cement by modifying existing ovens. Optimising full capacity may mean selling grey cement at low prices from time to time, but offers possibilities for cost-minimisation per unit which are difficult to achieve under constant changes in capacity use.

In this respect it is probably significant that Ålborg Portland has lower production per man-year than other Blue Circle production sites. A number of complimentary initiatives were taken by the new management over the 3–4 years to help increase the plant's productivity.

Major efforts have been put into reorganising the workforce's qualification structure, aiming at creating "multipurpose" or general employees, able to operate the core processes in the production system, which are today largely a matter of control room surveillance combined with reacting to any disturbances, using the back-up personnel present, but having other duties. This "process-operator" qualification is an industry-wide initiative and has been accepted as a formal apprenticeship, offered to young people as a 4-year course and to employees as a 2-year further education programme. The company has a collective agreement with its workforce and the General Workers Union, represented by its Cement Workers Branch, where responsibility for operating processes used to lie with craftsmen and/or technicians. A single-union approach seems appropriate for the "multipurpose" employee now established at other Blue Circle production sites.

The use of outsourcing for non-core activities, even non-core services, has a similar reason. From January 1998, internal company transportation was contracted out to a transport and construction company, moving 25 jobs out of company pay roll. A certain amount of technical services will always be bought in for peak loads and for tasks requiring specialist qualifications. This might lead into direct outsourcing of all technical services and maintenance, but this is not the case as yet.

4.2.2 The Trade Unions

Trade union involvement has been considerable over the entire two phases, starting with the lay-offs in 1980 and 1981, and becoming important again in the latter part of Phase III, with the new capacity optimisation strategy taking effect from 1993/94. The issues here have been employment levels, job content, working conditions and, particularly in recent years, new health and safety (H&S) issues related to alternative raw materials; the environment has been a minor issue for trade union representatives. The most important trade union, the Cement Workers

Branch of the General Workers Union (SID), has links to other regional branches and to the union's national level, but only with regard to employment and H&S issues only.

The main issues have been staffing levels, job content and job security. These themes have been important at all stages, although problems arose for different reasons. Working conditions (H&S) have also been an important issue throughout, but here the options have also changed and the influence of the unions has increased. We will look briefly into each of the topics across all phases, concentrating on the Cement Workers Union, especially when we examine the production side, as they represent the majority of employees,

Employment levels. The number of jobs in the company has fallen steadily, with the biggest single reduction coming during the crisis years of 1980/81, where in the range of 800 jobs were lost across the board, but mostly in the production sectors. Some 2 200 employees produced about 2.7 Mt cement in 1978, while by the end of 1997, some 750 employees were producing exactly the same amount. Job losses have occurred gradually over the years, with a period of stability at around 1 150 employees between 1990 and 1993, where the production was rising due to increased domestic consumption for the Great Belt bridges and the tunnel, but the fall resumed in 1994, with the new management taking office in late 1993.

The increased use of alternative raw materials in the clinker burning process, as well as alternative energy sources in the form of solid waste, means more pretreatment, but this is to a great extent automated and has in itself failed to create new jobs as yet.

With continued rationalisation and the push for higher production per man/year to match other Blue Circle factories, there is no reason to believe that employees (or their trade unions) have seen the end of job losses.

The union has had little opportunity for strong opposition to this development. In the 1980/81 crisis, job-sharing was introduced during what was an extraordinary situation, but this was discontinued after a year, as no improvement could be seen. The union is generally opposed to job-sharing, as their experience is that jobs are always lost in the end anyway, and the job-sharing period only makes life and conditions worse and less secure for all employees.

Job Content. The union has always made considerable efforts to enhance job content and the responsibilities for its members, who tended to be unskilled or semi-skilled manual labourers around the core process of cement production. With the new management in 1993 and the search for the "multipurpose" worker, together with new training for Process Operators, the union had a chance to bring its members in from the margins to the heart of the process, and to fulfil all functions except the skilled maintenance functions. The union exploited this opportunity to the full, bringing the process operator's functions into their part of the collective agreement, regardless of the employee's professional background. The union therefore has all main production activities, except maintenance, covered in its agreements. That offers its members the prospect of enhanced job content, with the adult further education option as Process Operator an especially promising element. The dilemma for the union is that it is encouraging the creation of a multipurpose grade of employee, thereby tending to remove one of the historical pillars of trade union influence (and job protection) – the monopoly of each trade and craft on its own area of work.

With the arrival of new management, and with it the new focus on customers and on lean and efficient organisation, each individual employee gained greater influence and enjoyed more self-determination. The hierarchical structure was flattened, and most foremen and some of middle management were removed to save wage costs, but also to allow for more influence at shop floor level, reflecting the quite considerable responsibilities placed on the shop floor worker in many operations in the complex, technology-based production process. The union has responded positively to this management initiative, and the convenor and union officials see definite improvements for members as well as for the union's influence on job content in the wake of these changes.

Job Security. The best guarantee of job security for the union lies in job content, giving their members a central position in all production functions and thereby also some control over their own work. Job security has therefore been an important factor for the union in its consideration of options and possible strategies in response to management policies since 1993. In addition, protection against or at least compensation for unfair dismissal is a traditional element in Danish collective agreements, with grievance procedures, also a part of the agreement with Ålborg Portland. In Denmark, however, these arrangements give no protection, against dismissal on the grounds of restructuring or reduction of business activity. And this is where the strategy of making employees indispensable becomes attractive, although it provides no absolute job protection either.

Health and Safety. Cement industry is heavy industry, which means the workplace is considered "risky". Cement-induced eczema was a major problem at the factory but also – and even more so – for workers in the concrete and the construction industries. This was handled by Ålborg Portland, who developed an anti-chromate treatment, to which all cement has been subjected since 1985. That process has turned out to offer a useful response to market dumping from CEE countries, as they do not produce chromate-free cement.

In Phase II, the H&S issue concerned the alternative raw materials, first of all fly-ash, which is a very fine powder and therefore considered dangerous when inhaled. Research is now being conducted into the amounts and the content of what is being inhaled. Various measures have already been taken in relation to storage (a huge shed with 100 000 t capacity) and feeding the ovens.

The union has had a number of traumatic experiences with toxic substances in production, albeit not at Ålborg Portland, but at the Eternit Roof Plating factory, also situated in Ålborg and owned by Ålborg Portland. Here asbestos was a key raw material until 1978, and a number of trade union members have died over the years from diseases caused by inhaled asbestos fibre. The union fought long and hard on this issue (as did unions world-wide), and the Cement Workers Union in Ålborg does not want to take any chances with the health of its members again – with fly-ash and with other possible raw materials now in the pipeline.

At factory level, the H&S system which the company is obliged to operate and which has full worker participation, is also dealing with environmental issues. Until now, however, there are only a few examples where environmental issues gain a high priority in the H&S system, or where these matters are handled with the same vigour as "pure" health and safety issues.

4.2.3 The NGOs

NGO involvement has been very modest in both phases, mostly through the Ålborg Branch of the Danish Nature Conservation Association. Their involvement is guaranteed in the Environment Act, as regionally registered organisations have information access rights. Their efforts were most evident in the use of alternative raw materials – fly-ash and used car tyres – because of the possible impact on the local and broader environment. Permission was granted to use both, despite the NGO's reservations and protests, but while fly-ash has become very important indeed, the Danish cement industry has not yet exploited its permission to burn tyres in the cement production process. There was no evident involvement at the NGO's national level, nor by other significant NGOs.

5 The Influence of Environmental Policy

5.1 Convergence Between Environmental Policy and Company Interests

Apart from their welcome involvement in the pricing system in early 1980s, at the peak of the problems, the company wanted the authorities to refrain from taking action on particular problems encountered with the slump in demand and the dramatic rise in energy costs in Phase II, and to focuse only on areas and issues where licensing and/or control was required by legislation. This applies for the environment, where all major changes in the production system, including the use of traditional and alternative raw materials, need a licence from the regional authorities, as cement production is the responsibility of the County Council. Other regulations concern factory inspections under health and safety legislation, but this is routine and did not change during the two phases considered here.

Extracting virgin material – including ground water – needs permission from the County Council. So does the discharge of waste water and the use of alternative, processed, raw materials, and various parts of the process need a licence to operate. The company obtained permission for all of this as required, without major battles at any point. As shown in the analysis above, the changes the company made, and therefore the licences they asked for, were oriented towards reducing consumption of energy and exploiting waste or other alternative materials, which would themselves represent environmental problems if not disposed of. Environmental authorities therefore view requests for such licences positively. The official opinion of the company's environmental performance was underlined when the County awarded the company their "Annual Environmental Award" in 1994, for sustained good environmental performance, in particular with regard to energy use.

The tax levied on landfill disposal of solid waste since 1989 has increased twice since then, providing the company with an economic incentive to involve itself in alternative materials other than fly-ash. That has led to the use of some 60 kt of waste paper per year, and now other waste with a high dry content and high calorific value, with a target of some 100 kt/a over the coming 2–3 years. Used tyres were also considered, but they went in the end to the Swedish cement production plant on the island of Gotland, which appeared cheaper to the tyre sector. This area of handling solid waste, i.e. supplying the cement production process with a parallel "by-process" – an incineration plant – has not much to do with the product itself or the process of making the clinker for cement. It is a question of earning money as a waste incinerator from incineration which would take place anyway – as long as it does not harm the quality of the cement produced.

On the energy side, policies exempted industry, in particular the major energy users, from all energy taxation, which had been going on since the 1970s in the wake of the first energy crisis in 1973/74, finally resulting in industrial energy prices which were 60% lower than those for households. With inflation, and the dramatic drop in fuel prices since the 1980s, this tax holiday has meant a substantial indirect subsidy to industry, in particular to major energy users like Ålborg Portland Cement.

5.2 The CO_2 Tax Represents a Divergence of Interests

CO_2 regulation started in 1990/91 with a parliamentary decision, forcing the then conservative led government to draft a law to come into effect for households by 1st January 1992, and for industry by 1st February 1993.

The "green" parliamentary majority in May 1991 consisted of the Social Democratic party, and the Social Liberal Party, the Socialist Peoples Party. The proposal came out of a long debate on the Greenhouse issue and the Toronto declaration to reduce CO_2 emissions by 20% relative to 1988 by the year 2005. The three parties in the green majority did not consider the goal realistic without political initiatives, like the CO_2 tax, creating incentives for industry, the biggest CO_2 "producer" anyway, to do something.

The government and its parties voted against the proposal – and against the act when it was passed in 1991. It brought household and trade, commerce and services under the tax at a level of 600 DKK/t CO_2, but left industry out, contradicting the original proposal. The final decision–making process, where Ålborg Portland was very active in trying to avoid inclusion under the act, finally exempted the processing industries, although the non-government majority had actually offered a compromise on how to include industry at this early stage. Accusations were aired afterwards that the conservative minister for energy, due to the poor drafting of the law and her basic opposition to it, was to blame for the final unclarity, and for exempting industry. This initial victory by the production sector created a rift within business, which emerged again in the 1995 amendment of the law.

The political scene had shifted when it came to the 1994–95 discussions on the review of the 1991 CO_2 tax law. Major parts of the green majority had become part of the government in January 1993. The Social Democrats were leading the government, with the Social Liberal Party as the main partner. But instead of the third party of the green majority of 1991 – the Socialist Peoples Party – a small centre-right party (CD), known for a lax attitude towards environmental protection and directly opposed to the CO_2- tax, was now in the government coalition.

The difficult political situation was fuelled by the fact that there was still a green majority in parliament for extending CO_2 regulation to industry, with the Socialist Peoples Party and a new Socialist Left Party favouring strong "green" policies, including tough CO_2 taxes, if the government dared to propose them.

Industry, in the shape of the CBI, was also divided, between the interests of the energy-intensive part of industry (12 factories consume some 40% of all industrial energy) and the remainder, which were going to shoulder most of the burden, as they would have little chance to avoid paying the full rates. As it became increasingly clear that, despite the government's political problems on the CO_2 issue, some action was likely to be taken on industrial processing, the CBI could not stand by the closure argument, which was obviously not valid for the greater part of Danish industry, with energy costs making up 3–7% of total production costs. They therefore became increasingly involved in the discussion about how to return tax revenue to industry. But Ålborg Portland put up a string case themselves, although they were not exempted from the CO_2 and the SO_2 tax altogether.

5.3 Ålborg Portland's Position

Ålborg Portland sees the CO_2 tax as unreasonable and unfair taxation, primarily because the firm has invested some 850 Mio DKK in energy saving technology, which over a period of 13 years (1979–92) has meant energy savings of some 33% (see Table 3, p. 334) In a 1991 report on the CO_2 issue, which was part of the company's lobbying efforts against the tax, the company stated that "There is no technology today which will be able to reduce specific energy consumption at Ålborg Portland, nor is anything currently under development".

Despite this, initiatives established in agreement with the Ministry of Energy in 1996 (see below) actually led to a further reduction of around 7% in energy consumption. A statement like the one above, where a major company claims that technological development has come to an end, cannot be sustained for long.

The real background to the protest, at the same time the key limit in the convergence of interests, is the competition issue: the problem of imported cement, some 15–20% coming from Spain and Central Europe, free of CO_2 taxes, but with higher transportation costs. The problem also applies for exports, as foreign competitors are not burdened by the CO_2 tax. The company overcame far greater energy cost problems in the 1980s than those represented by the CO_2 tax, proving that incentives from increased costs do work, and can produce substantial energy savings. How ever, the politically created cost curve is necessarily seen as "filling artificially created holes", with the end result for the company being a return to where they started, at best. In other words, the incentives provided are not attractive, because they do not contribute to company profitability.

In 1995, during the new campaign to avoid the CO_2 tax, the company reviewed and extended its 1991 report to reinforce its stance. Again, the merits of the company were listed in a letter to the Parliamentary Committee, sent while the Committee was reviewing the draft law for its 2nd reading in Parliament. The report now included the main social consequences of the CO_2 tax, should it led to the closure of the factory:

- 250 kt fly ash would need disposal
- 64 kt of waste paper from paper recycling would need disposal
- 75 kt of gypsum would need to be used elsewhere or disposed of
- 10 kt of waste sand from sand-blasting would need disposal
- 1 kt of iron-oxide sludge would need to be burned elsewhere
- 20 000 households in Ålborg, heated by recovered energy from Ålborg Portland cement production, would need to be heated by primary energy
- several thousand jobs would be lost in the region, starting with some 900 jobs at Ålborg Portland
- the national balance of payments would fall by 1 000 Mio DKK
- energy saving and environmentally beneficial plant worth some 850 Mio DKK would be lost
- research and development of cement technology would stop in Denmark
- the Portland group's (F.L. Smith Industries) development of cement production technology would relocate abroad
- some 1.2 Mt of cement would need to be imported annually, further damaging the balance of payments – and also increasing global CO_2 and SO_2 impact from cement production, as Ålborg Portland consider themselves a world leader in energy efficient cement production

This apocalyptic worst case scenario, put to parliamentarians at the eleventh hour, was designed to shatter the resolve of even the most experienced politicians. But it was overdone and thereby – again, like the former "end of technological development" – doubted by politicians and a large part of the media.

5.4 The 1995/96 Energy Tax Reform – a Complex Combination of a CO_2 and an SO_2 Tax

The 1994–95 discussions finally looked at bringing industry and other production sectors into the CO_2 tax scheme, based on two principles:

1. a distinction between energy used for 1) heating and office use (as in the 1991 law for households etc. and with the same 600 DKK/t CO_2), and 2) production processes, which were again divided into a) light processes, using fairly limited amounts of energy and b) heavy industries with high energy consumption. The proposed tax rate was 100 DKK/t. For both types of tax and rate, a gradual introduction was planned, with various forms of compensation or subsidy (see categories "heavy process", "light process" and "other" in Table 5).
2. giving the revenue back to industry and users, partly as an energy savings support scheme (30% of the revenue) and partly in easing indirect labour costs and an improved depreciation scheme for new investments. It was thought impossible that more than the 30% would be spent on subsidising energy saving projects, as that, along with private investment, took the total amount to some 6.000 Mio DKK (this has proven correct, as there has been a surplus in the first three years).

Furthermore, a SO_2 tax was proposed as part of the package, at a proposed level of 15 DKK/t and limited to a period of a few years, inspired by Swedish experiences, where similar policies had proved effective in providing an incentive to reduce emissions. It was linked with the CO_2 tax in the discussions, as it was expected to lead to reductions in energy consumption, and in that way contribute to CO_2 reduction as much as an additional 70 DKK/t on the CO_2 tax. This was also convenient because Denmark had signed an international agreement to reduce SO_2 emissions by 80% relative to 1980 by the year 2000, the 1994 ECE Sulphur Protocol.

Table 5. Differentiation in CO_2 tax and compensation for different industrial purposes according to the CO_2 tax reform of 1995

Purpose	Year	1996	1997	1998	1999	2000
with special agreement	compensation	97%	97%	97%	97%	97%
	actually paying tax[a] (DKK)	3	3	3	3	3
heavy process	compensation	50%	40%	30%	20%	10%
	actually paying tax[a] (DKK)	50	60	70	80	90
light process	compensation	95%	90%	85%	80%	75%
	actually paying tax[a] (DKK)	5	10	15	20	25
other	compensation	70%	35%	0%	0%	0%
	actually paying tax[a] (DKK)	180	390	600	600	600

[a] per ton CO_2

The 3-tier CO_2 tax was passed in May 1995 and took effect from 1st January 1996. But for the energy intensive industries, with Ålborg Portland heading the team, a further compensation system was introduced, consisting of two elements: an agreement with the Ministry of Energy on energy savings, and entitlement to further compensation (This scheme is also open for companies in the lighter process section, provided that they have energy costs above a certain level, according to a formula linked to the amount of V.A.T. they pay. The number of companies from this group is small).

The agreement with the Ministry of Energy has precondition that the company present an "energy check" on the whole production process, a report on energy management and a plan for reducing energy consumption. On this basis, negotiations on the agreement are taking place with the Ministry, with three elements: 1) energy saving projects, following up on the options found in the initial energy check, 2) projects on structural or other major change to permanently reduce the company's energy consumption 3) introduction of systematic energy management activities in the company, possibly combined with an overall environmental management system. The agreement can run for three years and must then be renewed, if the company wants to continue to receive the extra compensation.

The investment projects of the second type are permitted to pay for themselves within 4 years for the heavy process firms, while the firms with light processes and access to the scheme accept a period of 6 years for their share in the investment. These extended terms, compared to the 2–3 years normally accepted by industry, mean that a lot more energy saving projects will be implemented. For the investments included in the agreement, the company will receive a 30% subsidy of the total costs, provided the total investment exceeds a certain minimum threshold. This premium comes from a pool of 30% of the total CO_2 and SO_2 tax revenue, as specified in the law. The agreements can be made with branch organisations, groups of companies or with individual firms. For companies under this "special agreement" scheme, compensation will be 97% of the industry standard of 100 DKK/t until the year 2000, when the law will be reviewed (Table 5).

When the law was finally passed, the SO_2 tax was set at 10 instead of 15 DKK/t, due not least to lobbying by industry and the complicated political situation involved in getting the whole taxation package through parliament. Still, there is also an incentive for major users to do something about either 1) reducing energy consumption, or 2) using fuel with lower sulphur content or 3) improving the flue gas cleaning, as there is an "end-of-pipe" solution for SO_2 reductions which is not available for CO_2.

The introduction of the SO_2 tax was graduated, by giving all companies a basic exemption on SO_2 per ton of fuel (5.6 kg/t coal, 8 kg/t heavy fuel oil and 6.4 kg/t pet-coke), to be reduced by 25% of the initial rate each year after 1996, and therefore fully phased out by year 2000. This gives the companies time to evaluate their situation and the three options described above.

In short, the package with the two key reductions – CO_2 for major users at 3 DKK/t and SO_2 with a special deduction on Ålborg Portland's pet-coke consumption – made it very difficult for the small Centre-Right party to use this issue against the government, having to accept that it was vital for the other two – bigger – partners. At the same time, the party could abstain, as the Socialist Peoples party and the Socialist Left party were willing to support the law, although they found the final package unsatisfactory, for different reasons (see 5.6, p. 349). The majority for the law was therefore secured.

The Conservative and Liberal parties, supported by the small ultra-right party, have promised to repeal the law when next in office, and every year they seek to repeal the law in Parliament while in opposition – the latest attempt being in December 1998.

5.5 What Did the Political Outcome Mean for Ålborg Portland?

The economic interests in this law can be illustrated by the amount of taxes the company would have to pay if the 100 DKK/t applied (Table 6). At the same time it shows the environmental impact of cement production. This level of taxation amounts to some 15% of turnover, some 5 times higher than the net post-tax profit for 1994–95 (in relation to 1997 profits, only 1.5 times, as profits grew) and equals some 2/3 of the company's annual wage bill for its 800+ employees.

The tax actually paid by Ålborg Portland is shown in Table 7. It amounts to some 6% of the taxation level if the full 100 DKK/t CO_2 and the normal SO_2 tax level applies to Ålborg Portland, and shows that the law has a substantial "Lex Portland" component. Investments will have paid for themselves in 4 years, adding to company profits from then on, and from day one will mean reduced energy consumption and thereby a reduction in taxes paid – fully in line with the idea behind the investment support scheme, and indeed the tax scheme as far as its environmental perspective goes.

Thus, even at a company which has received extraordinary concessions in CO_2/SO_2 taxes, the policy actually works! The tax has led to an agreement which yields all the cost savings and reductions in environmental impact the parties could hope for, as far as its – limited – application has shown up to now.

5.6 Is the Environmental Tax About the Environment or Shifting Tax Burdens?

This tricky question was central to the parliamentary debate, and placed what is a general problem firmly on the political agenda. We shall limit discussion here to the core elements as they relate to the law in question, and leave both a detailed discussion and a wider elaboration of the general issues for another occasion.

The question raised was whether the CO_2 and the SO_2 taxes were intended for the benefit of the environment. If so, the entire revenue was available for environmental initiatives by companies, and also for the households and trades, businesses and services, paying the tax under the 1991 act. However, the fund set

Table 6. Hypothetical CO_2 and SO_2 taxes to be paid by Ålborg Portland

	amount of fuel (kt or GWh)	taxes (Mio DDK)
CO_2 tax[a]	all fuels	151.5
coal	271.6 kt	65.7
gas-oil	0.6 kt	0.2
heavy fuel oil	80.6 kt	25.8
pet-coke	87.4 kt	28.2
electricity	315 GWh	31.5
SO_2 tax[b]	all fuels	18.8
total	all fuels	170.3

[a] 100 DKK/ton CO_2, no reductions
[b] after reductions phased out

Table 7. The final result for Ålborg Portland of the CO_2 and SO_2 policy finally passed by Parliament in May 1995

Mio DKK	1996	1997	1998	1999	2000
Total tax	7.5	8.0	8.4	10.5	13.3
CO_2 tax	4.7	5.2	5.6	5.7	5.8
SO2 tax	2.8	2.8	2.8	4.8	7.5
Returned revenue from the tax on labour costs	-0.2	-0.5	-0.9	-1.1	-1.8
Net costs on tax scheme	7.3	7.5	7.5	9.4	11.5

aside by the act for subsidising company environmental investment has by no means been exhausted. As the subsidy can only amount to 30%, it requires companies to invest 70%, and as there are some 1 800 Mio DKK in the fund, company investment would have to reach 4 200 Mio DKK a year to exhaust it.

The Socialist Left party challenged this part of the concept. Their view was that if the full revenue is available for the benefit of the environment, the percentage of support for company investment could range from 25% to 100%, depending on the environmental importance or scope of the projects proposed. This would mean more projects, and all the money would be spent for the benefit of the environment, if not immediately then fairly quickly, and the tax and thereby the money available would be reduced as the environmental benefits were realised. Their point is that this policy could gain far more public support for green measures, especially when these measures hit industry and threaten jobs, as Ålborg Portland exploited in their report to Parliament.

An interview with representatives of the Socialist Left party revealed that they had actually visited Ålborg Portland during the political process and discussed the possibility of the company making more of their own electricity and including renewable sources like wind power. There has been no visible outcome of the talks, but the company had shown interest in the idea. The level of the possible grant would probably be crucial to the company's decision.

5.7 What is the Overall Impact of the Policy – and What Perspectives Are There for the Future?

The main reason for introducing a tax on CO_2 emissions by industry are international agreements (the latest in Kyoto has relaxed commitments, but the consequences for Danish policies on CO_2 are not clear yet) on a 20% reduction of CO_2 emission by 2005 compared with 1988. The Ministry set up a commission, which back in 1994 came to the conclusion that there would be a shortfall of some 5% without new political initiatives. An immediate goal of the tax is therefore to reduce CO_2 emissions by 5% – putting the whole debate with the Socialist Left in perspective. The ambitions were apparently quite modest.

In 1998, the 1995 act was evaluated. The evaluation has only covered the special section on the few heavy users covered by the agreement and the 3 DKK/t CO_2 scheme. The remaining sections have not yet been evaluated.

The evaluation of the agreement estimates that, with 1998 production levels, there will be a reduction of 440 kt CO_2 per year by 2005, which equals a 6% reduction in CO_2 emissions. However, more than half of this reduction is due to the expected effect of obligatory energy management by the companies. The evaluation is therefore ambiguous – suggesting that government aspirations have been met rather than supplying genuine documentation of environmental benefits. And in the meantime it has become obvious that the transport sector will increase its contribution to domestic CO_2 emissions substantially, instead of contributing to savings. The enormous infrastructure building programme Denmark has been experiencing since the early 1990s, with the bridge to Sweden to be completed by year 2000, explains part of this, although the picture is more complex than that.

Uncertainty is increased by the fact that companies are just about to introduce their management systems. The evaluation claims that it has taken this into account in the positive result. It also states that the results depend on the Ministry ensuring compliance by companies. Despite the uncertainties, the example from Ålborg Portland shows that this kind of policy can work and bring substantial environmental benefits. A discussion is needed on how to reap the full potential of a CO_2 tax policy. Thus, the discussion between the Socialist Left and the remainder of the green majority in the Danish parliament should be taken more seriously than has been the case so far.

References

Andersen, M. S. (1994): Governance by green taxes. Manchester.
CEMBUREAU (1997): Best Available Techniques for the Cement Industry. Bruxelles.

Cementarbejdernes Fagforening 1896–1996 (1996): (The Cement Workers Union 1896–1996, 100th anniversary 1996). Ålborg.
Clayton A., Spinardi G., Williams R. (1999): Cleaner Technology, a new agenda for Government and Industry. London.
Conrad, Jobst (1998): Environmental Management in European Companies. Amsterdam.
Danmarks Statistik (1972-1996): Foreign Trade and Industrial Commodities Statistics. København.
Dutch Ministry of Environment (1997): NL-BAT-Note on Cement Industry, Final Report. Deventer.
Energistyrelsen (1998): Evaluering af energiaftalernes effekt (EnergyAgency: Evaluation of the effect of the (CO_2-tax-related) Energy-agreements). København.
EU-Commission (1991–1996):Panorama of EC Industry. Several years.
Gouldson, A., Murphy, J. (1998): Regulatory Realities. London.
Holm, J., Kjærgård, B., Pedersen, K. (1997): Miljøregulering (Interdisciplinary Studies in Environmental Regulation). Roskilde.
Jänicke, M., Mönch, H., Binder, M. et al. (1992): Umweltentlastung durch industriellen Strukturwandel. Berlin: edition sigma.
Jänicke, M., Weidner, H. (1995): Successful Environmental Policy. Berlin: edition sigma.
Johansen, H.C. (1989). Gennem Forandring til Fremskridt - Ålborg Portlands udvikling 1972–1989 (Via Change to Progress – Ålborg Portlands development 72–89). Ålborg.
Johannsen O., Kjærgård B. (1986): Kalk – Produktion og Brancheudvikling. (Chalk – production and sector development). unpublished RUC.
Mol, Arthur P. J. (1995): The Refinement of Production. Utrecht.
Mez L., Jänicke, M. (1997): Sektorale Umweltpolitik. Berlin: edition sigma.
Miljø- og Energiministeriet (Environmental and Energy Ministry) (1999): Natur og Miljøpolitisk Redegørelse (Statement on policies for nature and the environment). København.
Ministry of Energy (1990): Energi 2000 – Handlingsplan for en bæredygtig udvikling (Energy 2000 – an action plan for sustainable development). Copenhagen.
UN, Statistical Office (1981 a.o.): Yearbook of Industrial Statistics. New York.
Ålborg Portland (1991–1998): Annual Reports, Green Annual Reports (1996–1998), Documents on CO_2-issue (1991, 1995), letters and reports to Parliament (1995), technical Reports on Cement and on Energy-Consumption etc.

Interviews

Ålborg Portland: environmental manager and energy manager
Cement Industry Technical Information Center: general manager
Danish Concrete Industry: statistical section
Danish Ready-mix Concrete Industry: statistical section
Cement Workers Union, Ålborg: chairman and Ålborg Portland convenor
Environmental Control Department, Ålborg County Council: head of industry section, environmental authority for Ålborg Portland A/S
power station *Nordjylland*: environmental manager (symbiosis with ÅP on chalk and gypsum)
Ålborg Municipal District heating: energy manager
Dansk Naturfredningsforening, Ålborg Afdeling: (Danish Nature Protection Association / Environmental NGO): chairmen for the period 1986–1998.

Oil Refineries in France

Samir Allal, Natacha Amorsi, Sylvie Faucheux and Julia Haake
C3ED, 47 Boulevard Vauban, 78047 Guyancourt Cedex, France

1 Introduction

After a 15-year period of regression, global oil refining capacity has begun to increase again since the end of the 1990s. This global change conceals considerable variance among the oil-refining countries of the world, since there are notable differences between local markets. While this growth is mainly to be found in the Asia-Pacific region, American refining capacity has remained constant, and Europe has entered a phase of reduction.

Oil refining in France is suffering from the same symptoms as its European neighbours, added to a number of specific handicaps. Its infrastructure and technological installations are outdated and unsuited to responding to the changed demand structure. In addition, the recent rise of environmental awareness in France is exerting strong influence on production processes via legislation or demand side environmentalism.

In Sect. 2 of this chapter we deal with structural change over the last twenty years in the French crude oil refining sector. The analysis is preceded by background information on the sector, concerning government influence, the companies, the economic importance and the environmental impact of this industry. Then we analyse the decline in production and consumption and compare it with developments in other countries.

In Sect. 3 we deal with the political decision-making process in relation to structural change in this sector, with special emphasis on the environment. One of the main ideas behind this study is to identify the key participants in decision-making in the oil refining sector, their goals, aims and strategies. This multi-stakeholder approach is supported with a series of face-to-face interviews[1] with the main actors, as well as on a thorough study of the relevant printed and electronic literature. Our approach enables us to explain why structural change in the refining sector over the last few decades has developed in a way significantly different to that in other European countries, and to hint at what the future might look like. In Sect. 3.1.1, we introduce the various stakeholders in decision-making on petroleum refining in France and their current constellation. Following this, we

[1] A series of eleven face-to-face, semi-guided interviews was conducted between June and August 1998, including the following actors: ADEME, DHYCA (environment division), Elf (environment division), Esso (environment division), Shell (external relations), Total (refining and distribution division), UFIP.

describe French energy policy, giving a brief overview of its three main phases in the 20th century (Sect. 3.1.2). Finally, we suggest some explanations for the fact that refiners have exerted relatively little influence on energy policy, although it has caused a decline in production by the sector (Sect. 3.1.3). Section 3.2 deals with environmental policy and its effects on structural change in the refining sector, in particular its potential to induce a decline in production. In Sect. 4, we deal with possible future scenarios for the refining sector, noting potential influences on structural change and discussing the positions of the refiners themselves.

2 Structural Change in the French Petroleum Industries

2.1 Market Structure and Refineries

The mineral oil refining industry in France is today composed of thirteen active mineral oil refineries. They are geographically concentrated in the South, the North and the East of France (Table 1). All refineries belong to vertically integrated oil-producing companies. Refining constitutes the most important section of these different groups: it functions as an interface between the exploitation/production sector and the distribution/petrochemicals sector.

The French mineral oil industry is characterised by heavy concentration. Six companies control the French market in refined products, although only the first two are French-owned: Elf Aquitaine, Total, Esso, BP, Shell, and Mobil (since January 1st 1997, BP and Mobil have merged their European refining, logistics and distribution activities in a joint venture).[2] The market for refined products is oligopolistic: a small number of producers serve several million consumers.

With the exceptions of the joint venture *Dunkerque,* which belongs to BP (75%) and Elf (25%), and of the refinery *Reichstett,* owned jointly by the major companies except Esso (65% belongs to Shell, 12% to BP, 10% to Elf, 8% to Total and 5% to Mobil Oil France), the mineral oil industry companies all own 100% of their refineries.

In spite of the heavy competition in mineral oil products on the French market, there are some co-operation agreements between competing oil-producing companies operating on the same market. One can, for example, cite the case of an agreement between BP (3rd largest European refiner with a market share of 7.4%) and Mobil (6th largest European refiner with a market share of 4.4%) on the joint venture merging their refining and marketing activities. This agreement relates to the common stock of pipelines, tanker terminals, trucks, refineries, depots, mineral oil stations etc. It covers almost all European countries and received approval from the European Union in September 1996. Another example comes from BP,

[2] Another important change has occurred since this case study was carried out: in 1999, Elf Aquitaine and TotalFina merged to form the world's 4th largest oil company. The new company, Elf-TotalFina, is currently (January 2000) developing common strategies and operations. In this chapter we therefore refer to the two as separate entities for historical correctness' sake.

which intends comparable co-operation with Shell and Exxon on the fuel market. Mobil will become market leader in lubricants (the refinery of *Notre-Dame de Gravenchon* being its main refinery), backed up with an estimated investment of $40 million and with the restructuring of three units in Europe (Germany, Austria, and the United Kingdom). The companies, facing over-capacity, are currently tending to negotiate jointly their regional capacity reduction. Shell is negotiating with Esso, BP and Total over a joint reduction in production capacity in the region of *l'étang de Berre*.

Table 1. Development of refining capacity (in kt) in French refineries in 1978–96

Region	Group	1978	1994	1995	1996
North		*14 800*	*6 210*	*6 210*	*6 210*
Flandres (Mardyck)	Total RD	6 300	6 210	6 210	6 210
Dunkerque	BP France	5 200	-	-	-
Valenciennes	Elf Aquitaine	3 300	-	-	-
Vallée de la Seine		*66 800*	*31 730*	*38 081*	*38 524*
Gonfreville	Total RD	22 300	10 350	16 215	16 215
Petit-Couronne	Société des Pétroles Shell	18 800	7 000	7 000	7 035
Port-Jérôme	Esso SAF	7 200	6 800	7 200	7 595
Notre Dame de Gravenchon	Mobil Oil française	3 600	3 100	3 186	3 195
Vernon	BP France	3 400	-	-	-
Gargenville	Elf France	6 500	-	-	-
Grandpuits	Elf France	5 000	4 480	4 480	4 484
L'Atlantique		*17 400*	*9 960*	*9 960*	*9 965*
Donges	Elf France	9 000	9 960	9 960	9 965
Vern-sur-Seiche (Rennes)	Elf France	1 500	-	-	-
Pauillac	Société des Pétroles Shell	4 000	-	-	-
Bordeaux	Esso SAF	2 900	-	-	-
Ambès	Elf France	-	-	-	-
Méditerranée-Rhône		*58 300*	*32 085*	*32 110*	*33 209*
Frontignan	Mobil Oil française	5 700	-	-	-
Berre	Société des Pétroles Shell	13 500	6 300	6 300	6 290
Lavéra	BP France	11 500	8 925	8 950	9 504
La Mède	Total RD	10 500	6 210	6 210	6 210
Fos-sur-Mer	Esso SAF	8 000	5 000	5 000	5 558
Feyzin	Elf France	9 100	5 650	5 650	5 647
East		*13 940*	*4 000*	*32 110*	*33 209*
Herrlisheim (Strasbourg)	Raffinerie de Strasbourg (SA)	4 610	-	-	-
Reichstett (Strasbourg)	Cie rhénane de raffinage	4 200	4 000	4 000	4 002
Hauconcourt (Lorraine)	Société de la raffinerie de Lorraine	5 130	-	-	-
Sum Total		*171 240*	*83 985*	*90 361*	*91 910*

Source: UFIP

The main justification for the aforementioned agreements are economies of scale. The last example shows a firm willingness to recreate a situation of under-capacity relative to demand for refined products and has resulted in the closure of the *Pernis* refinery in the Netherlands.

2.2 Economic Importance of Oil Refining in France

According to an estimate by the French mineral oil industry association UFIP (Union Française des Industries du Pétrole), the French refinery and distribution industry employed 29 000 people in 1995. Correcting this figure to include another 2 000 employees in research centres, total employment in the sector amounts to 31 000. In contrast, 52 700 employees were working in the French refining sector in 1975. Within two decades, the total workforce has therefore been almost halved.

Regionally, the mineral oil refining industry is of undeniable economic importance. For example, the Shell refinery in the region of *Petit-Couronne* contributed about FF 45 million to local tax revenue in 1996. The total wage bill amounted to FF 320 million and the orders carried out by other firms on the behalf of this refinery came to FF 50 million. On a national level, the same refinery has realised exploitation expenses worth FF 500 million. With respect to investment, the sector represents a significant sum of one FF 1 000 million francs.

As regards international competition, crude oil refining is of considerable economic importance, both "upstream" and "downstream".[3] In 1992, sales of refined products had risen to 3 130 Mt; net turnover amounted to $ 589 000 million.

Mineral oil is the primary energy resource for the transport sector, which consumes about 61% of refined products. This is also the sector which has experienced the strongest growth in energy consumption. The reason for this increase is essentially the rise in the total number of cars and the spread of mobility. The environmental impact of this development and particularly the impact on urban pollution is an increasing cause for concern.

Mineral oil is also an energy source for an important number of other key economic sectors, especially that of chemical fuels, which currently represent the most innovative and promising sector for the future.

2.3 Government Regulation

Since the mineral oil charter came into force in 1928, French refineries were somewhat strictly controlled by government in the form of (renewable) licenses issued for a period of six years. These licenses provided a certain number of rights, but also included relatively stringent obligations. The agreement was created in order to protect the new-born mineral oil industry.

---

[3] The "upstream" includes exploitation, surveying, etc. activities leading to input to the refinery, while "downstream" signifies the distribution activities originating in the refinery.

The first fracture came with the liberalisation of refined products in 1985. One of the government's prime constraints – extremely strict price control – was therefore eliminated. The 1992 Maastricht Treaty finally led to the revoking of the mineral oil charta which had for quite some time been obsolete, and which had intensified the issue of structural change by creating certain legal obstacles which led to less than competitive sales prices. Since then, the market has had to adapt to new competitive rules (some of which determined by the 1995 European directive on price liberalisation).

Since the creation of the Single European Market, the market for refined products has gradually become highly competitive. One of the main reasons for this is of course the free movement of products within the EU. Competition between refined products has thus become European competition. Every European country possesses several refineries producing increasingly standardised products. The French ELF corporation has, for example, completed several major acquisitions in the refineries and distribution sectors in Germany, Spain and Great Britain.

Certain constraints nevertheless remain on crude oil refining activity in France: refiners in France are obliged to fly the French flag and to pay harbour dues when bringing crude oil into French territory. This generates supplementary costs of about FF 300 million. Another restrictive factor is the relatively high local taxation connected with the crude oil refining industry.

2.4 Environmental Impact and Related Innovation

Refining is a prime contributor to the degradation of air and of water. It also gives rise to problems of noise pollution and to the creation of more or less hazardous waste, and technological risk plays a further important role. In addition, the refineries themselves are major energy consumers, with auto-consumption levels of about 4 to 5%.

In terms of polluting activities, the main problems are the following:

- discharge of mineral oil causing long term pollution of soil and the oceans
- inflammable vapour causing a risk of fire or explosion
- outflows of liquids or gases which might affect air or water quality
- dust emissions due to combustion and catalytic cracking, causing health problems and diminishing air quality

With regard to specific environmental media, about 79% of the total emissions by a refinery go into the air, about 20% are disposed of as solid waste and only 1% is emitted into water. Emissions of SO_2 are currently the most important form of environmental damage caused by refining. Taking for example the region of *Haute-Normandie*, about 65% of total SO_2 emissions are caused by the four refineries located in this region. Water pollution includes hydrocarbons, suspended matter and carbolic acid in waste water. Solid waste consists of, for example, sludge from tanks and processing plants, used catalysts and other biological solids. They are disposed of in landfill sites or commercial waste incinerators. Noise pollution can be kept relatively low in modern refineries, but remains a considerable local problem.

A number of innovations of some environmental importance are being made in the refining industry, although most of these innovations are mainly incremental and end-of-pipe. Both products and processes are affected. New products are:

- unleaded fuel and fuel containing less than 0.15g benzene per litre of fuel
- light fuels with less than 1% sulphur and particulates
- heavy fuels with lower particulate emissions
- biodegradable oil and environmentally friendly lubricant which can be decomposed by natural processes

Process innovation in refining includes:

- desulphurization units, used to decrease the amount of sulphur in refined products
- improvements in filtering and in the quality of waste water (from refineries using resin or nano-filters)
- prevention of pollutant emissions to air, especially from volatile components
- research on prevention of – and the response to – accidents

2.5 Structural Change in French Oil Refining

Refining capacity in France totalled 154 Mt/a in 1973 and peaked in 1976 with 172 Mt/a. Until 1989, capacity was steadily reduced, reaching 85 Mt/a, half its peak value. It has since remained at this level (Fig. 1). This 50% reduction in production capacity is associated with many refinery closures between 1977 and 1986: 24 refineries were in operation in 1974–76, only 13 since 1986 (DHYCA).

The rate of utilisation of refining capacity, almost 90% in 1973, had fallen to 59% by 1982. Since then, the rate has consistently risen and reached 90% once again in 1992. Gross production by refineries fell by 43% in twenty years, from 135 Mt in 1973 to 77 Mt in 1992, finally rising slightly to 83 Mt in 1996. During the same period, the structure of production has changed fundamentally: the share of light products (gasoline) had risen by 9 points to 22% by 1992. The share of domestic diesel oil and light fuel oil has risen by 3 points, reaching 37% whereas the share of heavy fuels / oils has fallen by 16 points, totalling only 16% in 1992.

Given the erosion of net profit margins, the ever-apparent over-capacity in production, as well as the steady increase in competition, quite a number of refinery sites have been closed down in France. However, if refineries with a "simple structure" are currently tending to disappear, others with a "complex structure" are emerging more and more frequently. In simple structure refineries, the treatment of crude oil is based on a distillation process using catalytic reforming with a low yield of light products. These refineries are not equipped with conversion units.

In the face of their associated environmental problems, the future of heavy fuels is uncertain at least, if not negative. There are presently six refineries with a com-

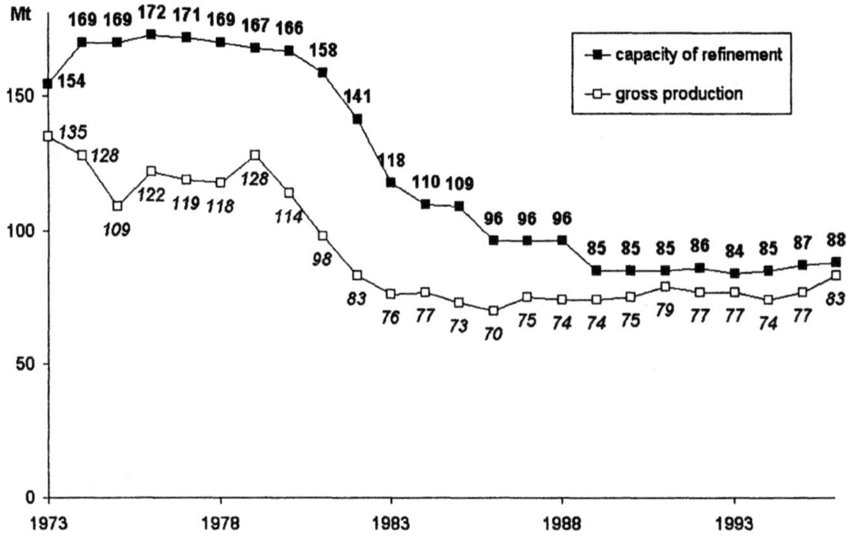

Source: CPDP

Fig. 1. Development of refining capacity and gross production in France 1973–96

plex structure in France. Their advantage is the ability to transform heavy layers obtained from the distillation process into light layers, which makes possible the maximum yield of light products.

The aforementioned closures went hand in hand with a complete restructuring of the mineral oil refining instrument. In order to finance the restructuring, a significant financial basis was required while, at the same time, profits were declining. The main social problems emerging from this economic situation were restructuring of the workforce, early retirement and some redundancies.

2.6 The French Situation in International Comparison

The structural change in France described above is fairly typical for European and North American countries, with some exceptions (such as the Netherlands; see *Jan van der Straaten* in this book). But the development of the refinery sector in Asia (Table 2) is characterised by a steady rise which may, in the near future, evolve into a strong competitive factor for not only the European, but also the North American industries. This development has been sustained by healthy economic growth and by a strong increase in consumption of mineral oil-related products.

The countries of South Eastern Asia in particular are currently experiencing strong growth in demand: South Korea, Malaysia, Taiwan, Thailand and even – to a certain extent – Japan. Finally, China has doubled its mineral oil consumption

Table 2. Development of global refinery capacity 1980–95

(Mt)	1980	1993	1994	1995
Western Europe	1 000.0	710.9	708.8	704.1
Near and Middle East	205.8	255.0	266.2	264.8
Africa	107.4	145.9	144.2	144.1
North America	1 025.0	851.1	861.3	860.1
Latin America	436.3	375.4	367.8	371.0
Far East	572.0	682.9	720.9	740.1
Eastern Europe / CIS	769.7	642.3	642.6	636.6
Total	4 116.2	3 663.5	3 711.8	3 702.8

Source: UFIP

since 1980. Asia currently possesses some of the most modern refineries, which makes it possible to produce vast quantities of refined products at attractive prices.

The Near East (the Middle East and the Persian Gulf) increased its crude oil treatment capacity by about 30% between 1990 and 1995, but remained relatively weak notwithstanding. Consumption also rose (almost doubled) over the same period, but still represents merely 6% of total world consumption.

The situation in Africa is almost the same as in the Near East: refining capacity is low, but has expanded in the last 15 years.

2.7 Driving Forces of Structural Change in French Oil Refining

The main reason for the structural change described here is not to be found "upstream" from the refinery (research, discovery and extraction). Given the technological developments of the last years, quite a number of oil deposits that were formerly inaccessible are now economically profitable. Offshore zones will, for example, become interesting in the future, in the wake of a number of significant discoveries, especially in Spain, Egypt and North Africa. The discovery of enormous deposits shows the importance of this region for future stocks of crude oil. In addition, the research/discovery/extraction sector is in the midst of a period of expansion, contributing considerably to the strong economic results from oil-producing companies.

The principal reasons for this structural change are located "downstream" from the refinery. Quantitative demand for refined products fell steadily – sharply following the second oil crisis – until the end of the 1980s, when demand rose again, albeit without ever reaching 1970s levels (Fig. 2).

Immediately after the first oil crisis, the French government launched an energy policy aiming at improving domestic energy independence. Two clear strategies were followed, bringing significant changes to the French energy situation: the development of nuclear energy for electricity generation on the one hand, and energy conservation on the other hand. France is one of the heaviest users of nu-

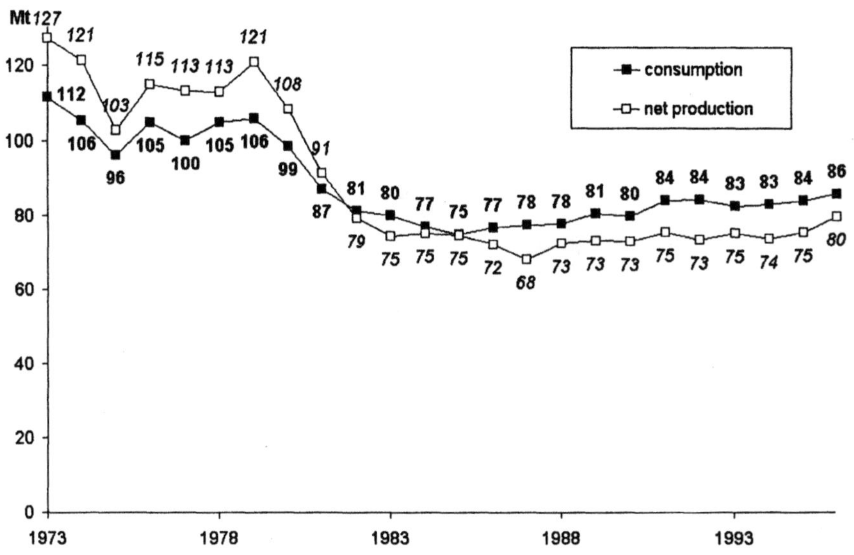

Source: CPDP

Fig. 2. Consumption and production of refined products in France1973–96

clear energy: 76.1% of electricity is generated in nuclear power stations. In Germany, nuclear energy makes up 29.6% of electricity production, in the Netherlands only 4.9%.

Demand for heavy fuel oil to be used in electricity production has therefore declined, while demand for light products like gasoline and for gas-oil has increased: this forced refiners to undertake a complete review of their development strategy and to adapt their production instruments, which had already become obsolete. But even with modification of the production structure, supply of refined products in France has still not adjusted to the change in demand. This development has thrown up the following paradox: France exports heavy fuel with a high sulphur content as well as lead-free fuel. At the same time, (principally) gas-oil is being imported at higher prices than the fuels exported (cf. Fig. 3).

This imbalance implies massive investment in new installations in order to transform heavy products for which there is less demand into light and middle-weight products experiencing a growth in demand. The French crude oil refining industry thus suffers from a chronic difficulty in balancing its financing capacity (because of falling net profit margins) with its investment requirements.

With respect to the quantitative fall in demand for refined products in France, the enormous costs associated with the closure of a refinery should also be mentioned. A number of French refineries are therefore still operational, even though their profit margins are extremely low.

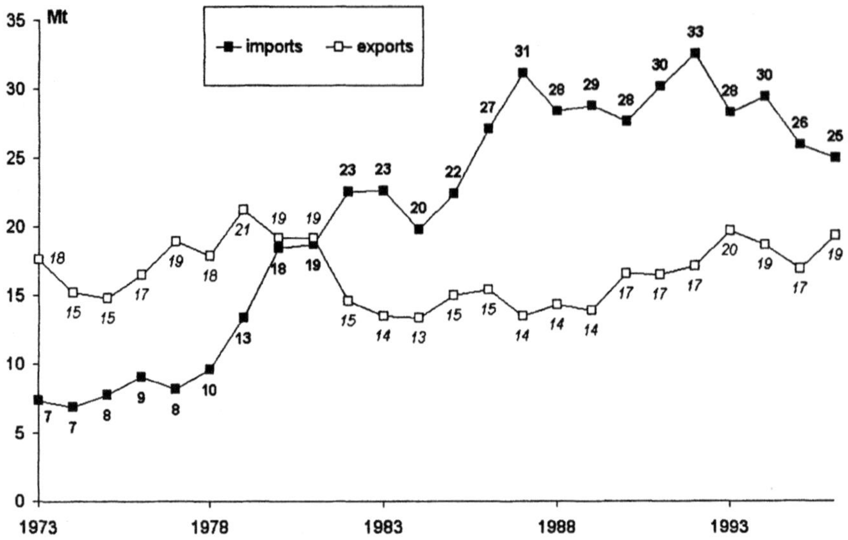

Source: CPDP

Fig. 3. Import and export of refined products in France 1973–96

In addition to the aforementioned factors influencing the refining sector in France, rising globalisation of the relevant markets must also be taken into account. This has important consequences for the market via limitations on, and the gradual alignment of, global prices for refined products. The recent deregulation of Asian mineral oil markets has contributed to the shift towards global trade in refined products. Another contributing factor in the near future might be the new production capacity in projects to be constructed in the Southern and Eastern Mediterranean regions.

Finally, environmental factors are important to structural change. Recent developments in environmental legislation, environmentally oriented demand, as well as general business practice in respect of environmental performance have influenced the structure of mineral oil refining in France. The results of the Kyoto conference will doubtless intensify the environmental debate on the long-term quantitative as well as qualitative evolution of refining.

3 Political Processes and Possibilities for Intervention

3.1 The Political Context of French Mineral Oil Refining

In order to better explain structural change and the decline in oil refining activity over the last twenty years, as well as to discuss possible future developments, this chapter will introduce the political context of mineral oil refining in France. We will start by presenting the current situation in the sector, the stakeholders involved and the interaction which forms the decision-making processes in French mineral oil refining. Secondly, we outline the main developments in French energy policy of the 20th century, with its crucial consequences for the evolution of French mineral oil refining. We finally seek an explanation for the relatively weak influence of the sector on (recent) developments in energy policy.

3.1.1 Stakeholders in the Decision-Making Process of the French Oil Refining Sector

Three broad categories of stakeholder can be identified: the first is industry, regrouping the oil companies as well as their representative associations at French and European level; the second group are the ministries and other political decision-makers concerned with the sector, the third group are the actors involved in research and ensuring energy supply. Fig. 4 gives an overview of the interaction between these actors.

Industry. The French mineral oil refining sector is composed of five companies, two of which are French in origin. The sector employed 20 000 in 1997 and produced 12 billion francs in added value. The oil companies are divided roughly into three types of activity: (1) the upstream, including exploration and production, (2)

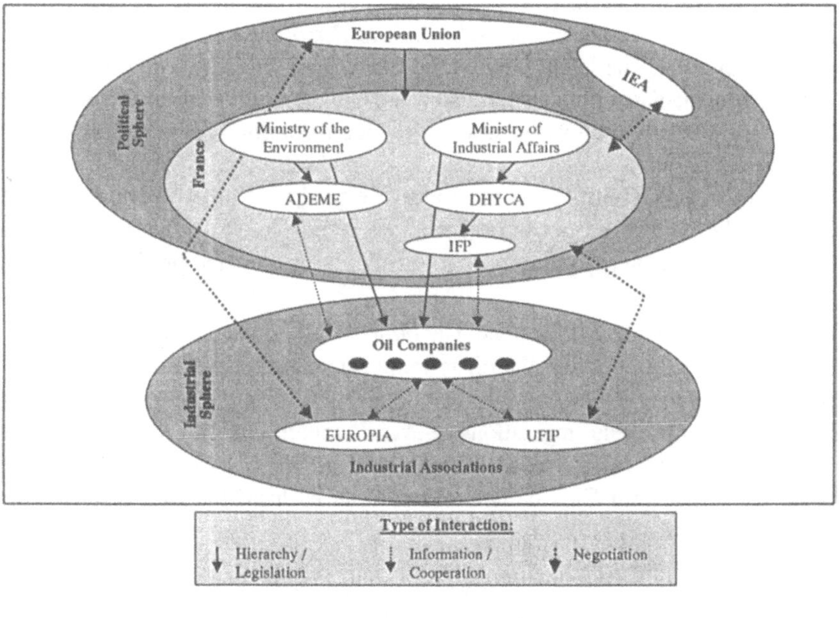

Fig. 4. Interaction between the different actors in mineral oil refining

the downstream, (refining and distribution) and (3) chemicals (the structure of each of the companies is slightly different, according to each of these categories). The economic objectives and interests of the different French oil companies are obviously coherent. They all aim to mobilise ways to better adapt to the European and French refining context, which is characterised by weak margins and a general over-capacity (Julien 1998; Le Tallec 1998; Legros 1997; Pétrole Informations 1997e). Specialists estimate that European refineries have an over-capacity of about 670 Mt, which means that about ten to fifteen refineries are "superfluous". This over-capacity, incompatible with the existing demand structure (especially in France) remains one of the main challenges. They did all achieve good results in 1997 however – their economic situation are therefore comparable, in their positive and negative features (BIP 1997e; Chabbi 1998; No Author 1998b).

For a couple of years, another type of constraint has been present in the French market: respect for and inclusion of the natural environment, essentially via environmental legislation. The oil refining sector is therefore called upon to incorporate environmental reasoning into their decision-making, while respecting their economic constraints. However, when comparing the stated objectives of the different oil companies, it is obvious that the environment plays only a secondary role and does not seem to have exerted explicit influence on the firms' strategies. One exception is Elf Aquitaine, which has clearly included environmental issues among its key targets and recently even embarked on some form of "eco-marketing".[4] Refiners have nevertheless remained rather hostile to environmental legislation, worrying about the investments imposed and higher costs (No Author 1998a; Rochard 1998a, 1998b; UFIP 1995b).

The interests of the oil companies are represented by industrial associations – the *UFIP (Union Française des Industries de Pétrole)* in France, and EUROPIA at European level. The UFIP informs the French public institutions on the state of refining and mediates between industry and government. It aims for example at influencing the tax system, in order to reduce disparities between gasoline and diesel (UFIP 1998b), or between natural gas and heating oil. *EUROPIA* advocates the interests of its members within Europe, e.g. against other industrial associations and the various European organisations responsible for legislation. Its main activities concern energy policy, environmental protection and product quality.

Ministries and Public Institutions. The *Ministry of the Environment* intervenes at two levels directly affecting the oil refining sector and structural change: first, it translates European legislation into French law and second, it monitors industry's compliance with these regulations. Several departments of the ministry are charged with these various tasks, in particular the SEI (*Service de l'Environment Industriel* – Industrial Environment Service), at its heart the Bureau for Atmos-

[4] *L'Environnement Magazine* (1998, p. 2) published an ecologically orientated advertisement of Elf (with the heading "Breathe – Elf invests so that you can drive ecologically"). It has to be mentioned that all forms of eco-marketing are very rare in France and only beginning to evolve since a relatively short time in comparison to the Northern European countries.

phere, Energy and Transport. Both of these are attached to the Directorate for Prevention of Pollution and Risks.

The Ministry of the Environment also interacts indirectly with industry, and thus with the oil refining sector, through the *ADEME*, the French Environment and Energy Agency established in 1992. Under the joint responsibility of the Ministries of the Environment, Industry and Research, the ADEME is involved in energy conservation, renewable resources, air, soil and noise pollution, and waste. With regard to the refining industry, the ADEME helps French refiners develop programmes and investment aimed at reducing SO_2 emissions. It also contributes to and assists research programmes on environmental improvements to the refining sector. The ADEME made a great contribution to the preparation of the Auto-Oil negotiations, for example.

The *Ministry of Industrial Affairs* interacts with the refining sector mostly through the *DGEMP (Direction Générale des Matières Premières)* and the *DHYCA (Direction des Hydrocarbures)*. The DHYCA's original job, before the liberalisation of the energy sector, had been to handle pricing, technical rules, import and export authorisations etc. for the French oil industry. This role has changed with the liberalisation measures taken in 1992, now focusing on defending the interests of the French oil industry in the context of European competition. In addition, the DHYCA is also concerned with ensuring (carbon) energy supply, thus playing a role similar to that of the International Energy Agency (IEA) at a national level (see below). Finally, environmental concerns do play an important part in the tasks of the DHYCA (DHYCA 1998)

Having evolved from the role of supervisor and decision-maker in the French oil industry, the DHYCA today acts as a kind of industrial association, not only representing the industrial sector, but taking into account national objectives and integrating the Ministries in order to reach a consensus at national level.

Another important actor is of course the *European Union*, since a large part of the member countries' legislation is drafted at this level today. As regards their intervention in the refining sector, the European Commission, the Council and the Parliament all have a similar goal: the implementation of a liberal industrial context while at the same time respecting the natural environment. The different actors are the Directorates General, especially the DG XVII (energy), charged with the environment problems of refining as well as with potential improvements, also the DG III (industry), DG VII (transport), DGXI (environment, nuclear safety, and civil defence), and the DG XII (science, research, and development).

***Institutions Concerned with Research on Mineral Oil and Ensuring Energy Supply*.** The *IFP (Institut Français du Pétrole - French Institute of Mineral Oil)* was created by the French government as a research organ working on improvements to mineral oil products, hierarchically subordinate to the DHYCA. Today, the French state no longer directly controls the oil companies, but is still concerned about the quality of their products and emissions by refineries. This concern is today realised by the IFP, which conducts basic research, often in co-operation with French and international oil companies. The IFP's main objectives are (a) to widen scientific and technological knowledge in oil refining and related domains, and (b) to ensure knowledge transfer between different companies, especially to small and medium-sized businesses. The knowledge transmission network ISIS (*Internationale de Services Industriels et Scientifiques*) was set up in this spirit, covering the full range of activities in the oil and gas industry.[5]

Following the first oil crisis, the *International Energy Agency (IEA)* was created in 1973 by several member states of the OECD. It was a response to an emergency, with energy supply being at risk in some of the member states. Its central aim is thus to guarantee energy supply for member countries (although there is official mission statement). This energy supply mainly consists of mineral oil based products (90%), which makes it an important actor for the refining industry.

3.1.2 A Brief History of French Energy Policy

French energy policy has gone through three main phases since the beginning of the century, the first two aimed at securing independent energy supply for the country, while the third has shifted towards environmental concerns.

At the end of the First World War, the French government decided to develop its own refining capacity, in order to reduce its dependency on other countries, and the state bought large parts of the *Compagnie Française des Pétroles* and the *Compagnie Française de Raffinage*, created in 1930. At this time, the French state monopoly in the refining sector was fixed by law on March 30th 1928. All imports of mineral oil were then controlled by state authorities.

After the end of World War II, and the exponential growth rates and economic transformations, the sector underwent deep restructuring to meet modified energy needs. This restructuring is essentially coupled with an important wave of nationalisation: in 1946, EDF (*Electricité de France*), GDF (*Gaz de France*) and *Charbonnages de France* were created. The nationalisation law of the same year granted important responsibilities to EDF, placing the company at the centre of French energy policy. The end of the Second World War also marked the start of a revival of nuclear energy research. The CEA (*Commissariat à l'Energie Atomique*) was created in 1945, and in 1955 the first studies began on a French nuclear Energy programme for the next twenty years. In the 1970s French energy policy shifted sharply following the oil price shocks provoked by OPEC. France, not unlike most of the other oil importing countries of the Western world, underwent a crisis which caused them to strive for independence from OPEC. French energy policy was thus determined by two main principles: energy efficiency and nuclear power. The first of these led to the creation of the AEE (*Agence pour les Economies*

[5] ISIS has been founded as a holding company by the IFP (57%), Elf (39%) and the French bank BNP (4%). See IFP 1998 or the website of GEP-France (*Groupement des Entreprises Parapétrolières et Paragazières*).

mies d'Energie – today ADEME), while the second started a massive programme of nuclear power stations, in comparison with other countries.

The energy strategy in France during the 1970s produced relatively low levels of CO_2 emissions in comparison to other countries, due to the decline of mineral oil in the French energy mix, from 70% to 40% over the last three decades, while imports have fallen from 135 Mt/a to 83 Mt/a. Still, these numbers should not mask the real extent of this independence, only about 50% of energy supply. This energy independence mainly took effect in the electricity market, replacing the need for heavy mineral oil products, while lighter refined products are still in great demand in the transport sector.

With the 1990s, French energy policy, like that of most other Northern countries, entered into a new phase, taking into account environmental issues, and in particular the problems of global warming and climate change. This represents an important shift in the main preoccupations of energy policy, and it will most probably lead to another shift in the French energy mix: less carbon energy sources, a focus on renewable energies, and environmental questions about nuclear energy. An official report to the Sénat on the future of French energy policy (Revol 1998) summarised the most important challenges for the coming century:

- environmental issues and the climate change negotiations
- the evolution of the European context and the consequences of European liberalisation of the energy market
- the future of French nuclear energy, given the need to renew the nuclear power plant stock from 2010 onwards, and the debate on abandoning nuclear energy in countries like Germany

3.1.3 French Mineral Oil Refiners and Energy Policy

The aim of French energy policy today is, as stated above, no longer simply to increase energy independence, but also to gradually reduce the environmental impact of technological choices. The main problem for the oil refining sector in the context of an evolving energy policy lies in the expansion of its activities. The first consequence for oil companies of previous French energy policy was that the market for heavy fuels was severely reduced. This not only led to a reduction of total oil production, but the oil companies also tried to adapt their production to remaining markets, i.e. to demand for light products, which required installing more complex refineries, with significant investment.

The influence of the refiners themselves on energy policy remained relatively weak, compared to other European countries (Boisson 1998) and their strategies seem to have been more or less reactive. One explanation for this fact is the market structure of the sector. While the tendency in the 1950s was to regroup and restructure enterprises under state authority, the 1970s, with the wave of privatisation of public enterprises, saw the opposite trend. In addition, quantitative regulations gradually disappeared, followed in the 1980s by the end of price management. The first step had been a partial freeing of prices in 1982, when they were determined with reference to Rotterdam. Since 1986, prices have been completely deregulated. The petrol charta of 1928 was finally abrogated in 1992 by a law ending the public monopoly in the oil industry. Its principal characteristics are:

- The oil companies operating on the French market are required to inform the government about their activities.
- They are forced to adapt the concept of strategic minimum stocks.
- They are obliged to fly the French flag.
- The ban on energy advertising is ended.
- A regulatory context is elaborated incl. sanctions in case of non-compliance.

Since the government is no longer involved in their management, the private multinational and French companies now compete with public energy companies like EDF or GDF in the sectors of nuclear energy or natural gas. While the mineral oil companies before privatisation had to adapt to national energy policy, the influence of the now private companies on government decision-making is equally slight in comparison to that of the large public companies like EDF.

Another explanation for the relatively weak influence of refiners in France is the fact that mineral oil based products fill the gap between nuclear energy and natural gas: the weight of mineral oil in the French energy mix is thus relatively small. Finally, it is surely the centralist nature of French political life itself that leads to the fact that industrial firms are less influential in comparison to other countries. In any case, the development of French energy policy shows that the influence of oil companies was not sufficient to defend their interests enough to stop the decline of the sector, especially given the fact that at the time of the great changes in the sector, in the 1970s, liberalisation and privatisation had not yet taken place. The state has thus even more influence on the sector.

Today, the French oil companies seem to oppose the slowing down of liberalisation in the energy sector provoked by public institutions like EDF and GDF, referring to the need to guarantee the quality of public service, and without doubt fearing the unions. From this point of view, the creation of the EU provides an advantage for refiners, since it promotes market liberalisation. The oil companies demand rapid change to the gas and electricity directives.

In all, the main characteristics producing this weak institutional influence of the oil companies on French energy policy seem to be their private ownership, the importance of nuclear energy, and the centralised institutional structures.

3.2 Environmental Policy and its Influence on Structural Change in French Mineral Oil Refining

Since the 1980s, respect for the natural environment has been becoming an important factor in political decision-making and French energy policy. The mineral oil refining sector will most likely be influenced more and more by environmental issues. The decline in mineral oil refining is also crucially affected by environmental problems, since stricter legislation and the shift towards alternative energy forms could cause further falls in the demand for refined products.

3.2.1 French Environmental Legislation and its Influence on Refineries

Four types of legislation can contribute to structural change in the refining sector:

- legislation on plants requiring authorisation
- legislation aimed at the environmental properties of products
- legislation on air quality[6]
- legislation on greenhouse gases

This legislation can emerge from environmental concerns at a French, European or international level, in the latter case through agreements between different countries. There is a series of protocols that influence the French legislative system: the Geneva Convention (1979) concerns crossborder pollution from SO_2 emissions; the Oslo Protocol (1994) has strengthened the Geneva objectives, with reduction targets of 74% in the year 2000, 77% in 2005 and 78% in 2010, based on emissions in 1980. Every year, emissions quotas are fixed for each country. NO_x emissions have also been subject of a protocol signed in 1988 in Sophia, where a reduction of 30% in emissions was fixed.

In all, French refining activity is influenced by legislation either of emissions by the refineries themselves or of the qualities of their products, the goal being the same in both cases: securing industrial development with more respect for the natural environment. The following sums up existing European and French legislation on emissions of refineries.

- *The Directive on Sulphur Content of Liquid Mineral Oil Products.* This proposal by the EC, representing a revision of directive 93/12/CEE on the sulphur content in gas and heating oils, largely aims to further restrict the sulphur content of heavy fuels (limited to 1% for January 1st 2000). The European sulphur directive will be implemented in French law by the DHYCA.
- *The Large Furnaces Directive.* This European directive, implemented in 1988, concerns the fight against air pollution and is currently being revised. It fixes permitted limits to combustion activities. The new values for SO_2 and NO_x proposed by the EC will be based on BAT (best available technology).
- *The Atmosphere and Rational Energy Use Act.* This law was promulgated on December 30th 1996. Its main aspects are supervision of air pollution and air quality (prevention), preparation of emergency measures, and technical measures ("end-of-the-pipe"). However, this law represents a list of measures aimed rather at decreasing urban air pollution than monitoring results (Pétrole Informations, 1997a).

[6] On French legislation concerning air quality see also Pétrole Information 1997a.

- *The Framework Directive on Air Quality and First "Daughter" Directives*[7]. The European directive 96/62/CE of December 27th 1996, concerning the evaluation and management of air quality represents the framework for future European air quality legislation. The four goals of this directive are: define and fix the objectives for air pollution in the EC in order to avoid the need for clean-up of health and environmental effects; develop common methods and criteria to evaluate air quality in the member states; provide information on air quality and inform the general public, e.g. via critical limits of air pollution; maintain air quality where it is good and improve it, where it is not sufficient. *The Fuel Directive*. The implementation of this directive has been delayed by administrative problems between the different institutions of the European Union. On June 29th 1998, the representatives of the Council and Parliament reached an agreement on measures aimed at reducing traffic emissions by 70% in 2010, based on 1990 levels (Reicherts 1998). By 2000, the sulphur content in gasoline should be cut to 150 ppm, that in diesel to 350 ppm. In 2005, sulphur concentration in transport fuels (gasoline and diesel) should not exceed 50 ppm.

3.2.2 The European Auto-Oil Programme

The European Commission, the automobile industry and the oil industry have launched a framework for co-operation: Auto-Oil. It was agreed in 1993, starting with a common research programme on air quality, emissions, motor technology and the quality of refined products. The main components considered are sulphur, benzine, aromatics, VOCs and NO_x. The principal participants in the programme are DG III (Industry), DGXI (Environment), DG XVII (Energy), the ACEA (the European car manufacturer association), and EUROPIA, the latter two representing their respective members in the automobile and mineral oil industries. An important part of the Auto-Oil programme is the research programme EPEFE (European Program Emissions Fuels Engines), which has led to a study of the effects of motor and fuel properties on traffic emissions and fuel consumption. More than 2 000 combinations have been studied. The costs of this programme are estimated at over 10 million ECU, shared between the ACEA and EUROPIA.

The basic goal of Auto-Oil is to identify measures to be considered in road transport to produce a respectable air quality and public health at a minimum cost for society. The experts have chosen seven European cities as reference points: Athens, Cologne, The Hague, London, Lyon, Madrid and Milan. The principle of the programme is to maximize cost effectiveness, with four main stages:

- establishing air quality norms and identifying possible necessary improvements
- evaluating the potential of motor/vehicle technologies and of fuel properties
- evaluating the cost effectiveness of the different possible measures
- selecting the best solutions, minimising the costs to the consumer as well as to society and the economy in general

---

[7] For further details, see the website http://www.europa.eu.int/index.htm.

Based on the results of the programme presented on June 18th 1996, the European Commission has made several recommendations. This has been the starting point for a long series of negotiations on the development of the different norms. In June 1997, a series of stricter norms was accepted by the European Council. In February 1998, the Parliament rejected the entire set of measures accepted in the year before, in order to adopt a maximalist position with new and stricter specifications for 2000 and even harsher norms for 2005. The costs to the member countries implementing this directive are estimated at 60 billion ECU.

The oil companies question the scientific justification for stricter standards. According to the oil industry, the environmental effects in terms of pollution reduction would hardly exceed 1%, while incurring relatively high costs (UFIP 1998a). To sum up the view of the oil industry, the environmental benefits of the strict norms are relatively low and emotional arguments have diminished their cost effectiveness, especially from an industrial point of view (cf. Legros 1998; No Author 1998a). The French Ministry of the Environment, on the other hand sees the strict norms as justified from a technological and scientific point of view.

3.2.3 Structural Change in Oil Refining due to Environmental Legislation or Considerations

Environmental regulation has had a considerable impact on the oil refining industry. The environmental laws and norms demand significant investment from the refiners, influencing their economic situation and their cost structure. The effects on the refineries and oil companies can take the following forms:

- changing product lines ("traditional" refined products can become obsolete)
- new (cleaner) production technologies
- change in the economic advantages of refining compared to their other activities (long term decline of refining?)
- further diversification of the oil companies' activities
- decreased demand for refined products
- change in the relationships and interaction between the different actors

As is the case for other industries as well, the environment now represents a new constraint. However, certain oil companies like Shell or BP (the two largest in the EU) aim to transform this constraint into an opportunity. This "win-win" strategy, as proposed by Porter and van der Linde (1995), means diversifying activities into renewable energies (biomass, wind, or solar energy) or building relationships with environmental NGOs in order to better anticipate (or even influence) environmental legislation. Others, like certain American firms (for example Exxon), tend towards an extremely aggressive position on the environment, trying to fight or influence environmental legislation. As for the French oil companies, they have adopted an intermediate position, which could be called a "follower" strategy, trying in particular to adapt to new environmental legislation without actually participating in the process of their formulation (Faucheux et al. 1997, 1998).

Even if environmental factors seem to restrict the French refiners today, it should be remembered that they are not the only influence on structural change. Low profitability and over-capacity were characteristics of refining before the environment became a major issue in legislation. French refiners have also been slow in modernising their refining tools over the last twenty years, compared to neighbouring countries. Restructuring is therefore necessary, even without environmental legislation and concerns. The first conclusion is therefore that refiners should consider the environment a "simple" parameter in decision-making. As the "Berger Report" of 1997 states, the environment is one of the three major pillars in energy policy, alongside competitiveness and guaranteed energy supply.[8] From the point of view of the European refiners, the environment represents a constraint primarily because taking it into account has been imposed on them. One aspect nonetheless turns this restriction into an opportunity, as competition is intensified and advantages can be gained from a pro-active strategy. This is no easy task, as the norms apply to all refineries and technological niches must be found.

In fact, the environmental constraint has become a part of the framework of action for French mineral oil refining, and has induced some structural change, either internal, inducing organisational or technological changes (products / processes), or external, inducing more radical changes, for example as alternative forms of energy become politically, economically and environmentally more advantageous than refined products. "Dematerialization"[9] of the mineral oil sector can thus be engendered, i.e. an absolute fall in production of refined products. Currently, there are no clear signs of such a development in the refining sector – the fall in production of refined products and the shift in activities towards production of natural gas, for example, is today still due more to French energy policy in the 1970s than environmental considerations or legislation.

However, environmental regulation is a major influence on the refining sector as regards internal structural change, as in product and process innovations which do not necessarily reduce absolute production. This integration of environmental parameters into refining requires significant investment, and the weak economic situation of the refiners does not make this any easier. ROCE (return on capital employed), which is considered to be at least 12% in order to assure profitability of investment, is currently about 4.4%. The director of Elf said that for ten years, all investment has been a response to constraints, without a real perspective of profitability (Legros 1998).

Apart from the low ROCE, refiners hesitate to invest because of uncertainty about future developments in the sector. To become profitable, investment re-

---

[8] The "Berger Report" has been accomplished by the German Consultants Roland Berger & Partner on command of the DG XVII and the European Union. This study formulated recommendations for stabilising the refining activity in Europe, precising that the actions should mainly be initiated by the companies themselves (see Berger & Partner 1997).

[9] ... as we use the term in this project, which does not totally correspond to other definitions that view this concept as the reduction of material flows through the economy. See for example Haake et al. 1999.

quires not just time, but a stable economic and legislative context, ensuring that structural change caused by the investment is not obsolete by the time it has amortised. A spokesperson from Esso said that it is uncertain whether a new concept, implemented over five years, will still be up to date by the time it is complete (Pétrole Informations 1997e). The investment cycle should therefore correspond to financing options in order to be technically feasible (BIP 1996b).

Given the importance of necessary investment and the lack of ROCE, refiners develop complex strategies, combining plant, operating methods and strategic alliances. The three main options refiners have are the following:

- *Partnership programmes in order to gain economies of scale:* Refiners marry with the head rather than the heart – industry considers partnership as a way of financing heavy investment rather than exchanging information or fostering sustainable development.[10]
- *Cutting refining capacity*
- *Closing down refineries:* This is a costly alternative. Decommissioning plants, cleaning up soil and social charges add up to a significant sum, often less tolerable than running a plant at a loss. Closing down a refinery producing 6 Mt/a can cost $ 100 million. In addition, closing a refinery will benefit competitors. This solution is nevertheless a feasible option in some economic situations.

In 1998, the majority of French refiners thought themselves ready for the environmental norms coming in 2000, although those in 2005 were considered more of a challenge. A spokesman from Esso emphasised that these necessitate technological changes, while the norms for 2000 simply demand improvements in the existing situation.

For the respondent from the UFIP, the European logic of inducing significant investment by environmental norms will lead either to the closure of less profitable refineries, or to the disappearance of certain products, as is the case for fuels with a high sulphur content. In all, this source believes the investments are feasible, if refiners are able to combine productivity gains, sources of low-sulphur crude oil, and the use of catalysts.

4 Future Perspectives for Mineral Oil Refining in France

4.1 Possible Factors in Future Structural Change for French Refining

4.1.1 Upstream Factors: Limited Resources of Crude Oil

According to one estimate, proven mineral oil reserves of about 139 billion tonnes will last for another 43 years at current rates of production (UFIP 1998a). Even if the UFIP cites this argument as the key potential danger to refining, the rest of the relevant public and private actors do not seem to see the future of the sector as

[10] On these kinds of firms strategies see also Faucheux and Nicolaï 1998.

realistically at risk. On the contrary, our various respondents seemed optimistic. The principal element on which this optimism is based is technological progress. The latter is seen as a double miracle, a solution to the problem of limited crude oil resources: on the one hand, alternative forms of energy will be found, and on the other hand, new oil resources. The 43 years predicted supply is made of reserves accessible to today's technologies and within today's cost structures. If higher costs were accepted, for example, for drilling in deeper parts of the ocean (which necessitates more developed technologies), far greater reserves can be predicted. The estimates must therefore be updated every year. In the 1970s, one estimate had predicted only 30 years more oil production, which had led some oil companies to consider shifting their activities by the turn of the century.

4.1.2 Downstream Factors: Decreasing Demand due to Environmental Policies and the Climate Change Negotiations

We have seen that environmental considerations and legislation represent an important new factor in the framework of mineral oil refining. One of the recurring issues in current environmental discussions with a possible effect on the refining sector is climate change. At the moment it is not yet clear how the climate change negotiations will influence the French and European environmental legislative and economic context, but it is clear that the six greenhouse gases at the focus of the discussions, and especially CO_2, are intimately connected with the oil refining sector. French energy policy in recent decades, based firmly on nuclear energy, has led to a relatively low level of CO_2 emissions (about 6 t/cap in 1990 against an average of about 12 t/cap in other OECD countries). As a consequence, France has managed to obtain a stabilisation target (i.e. stabilisation of emissions at 1990 levels in the period 2008–12) within the EU's internal burden-sharing – 8% of the EU target. So France (and thus French oil refining might end up requiring less action and thereby less negative effects for refiners. But even this is uncertain, as it depends strongly on decisions to continue the civil nuclear programme and on emissions growth in other sectors, in particular transport. The position of the French oil companies is vague, and harder to evaluate that those of companies like Exxon who, as leader of the Global Climate Coalition (an anti-climate change policy lobby mainly constituted of American oil and coal producers)[11], is openly hostile to climate change policy, or BP or Shell, who have clearly adopted proac-

[11] This business association presents itself as: "The Global Climate Coalition ('the Coalition') is an association of business trade associations and private companies formed to co-ordinate the active involvement of U.S. business in the scientific and policy debates concerning global climate change issues. Coalition membership represents a broad spectrum of virtually all elements of U.S. industry including the energy producing and energy consuming sectors. The Coalition is actively involved in the debate over global climate issues because of the concern with which its members share over the issues, and the potentially enormous impact which improper resolution of those issues may have on our industrial base, our customers and their lifestyles and the national economy". See http://www.globalclimate.org/gccdocs.html.

tive approaches, by developing their solar energy activities or setting up internal emissions reduction schemes, for instance. While Total claims to have no opinion on the subject, Elf had, just before the Kyoto conference, proposed a voluntary internal CO_2 reduction of 15%, which, according to some experts, might not be difficult nor costly to attain with the help of simple technological measures. Other environmental problems could be more crucial for this company.

The central negotiation point at the third Conference of the Parties to the United Framework Convention on Climate Change in Kyoto has been the definition of legally binding greenhouse gases emissions reduction commitments for industrialised countries (so-called Annex I countries). Although the question of participation by developing countries has been repeatedly pushed forward by some parties (notably the United States), the definition of reduction goals and the question of the instruments and policies to achieve the reductions have constituted the core of the Kyoto negotiations. The two main proposals as far as economic "flexibility" instruments are concerned are tradable emissions rights (tradable permits or project based emissions credits) and fiscal measures. The latter were strongly opposed by many parties to the negotiation, in particular the United States. As a result, the Kyoto protocol allows the use of the first type of instrument, tradable rights, in three "flexibility" instruments: tradable permits, joint implementation and the clean development mechanism (see e.g. Van den Hove et al. 1998). Taxes remain an option for parties but will not be internationally co-ordinated.

Even within France, differences can be found in acceptance of the flexibility instruments, which is mainly due to their innovative character. Emissions trading has only been applied sparsely in environmental policy, and only at national or lower level (e.g. the sulphur market in the United States). According to our respondent from the French Ministry of the Environment, these permits risk distorting competition: richer countries will more likely be able to buy pollution permits. However, from the point of view of industry, the permit solution is more welcome, since market mechanisms correspond better to the industrial mindset than imposed constraints, and taxes are always viewed with some degree of suspicion.

Even though the Kyoto protocol does not provide for any mandatory fiscal measures, the tax issue remains open in France. France may still rely on fiscal instruments to some extent to comply with its Kyoto commitments. This will depend both on the future design of French and European climate policy.

The future of these policies is linked to subsequent outcomes of the international climate change negotiations which will determine the rules and modalities for the use of flexibility instruments, which in turn will determine the extent to which these instruments will be used by the parties, as well as what other policies and measures the parties will need to implement to meet their commitments. Another crucial dimension will be compliance and enforcement mechanisms, which will determine the achievement of the Kyoto goals. But even before this, the first major signal will be the coming into force of the Kyoto Protocol, which will only take place when major players have ratified it. Until then, no commitment is binding. But the French oil refining sector now seems to have included the Protocol's coming into force as a very likely element of the framework in which to define their future strategy.

While it is clear that the climate change negotiations are of considerable importance for the future of the refining sector, other environmental considerations will also influence the sector strongly, through future legislation or social and economic change, which for an industrial company mean investment and higher costs. Apart from the examples already mentioned, the French refining industry seems to conform to the general French industrial attitude opposing environmental restrictions. An argument often brought forward against environmental policy is scientific uncertainty: it is not worth combating environmental harm which is not 100% proven. The environmental impact of acidification on forest diseases has for example recently been questioned, and might only be one of several factors, such as poorly adapted forestation, irrational soil exploitation or the development of certain fungi. There is further scientific uncertainty on the environmental effects of VOCs and particles, which are held to increase the ozone content of ground level air. This argument impedes the fight against greenhouse gas pollution: along with apocalyptic predictions for the economy, it is the central argument of the Global Climate Coalition. It is obvious that there is a difference between the arguments of industrialists and those of environmental government institutions. However uncertain the science may be, past experience has shown that the more obvious and visible the transborder and intertemporal character of pollution becomes, the more complex and influential environmental politics becomes. Global warming policy is particularly central because of its highly international dimension, and its potentially unprecedented influence on economic and social systems. And, as problem and differences of opinion are slowing down the political decision-making process at national and European level, the problems are greater at international level. Reaching international environmental agreements and regulations is not an easy task. However, the key issue in international climate policy today is not the scientific justification for action, but rather the choice of regulatory instruments and political control mechanisms, as confirmed by the Kyoto Conference in 1997.

There is another interesting point in the potential influence of current and future debates on the oil refining industry, which might possibly lead into a change in direction from the climate change negotiations and other aspects of environmental policy. We have already mentioned the fact that from about 2010 onwards, the stock of French nuclear power stations will need to be renewed (Aronsson 1998). The future of these power stations is at the moment still uncertain and being hotly debated. One of the main points of contention is the disposal and storage of radioactive waste, an issue of considerable environmental importance. These discussions (as well as recent developments in neighbouring countries like Germany) are likely to induce a revival of interest in classical energy sources, like mineral oil or natural gas, as well as for renewable energy sources. Such a development might contribute to a new increase in demand for refined products. However, from today's point of view, it seems more likely that the debates on climate change and emissions reduction will be of more importance for the sector than those on the future of nuclear energy; however the final outcome of the two developments cannot easily be predicted.

4.2 Reactions and Adaptation by Refiners

In our interview with an expert from the ENSPM (French national engineering school of mineral oil and motors), we heard a relatively optimistic view on the future of the refining sector: in order to be able to cope with future investment due to stricter environmental norms, the refiners will depend on refining margins between $3.50 and $4 per barrel. Currently, these margins are between $2.50 and $3. The gap of $1 between current and necessary margins is not, according to our interviewee, to be blamed on overly strict environmental norms. He feels that progress to $3 or $4 margins is possible, so long as stocks and sales remain constant. However, the competitive context remains important, as well as the fact that closing down less profitable refineries can guarantee sales. He sees two possible scenarios: if refineries are closed down, Europe's capacity will become more viable, but the related costs are immense and few people believe this solution to be a good one. The second scenario, considered more likely, is based on reducing capacity in refineries. In this case, investments are made and the current situation is maintained without an economic catastrophe.

The same "moderate optimism" came in an interview with a representative from the DHYCA as well as from different oil companies. This optimism is based partly on the strong results achieved by the companies in 1997 (although the results for 1998 were expected to be less good), and it is moderated by vestigial weakness in refining margins. Some of these actors also refuse to talk of a crisis at all, since demand is strong and France is currently reducing its over-capacity.

It is a broadly accepted view that hydrocarbons will decline relative to other energy sources. This would confirm the thesis that there will be further "dematerialization" of the oil refining sector. There are, however, differing views as to the quantitative development of demand – our correspondent at the IEA states for example that total energy demand will increase, which would lead to a quantitative freeze in fuel demand. This is largely due to growth in transport, current (road/air/water) transport technologies being highly reliant on hydrocarbons. In order to decrease today's negative environmental effects of transport, radical technological change would be necessary, which would of course threaten oil refining.

As stated above, it is mainly transport that is causing the growth in mineral oil products (about 1 to 2% per year). Fuel has become one of the most "undesirable" refined products, which French refiners either try to sell to foreign countries or else transform, which fits in with the objective of reducing the sulphur content of mineral oil products. One way in which refiners are responding to this objective is to drill in the North Sea, where crude oil is lighter (with a lower sulphur content). In this way, the installation of costly conversion units can be avoided. This choice will remain economically profitable as long the price of crude does not exceed $5 per barrel.

In all, the refining industry will respond to the evolution of the market, and the quality of products today depends importantly on legislation and on expertise coming from environmental research. Current developments indicate that LPG (Liquefied Petroleum Gas) will be able to gain higher market shares in future. Mineral oil products will be finer and better adapted to environmental needs. Domestic fuel oils will evolve to compete with other energy sources, like nuclear energy or natural gas. As for transport, research is being conducted on radical technological change. All these factors indicate that oil refining will be subject to internal structural change towards less harmful forms of refined products, and also that the activities of oil companies may as a whole evolve towards other forms of energy supply, not necessarily mineral oil.

It is thus important to distinguish between internal and external structural change due to environmental factors, as we already discussed above. Internal structural change means environmental improvements to the sector's activities and products, a process already taking place and likely to continue. External structural change on the other hand represents a threat to the existence of the sector, the negative environmental properties of mineral oil refining and its products suggest that alternatives to these rather than improvements may be necessary in order to attune the sector to the goals of sustainable development.

The refiners themselves are of course interested in the development which has the smalles negative impact on their activities, in both economic (investment) and absolute terms (a shift in global energy demand away from refined products). The oil companies will likely focus on diversification in the latter case, shifting their activities to other products, as is already the case with companies focussing on natural gas, and some of them also on renewable energy sources like solar or wind energy. However, the key measures aim to restructure refining internally, with a proposed stabilisation, removing the (negative) consequences of French activities in the sector, in order to establish an economic and environmental context which favours the future of mineral oil refining.

One proposal for improving the external conditions of oil refining is establishing a reliable fiscal environment. A first step in this direction is the recent decision to reduce the taxation gap between diesel and gasoline, since it cannot be justified environmentally (BIP 1997d). This reduction is to be spread over seven years, giving refiners enough time to anticipate changes in demand. Another financial influence on oil refining are the incentives favouring other energy sources, as is currently the case for natural gas, which is making French refiners uneasy (UFIP 1995b).

Another proposal for stabilising the sector is the abolition of anti-competitive factors. Two such distortions exist in France: the first is the fact that 51% of distributional market share is held by the largest firms. Public institutions could intervene in order to hinder these firms from fixing their own price policy, which could lead to imports of refined products (BIP 1997c). The DHYCA already intervenes, hindering large distributors from growing and selling at loss. Subsidies are the second type of distortion, reflecting the different interests of European member countries, all interested in restricting their industries as little as possible. As our respondent from the IEA puts it, it is the homogeneity of legislative constraints in Europe that conditions the positive, negative or neutral effects of environmental policy. Direct or indirect subsidies are thus counterproductive measures, creating competitive advantages for certain companies.

Finally, predictability seems to be an important factor for the future stabilisation of the oil industry. One of the components is anticipation of demand. In particular for gasoline, predictions of demand are subject to serious uncertainty. This is highlighted by the predictions made for demand for unleaded fuel at the time the product was launched – these predictions have proven false in the long run.

5 Conclusion

The case study presented in this chapter indicates that French mineral oil refining is in a state of decline and over-capacity, and is undergoing a qualitative structural change in its product lines. The main phenomenon distinguishing the French sector from its European competitors seems to be its "unadjusted" state, due to the gap between supply and demand of fuels and gasoline.

Current trends indicate that French refineries will face further legislative changes affecting demand structure in the near future, as well as incentives thrown up by proactive competitors at home and abroad. Among the most important causes of this change will be environmental factors, as present developments have shown. For example, the French government is currently discussing environmental legislation aimed at reducing consumption of gas-oil. European standards are also playing an important role here. Taking for example the Esso refinery at *Port-Jérôme*, one third of its investment spend is dictated by the need to adapt to European standards. These examples have shown that environmental factors are affecting and will continue to affect French refineries in different ways. On the one hand, they will intensify current problems and accelerate the drop in production by old plant. On the other hand, they favour innovation and qualitative change in the sector, which might help overcome present crises.

In this study we have also aimed to draw a thorough picture of the past and current decision–making processes in the French mineral oil refining sector. We have done so by summing up and evaluating the information we obtained from the various stakeholders and actors we interviewed and from an analysis of the relevant literature. Our main results are as follows:

The main objectives in French energy policy were traditionally oriented towards enhancing the country's independent energy supply, culminating in the incomparable growth of nuclear energy after the oil crises of the 1970s. Given the centralist nature of French politics, as well as the existence of large public monopolies in the electricity sector, the refiners had little significant influence on energy policy, and thus suffered major declines in production.

The aims of French energy policy have shifted since the late 1980s, towards the environment, which influences refiners' activities significantly, via legislation affecting their products and production sites. However, we have not yet identified an unambigous influence of environmental legislation on production volumes. It seems today as if internal structural change is the key strategy of refiners, focusing on product and process innovation in order to adapt these to the environmental demands of legislation.

It is nevertheless possible that this internal structural change will become external structural change in future, and thus lead to a further decline in production of refined products. The main factor likely to influence this are the current climate change negotiations, and their results will have considerable influence on the mineral oil refining sector. Given the fact that French CO_2 emissions are today lower than in other EU member countries, due to the large share of nuclear energy in the country's energy mix, the fixed emissions reduction goal for France is 0% relative to 1990 for the period 2008–12, for all six greenhouse gases. Nevertheless, the country will have to take action, since although the zero reduction goal is less urgent than in other countries, it is still difficult to achieve, especially if nuclear energy declines in the French energy mix. The development of French petroleum refining would thus converge with that of the refining sectors in the rest of Europe. The reaction of refiners to such an external structural change could lead to diversification. This would also include production of renewable energies, as certain oil companies like Shell or BP-Mobil have already begun, albeit to a very small extent in proportion to their overall activities.

References

Aronssohn, D. (1998): Energie: les quatre défis des années 2010. Alternatives Economiques, No. 162, September 1998, 38–41.

Berger R. & Partner (1997): Study on Oil Refining in the European Community. London.

BIP (Bulletin Industrie Pétrolière) (1996a): Shell annonce une série d'initiatives pour améliorer son raffinage en Europe. 11 December 1996.

BIP (1996b): Quelle est la capacité de l'industrie du raffinage à produire des carburants et à pourvoir aux futurs besoins et contraintes? No. 8241, 11 December 1996.

BIP (1997a): Elf veut rentabiliser ses activités en aval, un entretien avec Bernard de Combret, Directeur du Raffinage-distribution d'Elf. 4 March 1997.

BIP (1997b): Mobil France: perte économique de 149 MF en 1996. 4 April 1997.

BIP (1997c): La crise du raffinage français: est-ce la quadrature du cercle ? No. 8320, 7 April 1997.

BIP (1997d): Les contraintes fiscales et environnementales qui pèsent sur les fiouls handicapent le raffinage français. No. 8350, 22 May 1997.

BIP (1997e): Bons résultats semestriels des raffineurs distributeurs français qui sont sous la menace de la suppression envisagée de la PFC. 3 October 1997.

BIP (1997f): L'environnement au cœur des débats du Congrès Mondial du Pétrole à Pékin. 17 October 1997.

BIP (1998a): Un résultat net de 7,6 GF pour Total en 1997: record historique et doublement en deux ans. No. 8523, 29 January 1998.

BIP (1998b): Résultats 1997 d'Elf Aquitaine. No.8544, 27 February 1998.

BIP (1998c): Esso France veut développer son amont et son raffinage. 27 March 1998.

Boisson, P. (1998): Energie 2010 – 2020. Les chemins d'une criossance sobre, rapport du commissariat du plan, groupe plénier présidé par Pierre Boisson, rapporteur général: Patrick Criqui, La Documentation Française, Paris.

BP (1998): Rapport annuel de BP France 1997. Cergy-Pontoise.

Chabbi M. (1998): Bilan: L'industrie pétrolière a enregistré de bons résultats l'an dernier. La voix du Nord, 28 April 1998.

Comité professionnel du Pétrole (1995): Pétrole 1995 : éléments statistiques.

Cosnard, D. (1996): Esso envisage de supprimer 380 postes en deux ans. Les Echos, 29 November 1996.

Cosnard, D. (1997): Shell décide de fermer en solo la moitié de sa raffinerie de Berre. Les Echos, 10 February 1997.

DHYCA (Direction des Hydrocarbures) (1995): L'industrie pétrolière: rapport d'activité 1995. Paris.

DHYCA (1998): L'industrie pétrolière 1997, Rapport d'activité, Direction Générale de l'Energie et des Matières Premières. Paris.

Direction de la Prévention des Pollutions et des Risques (1995): Les composés organiques volatiles : Programme français et directives européennes. Elements de synthèse.

DRIRE (Directions régionales de l'industrie, de la recherche et de l'environnement), Alsace (1996): L'industrie et l'environnement en Alsace. Bilan 1996.

DRIRE, Pays de la Loire (1997): Industrie et Environnement: le temps du dialogue.

Elf (1995): Rapport annuel 1995.

Elf (1996): Rapport annuel 1996.

Enerpresse (1998a): Elf s'engage à faire sa "priorité" de la protection de l'environnement. No. 7073, 14 January 1998.

Enerpresse (1998b): Le chiffre d'affaire consolidé de Total en hausse de 8,2% en 97. No. 7021, 25 February 1998.

Enerpresse (1998c): Le bénéfice net de Total sera voisin de 4 milliards de francs au premier semestre 1998. No. 7074, 15 May 1998.

Enerpresse (1998d): Shell France espère susciter 500 créations d'emplois en Provence d'ici 2000. No. 7064, 29 April 1998.

ESSO (1998): Rapport annuel 1997. Rueil-Malmaison.

Faucheux, S. and Nicolaï, I. (1998): Les firmes face au développement soutenable: changement technologique et gouvernance au sein de la dynamique industrielle. Revue d'Economie Industrielle No. 83, 1er trimestre 1998, 127–146.

Faucheux, S., Haake, J., Nicolaï, I. (1997): Implications de la Mondialisation Economique sur la Relation Environnement-Entreprises. Research Report to the French Ministry of the Environment, rapport C3ED/DGAD/SRAE No. 95285.

Faucheux, S., Gowdy, J., Nicolaï, I. (eds.) (1998): Sustainability and Firms. Technological Change and the Changing Regulatory Environment. Cheltenham: Edward Elgar.

Haake, J., Kuhndt, M., Liedtke, C., Orbach, T. Rohn, H. (1999): Firms and Dematerialisation. In: Gowdy, J., Hinterberger, F., Köhn, J., Van der Straaten, J. (eds.): Sustainability in Question: the search for a conceptual framework. Cheltenham: Elgar.

Van den Hove, S., O'Connor, M., Faucheux, S. (1998): EU Cliamte Policy. The European Community Policy. Special Issue of: International Journal of Environment and Pollution 10 (3/4).

IFP (Institut Français du Pétrole) (1998): Rapport annuel 1997. Rueil-Malmaison.

Jolly, D. (1992): La politique environnemetale d'un groupe industriel: le cas d'Elf-Aquitaine. Révue française de gestion, June / July / August, 117–121.

Julien, A. (1998): A cause des coûts de production, un avenir incertain pour le raffinage français. Le Havre presse, 4 May 1998.

L'environnement Magazine (1998): No. 1570, September 1998.

Le Tallec M. (1998): La facture sera lourde pour le raffinage. PSE Professionnels du Sud-Est, March 1998.

Legros, E.J. (1997): Une convalescence exposée aux rechutes. Pétrole Informations No. 1727, April-May 1997.

Legros, E.J. (1998): Auto-Oil, Préparer la raffinerie de l'avenir. Pétrole Informations No. 1732, 1 February 1998, 24–26.

Mineralölwirtschaftsverband (1996): Mineralöl und Raffinerien.

MOBIL (1998): Rapport annuel 1997. Mobil Oil Française. Courbevoie.

No Author (1997a): Raffinage : Shell retarde l'annonce de son plan pour Berre. Les Echos, 3 February 1997.

No Author (1997b): Shell réduit de 20% sa capacité de raffinage en France. Pétrostratégies, 17 February 1997.

No Author (1998a): Union Européenne : nouvelles normes antipollution, Les raffineurs trouvent la note salée. Le quotidien de la réunion, March 1998.

No Author (1998b): France: forte hausse des résultats 1997 de l'industrie pétrolière. Agence économique et financière, 28 April 1998.

Pétrole Informations (1996): Le raffinage toujours négatif dans les comptes de la BP France. November 1996.

Pétrole Informations (1997a): L'air a sa propre loi. No. 1726, February–March 1997.

Pétrole Informations (1997b): Shell remodèle son raffinage. No.1726, February–March 1997.

Pétrole Informations (1997c): Chronique d'un raffinage en difficulté. No 1727, April–May 1997.

Pétrole Informations (1997d): BP/Mobil, joint-venture mode d'emploi. No.1729, August–September 1997.

Pétrole Informations (1997e): Raffinage: l'environnement, moteur des investissements. No.1730, October–November 1997.

Porter, M.E., Van der Linde, C. (1995): Green and Competitive: Ending the Stalemate. Harvard Business Review, September / October 1995, 120–134.

Reicherts, M. (1998): Agreement is Reached on Auto-Oil Package. Unpublished internal EC document, 30 June 1998.

Revol, H. (1998): Politique énergétique de la France, Rapport éffectué pour le Sénat, deuxième partie, accessible on the Internet: http://www.senat.fr/rap/197-4392/ 197-4392-mono.html.

Rochard, E. (1997a): Shell ferme la moitié de sa raffinerie de Berre. La Tribune, 10 February 1997.

Rochard, E. (1997b): Total compte investir massivement en Normandie. La Tribune, 25 September 1997.

Rochard, E. (1998a): Les raffineurs s'insurgent contre les normes sur les carburants. La Tribune, 20 February 1998.
Rochard, E. (1998b): Le raffinage s'inquiète de l'impact des nouvelles normes antipollution. La Tribune, 28 April 1998.
SHELL (1998): Shell en France, Les chiffres 1997. Rueil-Malmaison: Shell France.
SHELL (no date): Le site pétrochimique de Berre. Plaquette d'introduction du site pétrochimique de Berre.
TOTAL (1998a): Factbook Nanterre.
TOTAL (1998b): Rapport annuel 1997. Nanterre.
UFIP (Union Française des Industries du Pétrole) (1995): Rapport: l'industrie française du pétrole.
UFIP (1995a): Qualité de l'air: la démarche européenne. Union Française des Industries Pétrolières. La Lettre No.14, January 1995.
UFIP (1995b): Avenir du raffinage distribution en France, propositions de l'industrie pétrolière. Communiqué de presse, 3 July 1995.
UFIP (1995c): Contribution de l'industrie pétrolière à la préparation du projet de loi sur la qualité de l'air. Communiqué de presse, 29 November 1995.
UFIP (1995d): Les difficultés du raffinage français : les conditions de redressement. Communiqué de presse, 5 December 1995.
UFIP (1998a): L'industrie du Pétrole. Rapport annuel 1997. Paris.
UFIP (1998b): Un pas dans le bon sens pour le rééquilibrage de la fiscalité entre essence et diesel. La lettre No.45, January-February 1998.
Vignes, J.-L., André, G., Kapala, F. (1994): Données industrielles, économiques, géographiques sur des produits chimiques (minéraux et organiques). 6th edn. Paris.

Additionally, information and data for this case study have been obtained from the following organisations, publications and websites:
ADEME: http://www.ademe.fr
BMWI – Bundesministerium für Wirtschaft, Germany
BP – British Mineral Oil
CPDP – Comité Professionel du Pétrole, France
DG XII: http://europa.eu.int/comm/dg12/index.html
DRIRE – Directions régionales de l'industrie, de la recherche et de l'environnement, France
Elf: http://www.elf.fr
European Commission: http://www.europa.eu.int/index.htm
European Environmental Bureau: http://www.eeb.org
EUROPIA: http://www.europia.com
Friends of the Earth: http://www.foe.co.uk
GEP: http://www.gep-france.com
Global Climate Coalition: http://www.globalclimate.org
IEA: http://www.iea.org
IFP: http://www.ifp.fr
Ministry of the Environment: http://www.environnement.gouv.fr
MWV – Mineralölwirtschaftsverband e.V. Hamburg, Germany
Observatoire Méditerranéen de l'Energie
Pétrole informations
Total: http://www.total.com

Oil Refineries in the Netherlands

Jan van der Straaten
Tilburg University, P.O. Box 90153, 5000 LE Tilburg, The Netherlands

1 Introduction

The main aim of this paper is to describe and analyse the position of Dutch oil refineries on the process of industrial restructuring. This restructuring was triggered by changes in supply and demand in the European markets. The fact that oil refineries are seen as substantial polluters has also resulted in action by national and international authorities, aiming to reduce pollution by these industries. In the Netherlands, no special regulations have been implemented on pollution specifically by oil refineries; all measures affecting oil refineries are based on general environmental policy. Of course, these general principles must be "translated" into the specific terms of Dutch oil refineries.

The situation for Dutch oil refineries is in fact complicated by market change, which has forced refineries to change their production processes. However, the new technologies, implemented in response to market change, did not produce the same levels of emissions. This means that it is not always possible to show whether a measure has been taken as a result of changing markets or whether it is the result of implementing strict norms in order to be able to meet policy goals. This means that it is not possible to investigate market driven industrial restructuring in Dutch oil refineries as a distinct topic in this case study. As soon as we try to separate market driven restructuring from other influences such as environmental policy goals, we run into trouble. Therefore, we have chosen to discuss the process of restructuring as an inseparable and indefinable part of the changes which took place in the 1970s and the 1980s.

In Sect. 2 of this chapter, we will describe the actors involved in the restructuring process and the theoretical framework. In Sect. 3, the market position of the Dutch oil refineries is sketched out in an international context. Next, the cornerstones of Dutch environmental policy are discussed and, more specifically, policies to reduce oil refineries' emissions. We will pay special attention to their results and to the position of Dutch oil refineries compared with other economic sectors. We will then discuss the position of Shell, the market leader in this sector, in the environmental debate. This will make it possible to give more specific information regarding the complex process of industrial restructuring in a context of changing markets, a strict environmental policy and possible future developments. In the last section, conclusions will be drawn and we identify some lessons for the future.

2 Actors, Structural Change, and the Theoretical Framework

There are eight oil refineries located in the Netherlands, mainly in the Rijnmond area. Oil refining companies such as Shell, Esso, Mobil, and Total, which are the most significant, can be regarded as oligopolistic market actors. They have a strong position, giving them the opportunity to regulate the market to a certain extent. Oil refining is a capital-intensive process, where the total labour intensity is relatively low. This implies that labour unions played little part in restructuring. So on the production side, the oil companies were the relevant actors.

On the other side, the national authorities wanted to implement strict norms for polluting industries, including oil refineries. Environmental action groups have always played a significant role in the Netherlands, trying to mobilise public opinion and putting pressure on the political process to reduce pollution levels as much as possible. In addition, these groups have fairly good access to the media, which means that they cannot be ignored by the authorities.

Having identified the actors involved, we have to decide which theoretical approach is best suited to analysing the restructuring of the oil refining industry in the Netherlands. Various approaches to the political process in general, and environmental policy in particular, can be found.

The traditional approach is the "formal structural model". In this model, it is "maintained that Parliament represents and interprets the public will through its representatives who formulate executive policies which are faithfully implemented by civil servants" (Jones 1991, pp. 505–506). This model can be useful, but it does not provide important information on the political process itself, neither does the model explain or analyse what happens after policy is formulated.

The pluralist model can be viewed as a critical response to the structural model above. Defenders of the pluralist model argue that it is interest groups as such, and not the combined actions of individuals, which influence the outcome of the political process (Ham and Hill 1993, p. 28; Grant 1985, p. 21).

In the European political system, we can find many pressure groups able to articulate their interests in the bargaining which is the reality of the political decision-making system. In the pluralist model, the role of the government is vague. In the liberal corporatist model, more attention is paid to this aspect. The role of pressure groups in modern societies is seen as crucial; the government recognises this position and tries to use pressure groups, by inviting them to bargain with the government itself. However, the government expects some kind of co-operation from them. The advantage for the pressure groups is that they will be able to influence policy. The advantage for the government is that these pressure groups will, to a certain extent, keep their members under control.

In this model, there is always a certain amount of formalisation: pressure groups know what the government expects them to do and, on the other hand, pressure groups know how they can benefit from this position. This model describes "an alliance between Ministers, civil servants and the leaders of pressure groups". (Jones 1991, p. 506). Leaders of pressure groups can control and regulate their members only to a certain extent, which implies that if the government asks too much of the pressure groups'' leaders, both actors will lose control of the situation. So many checks and balances are in fact built into the process itself (Williamson 1989).

The corporatist model can be applied well to the case of Dutch oil refining. The situation we have here is quite clear. There are formal structures, in which the main actor is the government, while oil refining companies and environmental groups try to co-operate and to react to government actions. As the oil refining companies also react to changing markets, which force them to use new technologies with different levels of pollution, the government has to react to these initiatives as well.

3 The Position of the Dutch Refineries

In the Rotterdam Rijnmond region, five oil refineries are located. One refinery can be found in Vlissingen along the Scheldt river. One small plant is located in Amsterdam and another in Delfzijl. Otherwise, however, the significant oil refineries are located in Rotterdam. Approximately 35% of production is for domestic use, while 45% is exported, mainly to Germany. 20% of production is used either by ships or by aeroplanes; Kerosene for Schiphol airport is important in this respect. Since 1989, Schiphol's turnover has increased by more than 100%.

As can be seen from Table 1, the significant environmental problems of oil refineries are caused by the use of large quantities of fossil fuel. When coal or oil are burned, sulphur dioxides are emitted. Electricity plants where these sources are also used produce similar emissions. The combustion results in oxidation of nitrogen in the air, leading to emissions of nitrogen oxides. Motor cars and electric power plants are also responsible for similar emissions of nitrogen oxides. Nitrogen oxide and sulphur dioxide emissions lead to precipitation of acidic substances. In the Netherlands, where there is a high level of intensive agriculture, emissions of ammonia also play a role in acid rain.

Table 1. The significance of different economic sectors in acid rain and energy use

	SO_2	NO_x	NH_3	CO_2
oil refineries	+++	+++		+++
power plants	++	++		+++
motor cars	+[a]	++++		++++
agriculture			++++	++

[a] trucks

Source: based on data from Zwerver et al. (1984)

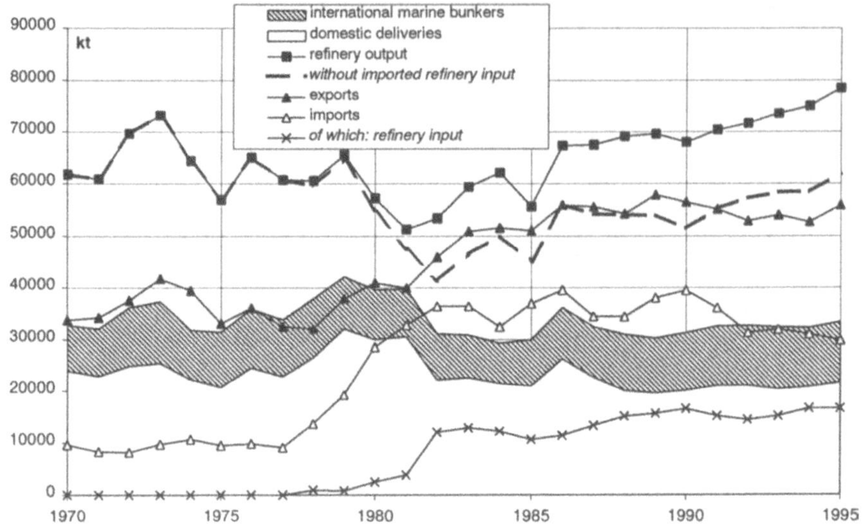

Source: IEA

Fig. 1. Dutch production, consumption and foreign trade of petroleum products 1970–95

Dutch production and consumption between 1970 and 1995 is shown in Fig. 1. As is usual in energy statistics, the petroleum products delivered to sea-going ships are not included in national energy consumption. Because they account for an abnormally large share of Dutch petroleum production, they are shown here separately, as "international marine bunkers". Those petroleum products which are imported for further processing in Dutch refineries are also indicated ("imports, of which: refinery input"). Domestic deliveries include total national consumption of petroleum products, with the exception of deliveries to the refineries themselves.

Dutch petroleum production is extremely and increasingly export-oriented. Thus, the decline in domestic consumption after the second oil crisis and its subsequent stagnation has not resulted in a lasting fall in production. This contrasts markedly with the French case study in this book, where a general fall in production is evident, even after a modest recovery in domestic demand. Particularly striking is the increase following the sharp decline at the start of the 1980s. Dutch oil refineries were presumably able to improve their competitive position during this period, but it should be noted that a remarkable share of this recovery is due to increases in the further processing of imported petroleum products, the final stage of production only. The broken line in Fig. 1 shows the difference between total production and imports of refinery feedstock, which result in almost the same quantities of end products. This is an indicator of the Netherlands' integrated production – from petroleum crude oil to products for final consumers – and shows a more modest increase, which has not yet brought production to 1970s levels.

The decline during the first and the second oil crises in the 1970s can also be made out from the graph, although this was not so important to the Dutch industry. More significant was the decline at the beginning of the 1980s, which was seen as a period of general economic crisis. This fall was of relatively short duration in the Netherlands and production has increased dramatically, by more than 50%, since 1981. There is a strong evidence that the sound development of the Dutch economy in the last decade has played a role in this respect.

There were no employment problems in the Netherlands until the economic crisis at the beginning of 1980s. At present, the employment situation in the Netherlands is not exactly favourable, but much better than in the majority of other European countries. The decrease in production levels at the beginning of the 1980s is not reflected in the employment figures of that period (Fig. 2). In the second half of the 1980s, employment rose while, later on, an irregular decline set in. In that period, new refineries were established and, additionally, there was an increase in the level of production which may have had a significant effect on the level of employment. The decline in employment in the sector took place during the 1990s, when the general employment situation in the Netherlands was relatively favourable. Thus, the employment situation in Dutch oil refineries could hardly have been a serious issue during that period.

Total production figures do not give sufficient insight into developments at Dutch oil refineries. Generally speaking, demand for light fractions such as naphtha, kerosene and motor fuels has been rising in recent decades. After the second

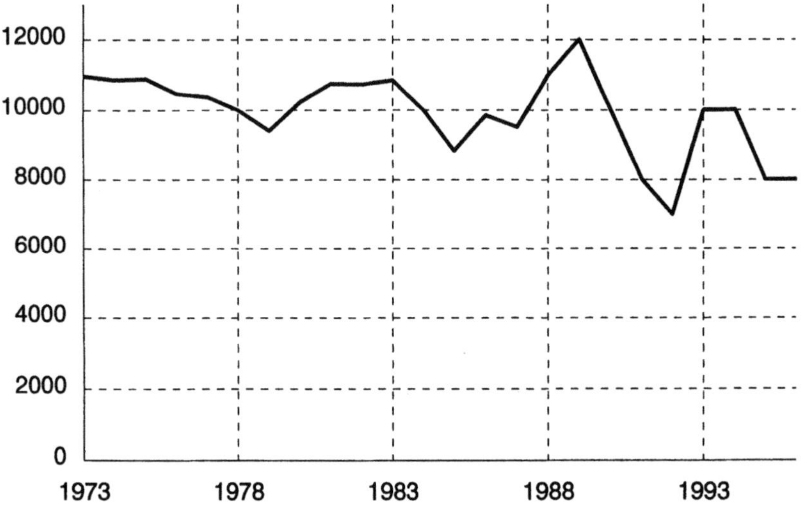

Source: Rijksinstituut voor Volksgezondheid en Milieuhygiëne 1997

Fig. 2. Total employment in Dutch oil refineries 1973–96

oil crisis in 1979, demand for fuel oil fell dramatically. It was broadly assumed that only oil refineries which produced a higher percentage of light products could survive. However, this could only be achieved with significant changes to the plants themselves. In the course of the 1980s, oil companies were therefore forced by the market to build expensive new refineries. Some of them were not able to do so. The Mobil refinery in Amsterdam was closed down in 1983, while the main parts of the Nerefco plant in Pernis shut down in 1989. Some companies, such as Esso, siezed the opportunity to introduce completely new technologies, enabling them to reduce emissions substantially. Others, such as Shell, refused to do so, and in fact built a fairly old-fashioned refinery, in which new environmental concepts found little or no resonance (De Bruin and Van Ooyen 1986; Barmentlo 1988).

The result of this modernisation was a fundamental change in the composition of the Dutch refineries' output (Fig. 3): While, up to the early 1980s, heavy fuel oil production remained low after a steep fall and production of gas oil (i.e. diesel and light fuel oil) did not change much throughout the period in question, quantities of light products increased substantially.

As shown above, the Dutch petroleum market is deeply integrated into international trade: both import and export flows exceed domestic demand. Imports of petroleum products (Fig. 4) originate to a great extent in Western Europe, especially neighbouring countries such as Belgium and the United Kingdom, but also

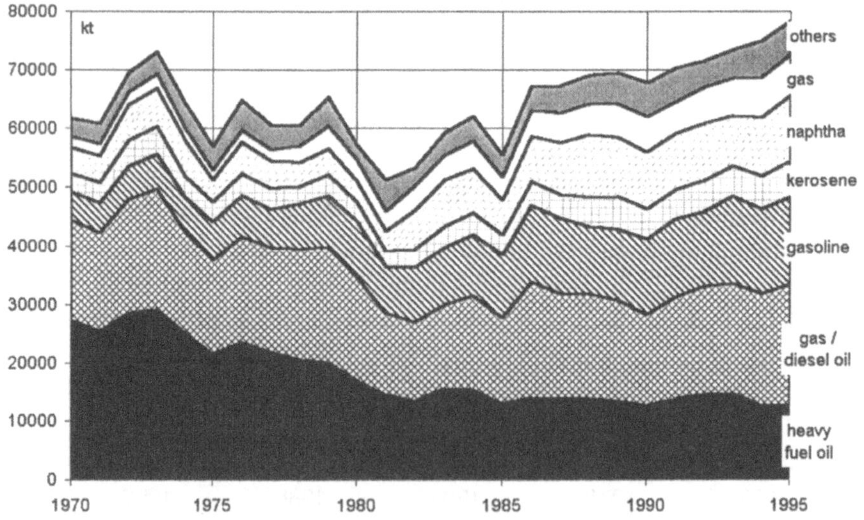

Source: IEA

Fig. 3. Total production of Dutch oil refineries by product 1970–95

in the former Soviet Union and Arab countries[1]. Dutch exports (Fig. 5) are almost completely destined for Western Europe, nearly half of them for Germany alone.

A closer look at domestic demand (Fig. 6) reveals developments which are fairly typical for industrialised countries:

- consistent but saturated demand for petrochemical feedstocks in the chemicals industry
- an at times drastic decline in other industrial or residential sectors, due to successful energy savings and substitution by natural gas and electricity, which resulted in stable but negligible market niches for petroleum producers
- a still growing demand from transportation sectors, mostly for road transport, but with higher growth rates on the still far smaller market in air transport
- a short-lived growth in demand for heavy fuel oils by large electricity and heat producers in the 1970s, which died out in the 1980s as a consequence of the second oil crisis, energy policies aimed at decreasing dependence on crude oil and successful competition from natural gas (or, in other countries, from nuclear energy or even coal)

4 Dutch Environmental Policy

4.1 The Political Climate in the Netherlands

It is obvious that the political climate in a country has a strong influence on its ability to solve political problems. This is especially true when the implementation of environmental goals with certain new instruments comes up for discussion. In many respects, the political climate in the Netherlands is the same as in other countries of Western Europe, but there are several important differences.

One of the most striking traditions in Dutch politics is the need for all political decisions to meet with general acceptance. This Dutch tradition has its roots in, among other things, an extremely open economy, which encourages tolerance in economic behaviour. The Netherlands is surrounded by countries which are among the most economically powerful in the world. When a country is accustomed to earning its money from trade with its neighbours, it is politically wise to give little public exposure to controversial issues.

Besides this, there is a long tradition of political and religious tolerance, which started during the long war against Spanish occupation (1568–1648). This climate facilitated the emergence of many political parties. Throughout Dutch history, there has never been a two-party system, as is rather common in other countries. Extreme opinions are not popular in Dutch politics. The large number of political parties necessitates coalitions in realising any policy.

[1] Imports from Arabia and the Soviet Union have experienced a considerable fall in recent years, but this may be a statistical illusion, since imports with an "unspecified" origin have increased at the same time.

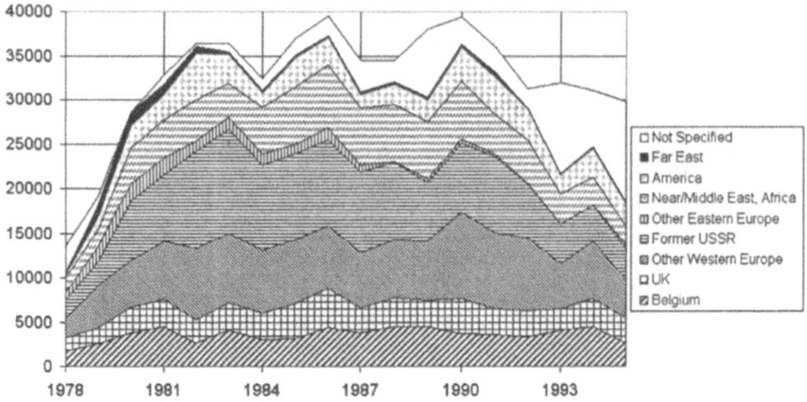

Fig. 4. Dutch petroleum imports by country 1978–95

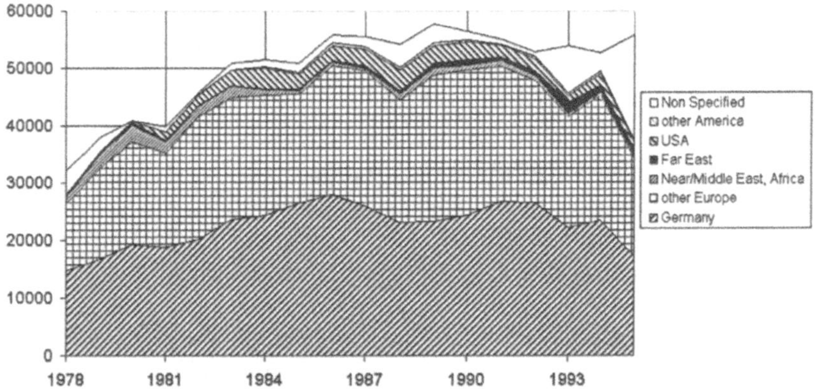

Fig. 5. Dutch petroleum exports by country 1978–95

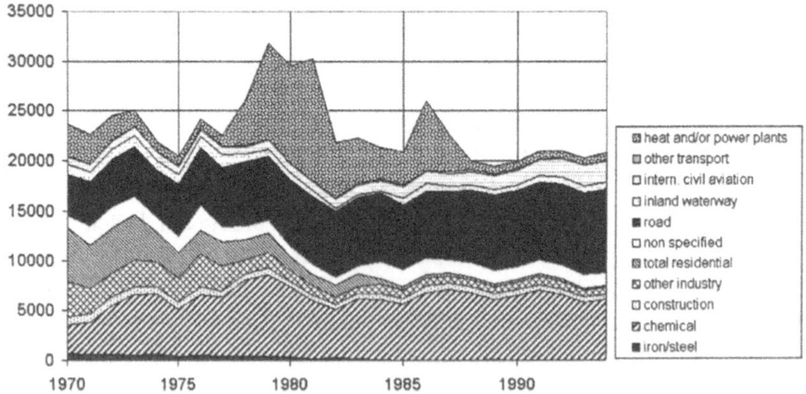

Fig. 6. Dutch petroleum consumption by sector 1970-94

This policy of tolerance and consultation has the advantage that a policy, when it comes to a decision, will be supported by all the parties involved in the problem. The disadvantage is that, in many cases, it is very difficult to reach this moment. Consequently, there is a strong tendency in Dutch politics to ignore the difference between the intent and the realisation of a policy. A ministry's written plan is often seen as "the" policy, not as a goal to be realised.

This is the picture in environmental politics as well. After the publication of the National Environmental Policy Plan (NEPP), 88 organisations in the Netherlands published their comments on it (Beenakkers, Van Esch, Lacroix and Van de Moosdijk 1990). Of course, it would be impossible to give all these reports and memoranda their due attention, but there is a general belief that this is the way things should be done.

4.2 The National Environmental Policy Plan

Till 1989, Dutch environmental policy implementation was relatively weak and fragmented. In 1989, The Dutch Minister of the Environment launched a completely new approach: the National Environmental Policy Plan. This plan is generally seen as a milestone in Dutch and European environmental policy.

What really strikes the reader of the NEPP is the painstaking analysis the government made of environmental problems. In many cases the reports are vague and lack a real framework. They are, generally speaking, political documents without any real expressive power. However, the analysis of environmental problems in the NEPP could have been written by environmental groups such as Greenpeace or Friends of the Earth. It is a clear sketch of the state of affairs in the environment in the Netherlands.

In the NEPP, environmental problems are described according to the scale at which they occur: local and regional scale such as noise, odour, local air pollution, household refuse; continental scale such as acidification; global scale such as CO_2 emissions, climate change, and the ozone layer; at the level of river basins, such as eutrophication. Environmental problems are seen to affect the social and economic functions of the environment negatively. These problems are not seen as isolated issues, "they are interconnected and share common causes: opening or changing substance cycles, more intensive energy use and the neglect of quality aspects in production processes and products. The environment's sustainability is a reservoir, as is the quantity of a raw material – whether renewable or not. " (VROM 1989, p. 9). These ideas are strongly related to arguments used in environmental economics (see among others Dietz and Van der Straaten 1992; Opschoor and Van der Straaten 1993).

Attention is paid to the place and level at which measures should be taken: "Source oriented measures are to be preferred to effect orientation, because the possibility of control is greatest at the source; one source can produce more than one effect, uncertainties can exist about the causal chain, irreversible effects can occur, and it generally costs less to intervene at the source" (VROM 1989, p.13).

Source oriented measures can be divided into emissions measures, such as add-on technology, volume measures and structural measures, which change production processes.

At global level, the change in the climate caused by emissions of CO_2 is seen as an important issue. Emissions of carbon dioxide should be reduced to the level at which the oceans and the biosphere can absorb them.

At continental level, attention should be paid to emissions of acidifying substances, hydrocarbons and non-degradable or poorly degradable substances, such as heavy metals. This is necessary to stop the effects of acid rain, such as dying forests and destruction of monuments. Measures should be taken at the source, in view of the damage which is already occurring. To achieve a viable situation for forests and monuments, a reduction of 80 to 90% is necessary.

Effect oriented measures are still necessary, due to legacies from the past and the delay in source oriented measures, which nonetheless have higher priority. Emissions measures, such as add-on technologies, should be avoided if possible, as this is not compatible with sustainable development. In many cases, these measures consume energy, and there is no closure of cycles. Nevertheless, we will not be able to avoid them in the near future.

4.3 The Real Policy Field for Oil Refineries

The main environmental problem of oil refineries has always been emissions of acidifying substances, in particular SO_2 and NO_x. More recently, CO_2 reductions have become an important issue in the Netherlands. In the National Environmental Policy Plan (VROM 1989), the use of fossil fuels is recognised as a serious global problem, but real policy goals have been formulated only in recent years, which means that it is not possible to investigate Dutch policy on the accelerated greenhouse effect in the 1970s, the 1980s and the 1990s. However, the use of energy and oil in particular was paid much attention during the oil crises. The main concern was not global warming, but the dependency of the Dutch economy on Arab oil. In that period, electric power plants shifted away from the use of oil, and coal was given a higher profile in Dutch power plants.

This shift was strongly encouraged by the Minister of Economic Affairs, who not only preferred lessening Dutch dependency on Arab oil, but was also trying to aid Dutch energy intensive industries by lowering electricity prices, as coal was cheaper than oil. The decrease in the use of coal would result in higher pollution, which could then only be minimised with a strong abatement policy. This shift towards coal influenced oil refineries' demand for oil products considerably.

Nevertheless, we can conclude from Fig. 1 (p. 388) that these policy goals did not negatively impact on production by Dutch oil refineries. Three factors are responsible for this. In the first place, it should be remembered that Dutch policy on reducing energy consumption in response to the greenhouse effect has been and still is ineffective. In fact, there is a lot of window dressing in this policy area. Many memoranda and documents from the Minister of the Environment are full of the need to reduce energy consumption, but few concrete measures have been taken. Furthermore, as far as a real Dutch policy goal has emerged in this respect, this was only the case after the restructuring of Dutch oil refineries took place in the 1970s and 1980s.

A second reason for the increase in production is the rise in fuel consumption by Dutch automobiles. During the period in question, the number of cars grew rapidly. On the one hand, there was a strong tendency to produce cars with lower fuel consumption; on the other hand, the percentage of small cars with relatively low fuel consumption decreased sharply. This resulted in a dramatic rise in fuel consumption by Dutch cars. A third reason for the rise in production is the development of air traffic and in particular the significant growth of Schiphol airport. The importance of kerosene to Dutch oil refineries rose sharply during this period.

From this discussion it becomes clear that market forces had the strongest effects on production levels in Dutch oil refineries. The Minister of the Environment was not able to neutralise market effects from higher energy consumption. The growth in production was, in fact, an intricate and unusual product of market forces, industrial restructuring in the sector, acid rain policies and energy reduction policies.

The polluting sectors affecting acid rain are closely interconnected. This is mainly caused by the fact that various sectors in the economy produce acidic emissions. The implication is that environmental policy on oil refineries relate to other sectors which are also responsible for emissions of acidifying substances. This means that it is not possible to discuss acid rain policy and oil refineries as a discrete item. This brings us to Table 2, with which we can analyse the different policy outcomes for oil refineries.

We have to be aware of the special position of environmental NGOs in the Netherlands, particularly in this case. Environmental groups in the Netherlands were, generally speaking, not outsiders during the research period, as was the case in many other European countries. This means that during the entire period, general environmental policy was strongly influenced by environmental groups. Furthermore, environmental groups were very active on special issues, including acid rain.

The acid rain issue is quite complicated, as it is caused by emissions of sulphur dioxide, nitrogen oxides, and ammonia. Major producers of these emissions are the oil refining industry, power plants, transport, and intensive cattle breeding. Furthermore, a large percentage of these emissions is transported by wind to other parts of Europe. On the other hand, a high percentage of the emissions in other European countries is also deposited cross-border, including in the Netherlands.

The deterioration in forests in the Netherlands began in the early 1980s. From this time on, the Dutch government became active in reducing acidic emissions and precipitation. Of course, the main question was how much the acid precipitation had to be reduced. In the beginning, the costs and benefits of an abatement policy were weighed up (Minister van Volksgezondheid en Milieuhygiëne 1982–83; VROM 1983–84). However, these calculations proved unusable, as too many

Table 2. Key actors and their actions in acid rain policy

		environmental NGOs	Dutch	EU environmental policy	international [a]
oil refineries	SO_2	+++	+++	+ [b]	+++ [c]
	NO_x	++	+		++ [d]
power plants	SO_2	++++	+++	+ [b]	+++ [c]
	NO_x	+++++	+	+	++ [d]
motor cars	SO_2	0	0	0	0
	NO_x	++	++++++	++++++ [e]	++ [d]
agriculture	NH_3	+++++++	++++	0	0

[a] This column reflects international policies not generated by the European Community or the European Union.
[b] EU directive
[c] Helsinki and Oslo protocol
[d] Sofia protocol
[e] catalytic converter
based on Van der Straaten (1990)

significant environmental effects could not be priced. Therefore, the Minister of the Environment decided to follow another line of reasoning. A functioning ecosystem was taken as the starting point for all environmental policy. In the National Environment Policy Plan, targets were redefined. For the long term, a level of 1400 was adopted, and for the short term 1800.

Acid precipitation is caused by emissions of sulphur dioxide, nitrogen oxides, and ammonia. Sulphur dioxide is emitted mainly by oil refineries and electric power plants. The nitrogen oxides emissions result from high temperature incineration, such as in electric power plants, oil refineries, and cars, while ammonia results from the manure produced in intensive cattle breeding. In this period, sulphur dioxide emissions were found to be the main problem. Approximately 70% of acid rain was caused by these emissions, which are effected by the sulphur content of the fuels used in oil refineries and power plants. Acid rain abatement policy in this period focused on these economic sectors.

It can be concluded that the Dutch abatement policy was quite successful in substantially reducing deposits. However, the targets as defined in the National Environmental Policy Plan of 1989 could not be reached. It is obvious, too, that ammonia and nitrogen oxides emissions did not decrease during the period. This implies that Dutch farmers in particular have been able to exploit more than their share of the environmental space concerning acid deposits. In other words, the agricultural sector has been able to neutralise any efforts to reduce emissions. Fig. 7 shows recent figures that make this abundantly clear.

In the 1980s, the acid rain problem became increasingly the problem of ammonia and nitrogen oxides. Thus, oil refineries contributed considerably in the policy

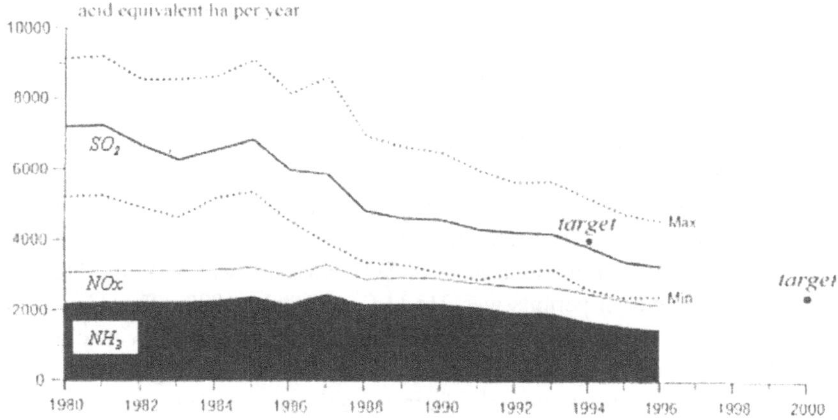

The dotted lines indicate regional differences.
Source: Rijksinstituut voor Volksgezondheid en Milieuhygiëne, 1997

Fig. 7. Mean deposits of potential acid in the Netherlands in 1980–96

goal of reducing substantially acidifying substances. Farmers and road traffic are the main targets in the current acid rain debate.

As was argued before, the aims of Dutch environmental policy are formulated broadly in the NEPP. Reduction targets for acid deposition and the time scale are set out. These targets relate to the total level of acid precipitation, which means that the civil servants of the Ministry of the Environment have to negotiate with all polluting sectors, including oil refineries, on emissions reductions needed across the different sectors. Of course, the total effect of all measures to be taken by all polluting sectors should equal the total set out in the NEPP. There is no *a priori* reason to suppose that the Ministry of the Environment will be successful in this respect. This depends on the results of negotiations with all polluting sectors.

5 Industrial Restructuring or Environmental Modernisation?

The figures might give the impression that oil refineries were, generally speaking, co-operative on reducing acidifying emissions. A closer analysis of the industry's reactions to stricter norms set by the Minister of the Environment in the 1980s, however, prove this to be only partially true.

The first initiatives were taken in 1984 (VROM 1984–85). This document stated that the German Grossfeuerungsanlagenverordnung (GFAV) would be used as a general reference point for policy measures. However, as regards oil refineries, Dutch policies have to differ from German measures, as "Dutch refineries are more dependent on exports, which makes these refineries far more vulnerable than refineries in Germany, in particular regarding capacity in France and Great Britain and increasing capacity in the Arab World" (VROM 1984–85, p. 59). This statement cannot be taken seriously, as much of Dutch production is exported to Germany, where standards are stricter than in the Netherlands. It has nonetheless played a significant role in the debate on the need to reduce oil refinery emissions, as will be demonstrated later on.

The Minister's claim to have reduced emissions can be viewed in different ways. According to documents in the Ministry of the Environment (VROM 1983–84, p. 25, Table 5.2.1), the total level of SO_2 emissions was 93 kt in 1983. However, the total level in permits was 215 kt (Van Ooyen 1985). The disparity was caused by the fact that oil refineries were not functioning at full capacity in that period. The permitted levels were based on full capacity. As regards reduction of emissions, the authorities can only, from a legal point of view, change permitted levels. During periods of lower capacity, the only result of lowering the defined permit levels would be a smaller gap between permits and real emissions. So, in fact the measures taken by the Minister in that period had no real effect on the level of emissions.

Nevertheless, the oil refineries were faced with increasing pressure from the Ministry of Environmental Affairs to substantially reduce emissions. The traditional way to realise this is either to reduce the content of sulphur in crude oil or to desulphurise the flue gas. Often, the heavy fractions are used as a source for heating in the refineries themselves or in electric power plants. However, Dutch policy aimed to reduce oil dependency in electricity production, and power plants were therefore adapted so that it was possible to use natural gas and coal as fuels. This meant that heavy fractions were no longer an appropriate fuel in electricity production. Oil refineries ran into difficulties, since they could not use the heavy fractions in their own plants, as this would increase the level of emissions. But on the other hand, there was less market demand for heavy fractions, as demonstrated in the previous paper in this volume. The market and the Ministry of the Environment's desire to substantially reduce emissions could only be realised by adapting the oil refinery plants themselves.

Restructuring of the oil refineries was achieved with various strategies. Esso, for example, was able to harmonise market developments with the implementation of strict environmental standards. Total emissions by Esso were 28 kt in 1980 (Openbaar Lichaam Rijnmond, December 1983). With the introduction of the completely new Flexicooker, they reduced these emissions to 10 kt. In fact, they were able to reduce this even further to 5–6 kt. However, these reduction measures did not increase costs. On the contrary, Esso has said that investment in the environment could be seen as profitable (De Bruin and Van Ooyen 1986). In other words, Esso was able to make a profit from compliance. This could only be realised by using the best available technology, developed by Esso itself. Pulles and Wiersma (1985) also came to the conclusion that a reduction of emissions could be realised without additional costs, by using an integrated technology which makes it possible to restructure the composition of output to meet current market demands.

Shell's response was completely different. It did not recognise the dangers in market developments; it requested a licence for a new refinery, the Hycon installation, as late as 1985. The new Esso refinery was already in place by then. This new plant could meet the changing demands of the market by making lighter products. However, Shell argued that no substantial progress could be made in reducing emissions. Emissions by the old refinery were approximately 70 kt (Openbaar Lichaam Rijnmond 1983; Franssen 1984). Shell claimed a new emissions level of 53 kt per year. When the figures from Shell and Esso are compared, a striking difference can be seen. In fact, Shell was unwilling to reduce emissions. On the contrary, it contacted the Minister of Economic Affairs and argued that if the authorities insisted on an emissions level below 53 kt, it would no longer invest in the Netherlands (Barmentlo 1988). This was essentially blackmail, as the norms for the same type of oil refinery in Germany were at the time between 4–8 kt per year (Franssen 1985). This "discussion" with the Minister got Shell what it wanted. The Minister of Environmental Affairs revoked the stricter standard and Shell was given a permit for 53 kt of emissions (Barmentlo 1988).

From a commercial point of view, it was Shell who took the wrong course. Hycon did not work adequately, as the new refinery was still based on old concepts and techniques. The need to reduce emissions was seen not as a challenge but as a danger. Shell's reaction was defensive, with negative effects on the design of the new refinery. Now, Shell is burdened with a "new" old-fashioned refinery which can neither meet the demands of the market nor the necessary reduction of emissions. In the end, Shell decided to build a genuinely new refinery based on modern principles. Only by doing so were they able to meet the new standards.

Shell's position in the Netherlands has been much debated in recent years. First came Brent Spar: in Germany and the Netherlands, the reactions of both the public and the authorities were very negative. We need not discuss this issue in detail here to conclude that Brent Spar was an environmental eye-opener for Shell. Another recent issue in the Netherlands is Shell's conduct in Nigeria, and the role it played in suppressing the environmental movement in a country ruled by a dictatorship. Again, reactions were very negative, even on the part of shareholders. More recently, Shell has given a press conference, in which it said that it had responded correctly to the many difficult issues. Additionally, it promised to incorporate environmental aspects in its company policy from then on. The public was surprised and somewhat sceptical.

Shell's current position is not far from green modernisation. However, when one takes their numerous negative reactions over a longer period into account, it is not to be expected that Shell will radically change its position. It needs time to incorporate these other practices and cultures of response.

6 Conclusions and Lessons

In the case of restructuring among oil refiners, the market was more important than environmental policy. The shift of the market towards light fractions forced refiners to build new refineries. Esso declared that, with their new refinery, very strict environmental norms could be introduced without an increase in total costs. Although still not the market leader, Esso proved to have a more up to date technology than market leader Shell.

The Minister of the Environment could have used the existence of Esso's technology as an argument against Shell. In fact, the minister was in a relatively strong position, because he could argue that Shell's complaints of excessive costs could not be justified, given the cheap Esso technology. The Minister of the Environment was badly informed about these different technologies, which made it impossible for him to resist pressure from the Minister of Economic Affairs.

Later on, it became clear that Shell would have to upgrade its plant again, due to the ongoing increase in demand for light fractions. It was then able to meet the stricter standards. Again, the influence of market change was stronger than that of environmental policy.

It makes sense for the Minister of the Environment to extend the quality and number of technical specialists at the ministry. In many cases, they will be able to provide the information necessary for exerting pressure on polluting industries.

The way in which policy issues are discussed and analysed in the Netherlands, in an open environment, where all groups, including NGOs, can contribute to the process, gives more flexibility to build up the necessary support within society for implementing strict standards. Polluting sectors are, in such an approach, always in a weaker position than they would be in a more rigid form of decision-making.

Although the results of environmental policy are quite substantial, it is notable that in the Dutch policy debate the vast majority of proposals and ideas remain within existing frameworks. Radical ideas and proposals are seldom put forward. Of course, this brings us to the crucial question of whether Dutch environmental policy results would not have been far better if the government had had the courage to come up with completely new ideas.

The distribution of access rights to use environmental space as a sink for acidifying substances was not paid sufficient attention by the Dutch authorities. This has led to a situation where farmers are currently using more than 50% of the total available to all polluting sectors. This brings other polluters, such as power plants, car owners, and oil refineries into a difficult position.

The difference to the French case study is striking. In France, no new technologies have been developed, while important information on policy aims and outcomes is also missing. This implies that NGOs are not able to obtain sufficient information to put pressure on the political system. In such a rigid and closed system, new developments in the field of environmental policy cannot be expected, as the necessary stimuli are lacking.

Threats by polluting industries that they would no longer invest in their own country if strict norms were implemented, and that they would expand in other countries, proved not to be serious. In Germany in particular, standards for oil refineries were and continue to be stricter than in the Netherlands. The Minister of Environmental Affairs should try to obtain all the relevant information on technologies and standards in polluting industries. This would provide him or her with a stronger position in negotiations with polluting industries.

Quite often, civil servants regard polluting industries' technologies as a given. The example of Shell and Esso makes it clear that this is often not the case. Technologies and their implementation are in a state of dynamic change. Furthermore, it should not be overlooked that this also applies to the market. Changes in demand for certain products can create new challenges requiring new technologies. Government should not therefore lay down the way in which pollution is to be reduced. The best way is to formulate targets to be reached within a certain period of time, and to let the polluting sectors themselves develop the new technologies and products necessary to meet political demands.

The Dutch energy conservation policy was based mainly on the need to reduce dependency on Arab oil. During the process of restructuring, the effects of policy aimed at reducing energy consumption in response to the greenhouse effect were either unnoticeable or non-existent. Market influences were far more important for the level of oil consumption than policy goals.

References

Barmentlo, J. (1988): Shell als Succesvol Lobbyist. Afstudeer-scriptie Organisatiesociologie (Master's thesis Sociology of Organisations). University of Leiden.

Beenakkers, M.F.M., Van Esch, S.A. , Lacroix, L., Van de Moosdijk, M.M.E. (1990): Het Nationaal Milieubeleidsplan Getoetst. Nijmegen: Katholieke Universiteit Nijmegen.

De Bruin, W. and Van Ooyen, D. (1986): Esso Helpt Beter dan Shell. Milieudefensie, December, 8–9.

Dietz, F. and Van der Straaten, J. (1992). Rethinking Environmental Economics: Missing Links between Economic Theory and Environmental Policy. Journal of Economic Issues, March, 27–51.

Franssen, J. (1985): Bezwaren tegen Vergunning voor Luchtverontreiniging door Shell, Natuur en Milieu 85/7–8, p. 25.

Grant, W. (1985): Introduction. In: Grant, W. (ed.); The Political Economy of Corporatism. London: Macmillan.

Ham, C. and Hill, M. (1993): The Political process in the Modern Capitalist State. Harvester Wheatsheaf, Hemel Hempstead.

IEA (International Energy Agency) (1997): Annual Oil and Gas Statistics (CD-ROM). Printed as: IEA: Oil Information, various issues.

Jones, B. (1991): The Policy Making Process. In: Jones, B. (ed.): Politics UK. London: Philip Allan.

Minister van Volksgezondheid en Milieuhygiëne (Minister of Public Health and Environmental Policy) (1982–83): Nota kosten en macro-economische gevolgen van het milieubeleid 16495, nrs. 1–2. SDU. The Hague.

Van Ooyen, D. (1985): Shell helpt de bossen om zeep. Milieudefensie, April 1985, pp. 4–6.

Openbaar Lichaam Rijnmond (1983): SO_2 Nota. December 1983.

Opschoor, J.B. and Van der Straaten, J. (1993): Sustainable Development, An Institutional Approach. Ecological Economics 7, 203–222.

Pulles, T. and Wiersma, D, (1985): Luchtverontreiniging door olieraffiinaderijen in een veranderende markt. IVEM rapport Nr. 9. Groningen: Rijksuniversiteit Groningen.

Rijksinstituut voor Volksgezondheid en Milieuhygiëne (1991): Nationale Milieuverkenning 2 1990–2010. Samsom H.D. Tjeenk Willink, Alphen aan den Rijn.

Rijksinstituut voor Volksgezondheid en Milieuhygiëne (1997). Milieubalans 97. Samsom H.D. Tjeenk Willink, Alphen aan den Rijn.

VROM (Minister van Volkshuisvesting, Ruimtelijke Ordening en Milieubeheer, i.e. Minister of Housing, Physical Planning and Environment) (1989): National Environmental Policy Plan. Session 1988–1989. No. 21137–1. SDU. Den Haag.

VROM (1983–84): Indicatief Meerjarenprogramma Lucht 1984–1988. 18100, Nos. 1–2. SDU. The Hague.

VROM (1983–84): De Problematiek van de verzuring. 18225, Nos. 1–2. SDU. The Hague.

VROM (1984–85): Indicatief Meerjarenprogramma Lucht 1985–1989. No. 18605, Nos. 1–2 and 8. SDU. The Hague.

Williamson, P.J. (1989): Corporatism in Perspective: An Introductory Guide to Corporatist Theory. London: Sage.

Zwerver, S., Bovenkerk, M., Mak, P.J. (1984): Verzuring: nationale en internationale ontwikkelingen. In: Adema, E.H. and Van Ham, J. (eds.): Zure Regen, oorzaken, effecten en beleid. Wageningen: Pudoc.

Nuclear Energy in Germany

Annette Piening
Free University of Berlin, FFU, Ihnestr. 22, 14195 Berlin, Germany

1 Introduction

Soon after the nuclear power plant industry reached relevant production levels in the late 1960s, serious environmental and safety concerns around the commercial use of nuclear energy were publicly brought forward in most Western nations, among them Germany, and have continued ever since. Internationally the industry's life cycle has passed its production peak. The trend for years has been an almost continuous decline in nuclear power plant construction. The number of orders for new reactors world-wide peaked in 1974 and since then has decreased to pre-1965 levels (fig. 1). In 1996, for the first time, more capacity was decommissioned than put on line, beginning a decline in nuclear electricity production that has continued until the present (fig. 2).

In studying the influence of environmental policies on the structural change of industries, the question therefore arises as to how much public policies and the industry's development were influenced by these early environmental concerns and to what degree these policies have been a decisive factor in the decline of the industry. In an attempt to contribute to an answer, the aim here is to analyse the German nuclear industry and the political and social framework within which the various actors have brought forward their interests and formed their strategies. Emphasis will be placed on the analysis of public policy with regard to the nuclear industry, the main determinants of public policy, and the role of environmental goals in shaping it. This will be preceded by a short description of the industry, its composition, and its economic status today.

2 The nuclear industry and market structure

2.1 General characteristics of the nuclear industry

Unlike most other industrial sectors, national nuclear industries are very much the result of specific national, often military defence policies. They are further influenced by such factors as a country's particular resource base or economic structure. This has led to a considerable variety in the structural features of the industries. Nevertheless, there are some general characteristics which hold true for nuclear industries world-wide.

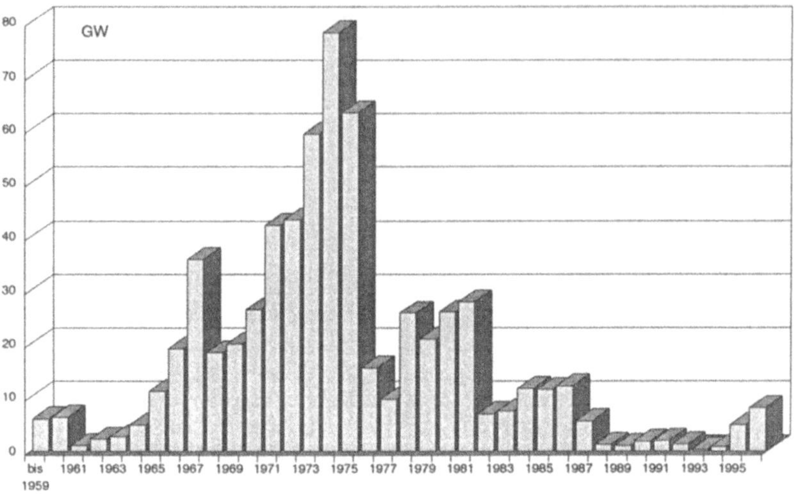

Source: Jahrbuch der Atomwirtschaft 1998

Fig. 1. Nuclear power stations ordered world-wide 1950–96

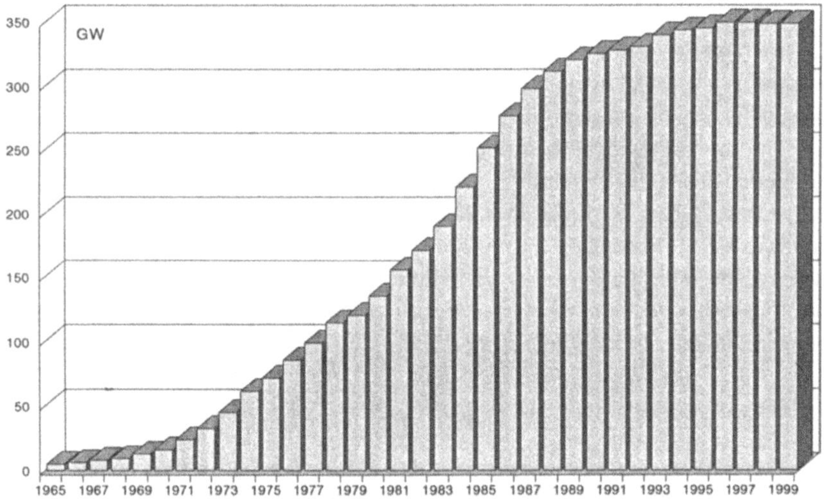

Source: IEAE, PRIS 1999/2000

Fig. 2. Nuclear capacity operational worldwide 1965–99

The nuclear industry is, in contrast to most basic industries, a young sector. Having started to take shape after World War II, it became commercialised in the mid-60s. It is rooted in nuclear weapon technology but, equally importantly, also in the chemical industry, on which it relies for nuclear fuel (re)processing and conditioning activities. Other engineering experience is concentrated in the electronics and machinery industries, i.e. those which are also technically involved in conventional power generation. At least in part, this explains the involvement of many companies from these sectors in early nuclear power enterprises.

The nuclear industry is characterised by the heavy involvement of state actors who have had a strong influence on the industry's development since its beginnings. There are three main reasons for this. First, the nuclear industry's close evolutionary links to nuclear weapon technologies has had serious implications for the aim of non-proliferation. Any activities in this market thus arouse international political attention and suspicion. Second, due to the high financial costs involved, governments have supported nuclear R&D with large sums compared to other industries. Third, the inherent risks and the relative lack of experience with nuclear technology have called attention to nuclear safety problems. In some nations, these factors have supported direct government ownership of main sections of the industry, as in France or Britain.[1] In Germany, as in the US, the privately owned industry must conform to strict regulations, while over long periods it has received a disproportionately large share of public research funds. In general, national policies, i.e. the explicit decision to promote the industry's build-up, are the determining factor for the existence of a relevant nuclear industry and the use of nuclear power in any given country. This has to be seen as an idiosyncrasy of the nuclear industry, albeit one that it shares with the aviation and defence industries (Moldenhauer 1975; Mez 1979, p. 335).

However, only those countries having actively researched already in the late 40s succeeded in developing a national, domestic reactor line, namely the USA, Soviet Union and Great Britain. Latecomers, like Germany and France, had the disadvantage of having to compete against second generation reactors (100 and more MW as opposed to 10–50 MW of the early years) and ultimately bought light-water reactor licences, which they further developed into national patents.

Despite the strong national policy influence, the nuclear industry today operates mainly at the international level and most national nuclear industries are affiliated or co-operate with companies of other nations. This holds true particularly for the German nuclear industry: International co-operation was elementary to its first success in constructing a nuclear power plant and even its early business strategies relied heavily on export prospects.[2] Nevertheless, only a few companies, predominantly US and former Soviet producers, have attained sizeable shares on the international market, while, for example, the German Siemens/KWU and French

---

[1] The other crucial reason for choosing a state managed system was that the UK and France have a well-established tradition of publicly managed provision of utility services.

[2] Cf. KWU 1969, 1970; Mez 1998, p. 115

Framatome have been most successful in their domestic markets. Until the take-over of the US-Westinghouse nuclear business by British Nuclear Fuels Ltd. in 1998 (Masters 1999, p. 20), only in the US and Japan did more than one company offer turnkey nuclear plants. Furthermore, links exist to the electricity sector, which constitutes the nuclear industry's demand side. Here, from a technological and economic viewpoint, nuclear energy encounters strong competition from alternative supply options – currently primarily from natural gas and coal. While nuclear power has a share of about 10% of electricity production world-wide, only 32 countries have decided to use nuclear power at all. In some of these, its share in electricity generation exceeds 75%.[3] Due to investment periods of over 20 years, nuclear power generation has had lasting effects on national electricity systems in most countries in which it is employed. They increasingly have been characterised by centralised supply structures and, consequently, supply-oriented planning. Moreover, the large investments required for nuclear power plant construction have contributed to an increased concentration in electricity supply industries (Kitschelt 1980, p. 130). In the case of Germany, all major electricity companies today meet a substantial part of their base load demand through nuclear power and the biggest companies are (financially) deeply involved in the nuclear power market's fuel cycle segment.

2.2 The German nuclear power industry

In 1998, 19 nuclear power plants with a capacity of 22 200 MW contributed 30% of the electricity generated in Germany (BMWi 1999, p. 11). The last plant was ordered in 1975 and put on stream in 1989; since then no new nuclear power plants have been planned.[4] With regard to total supply, nuclear power is, next to coal, the second most important power resource. For years the industry's efforts were concentrated on realising the ambitious nuclear programme and gaining a share in the international market. However, as in other countries, demand projections for nuclear plants had to be reduced in succeeding years. After the Chernobyl accident expansion goals were abandoned[5] and the focus since then has primarily

[3] The leading nuclear power producer are the US with 110 nuclear power stations, which produced 673.70 TWh in 1998. While this covered 18.7% of demand in 1996, France met 75.8% (1998) of its demand with nuclear energy. Leading in terms of dependency is Lithuania with 77.2% nuclear electricity share in 1998, however, only one nuclear power plant is operated (Masters 1999, p. 15).

[4] To the 19 plants, one further scandal-ridden plant, Mülheim-Kärlich, has to be added. It has been out of operation due to an invalid construction permit but has not been decommissioned yet. Known for its record construction period and costs, its commercial value is being revitalised in the context of the government-industry nuclear phase out negotiations by plans to include it with a 30 year credit in the shut off schedule (cf. also below 3.2.4.).

[5] A definite turning point was the accident at Three Mile Island in 1979, which set off a series of cancellations and refitting activities in the US and other countries. First cutbacks of German expansion plans, however, were already made in 1976 mainly due to decreasing growth of demand (Keck 1993).

been on improving the performance of existing reactors through upgrading, refurbishment and managerial changes. By comparison with other countries, the nuclear industry in Germany has been quite successful in this effort, and German nuclear plant operators have earned substantial profits. Nevertheless, the nuclear plant producing sector has been scaled back continuously over the last decade. Despite these difficulties, the nuclear sector was regarded as one of the most lucrative commercial sectors in German history (Der Spiegel 5.4.1993, p. 147).

In 1999 the industry employed about 38 000 people. Of these, 11 600 were highly qualified personnel at the nuclear power plants and ranked among the best paid workers in Germany. A further 10 000 people were employed to carry out plant revision, transport and fuel reconditioning activities. However, due to the high irradiation levels, the majority of these were employed on a temporary basis only. About 16 000 jobs were generated in other industrial sectors through the subcontracting of services (Mohr 1991, Majewski 1999, p. 16).

Structurally, the German nuclear industry[6] can be divided into three subsections: first, the construction of nuclear power plant and maintenance work; second, the operation of plant; and third, the nuclear fuel cycle business. Each of these segments is concerned with different activities which on the surface appear to be carried out by different companies or market actors. A more detailed analysis reveals close ties between the segments, ultimately presenting a picture of the nuclear industry as a closed and highly concentrated market.

2.2.1 Nuclear power plant production and construction

Initially, five German companies, some with foreign partners, took part in the federal "Eltville" R&D programme of 1957 to construct 100 MW nuclear reactors: Deutsche Babcock & Wilcox, Siemens, BBC/Krupp, AEG, Interatom.[7] While this programme failed, AEG and Siemens, who in their early R&D activities had benefited from their close co-operation with the American producers GE and Westinghouse, respectively, proved to be economically viable.[8] In 1969, in the aftermath of the 1966/67 recession and after AEG had experienced high losses in constructing the Würgassen reactor, AEG and Siemens merged their energy-related activities in the newly founded Kraftwerk Union (KWU). It was completely taken over

[6] The section described here concerns only those companies which have originated in the Western part of the country. The former German Democratic Republic had its own industry and research institutions. In the East, over years nuclear energy supply was regarded to be the only solution to the dire energy supply shortages, however, without much success. Following German unification, all nuclear reactors – which were of Soviet made – were shut down and construction projects terminated for safety and cost reasons (Matthes 2000). The following analysis is focused on the West German nuclear industry.

[7] Interatom is a consortium of the German Demag and North American Aviation. Other companies temporarily involved in follow-up programmes were MAN and GGH. (Keck 1984, pp. 52–90)

[8] General Electric held 25% of AEG at the time.

by Siemens in 1976.[9] Brown, Boveri & Cie AG (BBC) - held mainly by the Swiss BBC and largely relying on US Babcock & Wilcox technology - was active on the German market as a third company with a much smaller share. It built two commercial-sized reactors, though without the expected success. By 1980, KWU was the only German nuclear reactor producer (Mez 1979, pp. 33–35; Keck 1984, pp. 79–80; Matthes 2000, pp. 131–32).

Technologically, the German market has been dominated by light-water reactors, both heavy-water and pressurised water systems, for which the patents were bought from the US producers General Electric and Westinghouse. Already in the late 60s, licence provisions were escaped by doubling the reactor's size and a quasi national line was successfully developed by KWU. Besides this, persistent attempts were made at developing fast breeder (Kalkar) and high temperature reactors (Hamm-Uentrop) as part of the national fuel cycle project. They were largely financed by national nuclear research programmes and after many years of failures were finally abandoned at multi-billion losses in the early 1990s (AG Atomindustrie 1987; Keck 1993).

The costs of nuclear power plants averaged at 4 500 DM/kW[10] or DM 0.4-5 billion total construction cost for the large commercial nuclear power plants (AG Atomindustrie 1987). The plant core – the reactor – usually accounts for 10–20% of the total construction volume (KWU 18%). The remaining, larger, portion is provided by component producers, steel and machinery manufacturers, construction companies, and the electronics and chemical industries. However, since in Germany the producing company also acts as industrial architect, it is the responsible party vis-à-vis the permitting authorities as well as the electricity industry; component providers mostly act as sub-contractors (Keck 1993, p. 258; Mez 1979, p. 33; AG Atomindustrie 1987). All major German companies serve this segment. Some of the most important have been AEG, Siemens, Bosch, Klöckner, Krupp, Mannesmann, Thyssen, Phillip Holzmann, Hochtief, Bayer and Hoechst.[11]

Upward trends on the nuclear market did not last far beyond the first decade of commercial nuclear energy use. In spite of high performance ratings, turn-key plant sales by KWU has only for a short period passed the world market's middle section, when orders from Argentina, Brazil, Austria, the Netherlands and Iran filled the books.[12] Many of these plants however, were not completed or did not run as expected. Since then the emphasis has been on the component market. By

---

[9] Already in 1973, Interatom had been integrated in KWU. AEG's retreat from nuclear business confronted AEG shareholders with a net loss of 1.7 billion DM (Keck 1984 pp. 82–90). In 1987, KWU became part of Siemens' business section "Energy and automatisation technology" (Mez 1998, p. 115).

[10] This corresponds to 1 250 US-$/kW at an exchange rate of 3.25 DM. Other sources calculate costs at 4 000 – 4 500 DM/kW (OECD 1986 cit. Keck 1993, p. 235). All costs given here are pure construction costs excluding interest payments and later refurbishing.

[11] For a more complete list and the interrelations between the different firms cf. Mez 1979; Mez and Osnowski 1996; AG Atomindustrie 1987.

[12] The majority of these contracts were financially secured by national export guarantees (*Hermesbürgschaften*) and many of the costs ultimately had to be shouldered by taxpayers.

the second half of the 1980s KWU experienced a reduction in total volume of orders from DM 20.1 billion in 1987 to DM 7.7 billion in 1991. This resulted in far-reaching restructuring activities within KWU and the laying off of 8 500 of its 28 000 employees between 1987 and 1996. While by the mid-1990s new orders attained a volume of DM 11.9 billion, over the last decade the volume of new nuclear orders has stagnated in the range of DM 2.2 to 2.9 billion at the most. Thus the emphasis of the power business has shifted from nuclear to fossil-fuelled power plants, especially gas turbine technology, which was reinforced by the acquisition of Westinghouse Power Generation in 1998. The nuclear activities of KWU today comprise only a 20% share of its sales volume, having decreased from 70% in the late 1980s. At the same time, foreign clients (Central and Eastern Europe and Asia, esp. China) account for a growing share of KWU's nuclear business, rising to 50% of new nuclear orders in 1997/98 (Siemens 1999, p. 10).

In this period Siemens/KWU has transformed itself from a producer of turn-key nuclear power plants to a nuclear service enterprise. Current activities (in partnership with operating companies) are concentrated in the areas of components, refurbishing, monitoring, service and management for existing plants, as well as the fabrication and reloading of fuel assemblies (cf. below). (Hüttl 1992; Siemens 1999; 1994, p. 50; 1995, p. 25; Tagesspiegel, 16. Juni 1999; Breyer 1999). In late 1999, a joint venture with the French EDF's nuclear section was agreed.

Despite uncertain economic prospects, research activities continue. KWU has recently turned greater attention to international projects, i.e. the European Pressurised Water Reactor (EPR) and the SWR 1000, a boiling water reactor. Both projects are co-financed by the German ESI. The EPR is a joint venture of Siemens and Framatom and their common daughter-company Nuclear Power International (NPI) and is co-financed by the French utility EDF. Although there are, at this point, no new nuclear plants on order or any new orders anticipated, Siemens stands to be contracted by EDF should demand arise (Siemens 1998, pp. 32–35).

2.2.2 The operation of nuclear plant

Nuclear power plants in Germany are operated by the major electric utility companies and/or subsidiary operating companies. Initially, the German electricity service industry (ESI), i.e. the utilities, kept a sceptical distance from the government's nuclear programme because they feared another competitor, in addition to gas and oil, for their coal sources and shied away from the high projected costs involved (Kitschelt 1980, p. 38). As early as the late 1950s, however, two electricity companies, RWE and Bayernwerk AG, pursued an independent nuclear research project in co-operation with Siemens and AEG to gain operational experience. In 1962, RWE ordered the first commercial nuclear reactor after a change in management had dissipated the remaining internal resistance.[13] This set off a wave

[13] The 250 MW reactor was build by a German-American consortium (General Electric and AEG) in Grundremmingen and was co-financed by US-Euratom and the German government,

of plant orders and up until 1999, a total of 14 German companies owned nuclear plants, four of which – PreussenElektra, RWE, Bayernwerk, and EnBW – own more than 75% of the nuclear capacities.[14]

The operating companies also provide nuclear plant servicing and maintenance. At the same time, the large companies are co-owners of companies in the uranium mining and (re)processing business which operate internationally, often in co-operation with foreign companies. Thus they dominate the downstream but are also involved in upstream business segments. Operating companies are also responsible for decommissioning the power plants and for the subsequent site clean-up. Operating personnel as well as firms from other subsections of the industry participate in this task. Due to the long-term radioactivity of the substances involved, complete decommissioning will require several decades and, according to one estimate, cost roughly US-$ 300 million per reactor (U.S. GAO 1999). To finance this, the German operating companies put up non-taxable accruals of at least DM 72 billion until 1998 (Mez and Piening 1999; Reuters 1999).

While in other countries the high and often unpredictable costs of nuclear power plant construction have driven up the capital dept of a number of electricity companies, in Germany debts have been contained and high profits have been earned. This can be attributed to the leniency of German regulatory bodies towards the ESI, which has allowed high costs to be shifted to customers. Another reason is the financing schemes chosen for many power plants; in order to avoid burdening the electricity companies' balance sheets with large capital investments, since the mid-1970s construction and operation of power plant has increasingly been carried out by subsidiaries,[15] which raise the necessary financial capital from the mother company (and in part from financial institutions) but usually have no capital base themselves. The advantage is that the subsidiaries high investment costs do not have to be disclosed and do not appear in the utilities' balance sheets. Further, the company's capital is not invested and thereby dissolved, but employed to earn interest. As a result, most German nuclear power plants begin to earn a respectable profit even during the depreciation period (due to German tax regulations, usually 19 years) (Mez and Osnowski 1996, pp. 141–159).

In conclusion, the nuclear operating companies, i.e. the electric utilities, must be regarded as one of the most influential and lucrative sectors of Germany's economy. It has been a major employer, providing 213 000 jobs in 1991,[16] and for

---

despite initial government resistance against financing foreign industry products. It was not part of the Eltville programme and was motivated by the perceived need to gain operation experience (Keck 1984, p. 75).

[14] Further capacities are owned by VEW, HEW, NWS, the Deutsche Bahn AG, Elektromark, and the Munich municipal supplier, Münchner Stadtwerke (Mez and Piening 1999).

[15] Most common is the GmbH (limited liability company) solution. Another possibility is the leasing solution. Both models are applied to nuclear as well as conventional plant construction projects.

[16] Up until 1991 employment continuously increased, figures for West Germany only (BMWi 1993:22). Due to restructuring activities in the course of beginning deregulation and market liberalisation, employment was reduced by 20% until 1997 (BMWi 1999, p. 18).

decades has profited from the stable framework conditions of monopoly protection. With the onset of liberalisation of European electricity markets, this is likely to change.

2.2.3 Nuclear fuel cycle business and waste management

Another important section of the nuclear industry is the fuel cycle business: those activities that are necessary to produce reactor fuel from uranium, transport of the fuel to the reactor sites, installment, and the processing of spent fuel, either for recycling or in preparation for permanent disposal. Due to the extreme levels of radioactivity and the comparative lack of experience with safe treatment and permanent storage of nuclear wastes, this field is politically and economically highly sensitive. From the beginning, the German nuclear programme clearly aimed at gaining technological expertise and a lucrative share in this market. But despite enormous federal and private investment, the fuel cycle project has largely failed.

Initially, companies from the chemical industry together with the electric utilities were financially involved in German nuclear reprocessing and storage activities – represented by the DWK and its affiliates. In response to miscalculations and to negative experience in the US, the chemical industry sold most of their shares by the mid-70s to the major German utilities. Many other ventures, such as the fuel transportation company Transnuklear, are multinational enterprises. Among the most important is the Dutch-British-German venture for the enrichment of uranium, Urenco, of which Uranit, with its plant in Gronau, is the German partner. The ESI - especially RWE and VEBA/PreussenElektra via their subsidiary Nukem - holds major shares in most of these companies. Another important shareholder is Siemens/KWU with its uranium fuel assemblies in Lingen and Karlstein (Mez 1981; AG Atomindustrie 1987). The restructuring and regrouping of activities in this sub-sector has been constant, partly as a common response to economic and public relations problems, such as the Transnuklear (1987) and Castor contamination scandals (1998).[17]

The industry's venture into MOX fabrication at the Alkem facility in Hanau – owned by the large utilities and KWU – was finally abandoned in 1995. Its shutdown was preceded by legal contests over safety issues lasting for years, at the end of which the party responsible, the Hesse government, revoked the plant's permit and the ESI withdrew its financial support (Siemens 1996, p. 27; Jend 1996).[18] Siemens/KWU know-how and technology is now being employed in the construction of Cogema's MELOX plant in Cadarache, France, where the former Alkem director Jürgen Krellmann is the present manager (Lenail 1999, p. 41). Similarly, an attempt to build a fuel conditioning and reprocessing plant in Ger-

[17] One of the most controversial restructuring activities concerned NUKEM following the Transnuklear bribes scandal, which, however, did not fundamentally change the market structure (Mez 1988).

[18] From 1966 to 1995 total investments in the plant amounted to 1.1 billion DM, of which the German government contributed 384 million DM until 1991 (Jend 1996).

many was abandoned in 1989 when the German utilities decided to contract the fuel cycle services of the British BNFL and Cogema. At issue in all these cases – all of which were accompanied by heavy public protest – were safety issues and the cost increases they would bring about. As a consequence, services are very likely to be overpriced, due to the lack of competition within the sector and the fact that transporting firms are affiliated with and serve the ESI, which operates under monopoly conditions (Mez and Osnowski 1996, pp. 55–61).Overall, the German nuclear industry is extremely highly concentrated on the supply side as well as on the demand side. Except for the ties with French and British companies with regard to research and the fuel cycle business, the nuclear market in Germany is closed to domestic or foreign competitors. Competition and cost awareness is further stifled by the fact that virtually all operating costs incurred were guaranteed to be covered by captive customers. Thus the risk lay predominantly with the construction industry. Since the 1980s, nuclear plant production cannot be regarded a growing market, despite the continuing development of new reactor types.

3 The political context

3.1 Actors and alliances

Early nuclear policymaking was influenced by a number of actors who have since formed a stable alliance promoting the pro-nuclear energy stance of government policy. While many actors easily identified with the goals of this alliance, others did not enlist without internal struggles. However, a growing number of actors opposing the use of nuclear energy gained political influence – forming an anti-nuclear alliance - and have hampered the political impact the pro-nuclear alliance has had on the policymaking process and policy choices. This in turn has had consequences for the policy goals and strategies of the actors and alliances.

In the following, the dominating alliances and networks, composed of individual stakeholders, will be identified and the impacts these have had on the political process and policy choices will be discussed. Although they are presented here as two opposing blocks, they must be understood to be permeable to changing coalitions, particularly at their fringes.

3.1.1 The pro-nuclear alliance

The core of what here is referred to as the pro-nuclear alliance is comprised of the *nuclear industry*, of which Siemens/KWU and the large electric utility companies (RWE, PreussenElektra, Bayernwerk and EnBW) are most prominent in the political arena today. As the third largest industrial company in Germany, the leading electronics and engineering firm and a major employer, Siemens has a strong position in the political bargaining process, internationally as well as nationally.[19] The large electricity producers have a long tradition of co-operation in technical and economic matters, as national grid owners within the *Deutsche Verbundgesellschaft (DVG)* and other associations,[20] and politically, e.g. within the *Vereinigung Deutscher Eletrizitätswerke (VDEW)*. These companies more than others partook of the large profits that have been made from the nuclear business. As the development of nuclear energy was deemed to be in the national interest, the nuclear industry was accorded exceptional national political influence and reinforced its strong position in the electronics and engineering industry sectors and the electricity sector, respectively. Politically, these groups have been represented very persistently by the *Christian-Democratic Party (CDU)* and its Bavarian sister party *CSU*.[21]

Another central constituent of the alliance are the *nuclear research institutes*. The nuclear research community has been evolving since the 1950s on the basis of the federal research policy, which facilitated co-operation between the state and industry in the large state-funded nuclear research centres: *Kernforschungsanlage Jülich, Kernforschungszentrum Karlsruhe*, and some institutes of the *Max-Planck-Gesellschaft*. Here the great majority of projects was carried out.[22] To a large extent, their activities have been characterised by project-oriented "big science" research which involves scientists who, according to critics, did not adhere to traditional standards of scientific objectivity and political neutrality, but rather have shown great personal interest in leading their pet-projects to success, i.e. to secure continued funding. A condition of their success was to foster close ties to the state bureaucracy and the nuclear industry, since public R&D funding for individual projects legally requires co-financing from private sources. Over a period of many years this coalition supported even widely criticised projects, such as the fast breeder reactor (Keck 1984; Häusler 1988, pp. 91–94; Mez 1998, p. 127).

Thus, departments of the *federal and also Länder administrations* – those that were instituted during the formative years of German nuclear policy to oversee R&D, permitting procedures and safety requirements – have to be regarded as the

[19] A recent example of Siemens' international influence is the highly controversial debate on EBRD loans for two Ukrainian nuclear reactors to be refurbished as proposed by the G-7. Siemens together with Framatom and the Russian Nuclear Ministry would be the main beneficiary of this project which by European financial experts including from within the EBRD is judged to be highly uneconomic (cf. Münchmeyer 1999).

[20] One important association in the present context is Vereinigung der Großkesselbetreiber (association of large combustion unit operators), founded in 1973. Within this association RWE acted as co-ordinator and leader in communications between nuclear producer (KWU) and operating firms (EVU), thus structuring communications patterns (Keck 1993, p. 257).

[21] Support came also from other political parties, but not consistently (cf. Sect. 3.1.3). Even within the CDU, namely its workers branch CDA, resistance to nuclear energy use came up.

[22] Initially their basic budgets were co-financed by the Federal and *Länder* governments, but by the early 1970s 90% of their publicly financed budgets was provided by federal funds, 10% by Länder funds (Keck 1984, p. 59; BMFT 1975, pp. 229–42).

backbone of the pro-nuclear alliance.[23] However, in federal systems, administrations do not act as a homogenous block. Regulatory bodies in *Länder* administrations in particular have taken critical positions and questioned the judicial viability of legal provisions, due, among other things, to changes in the political composition of the respective governments. In its permitting tasks, the administration is supported by the certifying association *TÜV*, which officially operates independently from the industry but in reality is dependent on their fees for income and the consent of various industry board members (AG Atomenergie 1987, pp. 469–471).

A number of *financial institutions* comprise a further principal, if publicly rather invisible, constituent of the alliance. Large banks such as the *Deutsche Bank, Commerzbank, Dresdner Bank* and *WestLB* have influenced the development of nuclear energy use in Germany not only by providing major capital resources but also through their representation on utility company boards and as advisors in pivotal political negotiations, such as the *Jahrhundertvertrag* (cf. Sect. 3.2.2; Paulitz 1996). An equally important role must be attributed to the major German insurance companies, namely *Allianz Versicherung* and *Münchner Rück*, which hold large shares of the large utilities. These linkages gave rise to the concept of "alliance capitalism" (Shonfield 1968).

The interests of the pro-nuclear community are represented by a number of registered societies which fulfil important lobbying tasks, but also act to inform or form public opinion. The most prominent are *Deutsches Atomforum* (its subsection *Informationskreis Kernenergie*) and the *Kerntechnische Gesellschaft*, which are closely associated (AG Atomwirtschaft 1987, pp. 247–49), and the *Forum für Zukunftsenergien*. These associations are the main public forum for scientific and technical experts from the nuclear industry and research institutes. Other members also include bankers as well as government representatives and politicians in their membership and act as a central link between the political and commercial sector. They concentrate on public relations work through regularly held conferences, newsletters and lobbying. In general, the positive economic prospects for the industry are emphasised and its pivotal role in Germany's national economy and competitiveness are underlined, while critical assessments of nuclear energy use, e.g. after the Chernobyl accident, are refuted. Links exist between these organisations and such groups as the Deutsche Physikalische Gesellschaft and the major industrial associations (VDI, BDI, VIK). They are also part of the nuclear industry's international network, e.g. the European Nuclear Society (ENS).

While the various actors constituting the pro-nuclear alliance naturally do have different interests, they have all profited from the German nuclear programme, with respect not just to their economic standing but also their political influence. Generally, they have closely co-operated in working out solutions for the various technical, legal, political and financial problems the nuclear project has been con-

[23] This is underlined by the fact that a number of former administrators later took leading positions in the nuclear industry, such as Walter Hohlefelder, over years head of department in the federal government nuclear division and, since the change in government, member of the board of Preussen Elektra/E.ON.

fronted with over the years. Strategically, the alliance's focus has shifted since its beginnings: in the formative years, its aim was to establish nuclear energy as an integral sector of the economy. By the mid-1980s, as the further expansion of nuclear energy use became unrealistic, the alliance's efforts were redirected towards safeguarding the "politically undisturbed operation" of existing plant and, in the long term, keeping open the option for a future generation of nuclear plant, should the socio-economic parameters change.

3.1.2 The anti-nuclear alliance

The anti-nuclear alliance first formed when NIMBY-type local protests at planned reactor and research sites arose. Rooted also in the anti-(nuclear) re-armament movement of the 50s and 60s, they first caught national attention in the early 70s during the protests against the reactor plant in Wyhl. Originally a cause of *concerned citizens*, over time the ranks of the protesters were joined by renowned *scientists*, *nuclear energy experts* and *politicians*. The protests developed into an integral part of the "new social movements" (Brand et al. 1986) of the time – among them most importantly the environmental movement. As demonstrated by their successes in preventing nuclear construction projects like those in Wyhl and Gorleben, their activities had a palpable impact and by the end of the decade, they had built up a highly effective organisational infrastructure and, maybe even more importantly, acquired sufficient scientific, technical and legal expertise to seriously question the nuclear programme and discuss alternatives within the new public debate. Increasingly, the *media* provided the forum for this debate, and investigative journalism made important contributions. The anti-nuclear stance was initially founded on health, safety and environmental issues; later it included a broader-based critique of the supply-oriented national energy policy, and the promised economic benefits were called into question. Induced by this debate, a number of municipalities and some *Länder* governments launched programmes in support of alternative "soft" energy policy initiatives which were heavily based on the critical expertise of independent research centres like the *Öko-Institut* and others grown out of the movement.

This alliance's primary goal has been to put an end to the use of nuclear energy. More recently it has also been directed towards limiting the number of nuclear waste transports and raising policymakers' and industry's attention to the problems of nuclear waste disposal. As actors without a defined place within the German political system, early protesters took to unconventional forms of public articulation such as passive resistance, public information campaigns and festival-like events, and, as the conflict intensified, open confrontations with police forces. The national movement that had formed ultimately contributed to the rise and election of the Green Party. Representatives of the movement were thus integrated in the established political arena and their causes taken up in the traditional parties' programmes (Rucht 1980; Brand et al. 1986). With the national elections of

1998 and the new "red-green" coalition the movement's primary goal landed on the federal government's agenda.

3.1.3 Actors on the fringe

Between these two opposing blocks a number of social groups and political actors have influenced or been affected by the nuclear debate without always having taken a definite and clear position. Most important among these are the Social-Democratic Party (SPD) and the trade unions.

Just as the other major political parties, the SPD started out as a stern promoter of the nuclear project. After the first protests arose the SPD government attempted to dissipate the public's concerns.[24] At the same time, a number of prominent members of the SPD and the party's youth section *JUSO* revoked their support and enlisted themselves in the anti-nuclear alliance. Over the following years, several *Länder* administrations at least unofficially distanced themselves from the nuclear project,[25] while the federal leadership held to its pro-nuclear position. The internal breach and the resulting conflict within the administration inhibited the party's ability to serve as the "transmission belt" between the electorate or interest groups and the political administrative system and, according to one analysis, were elementary for the failure of nuclear policies at the time (Häusler 1988). The breach began to resolve when the liberal-conservative coalition of CDU and FDP took over the federal government in 1982 and the critical position of the SPD *Länder* governments, especially those governing in coalition with the Green Party, became much more outspoken. Finally, after the Chernobyl accident, the party convention voted, if re-elected, to pursue the phase-out of nuclear power within ten years (Ueberhorst 1992; Matthes 2000, pp. 138–40; Mez 1979).

In contrast, the liberal *Free Democratic Party (FDP)* has taken a much more outspoken pro-nuclear stance, especially some of its renowned legal experts. Nevertheless, party programmes of the 70s reflected the nuclear conflict and kept a sceptical distance to the nuclear programme, which later recurred in the demand to scale back public nuclear research funding.[26] But when in the 90s the coal-nuclear policy compromise was invalidated, the FDP pared with the pro-nuclear side of their coalition partner CDU/CSU. On the whole, the FDP took a sceptical view at times but the critique has never been made an issue of contention with its political partners.

---

[24] In the so-called "Bürgerdialog Kernenergie" in 1976, the federal government held public hearings on environmental issues pertaining to nuclear energy (cf. e.g. BMI 1976). While it was intended to soothe the protesters' "irrational" fears, the dialogue provided a public forum in which the concerns and critical arguments were spread to an even broader public and as a consequence stimulated further protests.

[25] Most outspoken were the Länder Hesse, Schleswig-Holstein and Lower Saxony, while the North-Rhine Westfalian SPD due to their traditional links to the energy industry remained silent as part of the compromise arrived at between the unions and government.

[26] It was the FDP Minister Riesenhuber who first cut research funds in the late 1980s.

The *trade unions* in Germany, united in the *Deutsche Gewerkschaftsbund* (DGB), have had a difficult position due to the various member unions' diverse and contradictory interests in industrial and energy policy (Grumbach 1986). In line with the SPD, the overwhelming majority of concerned branch unions initially supported the industrial policy goals underlying the first nuclear energy programmes and shared in the general optimism about nuclear power, especially its positive economic and employment effects up until the 1980s. While these leadership resolutions generally were not based on intensive internal debates, controversial debates developed and critical positions were at least internally supported, first in some smaller unions (education, printing industry), later also in the large *IG Metall*, despite the KWU workers' membership. Traditionally, conservative unions like the miners' *IG Bergbau und Energie* or the *IG Chemie* acted as outspoken promoters, partly also the *ÖTV*, where most nuclear plant employees were organised. Many of the conflicts can also be attributed to the fact that their leadership had strong ties to the SPD (as members of parliament) on the one hand, and to the nuclear industrial complex (as workers' representatives on the board of management) on the other, feeding the so-called *"Atomfilz"* (Brandes 1977). The political rifts – particularly between nuclear and coal mining interests – came to the open in the 1980s as electricity demand dropped and increased nuclear capacities were added to base load, with the result that coal as a source of fuel was faced with a new competitor (which was attempted to be solved by negotiating the *Jahrhundertvertrag*, cf. Sect. 3.2.2). Occupational safety issues remained on the agenda throughout the years, however, never became the decisive argument against nuclear energy. In parallel with re-evaluations within the SPD, a gradual change also within the *IG Bergbau und Energie* and the DGB leadership took place. After Chernobyl, the DGB announced its programmatic turnabout and since has supported the nuclear phase out and energy policy reforms.

It appears that the two opposing alliances have strengthened their respective positions over the years. Judged by their fundamentally different views on nuclear energy use and their opposing basic goals, this conflict does not seem likely to disappear. Its political and social virulence might, however, be calmed through decisions pursued and taken by actors at the fringes of the opposing blocks, who are confronted with polarisation within their own organisations, and therefore most likely to step into the role of negotiators.

3.2 Nuclear policies chosen

3.2.1 Early nuclear policies (1955–74)

Germany's nuclear policy was initiated in 1955[27] and was officially installed with the Eltville programme of 1957. It primarily aimed at the development of an internationally competitive, privately-owned nuclear sector, which was to include the production and operation of nuclear reactors. The focus of follow-up programmes shifted onto nuclear fuel cycle technologies, such as the fast breeder and reprocessing technologies. As the central element of nuclear policy, in the course of four research programmes from 1957–76 a total of 20 billion DM of research funds of the Federal Ministry for Research and Technology (BMFT) were spent.[28] Initially, the nuclear programme was regarded as an integral part of the general industrial policy, the so-called *"Modell Deutschland"*, to re-establish Germany as a viable force in international markets. Consequently, its co-initiators and the main recipients of public research funds were those industrial sectors mentioned above but also the new nuclear research centres. They constituted the majority in the *Deutsche Atomkommission*, which was the main advisory body of the newly founded Ministry for Nuclear Questions and the unofficial centre of nuclear policymaking[29] (Keck 1984, pp. 56–57; Häusler 1988, pp. 62–66). The Ministry and the needed legal framework were established with the goal to structure permitting procedures without interference from other ministries.

As the analysis of the nuclear industry showed, nuclear research policy completely failed with regard to the goal to establish a national reactor line. Its success lay in facilitating the constitution of a strong nuclear community in Germany: a corporatist alliance of policy makers, commercial enterprises and researchers. It laid the legal and institutional base for a quick and easy entrance of the electric utilities in the market when they finally recognised their advantages.

The early nuclear energy era in Germany, as in other countries, was characterised by an enormous technological optimism which until the early 1970s was based upon a general political consensus among the governing parties. By today's standards, it was a rather "silent" consensus, since at least in the formative years public or even parliamentary debate on the prudence or social consequences of

[27] In 1955, the Parisian Treaties were signed, which lifted the allied ban of any nuclear (civil and military) research from the losing parties of WW II. In the same year, the UN Conference on the peaceful use of nuclear energy in Geneva was held, which to the German delegates underlined the urgency of a nuclear programme to catch up to international standards (Keck 1984, pp. 52–56).

[28] According to historian Radkau this financial contribution was one of the preconditions for the success of Germany's nuclear industry, since many private investors by the late 50s had retreated from the nuclear venture because of low profitability expectations (Radkau 1983, p. 196).

[29] The commission was in place from 1956 until 1971. More than 200 delegates from industry, research centres and universities were members of its many working groups, and it thus brought together the leading representatives of the German nuclear community.

this policy was not pursued or even regarded opportune. The relative success of the industry – even though it grew largely as a consequence of electric utility decisions - initially seemed to reinforce the early political consensus: by the late 1960s, Germany had caught up to international nuclear industry standards; in 1974 Biblis A with a capacity of 1 300 MW started operation, the world's largest nuclear reactor at the time; and, consequently, KWU had advanced to rank third in the world nuclear industry. This success was reflected in the government's energy programme of September 1973, which adopted a 20-fold increase of nuclear capacity until 1985.[30] And since the commercial entrance of the first series of nuclear power stations happened to coincide with the first oil price crisis of 1973 and the arising fear of future energy shortages underlined the presumable prudence of the government's ambitious programme, nuclear energy proceeded to become a central topic on the energy policy agenda.

This concluded the first nuclear policy phase, which, in summary, was largely restricted to one main policy area, that of industrial policy.[31] As this phase was centred on research and development and did not have many publicly palpable economic and social effects, it involved a relatively restricted circle of political actors. Considering the results, it can be argued that the early nuclear energy policies of the German government were straightforward and quite successful with respect to supporting the inception of a German nuclear industry of acknowledged – albeit not indigenous – technical competence. However, the nuclear policy programme would have very likely failed without the independent financial involvement and dedication of the major German electric utilities.

3.2.2 Policies in the aftermath of the energy price crisis (1975–89)

Around 1975, Germany's nuclear energy policy entered a second phase which was characterised by both contention and stabilisation. Mainly induced by considerable public protest, the former consensus began to erode and nuclear energy policies increasingly came under public scrutiny for economic and safety reasons. The evolving anti-nuclear alliance published its first critical statements. And the public protests were proving effective: the reactor construction plans in Wyhl, where in 1975 the movement was "conceived", finally had to be given up and, in 1979, the prime minister of Lower Saxony had to declare the plans for a conditioning plant in Gorleben "impossible to enforce for political reasons." Calls for a paradigmatic change in energy policy were voiced. Over time, these and other factors contributed to a broad shift in public opinion, the formation of the Green Party, and finally its election to the German Bundestag in 1983 (Rucht 1980; Brand et al. 1986).

---

[30] This meant a total new nuclear capacity of 40–50 GW. The goal was to supply a share of up to 40% of electricity needs through nuclear energy. It was reconfirmed in 1974 (Rucht 1980, pp. 27–28; Deutscher Bundestag 1974).

[31] More detailed accounts of Germany's nuclear policy differentiate between a first "speculative phase" (1955–67) and a "demonstration phase" (1967–75) (Radkau 1983; Matthes 2000).

In substance, despite the well-founded criticisms raised and the nuclear programme's political setbacks, German nuclear energy policy of this period adhered to a pro-nuclear stance. In public statements and major policy decisions successive governments, irrespective of their political composition[32], underlined the status of nuclear energy as one of the strategic pillars of Germany's energy policy, namely the goal of sustaining a balanced mix of resources for the country's energy supply. And for many years persistent technical, economic and social obstacles were largely disregarded, even though these did not only occur in demonstration projects such as the fast breeder and high temperature reactors, but also in many operating commercial reactors. Successive federal governments agreed openly with representatives of the nuclear industry and consented to their political requests, even though the legitimacy of the critique had begun to be officially accepted.[33] By 1976 – the year Biblis B went on stream – the nuclear industry was judged to have achieved the economic status of "an important industrial sector" (Chancellor H. Schmidt on 16.12.1976, quoted in Mez 1981², p. 39), and, implicitly, abandoning the chosen strategy was expected to hurt employment policy goals. Major energy policy decisions were bargained for with the clearly-stated goal of expanding the use of nuclear technology. This continuity is reflected in key energy policy decisions of the following years.

The energy policy programme of 1977 was characterised by a shift in energy policies. First, the nuclear programme was not passed separately but became part of the general energy programme for the years 1977 to 1980. Second, beside the new important role of nuclear energy, the development of renewable energies and energy efficiency was underlined and added to the list of research projects. For the first time, environmental concerns were named as principal goals. Thus concerns brought forward by the new political actors were reflected. However, this shift did not end the preferred status of the nuclear industry: more than 75% of the total 6.5 billion DM in energy research spending was earmarked for nuclear energy (Rucht 1980, pp. 30–32; Mez 1981², p. 33). Up until 1986, spending levels were at least maintained, if not increased (Matthes 2000, p. 139).

Probably the farthest-reaching decision was the compromise arrived at between the interests of coal mining and nuclear industry, put down in the so-called "Jahrhundertvertrag" of 1980. This agreement quieted the miners' union protests against the new energy policy. The miners were confronted with two parallel developments: first, nuclear energy was about to become a serious competitor to coal as a power plant fuel; second, Germany's coal industry was subjected to increasing pressure due to falling world market prices for hard coal. Accordingly, influ-

[32] To demonstrate the fundamental agreement on nuclear policy issues among the major governing and opposition parties, in April 1982, party leaders of the Social Democrats and the Christian Democrats together visited the projected site of the reconditioning plant in Gorleben rallying in support of the project (Chronologie, p. 1).

[33] Important indicators of this are the reports of the „enquete commissions" on nuclear energy in the 1970s, which also took account of critical evaluations and alternatives to the nuclear course (Ueberhorst 1992; Matthes 2000, pp. 138–40).

ential trade union interests had threatened to withdraw their consent to pro-nuclear energy policies and the governing Social Democratic Party (SPD). In the "treaty" the ESI pledged to burn a minimum amount of German hard coal, purchased at a federally subsidised price, in exchange for the guarantee of uninhibited operation of nuclear power plants and continuation of the nuclear programme (Matthes 2000).

Meanwhile, the controversy over the use of nuclear energy continued, accompanied by public protests. Throughout the 1980s unresolved safety and policy issues were reflected in *Länder* administrations' restrictive interpretations of permitting provisions and related court decisions.[34] The federal government, since 1982 constituted by a liberal-conservative coalition, aimed at speeding up decisions by federal government administrative orders, but no serious attempts at achieving a consensus were made.

The Chernobyl accident of April 1986 and its serious consequences had the potential to bring about a re-evaluation of German energy policy as anti-nuclear protests caught the attention of principal political actors.[35] But official political reactions did not signal a fundamental policy change. The main response was, rather, institutional in nature: the German Ministry for Environmental Protection and Reactor Safety was established. Official statements took on a defensive tone, as expectations with regard to future capacity additions were slowly scaled back. But the nuclear energy programme as a whole – particularly its focus on nuclear fuel cycle-related technologies – was retained. The continued government support for the nuclear energy industry is also reflected in the way severe safety and financial irregularities, e.g. the Transnuklear bribery scandal, were dealt with. While the federal government announced that a fundamental restructuring of the nuclear industry was to be undertaken, the subsequent change of ownership of fuel cycle companies did not achieve the promised structural changes (Mez 1988, p. 56).

While the economic interests of the utility industry were continued to be supported, the utilities were confronted with a policy change in the field of environmental policy. In 1983, a new ordinance to combat acid rain emissions was passed (Großfeuerungsanlagen Verordnung, GFAVO). It prescribed strict emission limits for conventional power plants and set off DM 22 billion of utility investments in environmental technologies. Induced by broad environmental concerns, this ordinance for the first time allowed to retrieve licenses from the utility industry if certain plants would not meet emission standards. As such it marks a sea change (Mez 1995).[36]

---

[34] One underlying cause was the increasing political weight of the Green Party in some Länder, e.g. in Hesse and Lower Saxony, another was that the conflict with coal mining interests remained unresolved, especially in Northrhine-Westfalia (Matthes 2000, pp. 141–42).

[35] Most important the Social Democratic Party and the Deutsche Gewerkschaftsbund, cf. above.

[36] Trying to come to an evaluation of this issue, it should also be considered that financially the ordinance was not of disadvantage for the utilities. On the contrary, within the regulatory framework given, investments in environmental technology posed a justified need for investment which otherwise would have had to be reduced substantially in the face of large surplus

Throughout the 1980s, the schism within society around the use of nuclear energy remained unresolved, and even deepened as a consequence of the Chernobyl accident. Nevertheless, government response to public concerns usually did not go beyond symbolic political gestures. In summary, nuclear policymaking was influenced by three main factors: First, not only did nuclear policy become an integral part of energy policy, making policy reorientation and integration necessary, but, in the aftermath of the energy price crisis, public perception of energy policy itself changed. It overcame its former status as a "non-policy" (Lindberg 1977, p. 356) and became a widely and controversially debated subject. Second, environmental issues entered the energy policy arena and with the GFAVO showed first substantial results; one might even conclude that a paradigmatic change in the relation of the utility industry and the state as traditionally reluctant regulator set in. Third, new political actors, arising from the social movements of the time, proclaimed their interest in nuclear and energy policies and, in diverse and also unconventional forms, made a lasting impact on the formation of public opinion and discourse. Nevertheless, successive federal administrations adhered to the nuclear path and argued for the continuation of the dual strategy of coal and nuclear policies as part of an energy supply policy that "keeps all options open". Since this was considered illegitimate by large sections of the population and established political actors, disagreement on nuclear policy persisted. As a consequence, a viable, long-term energy policy strategy could not be formulated.

3.2.3 Policies of the nineties: government in retreat?

German nuclear policy finally entered a state of open crisis when, in April 1989, the ESI withdrew its plan to build a reprocessing plant in Wackersdorf. While this decision – mainly based on economic grounds – came as a complete surprise to government officials, it quickly led to the end of the nuclear fuel cycle programme and related projects such as the fast breeder in Kalkar. The programme was terminated in 1991 and the industry's fuel cycle activities were shifted to France and Great Britain. This policy change was also reflected in fiscal policies. Starting as early as 1986, nuclear research spending had gradually been cut back, and by 1990 had been reduced to DM 632 million from over DM 1.5 billion in 1985 (excluding expenditures for decommissioning and nuclear fusion). Primarily, publicly financed research and pilot projects for reconditioning and other fuel cycle-related activities were affected. The second factor influencing nuclear policy making was the German unification process and the consequent shift of responsibility for East Germany's nuclear reactors to the West German government in 1990. The detrimental state of the reactors and the fact that they were the same type as the Chernobyl reactor quickly led to the decision to shut down all of them.

in capacities. Consequently, the utilities managed to keep the rate base at profitable levels (Mez and Osnowski 1996).

Cutbacks and the more cautious approach at the time led some industry commentators to conclude that the political consensus on nuclear power use between the German federal government and the nuclear industry had been broken (Komorowski and Meuresch 1995). Indeed, on the one hand, nuclear policy apparently acquired a much more defensive character, even under the coalition government of Christian Democrats and Liberals. At least internally, a re-evaluation of the government's role vis-à-vis the industry began. Changes in German nuclear law were demanded by administration members, who called for the stated aim to fund and promote nuclear energy to be deleted from law, on the argument that nuclear energy by then had achieved commercial viability (Hohlefelder 1991). Official willingness for continued financial support was also stifled by the financial burden imposed by decommissioning the East German reactors, at costs up to 10 000 DM/kW. At the same time, the federal government's conflict with Länder governments, especially Lower Saxony, over the planning process for nuclear waste storage sites intensified, with no solution in sight. Thus, the nuclear industry clearly was on the defensive.

Fearing a slow but unavoidable retreat of the federal government, in 1992 two managers from RWE and VEBA (owner of PreussenElektra) together with Lower Saxony leader Gerhard Schröder suggested to join talks with all major political parties to agree upon an energy policy consensus and the future of nuclear energy. According to the managers, it was to include a solution of the waste storage problem and a guarantee of politically undisturbed operation of existing plants. Their idea initially was met with fierce resistance from a majority of utility company leaders including from PreussenElektra as well as the governing CDU/CSU, who did not agree to exit the nuclear energy path even in a long-term perspective. Another problem lay with the SPD which would have had to revise its resolution for a phase out. A way to compromise especially with respect to the trade union interests seemed to lay in the need to renew the "Jahrhundertvertrag", as the coal subsidy scheme was due to expire in 1994. After preparatory talks had lasted almost a year, in March 1993 Schröder invited to a first round of talks which also included environment organisations and industrial associations. It was characterised by much controversy and finally failed in October 1993, after the key negotiator from VEBA had tragically died and the pro-nuclear hardliners within the industry and the federal government had regained spokesmanship (Mez 1997).

At that point the federal government regained the initiative and in May 1995 parliament passed a bundle of energy law amendments. It included a federal subsidy for coal until 2002 and a number of changes of the nuclear law: safety requirements for new plant required inherently safe reactors which would preclude contamination of the environment in case of accidents, and direct storage of nuclear wastes as alternative to fuel reprocessing was allowed, thus, a lower cost option was introduced. Meanwhile, the need for a consensus was continued to be stressed by industry representatives. It had became virulent already in December 1994 when the constitutional court ruled the coal subsidy via rate payers to be against the constitution and thus invalidated the compromise of the "Jahrhundertvertrag". However, also talks taking place in 1995 and 1996 failed due to the industry/governments steadfast pro-nuclear position. The amendments of February 1998 allowed, for the first time, site-independent and plant type-specific permitting procedures and were designed to streamline the permitting process by reducing the chance of litigation.[37] The intention behind these amendments was to restore "legal security" with respect to current but also future nuclear energy use. And up until the end of the CDU/FDP government, in official statements, the use of nuclear energy was declared to make an important contribution to reducing greenhouse gases. Thus, despite a retreat from its former position as its principal promoter, the administration continued to be supportive of nuclear energy.

3.2.4 Current policy

A new phase of nuclear policymaking started in 1998, marked by a change in political leadership. In October of that year, the newly elected coalition of the SPD and the Green Party announced their goal of phasing out the use of nuclear power. However, practical policy formulation and implementation has proven to be a difficult task. As a first step, amendments to tax laws have extended the write-off periods for nuclear plant investments and reduced allowed accruals for decommissioning, so that over a period of ten years additional tax payments of about DM 16 billion will be collected from utility companies. Furthermore, vacant posts at institutions like the nuclear safety commission have been filled with experts critical of nuclear energy. Increases in liability for nuclear accidents from DM 500 million to DM 5 billion and other regulations scaling back the financial and legal advantages of the industry are planned to be introduced according to the coalition agreement.

As the central element of its new policy, in December 1998, the government resumed "consensus talks" on the phase-out of nuclear power with the four largest nuclear operators. As a precondition, the outcome of these talks was not to give rise to liability claims of the industry against government. Negotiations were overshadowed by serious disputes with industry over the initiated and planned legal changes. Disagreement among government members – namely between the Environment Minister and the Minister of Economic Affairs – has further complicated matters. After more than one year, on June 14, 2000, an agreement was initialled which is now in the process of being confirmed by amendments to nuclear energy law. The central purpose of the agreement is to limit the lifetime of the nuclear power plants and, at the same time, to ensure their operation undisturbed by regulatory activities. The operating licenses of the power plants – originally issued without limitation – are limited by fixing exact quantities of electricity which may be produced in the individual plants from January 1, 2000 on. The agreed quantity may be flexibly transferred to other plants and totals 2623.3 billion kWh. Roughly, this is equivalent to a total service life of 32 years at high utilisation.

---

[37] This demand of the nuclear industry had already been discussed since 1974 (BMI 1974, pp. 174–75; Kitschelt 1980, pp. 141–46).

While this agreement serves as an incentive to close down some nuclear reactors possibly within this administration's term, some plants will possibly operate until 2025. (see fig. 3).

The problem of waste disposal was a further crucial issue. On the question of nuclear waste reprocessing the administration finally backed away from its initial plan and agreed to postpone the immediate prohibition until July 2005. Since no final storage solution has been found yet, the signing parties agreed on the preliminary storage of wastes at or near plant sites, and to accept certification of preliminary storage as legally sufficient proof of secure waste disposal. From a legal, financial and safety point of view, this is very controversial. However, it has clear political advantages: first, long held industry demands are served and, from a mid-term perspective, substantial cost reductions would be attainable; second, the government hoped to avoid or reduce public protests against nuclear waste transports.

So far, the government's performance has not raised much support in the anti-nuclear alliance. On the contrary, leading figures of the movement handed in their Green Party membership cards at a party convention in early 2000. The general secrecy of the talks, and the fact that important issues such as permanent waste storage and the updating of safety requirements have not systematically been dealt with, underline the critique's impression that no comprehensive strategic change in nuclear policy matters is being pursued. And indeed, the phase out plan mirrors

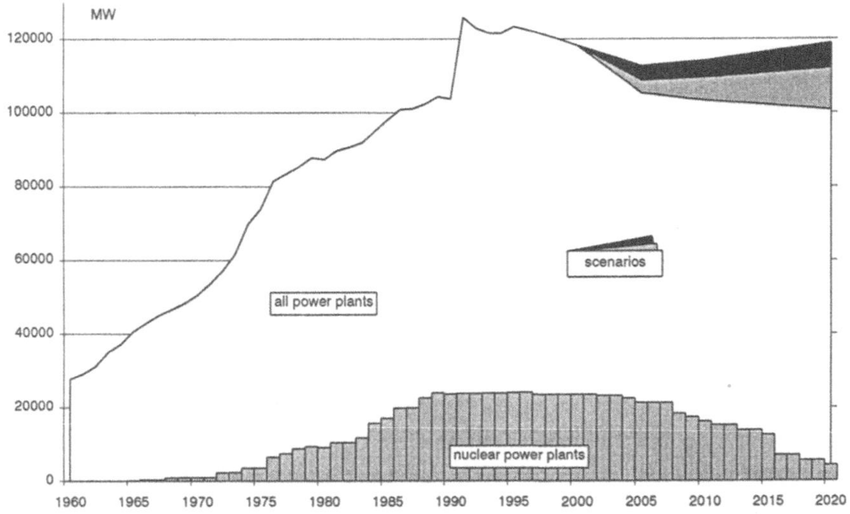

since 1990 unified Germany
Sources: VIK (1960–99); Öko-Institut 2000 (scenarios 2000–20)

Fig. 3. Maximum electricity capacity: all power stations and nuclear power stations 1960–2020

rather than pushes forward the economic-technical life cycle of existing reactors. In comparison, the reactions of the nuclear energy community have overwhelmingly been moderately critical and some prominent industry representatives have openly stated, that the agreement does not run counter to the industry's basic goals.

At the same time and rather unnoticed in the nuclear policy arena, the basic principles of energy policy have undergone fundamental changes in the course of the deregulation of the European electricity market. Since German liberalisation legislation of April 1998, dramatic changes in market structure have occurred characterised by the end of state-protected monopolies. So far, these events and their consequences do not appear to influence nuclear policy decisions. Furthermore, the continuing ecological problems, such as global warming, which are tied to electricity generation and supply-oriented structures might become a factor in the debate on nuclear policy.

4 Factors shaping nuclear policymaking

4.1 Factors leading to a decline of the nuclear industry

The statistical data, as well as the account of nuclear policies given so far, indicate rather clearly the general stagnation, if not decline, of the nuclear industry in the last decade. Some of the reasons for this development have implicitly been given above. They include economic, technical, organisational and informational, as well as political and social causes. The impact of the factors is weighted differently by different researchers, giving rise to contradictory interpretations. Often these interpretations are coloured by the personal views and political background of analysts writing in the context of an ideologically charged public debate on the use of nuclear power. This makes an objective assessment of causes and effects yet more difficult. With this in mind, the following factors can be identified as the main reasons for the nuclear industry's recent decline.

4.1.1 Planning and construction delays

One important factor was the delay in the planning and construction of nuclear plants, which became a serious problem soon after the first wave of orders. On the one hand, these can be seen as a direct effect of delays in the permitting procedures. With the expected construction wave of nuclear plants, the government had begun to promote research into safety issues in the late 1960s. Results of this research were integrated into official permitting procedures, thereby extending the lists of requirements and commenting authorities. This situation was further complicated by the novelty of the technical-scientific subject matter and the initial lack of qualified personnel at the permitting authorities of regional and Länder administrations, which by 1973 were clearly overburdened by the process.[38] In the mid-1970s safety issues were again pushed to the fore as public anti-nuclear protesters succeeded in the courts in a number of landmark cases (Kitschelt 1980, pp. 138–41; Mez 1998; Häusler 1988, p. 119).

On the other hand, as Keck (1993, pp. 239–46) points out, not all construction projects were affected by cost increases and/or delays, even though they had to meet the same strict regulations. According to his analysis, problems also resulted from mismanagement in project planning and construction. A lack of information and effective communication with respect to technical problems and requirements inhibited co-operation by producers and operators with the permitting authorities. By about 1980 these delays were partly overcome through a higher degree of standardisation in power plant construction (the so-called convoy concept in Germany) and the implementation of institutional and organisational changes (Keck 1993, pp. 258–59; for a more critical view of the industry's role cf. Kitschelt 1980, p. 139).[39] Both factors, administrative overload and industry mismanagement contributed to the fact that construction times extended far beyond the calculated five years, in most cases they added up to well over ten years.

4.1.2 Changing energy economic framework conditions

As a result of delays and the tightening of safety requirements, operators were confronted with severe cost increases. These inhibited the economic efficiency of nuclear power as a supply option. When, by the mid-1980s, the price of natural gas had started to decline and combined-cycle power plants entered the market as an economically and environmentally efficient generation technology, nuclear power lost further ground on the global market.[40] In contrast, generation in newer German nuclear reactors, especially those that had gone on stream after 1985, performed at higher costs than the alternative of burning hard coal imports.

The commercial entrance of the first series of nuclear power stations and the peak of orders world-wide coincided with the first energy price crisis, in 1973. While this seemed to boost the nuclear project, the crisis also marked the onset of economic recession. Induced by this and other factors, in Germany annual electricity demand growth rates decreased for the first time since 1950: from 7% in the early 1970s to 3–4% in the late 1970s and about 2% and less in the late 1990s. As demand projections were gradually adjusted, plant orders were cancelled. Nevertheless, some of the surplus capacity of the German electricity supply system of the 1980s and 1990s resulted directly from the unneeded

[38] Permitting procedures sometimes involved commenting by more than 44 different public authorities and in individual cases took up to 60 months (Kitschelt 1980, p. 138).

[39] For an example of an analysis by an open nuclear proponent which points out „impediments imposed on the nuclear industry by social, legislative and regulatory opponents“ as the main reasons for the large number of cancellations of plant order, cf. Simnad 1996, p. 1098.

[40] Even in Germany, only natural gas-fired generation capacities have increased since 1991, while total generation capacities declined (except renewable sources like wind energy) (BMWi 1998, p. 40).

the 1980s and 1990s resulted directly from the unneeded addition of nuclear plant.[41]

4.1.3 Public protest

Compared to the economic and managerial factors, it is difficult to pinpoint the impact of political and social factors on the development of the nuclear industry, and, for that matter, on nuclear policymaking. Clearly, the effect of the protests was not a political rejection of pro-nuclear policies at the federal level until very recently, but neither was the prior explicit and broad political support for the nuclear path upheld.

In the early years of nuclear expansion, the protests' effect on the nuclear industry was of an indirect nature. Due to the protests and the ambiguous position of the governing SPD, permitting authorities were confronted with many unanswered questions and lacked clear political directives which would have invalidated the critique. In the absence of a social and political consensus, jurisdiction on permitting requirements lacked unambiguous definitions for the many regulations left to the authorities' discretion. Consequently, the court decisions aggravated the authorities' difficulties in issuing "court-proof" permits. Political disorientation and uncertainty caused by the public opposition thus contributed to what has been called a "blockade of the administrative system" (Häusler 1988, p. 119). While this blockade was unintended and attempts were made to dissolve it, only recently, with the change in administration, has the government openly acknowledged the critique brought forward by the public protesters.

4.1.4 Liberalisation

A factor which has come to bear more recently is the liberalisation of electricity markets, which led to substantial price decreases for electricity of 30% for large consumers within one year. With respect to the nuclear industry, this has two consequences. First, a number of nuclear reactors are faced with increased economic pressure because they operate at costs too high to keep them competitive. Therefore, refitting and also safety related investments are being postponed.[42] While, as a consequence, some reactors may become candidates for early shut down, in other cases there may be increased incentive for continued operation because it guarantees that additional accruals can be set aside and employed to increase cash flow and that the point at which decommissioning costs fall due is postponed into the future. Second, investment in new reactors is becoming highly unlikely, since liberalised markets favour generation technologies with low capital

[41] In 1985, peak load was 58 800 MW at maximum electric capacity of 86 800 MW (Zängl 1989, p. 334). In 1994, the values (old *Länder* only) were 61 100 MW and 93 800 MW, respectively (BMWi 1996, p. 15).

[42] This is most evident in the case of the Biblis plant where unfulfilled safety related reconstruction requirements date from 1996 (Strohm 1999).

costs and short planning horizons. Therefore, modern combined-cycle gas-powered plants and flexible small-scale solutions have become increasingly attractive.[43] In the US, deregulation has given rise to a $200 billion support programme for nuclear power. Nonetheless, at least 40% of the reactors are expected to be shut down early for reasons of competition (Higley 1999). In Germany these effects are only slowly becoming visible, due to the lack of a public debate on cost structures as well as differences in supply and market structure which have prevented tougher competition up to now.

4.2 The role of environmental policy

What then has been the role of environmental policies or environmental concerns? Environmental policy as a policy area in its own right developed roughly at the same time as the nuclear industry was established. Because of the high environmental risks involved in nuclear energy, and the strong ties between the environmental and the anti-nuclear movements, these two policy fields soon overlapped. However, in retrospect, environmental concerns in the early years were just one factor that came to bear on nuclear policy making. As the analysis shows, they worked in concert with technical and economic factors to play a major role in shaping the nuclear industry's strategies. Environmental issues stayed on the nuclear policy agenda as risks to be considered, but only since the recent change in government have they worked their way to the top of the agenda. Nevertheless, one could argue that environmental issues did have an impact, namely at the level of policy implementation. They influenced requirements for occupational safety regulations – however insufficient they may be – and resulted in what has been described as the "administrative blockade". Furthermore, the industry as well as its public spokespeople within the political administrative system have stayed on the defensive and, despite continuous campaigning, have not succeeded in counteracting the deteriorating public image of the industry throughout the last three decades. Environmental and health concerns continue to be a predominant and persistent issue with regard to the use of nuclear energy.

Thus, ecological pressures did not pass the nuclear sector unnoticed. On the contrary, on the arena of energy policy as a whole environmental concerns, i.e. air pollution and acid rain problems, did have a lasting effect. Electric utilities were forced by federal legislation to carry out emission control retrofitting programmes with the result of a rapid decrease in emissions of classical pollutants from power plants. And, in the course of initiatives primarily at *Länder* level, the major utilities all set up energy saving campaigns. While these campaigns leave room for criticism, the changing policies and environmental concerns clearly induced the electric utility industry to publicly accept and emphasise its responsibility in safeguarding the environment – even though the industry continues to stress that the

[43] Capital costs of CCGT amounts to less than a third of those projected for the newest reactor line, the EPR, currently being developed by Siemens (Siemens 1998, p. 34).

risks from nuclear energy are not a defeating problem especially compared to global warming risks. In conclusion, policies over this period have contributed to a slow but evident change in the utility industry's relation to state actors. New policy tools which legally restrict formerly protected operational freedom of the utilities, such as the large combustion plant ordinance, have proved to be effective and within the frame of the consensus strategy is to serve as a model for a nuclear phase out legislation.

5 Future Perspectives

Meanwhile, the nuclear reactor industry maintains the stance that it is on the verge of a revival – one based on a new generation of self-containing, inherently safe reactors. Whether such an upswing in the industry will materialise in the near to intermediate future is impossible to predict for certain. But judging from experience to date with other technologies and their life cycles, it seems highly unlikely that a technology which has gone through such a distinct boom and downturn will recover so dramatically as to enter a second boom phase. Other less philosophical considerations also point toward a protracted decline. Economic conditions of liberalised electricity markets work towards the choice for supply sources which are inexpensive and flexible to demand changes. These characteristics are currently met by gas-turbine technologies and small-scale sources, while nuclear reactors with their high costs and long planning horizons are highly unattractive. Further, there are generation technologies such as the fuel cell which are poised on the brink of competitiveness and which promise to outlive not only nuclear power technologies, but all other central station technologies as well (Lönnroth 1989). From the perspective of policy making, the crucial question therefore is, whether a chosen policy should continue to support technology choices of the past – and pay subsidies – or whether it should facilitate the transition to this new energy market and safeguard social and environmental goals in the process.

The changes in nuclear policy over time illustrate a slow but steady retreat of state actors from their initial role as promoters and financiers to a more neutral position. Even under the conservative administration (CDU/FDP), certain members of the government pleaded for a more challenging stance towards the nuclear industry, which consequently ended the nuclear fuel cycle project and induced heavy cuts in public research funding. At the same time, the government's consensus finding activities - while aimed at securing and safeguarding the industry's interests in protecting the status quo - at least implicitly accepted the conflict potential inherent in the industry's demands. This attempt was later taken up by the SPD leadership, which had abdicated from the nuclear path after Chernobyl but due to conflicting energy interests never took on the radical stance of the Green Party. With these parties' election into government, a window of opportunity has opened for nuclear policies to be redirected toward greater independence of industry interests and broader consideration of social and environmental problems of nuclear issues (and their integration in energy and environmental policy) than in the past. Over the first year of their administration some progress was made. But the nuclear alliance has managed to keep ranks closed and was able to defend many of its main interests. In comparison, development of relations both, between and within, the government and the anti-nuclear alliance have been mixed and it has been difficult to adhere to a common strategy.

On the whole, it has turned out to be a very tricky task to solve the many problems the past forty years of nuclear policies have left unsolved. And while the policy change aimed for requires an assertive government actor, the ongoing process of electricity market deregulation and the underlying philosophy prescribes a policy pointing in the opposite direction: towards the retreat of government from its regulatory-intervening function vis-à-vis industry interests. However, this trend may be counteracted by the new, more independent relation evolving between the state and the former monopolistic utilities, in which the state may increasingly act as arbitrator. In principle, this holds the chance of extending the political debate over nuclear energy use – and, more generally, the role of public policy in the electricity sector - to include more critical views and to be led by changing actors' constellations.

References

AG Atomindustrie (1987): Wer mit Wem in Atomstaat und Großindustrie. Frankfurt a. M.: Zweitausendeins.

BMI (Bundesminister des Inneren) (1974): Sicherheitstechnische Einrichtungen und Strahlenschutz. Bonn.

BMI (ed.) (1976): Protokoll der öffentlichen Anhörung zu Fragen des Umweltschutzes am 9.6.76: Die Energiewirtschaft und die Entsorgung ihrer Kernkraftwerke. Bonn.

BMFT (Bundesminister für Forschung und Technologie) (1975): Bundesbericht Forschung V. Bonn.

BMWi (Bundesministerium für Wirtschaft) (1993): Energie Daten 92/93. Nationale und internationale Entwicklung. Bonn.

BMWI (1996): Energiedaten '95/'96. Nationale und internationale Entwicklung. Bonn.

BMWi (1998): Energie Daten `97/'98. Nationale und internationale Entwicklung. Bonn.

BMWi (1999): Energie Daten 1999. Nationale und internationale Entwicklung. Bonn.

Brand, K.-W., Büsser, D., Rucht, D. (1986): Aufbruch in eine andere Gesellschaft. Neue soziale Bewegungen in der Bundesrepublik. Frankfurt a. M., New York: Campus.

Brandes, V. (1977): Die Atompolitik der Gewerkschaften. In: Mez and Wilke 1977.

Brauch, H. G. (ed.) (1997): Energiepolitik – Technische Entwicklung, politische Strategien, Handlungskonzepte zu erneuerbaren und zur rationellen Energienutzung. Berlin: Springer.

Breyer, W., public relations manager of Siemens, interview, Südwestfunk/B3, 8 July 1999.

Deutscher Bundestag (1974): Erste Fortschreibung des Energieprogramms der Bundesregierung. BT-Drucksache 7/2713. 30 October. Bonn.

Deutsches Atomforum (ed.) (1991): Kernenergie im geeinten Deutschland. Bonn: Inforum.

Chronologie der Gorlebener Atomanlagen und des Widerstandes (1977–1997). http://www.oneworldweb.de./castor/diskurs/sonst/chronologie.html, retrieved 11.2.1999.

Grumbach, J. (1986): Energieforschung und Arbeitnehmerinteressen. Zielfindungsprobleme gewerkschaftlicher Forschungs- und Technologiepolitik. Frankfurt a. M.: Campus.
Hauff, V. (ed.) (1979): Argumente in der Energiediskussion. Band 7. Villingen-Schwenningen: Neckar-Verlag.
Häusler, J. (1988): Der Traum wird zum Alptraum. Das Dilemma einer Volkspartei: die SPD im Atomkonflikt. Berlin: edition sigma.
Higley, C. (1999): Statement on the future of nuclear energy before the Center for Clean Air Policy's conference on promoting clean power in a competitive market. Washington, DC: Public Citizen's – Critical Mass Energy Project. 2 December.
Hohlefelder, W. (1991): Die Pläne des Bundesumweltministeriums für eine Novellierung des Atomgesetzes. In: Deutsches Atomforum 1991, pp. 139–149.
Hüttl, A. (1992): Einleitungsreferat anläßlich der Jahrespressekonferenz des Bereichs Energieerzeugung (KWU) der Siemens AG Mühlheim/Ruhr, 18 February.
IEAE (International Atomic Energy Agency) (1999): Power Reactor Information System.
Jänicke, M. and Weidner, H. (1995): Successful environmental policy. Berlin: edition sigma
Jahrbuch der Atomwirtschaft (1986): Daten, Öffentliche Ausgaben und Haushalte für die Kernenergie.
Jend, R. (1996): Das Brennelementewerk Hanau. In: Jahrbuch der Atomwirtschaft 1996, pp. 40–50
Johansson, T.B., Bodlund, B., Williams R.H. (eds.) (1989): Electricity. Efficient end-use and new generation technologies and their planning implications. Lund.
Keck, O. (1984): Der schnelle Brüter. Eine Fallstudie über Entscheidungsprozesse in der Großtechnik. Frankfurt a.M., New York: Campus.
Keck, O. (1993): Information, Macht und gesellschaftliche Rationalität. Das Dilemma rationalen kommunikativen Handelns dargestellt am Beispiel eines internationalen Vergleichs der Kernenergiepolitik. Baden-Baden: Nomos.
Kitschelt, H. (1980): Kernenergiepolitik. Arena eines gesellschaftlichen Konflikts. Frankfurt a.M.: Campus.
Komitee für Grundrechte und Demokratie (ed.) (1988): Jahrbuch 1987. Köln.
Komorowski, K. and Meuresch, S. (1995): Stillegung und Rückbau kerntechnischer Anlagen. In: atw, Heft 4, p. 231
Lenail, B. (1999): Deutsch-französische Zusammenarbeit bei der Rezyklierung von Kernbrennstoffen. In: Jahrbuch der Atomwirtschaft, pp. 39–41.
Lindberg, L. N. (1977): Energy Policy and the Politics of Economic Development. In: Comparative Political Studies 10, pp. 355–382.
Lönnroth, M. (1989): The coming reformation of the electric utility industry. In: Johansson et al. 1989, pp. 765–786.
Majewski, O. (1999): Bedingungen der Kernenergie für Bayern und Deutschland. In: Jahrbuch atw, pp. 15–17.
Masters, R. (1999): World Survey, Station purchases and waste issues dominate. In: Nuclear Engineering International (June) pp. 15–22.
Matthes, F. Chr. (2000): Stromwirtschaft und deutsche Einheit. Eine Fallstudie zur Transformation der Eletrizitätswirtschaft in Ost-Deutschland. Norderstedt: Edition Energie + Umwelt. [LIBRI, Dr. und Vertrieb] (zugl. Berlin, Freie Univ., Diss., 1999).
Mez, L. (1979): Energiediskussion in Europa. Berichte und Dokumente über die Haltung der Regierungen, Parteien und Gewerkschaften in der Europäischen Gemeinschaft zur Kernenergie. In: Hauff 1979.

Mez, L. (ed.) (1981): Der Atomkonflikt. Berichte zur internationalen Atomindustrie. Atompolitik und Anti-Atom-Bewegung. 2nd edition. Reinbek: Rowohlt.
Mez, L. (1988): Transnuklear oder Urangate in Hanau. In: Komitee für Grundrechte und Demokratie 1988.
Mez, L. (1995): Reduction of Exhaust Gases at Large Combustion Plants in the Federal Republic of Germany. In: Jänicke and Weidner 1995.
Mez, L. (1997): Energiekonsens in Deutschland? Eine politikwissenschaftliche Analyse der Konsensgespräche. Voraussetzungen, Vorgeschichte, Verlauf und Nachgeplänkel. In: Brauch 1997.
Mez, L. (1998): Fortbildung Energiepolitik, Kapitel 5: Atomenergie in Deutschland. Manuskript.
Mez, L. and Osnowski, R. (1996): Ein Riese mit Ausstrahlung. Köln: Kiepenheuer & Witsch.
Mez, L. and Wilke, M. (ed.) (1977): Der Atomfilz – Gewerkschaften und Atomkraft. Berlin: Olle & Wolter
Mohr, M. (1991): Gewerkschaftliche Interessen von Beschäftigten in der Atomindustrie. Diplom-Arbeit, Freie Universität Berlin.
Moldenhauer, B. (1975): Die Atomindustrie in der BRD. In: Blätter für deutsche und internationale Politik, Heft 10.
Münchmeyer, T. (1999): Wenn grüne Minister den Bau von Kernkraftwerken absegnen. In der Ukraine sollen zwei neue Atommeiler errichtet werden – mit Bonner Unterstützung. In: Frankfurter Rundschau, March 15, p. 9.
Öko-Institut 2000: Energiewende 2020. Der Weg in eine zukunftsfähige Energiewirtschaft. Berlin: Heinrich-Böll-Stiftung.
Paulitz, H. (1996): Nachhaltige Einflußnahme. Die Deutsche Bank und der aufhaltsame Ausstieg aus der Kernenergie. In: Blätter für deutsche und internationale Politik, No. 41(2), pp. 223–230.
Radkau, J. (1983): Aufstieg und Krise der deutschen Atomwirtschaft 1945–1975. Reinbek: Rowohlt.
Reuters (1999): Steuerlast für Atomkonzerne umstritten. 28 July (18:22).
Rucht, D. (1980): Von Wyhl nach Gorleben. Frankfurt a. M., New York: Campus.
Shonfield, A. (1968): Geplanter Kapitalismus – Wirtschaftspolitik in Westeuropa u. USA. Köln: Kiepenheuer und Witsch.
Siemens (1994): Kraftwerk-Perspektiven '93. Jahresbericht des Siemens-Bereichs Energieerzeugung – KWU. Erlangen.
Siemens (1995): Kraftwerk-Perspektiven '94. Jahresbericht des Siemens-Bereichs Energieerzeugung – KWU. Erlangen.
Siemens AG (1996): Focus on Power Plants '95. Annual Report of the Power Generation Group – KWU. Erlangen.
Siemens (1997): Umfassende Partnerschaft für wirtschaftliche Energieerzeugung. Jahresbericht Bereich Energieerzeugung – KWU 1996. Erlangen.
Siemens (1998): Geschäftsbericht 1997. Erlangen.
Siemens (1999): Geschäftsbericht 1998. Erlangen.
Simnad, M. (1996): Has the nuclear power industry risen and fallen or will it rise again? In: Energy 21(12), pp. 1095–1100.
Strohm, H. (1999): Die stille Katastrophe. Berlin: Zweitausendeins.

U.S. GAO (U.S. General Accounting Office) (1999):Nuclear Regulation: Better Oversight Needed to Assure Accumulation of Funds to Decommission Nuclear Plants. RCED-99-75. Washington, DC. May.

Ueberhorst, R. (1992): Planungsstudie zur Bildung und Arbeitsplanung einer unabhängigen Kommission zur Förderung energiepolitischer Verständigungsprozesse in der Bundesrepublik Deutschland in den Jahren 1992 bis 1995. Elmshorn.

VIK (Verband der Industriellen Energie- und Kraftwirtschaft) (various volumes): Statistik der Energiewirtschaft. Essen.

Zängl, W. (1989): Deutschlands Strom. Die Politik der Elektrifizierung von 1866 bis heute. Frankfurt a. M., New York: Campus.

Printing: Saladruck, Berlin
Binding: H. Stürtz AG, Würzburg

GPSR Compliance

The European Union's (EU) General Product Safety Regulation (GPSR) is a set of rules that requires consumer products to be safe and our obligations to ensure this.

If you have any concerns about our products, you can contact us on ProductSafety@springernature.com

In case Publisher is established outside the EU, the EU authorized representative is:

Springer Nature Customer Service Center GmbH
Europaplatz 3
69115 Heidelberg, Germany

Batch number: 10704030

Printed by Printforce, the Netherlands